William H. (William Harris) Ashmead

A Monograph of the North American Proctotrupidae

William H. (William Harris) Ashmead

A Monograph of the North American Proctotrupidae

ISBN/EAN: 9783741126734

Manufactured in Europe, USA, Canada, Australia, Japa

Cover: Foto ©Klaus-Uwe Gerhardt /pixelio.de

Manufactured and distributed by brebook publishing software
(www.brebook.com)

William H. (William Harris) Ashmead

A Monograph of the North American Proctotrupidae

MONOGRAPH

OF THE

NORTH AMERICAN PROCTOTRYPIDÆ.

BY

WILLIAM H. ASHMEAD.

WASHINGTON:
GOVERNMENT PRINTING OFFICE.
1893.

PREFACE.

In the present work I have attempted to give a systematic description of the species of the Hymenopterous family *Proctotrypidæ*, found in North America, north of Mexico, at the same time systematizing and describing the genera of the world, as an aid to future students.

It represents several years study of the family, and although sensible of its incompleteness and many imperfections I shall be fully repaid for the labor involved if I have paved the way for a more thorough study of the family or stimulated others to collect these curious insects and study their remarkable habits.

In the preparation of this work my thanks are especially due to the following friends:

To Mr. W. Hague Harrington, of Ottawa, Canada, I am indebted for valuable contributions to my cabinet during many years and for types of some of the species described by the Abbé Provancher; to my friends and colleagues, E. A. Schwarz, Theo. Pergande, and Otto Heidemann, who have contributed by donating, from time to time, specimens taken during our entomological excursions in and around Washington and elsewhere; to Prof. E. A. Popenoe, for a few species collected in Kansas; to Dr. C. V. Riley, for the free use of his valuable notes on the rearings and habits of the species, and for other courtesies shown me during the progress of the work; to Mr. L. O. Howard, my friend and colaborer in the Microhymenoptera for like courtesies; to Dr. Gustav Mayr, of Vienna, Austria, for some typical European genera, and to Dr. C. W. Stiles, for making a microscopical section of the ovipositor.

Finally, to Prof. Dr. K. Möbius, director of the Royal Berlin Museum, and to Dr. F. Karsch, custodian of said museum, my warmest thanks are due for allowing me to study and describe the North American species contained therein.

I have also had the privilege of studying and describing the many new and interesting forms in the National collection and in the collection of the American Entomological Society, kindly sent by Mr. E. T. Cresson to the Museum for study and exchange, and which contains the types of Cresson, Patton, and Haldeman.

Dr. Riley has also permitted me to include from his manuscript, a few descriptions of species in which he has been particularly interested.

W. H. A.

Washington, D. C., *March 4, 1893.*

5

CONTENTS.

INTRODUCTION.

What a wide field, therefore, remains to be investigated before we shall become acquainted with the 600,000 or even 400,000 species supposed by Messrs Kirby and Spence to exist; and how absurd does it seem to consider our systems or rather system as firmly established whilst so little is comparatively known.—WESTWOOD, in 1833.

Although the classification of insects is in a more satisfactory condition than when Prof. Westwood wrote these memorable words, more than half a century ago, a fact due in great part to this Nestor of entomology and his contemporaries, our system of classification can not yet be considered firmly established. Instead of 600,000 species to deal with, later estimates place them at millions. Messrs. Sharp and Walsingham in 1889 placed them at two millions; while the latest authority, that of Dr. C. V. Riley, 1892, indicates that there are perhaps 10,000,000 species existing on the globe.

In these pages it is my province to treat of only a small proportion of this intricate and perplexing aggregate of forms, as found in America, north of Mexico, viz: those of the single family Proctotrypidæ.

The Proctotrypidæ, by some authorities, are considered to be closely allied to the Chalcididæ and, in a systematic arrangement of the hymenopterous families, usually follow them in our manuals and catalogues.

I consider, however, that they have but little affinity with the Chalcididæ and that this arrangement is unnatural. They are in every respect more closely allied to the *Hymenoptera aculeata*, the Chrysididæ, Scoliidæ, Mutillidæ, and Thynnidæ; while in the Terebrantia, I believe, they approach closest to the parasitic Cynipidæ (*Allotria, Eucoila,* and *Figites*).

In a natural arrangement, therefore, they should be placed at the head of the Terebrantia; for after the removal of the group Mymarinæ, [which I hold with Haliday forms a separate and distinct family allied to the Chalcididæ,] there is no relationship with the Chalcididæ.

With the Mymarinæ removed, there will be no difficulty in distinguishing, at a glance, a Proctotrypid from a Chalcid. In all true Proctotrypids the pronotum extends back to the tegulæ and the ovipositor issues from the tip of the abdomen, the sheaths, except in a few abnormal cases, being conjoined and forming a more or less cylindrical tube or scabbard for the reception of the two spiculæ and the ovipositor proper; whereas, in all Chalcids the pronotum never extends back to

9

dentate or pedicellate and verticillate. In some groups the number of joints is alike in both sexes, but in others they are more numerous in the males, while in at least one group, the Belytinæ, the females have the greater number.

The mandibles are in the majority of the groups bidentate or bifid, but in the Bethylinæ and Dryininæ they vary from 2- to 6-dentate: *Apenesia* has but 2; *Epyris* and *Mesitius* 5 or 6; *Scleroderma* 3 or 4, etc. Most of the genera in the Dryininæ have 3 or 4; the Scelioninæ 2 or 3; while in only one group, the Proctotrypinæ, are they edentate. The maxillæ, Pl. I, Fig. 2, terminate in one or two large, thin, membranous lobes, the galea and lacinia, while the mentum is small, horny, or coriaceous. The maxillary palpi vary in the number of joints from 2 to 6 and are usually conspicuous from beneath. The labial palpi, Pl. I, Fig. 3, on the contrary, are short and inconspicuous, usually 2- or 3-jointed, the terminal joint being the longest; while in one group, the Platygasterinæ, they are 1-jointed.

THE THORAX.

The thorax is, as a rule, rarely very much narrower than the head, variable in breadth and length in the different groups, and is of the greatest importance in classification. The three principal divisions are the pro-, meso-, and meta-thorax, and as it is essential that the component parts of each of these divisions should be thoroughly understood, they are here taken up separately, the parts being explained by letters on Pl. I, (Fig. 1, *T.*)

The prothorax (Fig. *pt*) is large and conspicuous from above, with but few exceptions, in the Bethylinæ, Embolemiinæ, Dryininæ, Proctotrypinæ, and in some of the Scelioninæ, while in the other groups it is more variable, usually short and inconspicuous, sometimes very small, and often not, or scarcely, visible from above (*Telenomus*, etc.). It supports the head and front legs, and its principal component parts are: (Fig. *pn*) the pronotum, (Fig. *p*) the pleuron, and (Fig. *sp*) the praesternum.

The mesothorax is the largest division of the thorax. It supports the front wings and the middle legs, and variations and peculiarities noticeable in its various sclerites have been found to be of excellent specific and generic value. The principal sclerites are: (Fig. *ms*) mesoscutum, which is frequently subdivided into three parts by longitudinal furrows (Fig. *pf*) called the parapsidal furrows; it is then said to be trilobed. Fig. *m* becomes the middle lobe, Fig. *p* the parapsides scapulæ or lateral lobes. Fig. *s* is the meso-scutellum, usually designated as the scutellum, and has usually at the lateral basal angles (Fig. *ax*) two subtriangular sclerites termed axillæ. Fig. *mps* is the mesopostscutellum, often visible only as a transverse band, carina, or fold, and again quite distinct and armed with one or more strong spines or

thorns. By some authorities it is considered a sclerite of the metathorax. Fig. *tg* is the tegula or wing scale of front wing. The mesopleuron (Fig. *mp*) is composed of episternum and epimeron. The mesosternum is undivided and is frequently not separated from the pleural
sclerites.

The metathorax, or third division of the thorax, supports the hind
wings, hind legs, and the abdomen, and it also is quite variable in
length, shape, and sculpture. The sides (Fig. *mtp*) are termed the
metapleura, and the sclerite (Fig. *sp*) extending obliquely forward
from the insertion of the hind legs to the base of the hind wing and
behind the mesopleuron is the metepisternum; the upper surface (Fig.
mn) is the metanotum. Fig. *sp* represents the position of the spiracles,
while *ihn* shows the insertion of the hind wings.

The other normal sclerites are not differentiated.

THE WINGS.

In shape and neuration, the wings exhibit the greatest diversity, and
are of primary importance in classification. As a rule, the hind wings
are veinless or the neuration is meager (Pl. 1, Fig. 5), the highest developed having but a single basal cell. They are rather broad and
with a distinct lobe in the Bethylinae, Emboleminae, and the Dryininae;
broad in the Proctotryinae, but without a distinct lobe; while in the
other groups they are much narrowed toward base, and not especially
widened toward apex, the apex being sometimes acute. The front
wings are entirely veinless in only a single tribe, the Platygasterini,
although another tribe, the Inostemmini, in the same subfamily, and a
few genera in another subfamily, the Scelioninae, are almost veinless,
having only a submarginal vein terminating in a small knob or stigma.

The neuration of the anterior wings reaches its fullest development
in the Bethylinae, Emboleminae, Dryininae, and Helorinae, although even
in these groups there are genera with almost veinless wings (*Cephalonomia*, etc.).

The neuration of the first three of these groups mentioned more
closely resembles that of the Chrysididae and Scoliidae; the Helorinae
and some genera in other groups that of the Braconidae; while some
genera in the Scelioninae closely approach that of the Chalcididae.

The great variation in the shape of the wings and the peculiarities
of venation are fully brought out in the plates; but as the neuration, in
connection with other characters, is of primary importance in a systematic study of the family, it is important that the horismology of
the wings be fully understood.

In Figs. 4 and 5, on Pl. 1, by the use of the front wing of *Pristocera
atra* Klug. and the hind wing of a belytid, as being the most special-
ized in the group, and, with the explanation given below, the technical
terms may be easily acquired.

Horismology of the wings.

FRONT WING.

Longitudinal nerves.	Transverse nerves.	Cells or areolets.
c. Costal.	b. Basal.	1. Costal (closed).
sc. Subcostal or submarginal.	tm. Transverse median.	2. First basal.
m. Median.	ftc. First transverse cubital.	3. Second basal.
sm. Submedian.	stc. Second transverse cubital.	4. Anal.
st. Stigma, or marginal nerve stigmated.	fr. First recurrent.	5. Marginal or radial.
	sr. Second recurrent.	6. First discoidal.
ps. Postmarginal or poststigmal.	d. Discoidal.	7. Second discoidal.
r. Radius or stigmal.		8. Third discoidal.
cbt. Cubital.		9. First submarginal.
sd. Subdiscoidal.		10. Second submarginal.
		11. Third submarginal.
		12. First apical or postanal.
		13. Second apical.

HIND WING.

c. Costal.	b. Basal.	1. Costal cell (open).
sc. Subcostal.		2. Basal.
cbt. Cubital		3. Anal.
st. Marginal or stigma.		

THE LEGS.

The shape of the legs, armature, tibial spurs, number of joints in tarsi, length, etc., also afford excellent characters in classification. In all groups, except sometimes in the Dryininae, the hind legs are the longest. The femora in the Bethylinae are usually very much swollen, their tibiae stout, often spiny or fossorial; in the Dryininae and Emboleminae they are, for the most part, obclavate; while in other groups they are usually clavate or but slightly swollen. The tarsi, except in the single genus *Iphetrachelus* Haliday, in which they are 4-jointed, are always 5-jointed. In all groups, except in the Dryininae, they are normal, but in this group the anterior tarsi in the females in most of the genera, are peculiarly modified, being chelate or furnished with a pair of pinchers or tongs, that evidently afford assistance in seizing and holding a living fulgorid, membracid, or jassid (on which in the larval stage they are parasitic), while in the act of oviposition.

The structure of the leg is explained in Pl. I, Fig. 1: *c c c*, coxae, anterior, middle and posterior; *tr*, trochanters, one-jointed; *fa*, femora; *ta*, tibiae; *tas*, tibial spurs; *tsi*, tarsi; *cl*, claws; *cls*, simple, *clt*, toothed, *clp*, pectinate; *ps*, pulvillus.

THE ABDOMEN.

The abdomen is usually composed of 8 visible tergites and 6 urites, although sometimes these are reduced in number to as low as 3. As in other families, it exhibits great variation in shape and size and in its attachment to the metathorax, the terms sessile, subsessile, or petiolate expressing the form of attachment. It is, however, never attached to the dorsum of the metanotum, as in the Evaniidæ and some Braconidæ.

In counting the abdominal segments, the basal one is usually referred to as the first segment or the petiole; sometimes, however, this segment is so short as to be invisible from above, or at least not visible until the abdomen has been detached from the metathorax; it is then usual to designate the second as the first.

The shape of the abdomen is generally ovate, ovate-conic, or oval, but often oblong-oval, broadly oval, fusiform, or linear. It is rarely greatly compressed, although frequently depressed or somewhat flattened.

In some genera, in the Platygasterinæ and Scelioninæ, the females are furnished with a peculiar horn-like structure at the base of the abdomen that projects forward over the metathorax, and, indeed, often as far forward as the head (Pl. XI, Fig. 6). It gives to these insects a singular appearance, but otherwise has no functional significance, being merely a sexual peculiarity.

The abdomen reaches its greatest length in proportion to the rest of the body, probably in the genus *Macroteleia* (Pl. IX, Fig. 6), and its most unique shape in *Sactogaster* (Pl. XII, Fig. 4).

Connected with the abdomen are the important organs of reproduction, the ovipositor, etc., which will now be described.

THE OVIPOSITOR.

In the differentiation of its parts, the ovipositor, in this family, agrees with other terebrant Hymenoptera, the only visible difference being that in the whole group, except in three or four abnormal individuals, the outer sheaths are conjoined and form a tube or scabbard at the tip of the abdomen that affords protection for the ovipositor proper and its two spiculæ when not in use. Distinct sutures are visible in this tube, and after death it can be readily separated into two plates.

This tubular formation of the sheaths reaches its highest development in the genus *Proctotrypes*.

On Pl. I two forms are shown. Fig. 6 represents that of *Proctotrypes caudatus* Say, our largest species. Here the tube is as long as the abdomen, slightly compressed and curving downward at tip. It is readily detached, and in the figure is represented partly detached, in order to show the internal structure of the ovipositor. The terebra, or ovipositor proper (*tba*), is the piece through which the egg passes; it is composed of three pieces, a broad upper plate and two spiculæ (*spc*) or lower plates, the latter united to the former by a dovetailed joint (see Fig. 9, A and B). These three pieces are dilated and lobed at base (*spl*), and are connected by strong muscles to a basal plate (*bp*) with the internal walls of the abdomen, their tips being sharply pointed, needle-like, and usually microscopically serrated. The spiculæ are used for piercing or boring, and move forward and backward on the upper plate as a saw in the hands of a carpenter.

Figs. 7 and 8 represent the terebra of *Epyris grandis*, one of the largest species in our fauna. The different parts are lettered identically as in *P. caudatus*. Here the tube (or outer sheaths conjoined) (*shs*) is short and conical, and is the form most commonly met with in the family.

At Fig. 9 two cross-sections of the ovipositor proper are shown—A from near the base, B from nearer the apex, the breadth and thickness varying slightly in accordance with the tapering of the ovipositor. The transverse diameter of the upper plate averages about 0.042^{mm}; of the basal plates or spiculæ, 0.015^{mm}, while vertically they average 0.018^{mm} in diameter.

It will be seen that the structure of this cross-section agrees fairly well with others that have been made of the terebrant Hymenoptera, the only real difference being the much larger canals through the center of the three pieces and the small additional canal in the lower inner angle of the spiculæ.

All the canals have apparently a membranous lining, and unless they have something to do with the control of the ovipositor their function is not apparent.

MALE GENITALIA.

It takes so much time to thoroughly study the organs of generation, on account of their minute size, in most groups of the Hymenoptera Parasitica, besides the almost absolute certainty of the destruction of valuable specimens, that most entomologists are content to depend upon external characters for the recognition of species, and consequently comparatively little is known of the male genitalia in any family composing this section, although there can be no doubt as to the specific and classificatory value these organs possess.

In order to give some idea of these organs in the Proctotypidæ, on Pl. I, I have figured the male genitalia of three different species.

Fig. 10 represents the male genitalia of *Epyris carbonarius*: A, the parts viewed from above; *p*, penis or penal sheath, strongly exserted, in outline narrow, harp-shaped, although strongly flattened like a leaf, bilobed at apex, the lobes with rounded margins; at the middle (*o*) is a longitudinal slit or opening, the orifice for the fleshy penis. On each side of the penis, attached to a swollen lobe at base, are two horny sheaths, a pair on each side, that may be termed respectively the upper and lower sheaths; *us* is the upper sheath. In this species it is slender and gradually acuminate, or lanceolate; *ls* is the lower sheath; it is much larger, broader, and more rounded at apex, although it is here divided by a longitudinal slit into two more or less distinct points, each point being surrounded by stiff hairs; outwardly the surface is convex and smooth, except at about two-thirds its length, where there is a tuft of long hairs. At B the same parts are shown as viewed from beneath, together with other parts not visible from above; *us*, the upper sheath; *ls*, the lower sheath; *p*, penis; *pc*, penal claspers, the outer margin of which are fringed with several long spines or hooks; *bl*, swollen basal lobes, to which the upper and lower sheaths are attached. All these organs, except the tips of the upper and lower sheaths, when not in use, are withdrawn within the apical abdominal segments and are only visible when exserted. The swollen basal lobes (*bl*) and the penal claspers (*pc*) seem not to have been noticed before, and I believe are now pointed out for the first time.

Fig. 11 represents the male genitalia of *Proctotrypes caudatus* Say, as seen from the side after the removal of the left ventral spine: *vs*, right ventral spine; *us*, upper sheath, slender and clavate; *ls*, lower sheath, very broad and flat and terminating in four chitinous lobes.

Fig. 12 represents the same organs in the male of *Scleroderma cylindrica*, after Westwood: *us*, upper sheaths, "the extremities of which are thin and incurved"; *ls*, lower sheaths, "broad and each with a broad, simple stipes (*dd*), and four terminal lobes (*e*), two of which are setose at the tips, and one at least more rigid than the other parts."

In Fig. 13, Pl. I, I represent the ovipositor of *Scleroderma ephippium*, after Westwood. In his explanation he says:

The parts of the ovipositor itself are vertically compressed, the recurved bases of the spiculæ (*e*), with their muscular angulated lobe or catch (*ff*), being represented as flattened. By strong protrusion of the spiculæ beyond the extremity of terebra, the curved basal portions of the former are straightened and brought forward to the base of the terebra, where their dilated angular form prevents them from further protrusion. The parts marked *e, e,* are the membranous plates connecting the base of the spiculæ and of the terebra itself with the interior of the abdomen.

HABITS OF THE PERFECT INSECTS.

The imagos are most frequently found wherever their hosts are most plentiful and their lives are of short duration, seldom extending beyond a few days. Those I have kept in confinement live but four or five days, although in freedom they probably live longer.

The favorite resorts for diapriids, bethylids, and proctotrypids are low, moist places, where there is a luxuriant growth of vegetation and a black or mucky soil, the decaying vegetation affording excellent food for their hosts—dipterous and other larvæ. The opening buds and newly formed leaves of plants and trees, and especially along the outskirts of a dense forest or wood, are particularly attractive to platygasterids and telenomids, while the bethylids, dryinids, and scelionids, as a rule, frequent the more open fields. In Florida, dry sandy knolls, where the scrub oak grows, are the favorite resort of the bethylids. Species in the genera *Epyris* and *Mesitius,* I have taken most frequently on trees badly infested with lepidopterous leaf-rollers and miners. Other genera frequent fungi, while from recent observations ant nests and their vicinity are good fields for some rare species in other genera.

As a rule, flowers are less frequently visited by the tiny species of this family than in the other hymenopterous families, except those flowers that afford protection or food for their hosts. Various platygasterids are the only members of the family that I can recall having captured on flowers, and in such cases the flowers were invariably infested with cecidomyiid larvæ, on which they were parasitic.

When captured, as with other Hymenoptera, some proctotrypids give off a waspy or pungent odor. Prof. Westwood has recorded the fact, taken from Mr. Saunders's MS. notes, that the female *Scleroderma linearis*, taken in a house September 25, 1849, stinging his neck, when captured "threw out a pungent fœtid odor."

DIMORPHISM AND PARTHENOGENESIS.

There is scarcely any doubt but that many of the wingless forms to be found in various genera of this family are only dimorphic forms of winged species, although comparatively little is positively known on the subject. Until such forms are bred from generation to generation, however, as has been done in the Cynipidæ by Dr. H. Adler, we must be content to describe these forms as distinct species, for any other course would be unscientific or guesswork. In this monograph I have not hesitated, therefore, to give these apterous individuals a separate specific name. Doubtless also, as in the Blastophagæ, or fig-insects, trimorphism occurs.

Prof. Westwood,[1] in speaking of the genus *Scleroderma* says:

Some of the species of this genus exhibit a remarkable instance of dimorphism and even trimorphism in the females, some of which are furnished with certain of the characteristics of the opposite sex. Sir S. S. Saunders, in addition to the ordinary female of *S. ephippium* destitute of wings and ocelli, found one agreeing in the general form of the body with the females, but possessing the fully developed wings and ocelli of the male. Remembering the diversity in the size of the heads of different individuals among some of the species of ants, I was curious to ascertain how far this character was to be found in these different individuals. Their heads were, therefore, drawn of a large size by the camera lucida, which was then carefully reduced by measurement, when it appeared that the heads of the normal females were slightly longer than wide, whereas the head of the winged and ocellated females was considerably broader than long, although its length agreed very nearly with that of the normal female. Their relative proportions may be expressed by the following figures:

	Long.	Wide.
Head of normal female without wings or ocelli	77	75
Head of female with wings and ocelli	78	90

[1] Trans. Ent. Soc. Lond., 1881, p. 120.

These measurements may be contrasted with those of the normal male of *S. cylindrica* from Prevesa, drawn to the same scale:

	Long.	Wide.
That of the winged and ocellated male being	61	60
That of the wingless and unocellated female being	97	78

The anomaly is carried still further in the Ceylonese *S. rigilans*, of which I have only seen two female specimens, one of which had fully developed wings and ocelli, whilst the other was wingless, but possessed ocelli. But the anomalous characters of the genus are not confined to the females, since Sir S. S. Saunders captured a wingless male provided with slender antennae and with three large ocelli, but entirely destitute of wings.

Mr. Haliday's observation on his *Labolips innupta* is the only case bearing directly upon parthenogensis in the Proctotrypidae, known to me. He found the ovaries form each an oblate spheroid, entirely covered with regular small protuberances, as if they were composed of an agglomeration of globular cells; the separate oviducts, in the short axis of the ovary, of considerable volume, and nearly as long as the transverse diameter of the ovary, united into a short excretory canal; no seminal receptacle was discovered, while in both specimens examined there was apparently a malformation of the malpighian vessels, so that he could not determine their number with absolute certainty.

Mr. Haliday considered the form of the ovaries without a parallel among the rest of the Hymenoptera, and the absence of a seminal receptacle, if his observation could be depended upon as exact, most singular, as this appendage is found in form even in the agamous Cynipidae.

TRANSFORMATIONS OR LIFE HISTORY.

THE EGGS.

The eggs of the Proctotrypids known to me are ovate or oblong in shape, with a more or less distinct peduncle at one end, and agree well in general with many in the family Ichneumonidae, although those in the subfamily Platygasterinae, on account of the longer peduncle, more closely resemble those in the family Cynipidae.

EMBRYOLOGICAL DEVELOPMENT.

Ganin,[1] 1869, was the first biologist to study the embryological development of certain proctotrypids—*Platygaster*, *Teleas*, etc., while an American biologist, Prof. Ayers,[2] has given the embryological development of a Scelionid, in his paper entitled "On the Development

[1] Ueber der Embryonalhülle der Hymenopt. und Lepidopt. Embryonen, St. Petersburg, 1869.

[2] Mem. Boston Soc. Nat. Hist., vol. III, 1884. This insect is not a true *Teleas* and is either one of two insects described in this memoir, viz., *Cacus acanthi* or *Baryconus acanthi*.

of *Œcanthus niveus* and its Parasite *Teleas*." Both authors may be consulted to the advantage of the student.

In lieu of original work of this kind I have deemed it advisable to reproduce here a summary of Ganin's work, by Balfour, together with his figures, in order to illustrate the remarkable changes the embryolarva must undergo before assuming the normal larval stage.

Balfour says:

The very first stages are unfortunately but imperfectly known, and the interpretations offered by Ganin do not in all cases appear quite satisfactory. In the earliest stage after being laid the egg is inclosed in a capsule produced into a stalk (Fig. 1, A). In the interior of the egg there soon appears a single spherical body, regarded by Ganin as a cell (Fig. 1, B). In the next stage three similar bodies appear in the vitellus, no doubt derived from the first one (Fig. 1, C). The central one presents somewhat different characters to the two others, and, according to Ganin, gives rise to *the whole embryo*. The two peripheral bodies increase by division, and soon appear as nuclei imbedded in a layer of protoplasm (Fig. 1, D, E, F). The layer so formed serves as a covering for the embryo, regarded by Ganin as equivalent

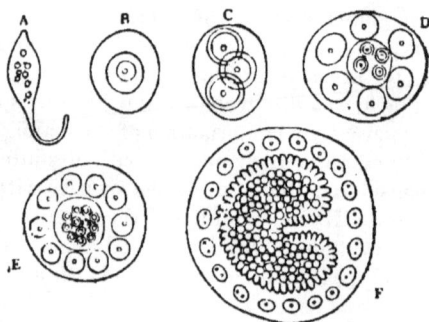

Fig. 1. --Embryonic development of platygaster: A, egg; B, primitive cell; C, additional cells developed from the first; D and E, further cellular development; F, final embryonic stage. (After Ganin.)

to the amnion (? serous membrane) of other insect embryos. In the embryo cell new cells are stated to be formed by a process of endogenous cell formation (Fig. 1, D, E). It appears probable that Ganin has mistaken nuclei for cells in the earlier stages, and that a blastoderm is formed as in other insects, and that this becomes divided in a way not explained into a superficial layer which gives rise to the serous envelope and a deeper layer which forms the embryo. However this may be, a differentiation into an epiblastic layer of columnar cells and a hypoblastic layer of more rounded cells soon becomes apparent in the body of the embryo. Subsequently to this the embryo grows rapidly, till by a deep transverse constriction on the ventral surface it becomes divided into an anterior cephalothoracic portion and a posterior caudal portion (Fig. F). The cephalothorax grows in breadth, and near its anterior end an invagination appears, which gives rise to the mouth and œsophagus. On the ventral side of the cephalothorax there is first formed a pair of claw-like appendages on each side of the mouth, then a posterior pair of appendages near the junction of the cephalothorax and abdomen, and lastly a pair of short conical antennæ in front.

At the same time the hind end of the abdomen becomes bifid, and gives rise to a fork-like caudal appendage; and at a slightly later period four grooves make their appearance in the caudal region, and divide this part of the embryo into successive segments. While these changes have been taking place in the general form of the embryo, the epiblast has given rise to a cuticle, and the hypoblastic cells have become differentiated into a central hypoblastic axis—the mesenteron—and a surrounding layer of mesoblast, some of the cells of which form longitudinal muscles.

With this stage closes what may be regarded as the embryonic development of platygaster. The embryo becomes free from the amnion, and presents itself as a larva, which from its very remarkable characters has been spoken of as the Cyclops larva by Ganin.

The larvæ of three species have been described by Ganin, which are represented in Fig. 2, A, B, C. These larvæ are strangely dissimilar to the ordinary hexapod

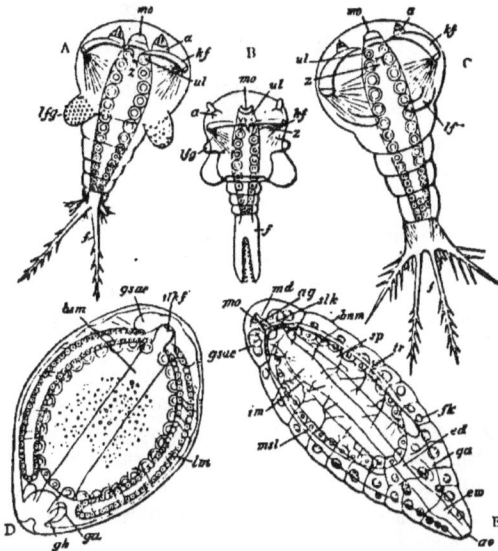

Fig. 2.— Embryonic larval development of platygaster: A, B, C, Cyclops larvæ of three species of platygaster; D, second larval stage; E, third larval stage; *mo*, mouth; *a*, antenna; *kf*, hooked feet; *lfg*, lateral feet; *f*, branches of tail; *ul*, lower lip; *slkf*, œsophagus; *sac*, supra-œsophageal ganglion; *bsm*, ventral epiblastic plate; *lm*, lateral muscles (the letters also point in D to the salivary glands); *gh*, proctodæum; *ga*, generative organs; *md*, mandibles; *ag*, ducts of salivary glands; *sp*, (in E) salivary glands; *mls*, stomach; *ed*, intestine; *ew*, rectum; *ao*, anus; *tr*, trachea; *fk*, fat body. (After Ganin.)

type, whether larval or adult. They are formed of a cephalothoracic shield with the three pairs of appendages (*a*, *kf*, *lfg*), the development of which has already been described, and of an abdomen formed of five segments, the last of which bears the somewhat varying caudal appendages. The nervous system is as yet undeveloped. The larvæ move about in the tissues of their hosts by means of their claws.

The first larval condition is succeeded by a second with very different characters, and the passage from the first to the second is accompanied by an ecdysis.

The ecdysis commences at the caudal extremity, and the whole of the last segment is completely thrown off. As the ecdysis extends forwards the tail loses its segmentation and becomes strongly compressed, the appendages of the cephalothorax are thrown off and the whole embryo assumes an oval form without any sharp distinction into different regions and without the slightest indication of segmentation (Fig. 2, D). Of the internal changes which take place during the shedding of the cuticle, the first is the formation of a proctodæum (*gh*) by an invagination, which ends blindly in contact with the mesenteron. Shortly after this a thickening of the epiblast (*bsm*) appears along the ventral surface, which gives rise mainly to the ventral nerve cord; this thickening is continuous behind with the epiblast, which is invaginated to form the proctodæum, and in front is prolonged on each side into two procephalic lobes, in which there are also thickenings of the epiblast (*gsae*), which become converted into supraœsophageal ganglia, and possibly other parts.

Toward the close of the second larval period the muscles (*lm*) become segmentally arranged, and give indications of the segmentation which becomes apparent in the third larval period. The third and last larval stage (Fig. 2, E) of *Platygaster*, during which it still remains in the tissues of its host, presents no very peculiar features. The passage from the second to the third form is accompanied by an ecdysis.

THE LARVÆ; THEIR HABITS AND PUPATION.

The larvæ are all, apparently, internal feeders, apodous and with minute mandibles. Those living in eggs transform into pupæ within the empty egg-shell of their hosts; sometimes a half dozen or more being found within a single egg-shell, although of the larger scelionids —*Scelio, Macroteleia*, etc.—only a single specimen is found in each egg. Those species that affect the larvæ or pupæ of other insects either gnaw their way out and spin small silken cocoons (*Cephalonomia* and other bethylids), or else weave silky or parchment-like cocoons, often arranged in parallel rows, side by side, within the empty skin of their host (*Platygaster* and others).

Lygocerus and allied genera, living in the Aphididæ, gnaw a hole through the ventral surface of the aphis, and after securely fastening the aphid by a silk-like secretion to the leaf or twig upon which it has been feeding, pupate within the body of their host, which, in lieu of a cocoon, affords ample protection for the larvæ to undergo their transformations. I know of no proctotrypid that transforms loosely into a pupa without a covering or protection of some kind, as is frequently, if not invariably, the case amongst the Chalcididæ.

DISTRIBUTION.

The Proctotrypidæ are apparently widely distributed over the whole world, although outside of Europe little is as yet known of the exotic forms, and it is not possible, therefore, to generalize upon the genera and their distribution. From an examination of various exotic collections of Hymenoptera, it is safe to predict the species will be found to be numerous and widely distributed, but far less numerous than the Chalcididæ; judging from my own collecting I should say less than one-fiftieth in number. Only a very small percentage of the species is yet described.

Our species, except in a few cases, fit very well into the established European genera; but from South America it was found necessary to erect many new genera, and doubtless this will be found to be necessary with collections brought from other regions. Very few species have been described from Asia, Africa, and Australia; only a few isolated descriptions of species, scattered here and there through the literature, could be found; and undoubtedly many remarkable and interesting forms will be discovered in the rich insect faunas of those countries.

CLASSIFICATION.

The name of the family, Proctotrypidæ, is derived from the genus Proctotrupes or Proctotrypes, established by Latreille in his Précis des Caractères Génériques des Insectes, published in 1796, and there placed with the Ichneumonidæ. It is derived from two Greek words πρωκτός, the anus, + τρυπάω, bore, pierce through, in allusion to the structure of the ovipositor, and I here follow late authorities in spelling the word Proctotrypes, although as a rule I strongly oppose changing a well-established zoölogical name.

From 1796 down to 1839 additional genera, belonging to this family, were established by Latreille, Dalman, Klug, Jurine, Spinola, Nees, Westwood, Walker, Haliday, and others, and various groups or families were proposed for their reception, according to the views of the different authors, the affinities of the genera not being fully understood. Some of the genera were classed with the Cynipidæ, Scoliidæ, Mutillidæ, Chalcididæ, or Braconidæ, or were considered to be entitled to family rank, and the greatest confusion existed as to their affinities and proper position.

An excellent historical sketch of the development of the family has been given by Dr. Arnold Förster, in his Hymenopterologische Studien, Heft II, and it will suffice here to call attention only to the two authors who wrote just previous to his work. To them we are mainly indebted for a proper conception of the family, as now understood, as well as for bringing order out of the chaos that existed up to their time. I also call attention to a few authors who wrote subsequent to the appearance of Förster's work.

In 1839 Haliday, in his "Hymenopterorum Synopsis," properly defined some of the groups as now understood, although he considered them to be of family rank, placing some among the Terebrantia and some among the Aculeata.

In order to illustrate his method, I give below such portions of his scheme of arrangement as refer to them:

Suborder 2.—Petioliventres.

Stirps 3.—Terebellifera.

cc. Trochanters exarticulati. Terebra abdominis apice exerenda per foramen apicale vel rimam transversam, aut exserta. Antennæ articulis ad summum 15. Alæ disco ferme exareolatæ, radio et cubito distantibus aut obsoletis.

Tribe 2.—OXYURA.

f. Tibiæ anticæ calcari unico.
    g. Mandibulæ edentulæ.
        Fam. 10.—*Proctotrupidæ.*
    gg. Mandibulæ dentatæ.
      h. Abdomen *fem.* longissimum, filiforme, articulatum.
        Fam. 9.—*Pelecinidæ.*
      hh. Abdomen petiolatum, immarginatum.
        Fam. 11.—*Diapriadæ.*
      hhh. Abdomen marginatum, subsessile aut subpetiolatum.
        Fam. 12.—*Scelionidæ.*
ff. Tibiæ anticæ bicalcaratæ.
        Fam. 13.—*Ceraphronidæ.*

Tribe 4.—*Hallicoptera.*

ff. Os epalpatum. Alæ angustissimæ, nervo brevissimo.
        Fam 17.—*Mymaridæ.*

Stirps 4.—*Aculeata.*

c. Antennæ utrinsque sexus articulis eodem numero. Pedes sublæves. Alæ posti-
    cæ incisæ.
        Tribe 6.—*Cenoptera.*
    d. Antennæ 10-articulatæ. Caput deflexum.
        Fam 19.—*Dryinidæ.*
    dd. Antennæ 12-13-articulatæ. Caput porrectum. Rapacia. Larvæ entomo-
    phagæ.
        Fam 20.—*Bethyllidæ.*

Chrysidis inde connexi, hinc *Scoliadis.*

It will be seen from the above "scheme of arrangement" that Haliday widely separates some closely allied groups, interpolating other
families between them; and, moreover, with our larger and better acquaintance with these groups, the arrangement is unnatural and the
characters he has employed, in defining them, will not now always
hold good.

About the same time, 1840, Prof. Westwood, in his memorable work,
"Introduction to the Modern Classification of Insects," brought all of
these families together and properly united them as one, under the
name Proctotrupidæ. He then separates the family into six subfamilies, as follows:

(1) MYMARIDES. Head transverse areolate; antennæ inserted above the middle of
    the face, ♂ long, slender, and elbowed, ♀ clavate; mouth destitute of palpi;
    wings narrowed, densely ciliated, with a very short subcostal nerve.
(2) PLATYGASTERIDES. Abdomen sessile, depressed, first segment not campanulate;
    antennæ elbowed, 10-12-jointed, inserted near the mouth.
(3) CERAPHRONTIDES. Abdomen subsessile, campanulate, terminal and ventral seg-
    ments carinated; antennæ elbowed, inserted near the mouth; wings nearly
    exareolate.

(4) GONATOPIDES. Abdomen convex, not campanulate, last ventral segment carinated; antennæ porrected, 10-jointed; hind wings lobed; mandibles toothed.

(5) PROCTOTRUPIDES. Abdomen subsessile, campanulate; antennæ porrected, 12-jointed, inserted beneath the front; maxillary lobe tripartite.

(6) DIAPRIDES. Abdomen petiolate, campanulated; antennæ inserted on the front, 12- or 15-jointed; maxillary palpi long, 5-jointed.

His group (2) included the Scelioninæ with the Platygasterinæ; group (4) the Bethylinæ, Embolemiinæ and Dryininæ; while group (6) included the Belytinæ and the Diapriinæ. All the groups are undoubtedly closely related, and Westwood's scheme was the most natural one formulated up to the time of Förster.

It was just sixteen years after the publication of Westwood's work that Dr. Förster's studies in the "Chalcidiæ und Proctotrupii," so often quoted in these pages, appeared. After an excellent historical review of the two groups, already referred to, he makes use of the following characters to distinguish them:

CHALCIDIÆ. ♂. Fühler immer gebrochen, mit ein oder mehreren kleinen Ringeln (Zwischengliedern) zwischen Stielchen und Geissel.
♀. Der Legebohrer unterhalb (d. h. vor) der Spitze des Hinterleibs entspringend.
PROCTOTRUPII. ♂. Fühler gebrochen oder ungebrochen, zwischen dem Stielchen und der Geissel keine Ringel (Zwischenglieder), selten ungebrochen mit einem kleinen Ringel.
♀. Der Legebohrer aus der Spitze des Hinterleibs entspringend.

After tabulating and defining the families of the Chalcidiæ, he separates the Proctotrupii, on page 20, into 9 families, according to the following scheme:

a. Hinterflügel mit einem lappenförmigen Anhang, oder, wo die Flügel beim ♀ fehlen, mit Raubfüssen versehen....................Dryinoidæ.
aa. Hinterflügel ohne den lappenförmigen Anhang.
 b. Die vordersten Schienen mit zwei Dornen ....Ceraphronoidæ.
 bb. Die vordersten Schienen mit einem Dorn.
  c. Mandibeln ungezähnt...................................Proctotrupoidæ.
  cc. Mandibeln gezähnt.
   d. Der Hinterleib seitlich gerandet; Fühler über dem Mundrande eingefügt.
    e. Flügel mit einem Ramus marginalis und bisweilen auch einem Ramus stigmaticus; die ungeflügelten Gattungen ohne Nebenaugen ....Scelionoidæ.
    ee. Flügel ohne Ramus marginalis und stigmaticus; alle Gattungen mit Nebenaugen ....Platygasteroidæ.
   dd. Hinterleib seitlich nicht gerandet; die Fühler weit über dem Mundrande eingefügt.
    f. Hinterflügel ohne Spur einer Mittelader.
     g. Die Hinterflügel sehr schmal, fast linienförmig..Mymaroidæ.
     gg. Die Hinterflügel breiter, nicht linienförmig......Diaprioidæ.
    ff. Die Hinterflügel mit einer Mittelader.
     h. Vorderflügel entweder ohne oder mit einer regelmässigen Grundader; Geissel ohne Ringel (annellus)....Belytoidæ.
     hh. Vorderflügel mit einer unregelmässigen, die Unterrandader nicht erreichenden, gekrümmten Grundader, welche eine unregelmässige Zelle in der vorderen Schulterzelle bildet; Geissel mit einem Ringel ....Heloroidæ.

Subsequently, farther on in the body of the work, Förster again subdivides his family Dryinoidæ into three families: Dryinoidæ, Embolemoidæ, and Bethyloidæ. He then follows with his generic tables, in which, including the *Mymaroidæ*, no less than one hundred and twenty-one genera are defined, about seventy being new.

Dr. Förster's work is by far the most philosophical, satisfactory, and important contribution published on this intricate family, and has formed the basis of my own work. Since his day, not taking into consideration mere descriptive work, no systematic work has been attempted except a short paper by A. H. Haliday, published in the Dublin Natural History Review, 1857; a series of articles by the Swedish entomologist C. G. Thomson, published in Öfversig af Kongliga Svenska Vetenskaps-Akademiens Förhandlingar, from 1857 to 1859, and an excellent contribution by Prof. J. O. Westwood on the subfamily Bethyllides, in his Thesaurus Entomologicus Oxoniensis, 1874.

Haliday, in his paper, proposed some new genera in the Diapriidæ (probably before seeing Förster's work) that unfortunately prove to be synonyms of some of those characterized by Förster.

Thomson's contribution is much more elaborate; besides describing some new genera he separated the family into eleven tribes (defined, but without a synoptic table), arranged as follows:

Tribus   I. Proctotrupini.
         II. Belytini.
        III. Ceraphronini.
         IV. Diapriini.
          V. Ismarini.
         VI. Helorini.
        VII. Scelionini.
       VIII. Platygasterini.
         IX. Telenomini.
          X. Dryinini.
         XI. Epyrini.

It is also quite evident that Thomson, in the beginning of his studies in the family, was unacquainted with Förster's work; and, considering this fact, his tribes agree quite closely with the families proposed by Förster. His tribus XI, Epyrini, equals the Bethyloidæ of Förster. In separating the Ismarini from the Belytoidæ and the Telenomini from the Scelionoidæ and elevating them to an equal rank with his other tribes it appears to me Thomson gave too much importance to trivial characters.

The arrangement of his tribes also does not show the relationship they bear to each other. I consider the Ceraphronini much more closely allied to the Epyrini and Dryinini, while he has them widely separated. The Belytini and the Diapriini are unquestionably closely allied, and he has interpolated the Ceraphronini between them; while the Helorini he places next to the Scelionini, with which they have few characters in common, their nearest allies being the Proctotrypini and the Belytini.

Prof. Westwood's contribution, mentioned above, is a most valuable one, in which some new genera and many new species are described and illustrated. It is of especial value for clearing up many obscure points in the old, imperfectly described genera, for the very full diagnoses of all the genera in the Bethylinæ known to him, and for the admirable illustrations of the mouth parts and other salient characters.

Other European contributors to a knowledge of the family are Walker, Mayr, Ruthe, Cameron, Marshall, Mik, Giard, etc.

In America no systematic study of the family has been before attempted.

The American writers on the family, besides myself, are Say, Haldeman, Cresson, Patton, Packard, Provancher, Howard, Riley, and Forbes, and, excepting an admirable translation of Förster's tables by Mr. L. O. Howard, the work thus far done has been purely descriptive, confined to isolated descriptions of genera or species. These will be found recorded in their proper place in this work and require no special mention here.

Förster separated the Proctotrupii into eleven families. In the present work, ten of these are recognized as subfamilies. His family Mymaroidæ (=Mymaridæ Haliday) I have not included, as I believe with Haliday, who first proposed the name, that it is a group entirely distinct from this family, more closely allied to the Chalcididæ than to the Proctotrypidæ and entitled to rank with other families of the Hymenoptera Terebrantia.

In the arrangement of the ten subfamilies or natural groups recognized in this family I have attempted to show their true relationship, and, although full diagnoses of them are given in their proper place in the body of the work, I believe the following table will be found all that is necessary to facilitate their recognition:

TABLE OF SUBFAMILIES.

Wingless forms..6
Winged.
  Posterior wings not lobed...2
  Posterior wings lobed; anterior wings with 1 or 2 basal cells.
    Head oblong; antennæ inserted at the clypeus, 12- or 13-jointed in both sexes.
        Subfam. I, BETHYLINÆ.
    Head not oblong.
      Head globose; front wings with a lanceolate stigma; anterior feet never chelate; antennæ in ♀ 13-jointed, in ♂ 10-jointed.
        Subfam. II, EMBOLEMINÆ.
      Head transverse; front wings usually with a large or semicircular stigma; anterior feet often chelate; antennæ 10-jointed in both sexes.
        Subfam. III, DRYININÆ.
2. Antennæ inserted on the middle of the face, often on a frontal prominence....4
  Antennæ inserted at the clypeus.
    Abdomen acute or margined at the sides, sessile or subpetiolate............3

Abdomen not margined at sides, sessile or subsessile. Anterior tibial spur strongly forked; antennae in ♀ usually 10–11-jointed, in ♂ 11-jointed; front wings never with a post-marginal vein, the stigmal vein long, the marginal often stigmated......Subfam. IV, CERAPHRONINÆ.

3. Front wings most frequently with marginal and stigmal veins; antennae in both sexes usually 12-jointed, in ♀ sometimes 11-jointed, or 7-jointed when the claval joints coalesce...Subfam. V, SCELIONINÆ.

Front wings always without marginal and stigmal veins and most frequently entirely veinless, at the most with only a clavate submarginal vein; antennae never more than 10-jointed, usually alike in both sexes, rarely only 8- or 9-jointed....Subfam. VI, PLATYGASTERINÆ.

4. Front wings with the marginal vein linear, not stigmated.................. 5

Front wings with the marginal vein stigmated, the costal cell closed.

Mandibles dentate; antennae 15-jointed; claws pectinate,
Subfam. VI , HELORINÆ.

Mandibles edentate; antennae 13-jointed, with a ring joint; claws simple,
Subfam. VIII, PROCTOTRYPINÆ.

5. Front wings with a distinct basal cell and usually with a distinct marginal cell (the latter rarely entirely wanting); hind wings always with a basal cell; antennae 11 to 15 jointed; labial palpi 3-jointed...........................Subfam. IX, BELYTINÆ.

Front wings with the brachial cell seldom distinct, the median nervure being usually obsolete, the marginal cell never complete, usually entirely wanting; hind wings always without a cell; antennae 12-, 13-, or 14-jointed; labial palpi 2-jointed,
Subfam. X, DIAPRIINÆ.

WINGLESS FORMS.

6. Abdomen acutely margined at sides............................. 7

Abdomen not margined at sides.

Fore femora much swollen.

Head oblong; antennae 12- to 13-jointed in both sexes, inserted near the mouth...................................Subfam. BETHYLINÆ.

Head transverse or globose.

Anterior feet normal; antennae in ♀ 13-jointed......Subfam. EMBOLEMINÆ.

Anterior feet chelate; antennae in ♀ 10-jointed........Subfam. DRYININÆ.

Fore femora not much swollen.

Mandibles edentate.................................Subfam. PROCTOTRYPINÆ.

Mandibles dentate.

Antennae inserted near the mouth.

Anterior tibial spur 2-pronged...................Subfam. CERAPHRONINÆ.

Antennae inserted on a frontal prominence.

Labial palpi 3-jointed...................... ....Subfam. BELYTINÆ.

Labial palpi 2-jointed.............................Subfam. DIAPRIINÆ.

7. Antennae 10-jointed; labial palpi 1-jointed...........Subfam. PLATYGASTERINÆ.

Antennae 12-jointed or with a solid club 7-jointed; labial palpi 2- or more, jointed...................................Subfam. SCELIONINÆ.

Subfamily I. BETHYLINÆ.

Head oblong, horizontal, the occipital line complete. Ocelli often absent; when present 3, triangularly arranged. Mandibles usually large and broad, from 2- to 6-dentate. Maxillary palpi 4- to 6-jointed; labial palpi 2- to 3-jointed. Antennae not elbowed, filiform or setaceous,

12- to 13-jointed, inserted in a clypeal fovea, the clypeus being more or less distinctly carinated. Pronotum large, well developed, narrowed before; mesonotum short, transverse, the scapulæ rarely entirely separated; scutellum subtriangular, rarely entirely wanting, the axillæ not separated; metathorax large, quadrate. Front wings most frequently with a small, stigmated marginal vein (sometimes with a parastigma), a radial and two basal cells, the radial cell rarely distinctly closed; occasionally the marginal and radial nervures are wanting or abbreviated; sometimes there is but a single basal cell, rarely entirely wanting. Hind wings distinctly lobed, veinless, except along the costa at base. The apterous forms are usually without ocelli and with minute eyes. Abdomen subpetiolate, ovate, or conic-ovate, depressed, composed of 8 segments. Legs rather short, the femora and tibiæ stout, most frequently swollen, tibial spurs 1, 2, 2, well developed, the tarsi 5-jointed, the claws often with a tooth within.

The *Bethylinæ*, so far as we at present know, are parasitic on coleopterous and lepidopterous larvæ, and before pupating most of them spin cocoons. The majority of the genera comprise species winged in both sexes; but in a few genera, viz, *Pristocera*, *Isobrachium*, *Scleroderma*, *Dissomphalus*, *Apenesia*, and *Cephalonomia*, the male alone is winged, the female, except in *Cephalonomia* being always apterous. Usually the disparity between the sexes in size and appearance is such, that, when captured at large, they cannot be satisfactorily correlated. In *Cephalonomia* a dimorphic form occurs; for, whereas the female is usually wingless and without ocelli, at times a winged form with ocelli is produced.

The European type of the genus *Cephalonomia formiciformis* Westw. has been bred in England and Germany from mushrooms infested with coleopterous larvæ. In this country species are reared from cynipidous galls, infested with coleopterous, lepidopterous, and other inquilines, and it is believed the genus is parasitic on coleopterous larvæ. Species belonging to the genera *Isobrachium*, *Apenesia*, and *Pristocera* have been taken in ant nests, and it is presumable they are parasitic on myrmecophilous Coleoptera. *Anoxus* is parasitic on the coleopterous genus *Cis* living in fungi, and I have reared *Ateleopterus tarsalis* from raisins affected with *Silvanus surinamensis*.

The new genus *Lælius* is parasitic on the coleopterous family Dermestidæ, while species in the genera *Bethylus*, *Perisemus*, and *Goniozus* prey upon microlepidopterous larvæ.

Mr. A. H. Haliday, in Ent. Mag. vol. II, p. 219, has recorded the following observation in regard to an unknown *Bethylus*:

The insects of this genus seem fond of the flowers of *Syngenesia*, but their principal haunts are in dry sandy districts near the sea. The low tufts of *Rosa spinosissima*, flourishing among the sand cliffs, support numerous larvæ of Tineidæ, which when full fed often fall into the little pits of loose sand formed at the foot of the cliffs by the gradual scaling of the bank and the eddies of wind. These pits are complete traps for various insects, to which *Myrmica rubra* and other predaceous species resort, and among these our Bethyli will be seen prowling. On the 5th of last June I observed a female of the largest size occupied with one of those larvæ which was full fed, and I should think about six times its own weight. It had seized this by the mouth, and was with great perseverance endeavoring to transport it up the sides of the pit. Perceiving that, though apparently not discouraged after ten minutes' ineffectual exertion, it had no chance of succeeding, and wishing to trace its proceedings, I placed a fragment of straw in the hollow within its reach. The moment it had touched this railway the state of affairs was changed; taking a firm hold with its hind feet, it swung its prey round and set off with it at a smart pace, walking backwards and dragging the body after it. From this time it was constantly endeavoring to ascend the face of the sand cliff, availing itself with admirable adroitness of morsels of grass, twigs, etc., embedded in it, not seeming to care how obliquely they lay, if they enabled it to gain a little elevation; so that its track was a zigzag. Frequently it chose stems, which, rising nearly erect, receded from the bank above. I first thought it was losing its labor, but it was at no loss how to act. After ascending a few inches with the whole weight suspended in the air from its mandibles, it would poise itself and its burden across the stem with its head towards the bank, then throw itself off, at the same time extending its wings, which, though incapable of raising it from the ground, were able to give it some impulse toward the bank, on which it thus alighted at a spot someway above the springing of the stem. If, on ascending one of these twigs, it discovered that it was bent the wrong way or receded too far from the cliff, it lost no time in hesitation, but, stopping short, of a sudden commenced the descent again. It may be guessed that, dragging a gross slimy body over twigs, etc., close to or half buried in the sand, frequent impediments would occur, which its extreme activity in walking indifferently sideways or backwards, and main exertion of muscular force, generally enabled it to overcome; but sometimes it had drawn its burden under or between two twigs, which arrested its course. After a violent tug or two without effect, it would retrace its steps, dragging the larva in the opposite direction till it was extricated, disposing it so as to keep clear of obstacles, and start again. On every occasion when it had left its hold it made for the same part, and spent some time in fastening its mandibles on the mouth of its prey beneath, so that the larva should be dragged on its back; once where this was not the case it was impeded by the latter grappling with its feet the twigs over which it was drawn, and its captor quickly finding the error, let go and took a new hold in the usual position.

When it ascended about two feet, it came upon a fragment of reed partly embedded in the sand, the stem of which was broken off and open below, a few dry elastic shreds of the leaf only remaining. Having reached the part where these grew, it by a strong pull drew its burden about half through, till its body was grasped between two of these as in a vice; then letting go, it began to explore the bank on each side to some distance, tapping with its antennæ the conspicuous objects. In a few minutes, seeming to be satisfied, it hastily descended the reed, and entered its stem at the lower end; it did not remain long in the interior, and on its reappearance set off for the spot where it had left the larva, which, after pulling it out of the holdfast, it seized by the mouth as usual and began to descend the reed again; it did not complete the journey this time, but taking advantage of the same kind of security to detain its prey, it repeated the reconnaissance, then returning, dragged it to the opening, and leaving it there plunged in itself, but immediately reappearing drew in the larva head foremost, speedily disappearing in the interior; so that I could not observe its subsequent proceedings, and being obliged to turn homewards, I left them undisturbed. I think, however, it will seem probable that the bore of the reed was employed instead of an artificial funnel for the cells which should contain the progeny of the Bethylus, with its store of provision.

This interesting observation of Haliday's on the habits of a species in this group is quoted at length as being quite unique and contrary to the observations made on the habits of these insects by all subsequent writers. It probably influenced Haliday into placing the Bethylidæ among the Aculeata; but we can see from this observation that they are not true Aculeata, for the insect observed by Haliday would unquestionably have stung the larva into insensibility [as all Aculeata do], before attempting to carry it to its cell.

The genera and species are numerous and are widely distributed, being found in all parts of the world.

It is believed that the following synoptic table, prepared with great care, will enable the genera of this difficult group to be readily determined by the student without recourse to the full generic description:

TABLE OF GENERA.

FEMALES.

```
Winged ....................................................................3
Wingless.
  With ocelli .............................................................2
  Without ocelli; antennæ 13-jointed.
    Metathorax much contracted or strangulated at base.
      Scutellum present.
        Maxillary palpi 6-jointed; mandibles 3- or 4-dentate...PRISTOCERA Klug
      Scutellum wanting.
        Maxillary palpi 6-jointed; mandibles 3-dentate....ISOBRACHIUM Förster
    Metathorax not much contracted at base, quadrate.
      Scutellum present.
        Maxillary palpi 5-jointed; mandibles 4-dentate.....SCLERODERMA Latr.
        Maxillary palpi 4-jointed; mandibles 3-dentate,
                                       DISSOMPHALUS Ashm. gen. nov.
        Maxillary palpi 4-jointed; normal; mandibles 2-dentate.
          Labial palpi 3-jointed .........................ATELEOPTERUS Förster
        Maxillary palpi 4-jointed, deformed; mandibles 2-dentate.
          Labial palpi 3-jointed ..............................APENESIA Westw.
2. Antennæ 12-jointed; maxillary palpi 4-jointed..........CEPHALONOMIA Westw.
3. Anterior wings with a stigmated marginal vein and a marginal cell, the radius
      always well developed ..................................................4
   Anterior wings without stigmated marginal vein and marginal cell, the radius
      not, or but slightly, developed.
     Anterior wings without marginal and stigmal vein.
       Antennæ 12-jointed; no basal cell.................CEPHALONOMIA Westw.
       Antennæ 13-jointed; one basal cell................ATELEOPTERUS Förster
     Anterior wings with a short linear marginal vein and a short radius.
       Two basal cells about equal in length.
         Antennæ 13-jointed ............................LÆLIUS Ashm. gen. nov.
         Antennæ 12-jointed .................................BETHYLUS Latr.
4. Basal nervure with a branch directed backwards, sometimes forming a small
      cell; a parastigma ....................................................6
   Basal nervure without a branch directed backwards; no parastigma.
     Anterior wings with an incomplete marginal cell .......................5
```

Anterior wings with a complete marginal cell.
  Antennæ 13-jointed.
    With 2 discoidal cells..................................EUPSENELLA Westw.
    With 1 discoidal cell..................................SIEROLA Cameron
5. Mesonotal furrows distinct.
  Antennæ 13-jointed; maxillary palpi 6-jointed; labials 3-jointed; fore femora
    much swollen.
    Scutellum with a transverse grooved line at base.
      Maxilla trilobed at apex ....CALYOZA Westw.
      Maxilla bilobed at apex................................EPYRIS Westw.
    Scutellum with 2 foveæ at base.........................MESITIUS Spinola
  Mesonotal furrows wanting or indistinct.
    Antennæ 13-jointed; fore femora much swollen.
      Scutellum with a transverse grooved line at base ....(EPYRIS)
      Scutellum with 2 foveæ at base...............................(MESITIUS)
    Antennæ 12-jointed..ANOXUS Thoms.
6. Antennæ 12-jointed ....PERISEMUS Förster
  Antennæ 13-jointed..GONIOZUS Förster

MALES.

1. Anterior femora much swollen..2
  Anterior femora not swollen; anterior wing with 1 or 2 discoidal cells, usually
    open.
  Mesonotal furrows distinct, usually 4; antennæ 13-jointed; no parastigma.
    Mandibles 4 or 5 dentate; maxillary palpi 6-jointed.
      Abdominal segment 2 without umbilicate tubercles.
        First discoidal cell distinct; stigma large ....PRISTOCERA Klug
        First discoidal cell indistinctly defined, always open; stigma small,
                                ISOBRACHIUM Förster
    Mandibles 3-dentate; maxillary palpi 4-jointed.
      Abdominal segment 2 with 2 umbilicate tubercles,
                          DISSOMPHALUS Ashm. gen. nov.
2. Anterior wings with a distinct marginal cell, the stigmal nervure always long..4
  Anterior wings without a marginal cell, the maginal and stigmal nervures want-
    ing, or very short.
  Anterior wings with a short linear marginal nervure and a short radius......3
  Anterior wings without marginal and stigmal nervures.
    Antennæ 13-jointed.
      With 2 basal cells..................................SCLERODERMA Latr.
      With 1 basal cell..................................ATELEOPTERUS Förster
    Antennæ 12-jointed..................................CEPHALONOMIA Westw.
3. Antennæ 13-jointed....................................LÆLIUS Ashm. gen. nov.
  Antennæ 12-jointed ....BETHYLUS Latr.
4. Anterior wing with an incomplete marginal cell, the stigmal vein long.........5
  Anterior wing with a complete marginal cell; antennæ 13-jointed.
    With 2 discoidal cells..................................EUPSENELLA Westw.
    With 1 discoidal cell ....SIEROLA Cameron
5. Basal nervure with a branch direct backwards, often forming a closed cell; a
    parastigma 6
  Basal nervure without a branch directed backward; no parastigma.
    Mesonotal furrows distinct.
      Antennæ 13-jointed, ramose......................CALYOZA Westw.
      Antennæ 13-jointed, simple, filiform.
        Scutellum with a transverse fovea at base; maxilla terminating in two
          lobes ....EPYRIS Westw.
      Scutellum with 2 foveæ at base; maxilla terminating in 3 lobes,
                                MESITIUS Spinola

```
Mesonotal furrows wanting or indistinct.
  Antennæ 13-jointed.
    Mandibles long, slender, bifid .....................APENESIA Westw.
    Mandibles broad, 4- to 6-dentate.
      Scutellum with a transverse fovea at base...................(EPYRIS)
      Scutellum with 2 foveæ at base............................(MESITIUS)
  Antennæ 12-jointed.
    Eyes hairy ........................................ANOXUS Thoms.
6. Antennæ 12-jointed .........................................PERISEMUS Förster
   Antennæ 13-jointed.........................................GONIOZUS Förster
```

PRISTOCERA Klug.

Weber u. Mohr. Beitr. Naturk. II, p. 202–6, pl. iv (1810), Westw. Thes. Ent. Oxon, p. 162 (1874).

(Type *P. depressa* Fabr.)

Head, in ♂ oblong-oval, convex with 3 ocelli; in ♀ large, oblong quadrate, the ocelli wanting; the eyes very small.

Antennæ in both sexes 13-jointed, in ♂ long, tapering toward tips; the scape stout, curved, punctate, the pedicel minute; the flagellar joints are filiform or subfiliform, truncate at tips, three or four times longer than thick, hairy or pubescent; in ♀ much shorter, submoniliform, the last joint somewhat longer than the penultimate.

Maxillary palpi 6-jointed, long; labial palpi 3-jointed.

Mandibles in ♂ dilated at apex and crossing each other, 5-dentate, teeth acute, conical, the outer tooth large; in ♀ stout, curved, the apex 3- or 4-dentate, the outer tooth large, obtuse.

Thorax: Prothorax in ♂ triangular, anteriorly subtruncate, deeply transversely impressed; mesonotum with 2 to 4 distinct furrows; metathorax convexly rounded posteriorly; in ♀ the prothorax is subquadrate, rounded anteriorly; mesothorax convexly rounded, with an impressed line on each side; the scutellum present; metathorax usually more or less strongly contracted at sides towards the base.

Front wings in ♂ with an oblong or oval stigma, an incomplete marginal, two basal, and one or two discoidal cells; and sometimes by holding the wing up to the light other cells may be seen; the second basal cell is a little longer than the first along the median nervure. The ♀ is always wingless and very much smaller than the ♂.

Abdomen in the ♂ ovate, depressed, subpetiolated, the second and third segments longer than the others; in ♀, it is variable, usually greatly elongated, pointed at apex.

Legs in ♂ rather long and not nearly so much swollen as in *Epyris*, *Mesitius*, etc.; the anterior femora a little thicker than the posterior, the anterior and middle tibiæ shorter than their femora, armed with rigid bristles anteriorly; tibial spurs 1. 2. 2, the outer spur of hind tibiæ very long, the claws long curved, with a tooth within near base; in ♀ shorter, stouter, the tibiæ outwardly spinulose, the claws simple, not toothed.

A well-defined genus, but the males easily confused with *Epyris*. The more strongly developed venation, the distinct, rhomboidal discoidal cell, the transverse medial nervure being always oblique and not so much curved as in *Epyris;* the 4 distinct mesonotal grooves, coarser sculpture, and the less distinctly swollen femora will, however, readily distinguish the genus. The females have frequently been mistaken for those of other genera, but the characters given in the table readily separate them. Many species described under the genus *Scleroderma* belong here.

Mœsary's work "Magyar Fauna, etc.," in which this genus is placed with the Sapygidæ as a subfamily Pristocerinæ, I have not been able to see.

The species in the United States are not numerous, but two being known to me, separated as follows:

Females wingless 2
Males winged.
    Black; the head coarsely rugose, with large punctures.
        Large, 10-11ᵐᵐ.; flagellar joints at apex more or less serrate..P. ATRA Klug
        Moderate, 7-8ᵐᵐ.; flagellar joints at apex not serrate......P. ARMIFERA Say
2. Large, 7 to 8ᵐᵐ. Rufo-piceous, the abdomen paler, the head and thorax
        with coarse scattered punctures.
    Mandibles 4-dentate...P. ATRA Klug
    Smaller, 4ᵐᵐ. Rufo-piceous, the abdomen darker, smooth, shining, impunc-
        tured.
    Mandibles 3-dentate..P. ARMIFERA Say

Pristocera atra Klug.

(Pl. II, Figs. 1, ♂ ; 2, ♀ .)

Web. u. Mohr Beitr. zur Natur II. 202, ♂ ; Westw. Thes. Ent. Oxon., p. 163, pl. XXXI, f. 5; Cress. Syn. Hym., p. 247.

Scleroderma thoracica Westw., Trans. Ent. Soc. Lond., II, p. 167, ♀ ; Cress. *l. c.*

♂.—Length 10 to 11ᵐᵐ. Black, shining, punctate, sparsely covered with a white pile. Head quadrate, coarsely reticulately punctate. Mandibles black. Palpi piceous. Antennæ 13-jointed, very long, longer than the head and thorax together, tapering toward tips, the flagellar joints about four times as long as thick, truncate at tips and slightly produced toward one side, appearing somewhat serrate. Collar rather short, rounded before, rugose and hairy. Mesonotum smooth, with some scattered punctures and 4 parallel grooved lines. Metathorax rugose, with raised lines and a slight median carina; the raised lines at base are short, longitudinally directed, those beyond transverse and irregular. Abdomen ovate, depressed, subpetiolate, highly polished, black, shorter than the thorax, and with a deep longitudinal furrow at base. Wings fusco-hyaline, the costa and stigma black, the other nervures brown; the stigma is somewhat ovate, the second basal cell (submedian) is a little longer than the first, the first discoidal cell distinct, while two other discoidal cells are plainly discernible by water-colored nervures.

♀.—Length 8 to 9ᵐᵐ. Rufo-piceous, the abdomen paler; head and thorax with some scattered punctures: mandibles 4-dentate; palpi pale. The prothorax is one and a half times as long as wide, the sides impressed; scutellum flat, rounded: metathorax elongate, laterally toward base very strongly contracted, so that the base is not half the width of the apex. Legs brownish yellow, the tibiae strongly spinulose. Abdomen elongate, ovate.

HABITAT.—Georgia, Florida, Texas, Mississippi, and Maryland.

Types in Berlin Museum.

The types of this species I have seen in Berlin. Westwood's type of *Scleroderma thoracica* is labeled "Carolina, Zimmermann" while it is accompanied by another specimen bearing label "Baltimore, Md." Klug's type of *Pristocera atra* is in good condition. As suggested by Prof. Westwood, these insects are sexes of a single species. It is the largest species known, and the male might easily be mistaken for a Tiphiid. My collection contains specimens compared with the types in Berlin. I have seen specimens from various parts of the country, the ♂ not being rare. The ♀ is extremely rare. The National Museum contains a single ♀ specimen collected by Mr. E. A. Schwarz, under bark, at Jackson, Miss., February 2, 1879.

Pristocera armifera Say.

Bethylus armiferus Say; Lec. Ed. Say, I, p. 383, ♂.
Epyris laevventris Cr.; Trans. Am. Ent. Soc., IV, p. 193, ♂; Ashm. Ent. Am., III, p. 76; Cress. Syn. Hym., p. 247.
Scleroderma contracta Westw.; Trans. Ent. Soc. Lond., II, p, 169, pl. 15, f. 11. ♀

♂.—Length 7 to 8ᵐᵐ. Black, shining, covered with a sparse, glittering white pile; head quadrate, rugose, with dense coarse punctures; mandibles rufo-piceous; palpi rufous. Antennae 13-jointed, setaceous, long, covered with a fine sericeous pubescence, the basal flagellar joints about three times as long as thick, toward the apex the joints relatively longer on account of the antennae tapering off at apex. Pronotum rugose; mesonotum smoother, with sparse deep punctures and 4 grooved lines; scutellum polished, with few punctures; metathorax longer than wide, abruptly truncate at tip, transversely rugulose; pleura coarsely punctured. Legs black, tarsi rufo-piceous. Wings subfuscous, the venation similar to *P. atra*, but without the water-colored nervures.

♀.—Length 4ᵐᵐ. Reddish piceous, smooth, shining, impunctured; abdomen darker, the apical margins of the segments pale; antennae and tarsi, yellowish; mandibles 3-dentate. Antennae 13-jointed, the pedicel more than twice longer than the first flagellar joint, the following joints to the thirteenth not longer than wide, the thirteenth twice longer than wide. The apical margins of the third, fourth, fifth, and sixth abdominal segments are roundedly emarginated.

HABITAT.—United States.

TYPE: ♀ in Berlin Museum; ♂ in National Museum and Coll. American Ent. Soc.

The type of *Scleroderma contracta* Westw. I have seen in the Berlin Museum; it has 3-dentate and not 4-dentate mandibles, as described by Westwood.

The species is widely distributed over the United States, but the ♀ is rare. Specimens are in my collection from Florida and Iowa, while the National Museum and the American Entomological Society have specimens from Texas, Georgia, Virginia, Maryland, and other States.

ISOBRACHIUM Förster.

Hym. Stud., II. p. 96 (1856).

(Type *I. (Omalus) fuscicornis Nees.*)

Head in ♀ much elongated, more rounded in ♂, the cheeks posteriorly delicately margined; eyes in ♀ wanting or very minute, in ♂ rather large, reaching to the base of the mandibles; ocelli absent in ♀, present in the ♂.

Antennæ 13-jointed, filiform, inserted in a clypeal fovea, the scape curved.

Maxillary palpi 6-jointed; labial palpi 4-jointed.

Mandibles broadened and truncate at apex, in ♂ with an acute outer tooth and with usually 4 small blunt dentations within; in ♀ with 3 teeth, the outer two most frequently equal.

Thorax in ♂ similar to *Epyris*, the prothorax rounded anteriorly, the mesonotal furrows rarely complete, most frequently indicated only anteriorly, sometimes entirely wanting; there is also a distinct groove on the shoulders; metathorax long; in the ♀ the thorax between the meso- and metathorax is strongly constricted and the scutellum is wanting.

Front wings in ♂ very similar to *Epyris*, only the stigma is smaller, more quadrate, the transverse median nervure usually more oblique, not so distinctly curved, while the discoidal nervure is present and the second discoidal cell is more or less present; the ♀ is always apterous.

Abdomen ovate, depressed, the segments unequal; in ♀ elongate, conic-ovate.

Legs as in *Pristocera*, not much thickened in the ♂; tarsi longer than their tibiæ, the basal joint of posterior tarsi as long as the remaining joints united; in the ♀ the middle tibiæ are spinous.

This genus bears a superficial resemblance in the male sex to both *Mesitius* and *Epyris*, but the much slenderer legs, eyes extending to base of mandibles, and the 4-jointed labial palpi readily distinguish it from those genera. The female comes nearest to *Pristocera*, but the absence of a scutellum and the strongly constricted metathorax sufficiently separate it from it and other wingless forms.

In stating this genus to be synonymous with *Mesitius* Spinola and *Heterocœlia* Dahlbom, Prof. Westwood erred and added not a little to the confusion respecting it, as all these genera are quite distinct. *Heterocœlia* is, as Dahlbom indicated, a genus in the family *Chrysididœ*, while *Mesitius* is a genus allied to *Epyris*. For an explanation of this statement see my remarks under the genus *Mesitius*.

The genus is found associated with various ants; it may be parasitic upon the ants, or upon the myrmecophilous Coleoptera so frequently found in their nests.

Our species may be recognized by the aid of the following table:

TABLE OF SPECIES.

MALES.

```
Mesonotum without furrows......................................................... 3
Mesonotum with furrows.
   Transverse median nervure interstitial......................................... 2
   Transverse median nervure evected.
      Antennæ and legs pale ferruginous...................... I. MAGNUM sp. nov.
2. Wings subhyaline.
      Abdomen piceous, paler at sutures, rufescent beneath.
         Antennæ and legs pale ferruginous or brownish yellow.
            Mandibles with small teeth within......... I. MYRMECOPHILUM, sp. nov.
            Mandibles very large and broad without teeth within.
                                                         I. MANDIBULARE, Ashm.
   Wings hyaline.
      Abdomen rufous, sometimes more or less piceous toward base above.
         Antennæ and legs pale ferruginous.
            Abdomen not longer than the thorax............... I. RUFIVENTRE, sp. nov.
      Abdomen brownish-yellow.
            Abdomen elongate, much longer than head and thorax together.
                                                         I. MONTANUM sp. nov.
3. Wings subhyaline.
      Black: legs pale ferruginous, the femora obfuscated above.
                                                         I. FLORIDANUM Ashm.
```

FEMALES.

```
Head black, with large punctures.
   Thorax dark brown, or mahogany.
      Abdomen not longer than the thorax, rufo-piceous....I. MYRMECOPHILUM
   Thorax rufous.
      Abdomen brownish-yellow, much longer than the head and thorax united.
                                                         I. MONTANUM
Head piceous.
   Thorax rufopiceous.
      Abdomen brownish-yellow, longer than the head and thorax together.
                                                         I. MANDIBULARE
```

Isobrachium magnum sp. nov.

♂.—Length nearly 5ᵐᵐ. Black, punctate: head a little longer than wide, rather strongly punctate, the surface alutaceous. Antennæ 13-jointed, brownish-yellow, the flagellar joints from 2 to 2½ times as long as thick, the last joint twice as long as the penultimate. Thorax anteriorly narrowed, the collar brownish, the mesonotum with 4 grooves, the middle furrows more or less indistinct and abbreviated posteriorly, the scutellum with a transverse furrow at base, the metathorax twice as long as wide, rugulose, with a median carina. Wings subhyaline, the stigma and nervures pale brownish-yellow, the transverse median nervure evected. Legs brownish-yellow, the posterior coxæ somewhat dusky. Abdomen oval, piceous-black, the apical edges of the segments testaceous.

Habitat.—Spokane Falls, Wash.

Type ♂ in Coll. Ashmead.

The largest species in the genus known to me, and readily distinguished by the evected transverse median nervure.

Isobrachium myrmecophilum sp. nov.

(Pl. ii, Figs. 3, ♂; 4, ♀.)

♀.—Length 3ᵐᵐ. Head twice as long as wide, black or piceous, shining, punctate, and sparsely covered with a dusky pubescence. Eyes and ocelli entirely wanting. Mandibles rufous, with 3 teeth. Antennæ short, 13-jointed, rufous, the scape nearly one-half as long as the flagellum; pedicel small, rounded; the flagellar joints, except the last, all transverse, the last oblong, more than twice as long as the preceding. Thorax brownish-piceous, highly polished, with a few scattered punctures; the prothorax is greatly elongate, 2½ times as long as wide, with nearly parallel sides rounded off anteriorly, the anterior depression of the collar pale rufous, finely transversely aciculate; prosternum black; mesonotum separated from the pronotum by a distinct furrow; the mesothorax is separated from the metathorax by a strong constriction, trilobed but without a scutellum; the metathorax is scarcely larger than the middle mesothoracic lobe. Legs pale brownish-yellow, stout, the middle tibiæ strongly spinous; all tarsi longer than their tibiæ. Abdomen oblong oval, brownish-piceous, much shorter than the thorax, smooth, impunctured, sparsely pilose.

♂.—Length 3 to 3.5ᵐᵐ. Black, punctate; head oblong, with a thimble-like punctuation, mandibles rufous, antennæ and legs pale brownish-yellow, collar pale anteriorly; abdomen piceous, paler at sutures.

Antennæ 13-jointed, extending to base of scutellum, the flagellar joints not quite twice as long as thick. Thorax with 4 grooved lines, the middle pair parallel, abbreviated posteriorly; those on the shoulder abbreviated anteriorly. Metathorax long, minutely shagreened with a central longitudinal carina. Wings hyaline, the stigma and subcostal nervure pale brownish, the other nervures subhyaline; the radius is very long, almost reaching to the apex of the wing, the transverse median nervure, oblique, interstitial with the basal.

Habitat.—Pennsylvania, District of Columbia, Montana, and Arizona.

Types in Coll. Ashmead.

This species has been taken from the nests of the ant *Formica rufibarbis* by H. G. Hubbard, at Ft. Assiniboine, Mont., and is also associated with ants under stones, at Washington, D. C., and at Beatty, Pa.

Isobrachium mandibulare sp. nov.

♀.—Length, 3.5mm. Closely allied to *I. rufiventre* and closely resembling it: Head brown, not so distinctly punctate, the outer tooth of mandibles longer, the inner tooth subobsolete; eyes entirely wanting; legs yellow, the anterior tibiæ at base dusky; middle tibiæ strongly spinous. The abdomen is long and measures 2mm; it is brownish-yellow, except the short petiole, which is black. The 13-jointed antennæ are nearly twice the length of the oblong head; scape half the length of the head, much narrowed at base; last flagellar joint twice as long as the penultimate, rounded at tip.

♂.—Length 3.6mm. Black and agreeing in color and structure with *I. myrmecophilum*, except as follows: The head is a little broader; the antennæ a little stouter and shorter, the flagellar joints scarcely 1½ times as long as thick: mandibles very large and broad, brownish-yellow; there are no teeth within after the second tooth which is small, while the dorsum of the metathorax, except at base, is smooth and shining.

HABITAT.—District of Columbia and Fort George, Fla.

Types in Coll. Ashmead.

Both sexes of this species were taken together in the nests of *Camponotus pennsylvanicus*, May 27, 1883, by Mr. Theo. Pergande.

Isobrachium rufiventre sp. nov.

♀.—Length, 3.5 to 4mm. Head variable, from piceous to black, about twice as long as wide, with scattered punctures; eyes wanting; thorax brown; antennæ, legs, and abdomen brownish-yellow. The mandibles are large, crossing each other at tips, and terminate in 3 black teeth, the two outer large, equal, the inner one small. Antennæ 13-jointed, 1½ times as long as the head; scape stout, not quite half as long as the head; pedicel a little longer and stouter than the first flagellar joint, the following joints to the last, wider than long, the last conic. Prothorax narrower but longer than the mesothorax, rounded before, more than twice as long as wide, separated from the mesonotum by a deep groove; mesothorax subcordate, as wide as long, trilobed, without a scutellum, the lateral lobes obcordate. Middle coxæ, much larger than either the front or hind coxæ, pilose beneath, the tibiæ spinous; anterior tibiæ with only a few spines. Abdomen long, pointed at apex, about 1.6mm in length.

♂.—Length, 3.6 to 4mm. Head and thorax black, shining, sparsely punctuate, the posterior margin of collar tinged with rufous. Antennæ and mandibles pale rufous; legs, including coxæ, pallid or whitish.

Abdomen entirely rufous. Wings salty white, the stigma and nervures hyaline.

HABITAT.—Yuma, Ariz., Montana, Texas, and Virginia.

Types in Coll. Ashmead, National Museum, and American Entomological Society.

The males of this species are common, the females rare.

Both sexes were taken together under stones near an ant's nest, by Mr. Pergande, August 5, 1885, at Occoquan Falls, Virginia, and a single ♀ was taken from the nest of *Formica obscuripes* at Helena, Mont., by Titus Ulke.

Isobrachium montanum sp nov.

♀.—Length, 5 ᵐᵐ. Head oblong, twice as long as wide, piceous black, punctate, very faintly pubescent. Eyes and ocelli wanting. Mandibles rufous ending in 2 black teeth. Antennæ 13-jointed, brownish-yellow, the scape less than half the length of the flagellum; pedicel longer than the first funiclar joint; flagellar joints, except the first and last, a little wider than long, the last joint oblong. Abdomen much longer than the head and thorax united, pale brownish-yellow.

♂.—Length, 4 ᵐᵐ. Head and thorax black, the collar pale; antennæ, mandibles, legs, and abdomen brownish-yellow; wings salty white. The thorax is smooth, impunctured, while the middle grooves of the mesonotum are only slightly indicated anteriorly. Metathorax long, smooth, impunctate, the central carina abbreviated. Abdomen longer than the head and thorax together.

HABITAT.—Montana and Arizona.

Types in Coll. Ashmead.

The ♀ of this species was taken from the nests of *Formica rufibarbis* by H. G. Hubbard at Helena, Mont., while the male, here associated with it, was obtained from a collection in Arizona; still the structure and color of the abdomen are so similar that there is no doubt left in my mind that they are sexes of the same species.

Isobrachium floridanum Ashm.

Ent. Am., III, p. 76; Cress. Syn. Hym., p. 247.

♂.—Length, 2.5 ᵐᵐ. Black, subopaque, finely punctate and sparsely covered with a whitish pubescence. Head oblong, closely punctate; eyes pubescent, mandibles long, crossing each other at tips. Antennæ 13-jointed, ferruginous, covered with a dense pubescence; scape clavate, about half the length of the head; pedicel not longer than wide at tip, narrowed at base; first flagellar joint a little shorter, the joints beyond longer. Prothorax trapezoidal, longer than wide across the base; mesonotum very short without furrows; scutellum with a transverse impressed line at base; metathorax quadrate with numerous longitudinal raised lines at base. Wings fusco-hyaline, the nervures brown;

stigma small. Legs pale ferruginous, the femora with a dusky streak
above. Abdomen black, highly polished, oval, shorter than the thorax.

HABITAT.—Jacksonville, Fla.

Type in Coll. Ashmead.

Distinguished at once by the absence of the mesonotal furrows and
the longer and different shape of the prothorax. It is scarcely congeneric with the others; but as I have but a single specimen and the
mouth parts can not be examined I must be content to allow it to remain here for the present. In general appearance it closely resembles
the ♂ in *Mesitius*.

SCLERODERMA Latreille.

Gen. Crust. et Ins., IV, p. 119 (1809); Westw. Trans. Ent. Loc. Lond., II, p. 164
(1839) and 1881, p. 17; Thes. Ent. Oxon., p. 169.

Sclerochroa Förster Hym. Stud., II, p. 168 (1856).

Head large, subquadrate, above convex; eyes in ♀ small, the ocelli
wanting; in ♂, eyes and ocelli normal.

Antennæ 13-jointed, longer in the ♂, shorter and more incrassated
in the ♀.

Maxillary palpi short, 5-jointed, the apical joints slender; labial palpi
3-jointed.

Mandibles small, obtuse at apex, 4-dentate.

Thorax elongate, the prothorax large, triangular or semiovate; mesonotum scutelliform; metathorax oblong, in ♀ narrowed basally.

Front wings in ♂ without a stigma or stigmal vein, and with only 2
short basal cells, the other cells entirely obliterated; ♀ apterous.

Abdomen in ♂ ovate, in ♀ greatly elongated, cylindrical and pointed
at apex.

Legs short, thick, the femora much swollen, tibial spurs 1, 2, 2; claws
simple.

The characters of the male sufficiently separate this genus from other
forms, but to distinguish the females from those in closely allied genera
is quite difficult. I am convinced that many females described under
this genus, by European authorities, do not properly belong in it, but
will be found to belong to other genera, i. e., *Pristocera*, *Isobrachium*,
and *Dissomphalus*, as the mouth parts can not always be satisfactorily
examined.

Only a single species is known in our fauna, as follows:

Scleroderma macrogaster Ashm.

(Pl. II, Fig. 6, ♀.)

Sclerochroa macrogaster Ashm. Ent. Am., III, p. 75, ♀ ; Cress. Syn. Hym., p. 247.

♀.—Length, 3ᵐᵐ. Head oblong-quadrate, black, polished; thorax
and legs rufo-piceous, the knees and tarsi, honey-yellow; metathorax
honey-yellow; abdomen black. Antennæ 13-jointed, honey-yellow, one-
fourth longer than the head, the scape about one-third the length of

the flagellum, curved, narrowed at the base, the joints of the flagellum not longer than thick. Abdomen elongate, pointed-ovate, two and a half times longer than the thorax, black, highly polished, with a few hairs at tip, the first and third segments about equal, the second slightly shorter.

HABITAT.—Jacksonville, Fla.

Type in Coll. Ashmead.

DISSOMPHALUS Ashmead gen. nov.

(Type *D. xanthopus.*)

Head in ♂ oval, very little longer than wide, with a long frontal carina; eyes oblong-oval, extending to base of mandible, nearly bare, or with few bristles; ocelli 3, distinct: in ♀ head oblong; eyes small oval; ocelli wanting.

Antennæ inserted rather far apart, 13-jointed, filiform, submonili-form, the first flagellar joint always smaller than the second.

Maxillary palpi 4-jointed; labial palpi 3-jointed.

Mandibles not very broad at tips, truncate, with 4 teeth.

Thorax in ♂ subovate, the prothorax not very long, constricted into a collar anteriorly, the promotum usually very short; mesonotum with or without furrows; scutellum with a transverse impressed line at base; metathorax subquadrate, truncate behind but the angles rounded; in ♀ with the prothorax long, narrowed anteriorly; scutellum distinct; metathorax quadrate, very slightly contracted at base.

Front wings in ♂ with two basal cells of an equal length, the transverse medial nervure being straight, an oblong-quadrate stigma and a long stigmal vein; ♀ always apterous.

Legs in ♂ slender, the femora not much swollen; tibial spurs 1, 2, 2; tarsi a little longer than the tibiæ, the basal joint nearly as long as the remaining joints united; in ♀ with the femora swollen, the tibiæ not spinous.

Abdomen in ♂ oval or rotund, depressed, subpetiolate, the second body segment always the longest and with two very small, round, wart-like tubercles, usually covered with short hairs, sometimes placed in a fovea or depression and variously situated, being sometimes close together or widely separated; in ♀ long, cylindric-ovate, the apical margins of segments 3 to 5 more or less emarginate or sinuate.

This new genus can always be distinguished in the male sex by the two minute wart-like, pubescent tubercles on the second abdominal segment and by the straight transverse median nervure: otherwise it bears a close resemblance to *Isobrachium*. It agrees with that genus, in having the eyes extending close to the mandibles, in its mesonotal characters, and in having (comparatively) slender legs: but besides the differences already mentioned, which are sufficient to distinguish it, it differs in the more slender, 4-dentate mandibles, in having 4-jointed maxillary palpi, and a shorter metathorax.

The female resembles *Pristocera* and *Scleroderma*, but the maxillary palpi are 4-, the labial 3-jointed, the scutellum distinct, while the metathorax is quadrate, not or scarcely contracted at base, the middle tibiæ not spinous.

Four species are known to me in the male sex from St. Vincent, West Indies, and two from the United States.

The species in our fauna may be distinguished as follows:

Wingless ♀ 2
Winged ♂.
Mesonotum with 2 distinct furrows.
    Legs, including coxæ, yellow; head opaque, finely, densely punctulate, with no frontal furrow..........................D. XANTHOPUS sp. nov.
    Legs yellow, the hind coxæ black or black basally; head shining, feebly punctulate, with a frontal furrow ....D. CALIFORNICUS sp. nov.
2. Abdomen black.
    Thorax, legs and antennæ testaceous; head rufo-testaceous.....D. XANTHOPUS
    Thorax, legs, and antennæ brownish-yellow; head blackish....D. CAROLINENSIS

Dissomphalus xanthopus sp. nov.

(Pl. II. Figs. 7, ♂; 8, ♀.)

♀. Length, 3.5ᵐᵐ. Head rufo-testaceous; thorax, legs and antennæ flavo-testaceous; abdomen black.

Antennæ 13-jointed about the length of the head; the scape clavate, one-third the length of the head; flagellum subclavate, involuted; pedicel longer than the first flagellar joint; first flagellar joint the slenderest, a little longer than thick, narrowed basally, the following joints very gradually becoming stouter, but not, or scarcely, longer than thick.

♂. Length, 2.8ᵐᵐ. Black; the head opaque, finely punctulate; thorax shining but still microscopically punctate, the mesonotum with two furrows and a grooved line on the shoulders; mandibles pale rufous with 4 nearly equal teeth; palpi white; legs, including the coxæ, yellow. Antennæ 13-jointed, the scape, pedicel, and first flagellar joint, yellow, remaining joints fuscous; the flagellar joints except the first, are about one and a half times as long as thick, densely pubescent, the last a little longer than the penultimate. Metathorax subquadrate, a little wider than long, almost smooth, being only slightly rugose at base and with a median carina extending to the superior edge of the truncature. Wings hyaline, the venation yellowish; the stigma is 3½ times as long as thick, a little narrowed at base and truncate at apex. Abdomen oblong-oval, not as long as the thorax, black, highly polished, the anus yellowish; the two small tubercles on the second segment are widely separated, situated at the basal angles.

HABITAT.—Arlington, Va., and Cedar Point, Md.

Type ♂ ♀ in Coll. Ashmead.

Described from 3 specimens.

Dissomphalus californicus sp. nov.

♂.—Length 3ᵐᵐ. Black, shining; the head but feebly punctured, with a frontal depression before the front ocellus: thorax with 2 furrows; mandibles rufous; palpi whitish; legs brownish yellow; the posterior coxæ black basally: the hind femora dusky above. Antennæ 13-jointed, fuscous, the scape and pedicel yellow; the first flagellar joint very small, only half the length of the pedicel; the three or four following scarcely longer than thick; those beyond longer, all densely covered with a short pubescence. Metathorax subquadrate, dorsally, except just at tip, rugose, with a median carina. Wings hyaline, the venation pale yellowish: the stigma is thrice as long as thick, truncate at apex. Abdomen oval, scarcely half the length of the thorax, black, shining: the two small tubercles on the second segment are wide apart, placed near the basal angles.

HABITAT.—California.

Type ♂ in Coll. American Entomological Society.

Described from a single specimen.

Distinguished from xanthopus by the sculpture of the head, color and relative lengths of the joints of the antennæ, longer thorax, color of the hind legs, shorter, oval abdomen, and in the position of the two tubercles on the second abdominal segment.

Dissomphalus carolinensis sp. nov.

♀.—Length 2.6ᵐᵐ. Polished, impunctate; head oblong-quadrate, piceous black; antennæ 13-jointed, not longer than the head, brownish yellow, the flagellum slightly dusky toward apex and very slightly incrassated; the scape is less than half as long as the flagellum and stouter; pedicel larger than the three following joints united, which are small and not longer than thick; the joints beyond these to the last are wider than long; the last joint longer than the penultimate. Thorax and legs brownish-yellow; the tarsi pale; metathorax quadrate, impressed laterally to the middle coxæ. Abdomen black, polished, conic-ovate, as long as the head and thorax united, the incisions pale, the apical margins of segments 3 to 6 slightly emarginate.

HABITAT.—North Carolina.

Type in Berlin Museum.

Described from a single specimen, labeled "Carolina, Zimmermann."

ATELEOPTERUS Förster.

Verh. d. Naturh. Ver. Preuss. Rheinl., 1851. p. 5. Tab. i. f. 1. alæ.

(Type A. Försteri Kirchner, ♂.)

Head in apterous ♀ oblong-quadrate, flattened subconvex above, the eyes small, oblong-oval, the ocelli wanting; in winged ♀ with ocelli; in ♂ shorter, the eyes large, prominent, slightly hairy.

Antennæ 13-jointed, the scape about one-third the length of the head, curved, subclavate, shorter in the ♀ ; pedicel small, rounded.

Maxillary palpi short, 4-jointed; labial palpi 2-jointed.

Mandibles in ♂ 4-dentate, the outer tooth acute.

Thorax elongate, the prothorax long, triangular, rounded in front; mesonotum short without furrows; scutellum rounded behind, with a transverse furrow at base; metathorax subquadrate, roundedly truncate in ♂ ; longer abruptly truncate posteriorly in ♀, with squared sides.

Front wings with only a single basal cell and a punctiform marginal vein; no stigmal vein, the costal cell open; ♀ sometimes apterous.

Abdomen subpetiolate, elongate, cylindrical, pointed at apex, the second and third segments the largest, the apical margins of segments 3, 4, and 5 roundedly emarginate in ♀, in ♂ straight.

Legs rather short and stout, the femora swollen, tibiæ not spinulose, the claws simple.

A genus closely allied to *Scleroderma* and the females apt to be mistaken for those of that genus; but the 4-jointed maxillary palpi and the simple, not spinous, tibiæ sufficiently distinguish them. The male is recognized at once by the single basal cell in the front wings.

Förster characterized the genus from a single male, the female being unknown to him. I have been enabled to perfect his diagnosis by the rearing of both sexes from a cluster of cocoons found in an old stump and from specimens of another species reared from the cosmopolitan beetle *Silvanus surinamensis*.

The species known to me in our fauna may be tabulated as follows:

TABLE OF SPECIES.

Wingless ♀ 2
Winged ♂ and ♀ .
    Black.
        Wings dusky; legs brownish yellow...............A. NUBILIPENNIS, Ashm.
        Wings hyaline.
            Legs rufo-piceous; sutures of abdominal segments rufous............
            ....A. VIRGINIENSIS sp. nov.
            Legs black or piceous black; tarsi honey yellow; sutures of abdomen
                not rufous ....A. TARSALIS sp. nov.
2 Head and thorax rufo-piceous; abdomen longer than the head and thorax
        united, black, the tip rufous.
        Antennæ and legs brownish-yellow ....A. VIRGINIENSIS, sp. nov.

Ateleopterus nubilipennis Ashm.

Ent. Am., III, p. 97 ♂ nec ♀ ; Cress. Syn. Hym. p. 312.

♂. Length 3ᵐᵐ. Black, shining; very faintly closely punctulate, but with a few coarser scattered punctures over the surface and sparsely covered with a black pubescence. Antennæ 13-jointed, the scape brownish yellow, the flagellum fuscous; the scape is one-third the length of the head, clavate, a little bent; the pedicel small; the flagellar joints a little longer than thick. Prothorax a little longer than its width at base, rounded before. Mesonotum short, transverse; scutellum subconvex, with a transverse impressed line at base. Metathorax a little longer than wide, finely rugulose. Wings dark fuscous. Legs brownish yellow. Abdomen black, shining, as long as the head and thorax together, with sparse long hairs, thicker at the pointed apex.

HABITAT.—Jacksonville, Fla.

Type ♂ in Coll. Ashmead.

Ateleopterus virginiensis sp. nov.

(Pl. III, Figs. 1, ♂ ; 2, ♀.)

♂. Length 1.75ᵐᵐ. Black, smooth, shining, varied with rufous or piceous. Antennæ 13-jointed, pale brown or yellowish; the scape one-third the length of the head.

Prothorax long, narrowed in front twice the length of the width at base. Metathorax a little more than twice as long as wide.

Wings hyaline, the veins of the single basal cell pale brownish-yellow. Legs rufous, the femora sometimes dusky above. Abdomen black, the sutures tinged with rufous; sometimes the rufous color extending on to the segments; the posterior margin of the second and third segments emarginate at the middle.

♀. Length 2.2ᵐᵐ. Head and thorax rufous; abdomen piceous-black, the pointed tip rufous, the sutures pale, the posterior margin of the third, fourth, and fifth segments emarginate.

HABITAT.—Arlington, Va.

Types ♂ and ♀ in Coll. Ashmead.

Described from several specimens in both sexes, reared from a cluster of cocoons taken under the bark of an old stump and which have enabled me to complete the generic diagnosis.

Ateleopterus tarsalis, sp. nov.

♀. Length, 1.8ᵐᵐ. Black, polished, impunctate; antennæ dark brown; legs piceous to black, articulations somewhat reddish; tarsi honey-yellow. The head is oblong, a little longer than wide, squarely truncate at base and apex, the sides convex; eyes rather large, oblong oval; antennæ 13-jointed, the scape clavate not quite one-half the length of the head; pedicel 1½ times as long as thick, and much larger than any of the following joints, the flagellar joints closely united, wider than long. Pronotum a little longer than wide, narrower than the head and longer than the mesonotum and scutellum united; mesonotum about half as long as the scutellum, smooth without furrows; scutellum with 2 foveæ at base; metanotum long, quadrate, with the sides margined and a central carina. Wings hyaline, the venation brownish. Abdomen conic-ovate, much wider than the thorax, but not or scarcely longer, perfectly smooth and shiny, with a fine sparse pubescence.

The ♂ measures scarcely 1.5^{mm}, and differs from the ♀ as follows: The head is more rounded; the flagellar joints are a little longer than wide, while the abdomen is shorter and more nearly oval, or short ovate.

HABITAT.—Washington, D. C., and Lafayette, Ind.

Types in Coll. Ashmead.

I first reared this species from raisins infested with *Silvanus surinamensis* and other Coleoptera; subsequently Prof. F. M. Webster sent me several specimens which he reared in Indiana from *Silvanus surinamensis* infesting stored grain. It comes quite close to *A. Försteri* Kirch., of Europe, but differs in being perfectly smooth, impunctured.

APENESIA Westwood.

Thes. Ent. Ox., p. 170.   Trans. Ent. Soc. Lond., 1881, p. 130.

(Type *A. amazonica* Westw.)

(Pl. III. Fig. 1, ♀.)

Head large oblong or subquadrate, flattened; eyes in ♀ minute and composed of only a few hexagons; ocelli wanting; clypeus anteriorly trituberculated; in the ♂ the eyes are normal and the ocelli are present.

Antennæ 13-jointed, short, the first joint long, the second a little longer than the third, the following short; in the ♂ the flagellar joints after the second are a little longer.

Maxillary palpi deformed, 4-jointed; labial palpi 3-jointed.

Mandibles slender, curved, armed at apex with two large teeth.

Thorax in ♀ elongate, the prothorax large, suboval; mesothorax short, the dorsum subcordate; metathorax oblong, contracted near the base, the angles rounded; in ♂ elongate, the pronotum very long, narrowed anteriorly, the mesonotum with one distinct furrow, the metathorax quadrate.

Wings in ♂ similar to *Epyris*, the stigma very minute, the radius very long and slender.

Abdomen in ♀ elongate oval, in ♂ ovate.

Legs in ♀ short, thick, fossorial, middle tibiæ compressed and spinous, the anterior tibiæ terminating in two spurs and the basal joint of the tarsus is lunate and armed on the under side with a row of very fine short bristles; the middle legs, on the contrary, are very robust, the tibiæ strongly serrated on the outer margin and the spurs finely spined; in the middle legs the tarsi are longer than the tibiæ and have the under side of the three basal joints finely spined.

The slender, bidentate mandibles and 4-jointed deformed maxillary palpi, sufficiently distinguish the female of this genus from other wingless genera. The male, which is here described for the first time, comes nearer to the genus *Dissomphalus* than to any other, but the mandibles are longer and more slender, bidentate and cross each other at tips; the head is quite differently shaped, being much longer, a little wider anteriorly than posteriorly: the pronotum wholly different: the meso-notum much shorter and without furrows; the abdominal segments similar, but without the warty-like tubercles on the second segment: while the venation of the wing is quite different, the stigma being minute, the radius very long and slender, the basal cells slightly sub-equal in length.

Apenesia coronata, sp. nov.

(Pl. III, Fig. 3, ♂.)

♂. Length, 3ᵐᵐ. Black, shining, alutaceous: mandibles long, slen-der, crossing each other at tips, bidentate, rufous; antennæ 13-jointed brown; pedicel longer than the first flagellar joint, joints 1 and 2 of flagellum equal, a little longer than thick, joint 3 a little longer, those be-yond somewhat longer than the third; head nearly twice as long as wide, a little wider before than behind, the vertex with several blister-like elevations. Thorax smooth, without furrows, the pronotum nearly three times the length of the mesonotum, narrowed before: scutellum sepa-rated from the mesonotum by a delicate transverse furrow at base; metathorax quadrate, with a longitudinal median carina. Wings hyaline, the venation pale brownish, the stigma minute, the stigmal nervure very long and slender: basal cells two, nearly equal. Legs black, the tibiæ, except at tips, piceous, tips of tibiæ and tarsi pale or whitish. Abdomen ovate, depressed, polished black, about as long as the thorax.

HABITAT.—Bladensburg, Md.

Type in Coll. Ashmead.

For the single specimen of this rare insect, the first male to be de-scribed in the genus and the first species to be detected in the United States, I am indebted to my friend Mr. E. A. Schwarz, who captured it at Bladensburg, July 20, 1890.

CEPHALONOMIA Westw.

Lond. Mag. Nat. Hist., VI. p. 420. [1833]; Syn. Holopedina, Först., Verh. naturh. Ver., preuss. Rheinl., 7. Jahrg., p. 502 (1850); Först., Hym. Stud., II. p. 125, 1856.

(Type C. formiciformis Westw.)

Head in ♀ large, oblong-quadrate, flattened, in ♂ more rounded; the ocelli in the winged form, in both sexes, present: in the wingless female absent.

Antennæ 12-jointed, the pedicel larger than the first flagellar joint, in ♂ filiform, nearly the length of the thorax, in ♀ not or scarcely longer than the head, submoniliform, not incrassated toward tips.

Maxillary palpi 4-jointed; labial palpi 3-jointed.

Mandibles 3-dentate.

Thorax elongate ovate, the prothorax large and triangular, narrowed or rounded in front; mesonotum short, without furrows; scutellum with a transverse furrow at base; metathorax subquadrate, rounded behind.

Front wings with a very short submarginal vein, terminating in a callous spot and a small stigma; costal cell open; the stigmal vein and basal cells wanting. The ♀ is usually apterous without ocelli, although sometimes winged with the venation as in the male.

Abdomen ovate or oval, subpetiolate, the second and third segments large, about equal, the following short, the posterior apical margins of the segments usually straight.

Legs short, stout, the femora swollen, the anterior pair larger than the others, tibial spurs, 1, 2, 2, claws simple.

For many years the true position of this genus was obscure. Prof. Westwood in his original description stated it was allied to *Teleas*, and Förster, as late as 1856, in his Hymenopterologische Studien, without a personal acquaintance with it, incorporated it among the *Diapriinæ*. It has been fully elucidated since, however, and properly placed by Westwood in the *Bethylinæ*, details of which will be found in the Proceedings of the London Entomological Society for 1881, p. 125.

The degraded type of venation and the absence of basal cells readily distinguish the winged form from other genera in the group; but with the apterous forms it is quite different, and great care must be taken to examine the mouth parts for the characters laid down in the table, or one is apt to confuse them with *Scleroderma*.

Only three species are known in our fauna, which may be tabulated as follows:

```
      Winged........................................................ 2
Wingless, without ocelli.
      Entirely brownish yellow or honey-yellow...............C. GALLICOLA Ashm.
      Head, thorax, and legs rufo-testaceous, the collar anteriorly and the meta-
            thorax waxy white.
      Abdomen black.........................................C. CYNIPIPHILA Ashm.
2.  Polished black; legs piceous.
      Wings hyaline .....................................C. HYALINIPENNIS sp. nov.
   Rufo-testaceous, collar and base of abdomen yellowish.
      Wings subhyaline...................................C. NUBILIPENNIS Ashm.
```

Cephalonomia gallicola Ashm.

(Pl. III, Fig. 6, ♀.)

Sclerochroa gallicola Ashm., Ent. Am., III, p. 75, ♀; Cress., Syn. Hym., p. 247.

♀.—Length, 1.8mm. Honey-yellow or pale brownish-yellow, smooth, polished. Eyes small, round. Antennae 12-jointed, about as long as the head; scape one-third the length of the head; pedicel a little longer than thick; the flagellar joints all small, submoniliform. Abdomen pointed, ovate, a little longer than the head and thorax united; the apical margins of the second and third segments slightly emarginate.

HABITAT.—Jacksonville, Fla.

Type ♀ in Coll. Ashmead.

My specimens were bred from a cynipidous oak gall (*Andricus foliatus* Ashm.), and is probably parasitic on some coleopterous larva inhabiting the gall.

Cephalonomia cynipiphila Ashm.

Sclerochroa cynipsiphila Ashm., *loc. cit.*, II, p. 75; Cress., Syn. Hym., p. 247.

♀.—Length, 2^{mm}.   Head, thorax, and legs rufo-testaceous, the collar anteriorly and the metathorax waxy white, the abdomen black, polished.   Antennæ 12-jointed, a little longer than the oblong head; scape slightly longer than one-third the length of the head, bent, narrowed at base; pedicel twice as long as thick; flagellum very slightly thickened at the middle, the first two joints very small, not longer than thick, the following wider than long.

HABITAT.—Jacksonville, Fla.

Type in Coll. Ashmead.

Bred from a cynipidous oak gall (*Holcaspis omnivora* Ashm.).   It is not necessarily parasitic on the cynipid, as other insects, dipterous, lepidopterous and coleopterous, were also reared from the gall.

Cephalonomia hyalinipennis, sp. nov.

(Pl. III, Fig. 5, ♂.)

♂ ♀.—Length, 1 to 1.5^{mm}.   Polished black; scape and pedicel rufous; flagellum brown-black; legs piceous; wings hyaline; tegulæ rufo-piceous.

The female in this species agrees in all respects with the male except in having a much longer and broader abdomen, it being broadly ovate, fully twice as wide as the thorax, and in having a much longer head.

HABITAT.—Jacksonville, Fla.

Type in Coll. Ashmead.

The color and the female in having wings render the species easy of identification.

The ♂ was reared March, 1887, from the galls of *Amphibolips cinerea* Ashm., while the ♀ was bred April, 1887, from *Holcaspis omnivora* Ashm.; but notwithstanding they came from different galls, I believe they are sexes of one and the same species.

Mr. F. H. Chittenden has recently shown me several specimens of what is evidently this same species, reared by him from the coleopteron *Hypothenemus eruditus*, living in the dead twigs of the cultivated fig. It differs from my type in having black femora.

Cephalonomia nubilipennis Ashm.

Holopedina nubilipennis Ashm., *loc. cit.*, III, p. 97.

♂.—Length, 1.8^{mm}.   Rufo-testaceous, smooth, impunctured; there is a dusky blotch across the scutellum and on the disk of the abdomen; collar and base of abdomen yellowish.   The antennæ are 12-jointed, pale brown, $1\frac{1}{2}$ times as long as the head; the scape a little more than one-third the length of the head, the pedicel stouter than the flagellar

21899—No. 45——4

joints; the flagellar joints are longer than thick, the last twice as long
as thick, pubescent. Legs rufous, the femora and tibiæ, at the middle, piceous. Wings subfuscous, with only an open costal cell.

HABITAT.—Florida.

Type ♂ in Coll. Ashmead.

LÆLIUS Ashmead gen. nov.

(Type *L. trogodermatis.*)

Head oblong, nearly as wide across the eyes as long; in ♂ wider
across the eyes than long; eyes large, oval, hairy; ocelli 3 in a triangle,
and larger in the ♂ than in the ♀.

Antennæ 13-jointed, filiform, more than twice as long as the head in
the female, and in the male much longer; the scape is about one-third
the length of the head, cylindrical, very slightly bent, and a little
stouter than the flagellum; the pedicel in the ♀ is a little longer and
stouter than the first flagellar joint, in the ♂ slightly shorter; the
second flagellar joint slightly shorter than the first.

Maxillary palpi 4-jointed: labial palpi 3-jointed.

Mandibles curved, broadened, and truncate at tips, 5-dentate, the
outer tooth acute, followed by small teeth.

Thorax elongate, the prothorax very long, triangular, rounded in
front; mesonotum very short, without furrows; scutellum with a transverse impressed line at base; metathorax oblong quadrate, abruptly
truncate behind, finely transversely rugulose, usually with 3 delicate
longitudinal carinæ.

Front wings with two distinct basal cells, about equal in length, a
very short marginal and stigmal vein, and with the costal cell closed,
a vein running all along the outer margin.

Abdomen subpetiolate, pointed ovate, the third segment the longest,
the fourth about half the length of the third, the following short.

Legs moderate, pilose, the femora swollen, the anterior pair the
stoutest, the tibiæ subclavate, tibial spurs 1, 2, 2, the tarsi longer than
the tibiæ, slender, claws simple.

In venation this genus resembles Bethylus, but the 13-jointed antennæ and the difference in the palpi readily separate the two; while
from *Ateleopterus*, which also has 13-jointed antennæ, it is at once distinguished by the two basal cells and the short but distinct stigmal
vein.

Three species are known to me, which may be separated as follows:

Coxæ and femora, except tips, black.
  Wings hyaline ....L. TROGODERMATIS, sp. nov.
Legs rufous or reddish-yellow.
  Anterior coxæ black, the others usually dusky.
    Metathorax with 3 carinæ on disk..................L. TRICARINATUS, sp. nov.
  All coxæ pale.
    Metathorax with 4 carinæ on disk, transversely rugulose..L. RUFIPES, sp. nov.
    Metathorax with 1 carina on disk, finely punctulate..L. NIGRIPILOSUS, sp. nov.

Lælius trogodermatis, sp. nov.

♀. Length, 2.4ᵐᵐ. Black, highly polished, impunctured, but under a high power exhibiting a fine alutaceous sculpture. Head very little longer than wide, without pubescence; a short keel between antennæ; mandibles 4-dentate, honey-yellow; palpi pale. Antennæ 13-jointed, extending to tegulæ, pale brownish; scape basally black, about as long as the pedicel and first two flagellar joints united; pedicel longer than the first flagellar joint; second flagellar joint slightly shorter than the first; those beyond nearly equal, quadrate. Thorax elongate, the pronotum three or more times as long as the mesonotum, the scutellum with a transverse impressed line at base. All coxæ black; all femora, except tips, black or piceous brown, the trochanters, tibiæ, and tarsi, honey-yellow. Wings hyaline, the venation yellowish. Abdomen a little longer than the thorax, pointed ovate, black, shining, impunctate, with sparse black hairs at apex. In the ♂ the head is slightly wider than long; the antennæ longer; the pedicel is a little shorter than the first flagellar joint, the flagellar joints, after the first, being about twice as long as wide, while the basal joint of the maxillary palpi is slightly swollen.

HABITAT.—District of Columbia (?)

Types in National Museum.

Described from 1 ♂ and 1 ♀ specimen, reared Nov. 1, 1884, from the larva of *Trogoderma tarsale* in the Belfrage collection.

Lælius tricarinatus. sp. nov.

♂ ♀. Length 2.5 to 2.9ᵐᵐ. Black; legs, except coxæ, and antennæ, brownish yellow. Antennæ 13-jointed, twice as long as the head, the pedicel and first flagellar joint about equal, the following shorter but still all longer than thick. Mesonotum very short, without furrows. Scutellum with a transverse fovea across the base. Metathorax longer than wide, transversely rugulose, with 3 carinæ on the disk. Wings hyaline, the venation pale yellowish; marginal and stigmal nervures very short, scarcely developed.

HABITAT.—District of Columbia and Florida.

Types in Coll. Ashmead.

This species is readily distinguished by the three longitudinal metathoracic carinæ and by the anterior coxæ being always black. It comes nearest to *L. nigripilosus*, and like that species is sparsely covered with long black hairs.

Lælius rufipes. sp. nov.

♀. Length 3ᵐᵐ. Black, shining, alutaceous, sparsely pilose; antennæ 13-jointed, brown; three basal joints, mandibles, palpi, and legs rufous. Metathorax with 4 longitudinal carinæ on the disk, the surface rugulose. Abdomen pointed ovate, shorter than the thorax, polished black, sparsely pilose. Wings hyaline or but faintly tinged, the venation pale, the marginal and stigmal nervures but slightly developed. The antennæ are rather long, the scape 2½ times the length of the pedicel; first flagellar joint shorter than the pedicel, the joints beyond all longer than thick.

HABITAT.—Western States.

Types in National Museum and Coll. Ashmead.

Lælius nigripilosus, sp. nov.

(Pl. III. Fig. 7, ♀.)

♂ ♀. Length 1.8 to 2.5ᵐᵐ. Black, shining, faintly alutaceous, the head in the ♀ much wider than in the ♂, similar to the head in *Goniozus*. Mandibles in ♀ black, in ♂ rufous. Antennæ 13-jointed, pale brown; flagellar joints 1 to 3 nearly equal, about twice as long as thick, the following shorter; in the ♀ the flagellum is stouter and the joints shorter than in the ♂. Mesonotum with 2 delicate furrows, indistinct or wanting in the male. Scutellum with a distinct transverse furrow at base. Metathorax subquadrate, the dorsum with a single central carina and with a scaly punctuation. Wings grayish-hyaline, the nervures yellowish or pale, the marginal and radius very short, scarcely developed. Legs ferruginous or reddish-yellow. Abdomen in ♀ conic-ovate, about as long as the head and thorax united, subpetiolate; in ♂ ovate, much shorter than the thorax. Both sexes are covered with long, sparse black hairs, more apparent in the male, and denser on the head and towards apex of abdomen.

HABITAT.—Jacksonville, Fla.

Types in Coll. Ashmead.

BETHYLUS Latreille.

Hist. Nat., III, 315 (1802); Förster, Hym. Stud., II, 96 (1856).

(Type *B. cenopterus* Panz.)

Head oblong, smooth; eyes broadly oval or rounded; ocelli 3, distinct, prominent.

Antennæ 12-jointed, filiform, moniliform, or setaceous; scape very short; pedicel in ♀ a little longer than the first flagellar joint, in ♂ shorter.

Maxillary palpi 6-jointed; labial palpi 3-jointed.

Mandibles short, curved, not very broad at tips, truncate, with 3 or 4 small teeth, the outer acute.

Thorax subovate; the prothorax triangular, rounded in front; mesonotum most frequently without furrows; short, transverse; metathorax usually with delicate longitudinal keels, the apex subtruncate.

Front wings with a very small marginal vein, a short stigmal vein, and two basal cells of an equal length; no traces of other veins or cells.

Abdomen ovate, pointed at tip, the third segment the longest, the apical margins of the segments straight.

Legs short, stout, the femora much swollen, the anterior pair the stoutest, the tarsi not or scarcely longer than the tibiæ, slender, claws simple.

This genus, the type of the subfamily, is of small extent, few species in it having been discovered either in Europe or America.

It resembles my genus *Lælius* in venation, but the 12-jointed antennæ and the 6-jointed palpi readily distinguish it.

Our species may be tabulated as follows:

TABLE OF SPECIES.

FEMALES.

Abdomen with a strong constriction between the third and fourth segments.

    Legs, except coxæ, reddish-yellow..................B. CONSTRICTUS, sp. nov.

Abdomen normal.

    Legs honey-yellow...B. PEDATUS, Say.

    Legs black, tibiæ and tarsi brownish,.....................B. CENTRATUS, Say.

Bethylus constrictus, sp. nov.

♀. Length 3^{mm}. Black, shining, impunctured, with some sparse black hairs. Head a little longer than wide. Antennæ 12-jointed, less than twice the length of the head, brown, the pedicel very small, the flagellar joints very little longer than wide. The mesonotum shows traces of furrows posteriorly. Metathorax finely, closely punctate, subopaque, with raised longitudinal lines towards the base. Legs reddish-yellow, the coxæ black. Wings dilute fuscous, the veins brown. Abdomen black, highly polished, as long as the thorax, the petiole very short, the second and third segments long, the latter the longer, with a strong constriction between it and the fourth, the apex produced into a stylus-like point, with long, sparse black hairs.

HABITAT.—Jacksonville, Fla.

Type in Coll. Ashmead.

Described from a single specimen, which is remarkable for, and easily distinguished by, the constriction between the third and fourth abdominal segments.

Bethylus pedatus Say.

(Pl. III, Fig. 8, ♀.)

Bost. Jour., I, p. 279; Lec. Ed. Say's Works, II, p. 727; Ashm. Ent. Am., III, p. 97; Cress. Syn. Hym., p. 247.

♀. Length 2.5^{mm}. Polished black, with sparse hairs. Head hardly longer than wide. Antennæ 12-jointed, 1½ times as long as the head, honey-yellow or pale brownish-yellow, fuscous at tips; the scape is a little more than thrice as long as thick, the joints of the flagellum scarcely longer than thick. Mesonotum without furrows. Scutellum with a transverse impressed line at base. Mesonotum finely punctulate, with longitudinal raised lines or carinæ. Wings hyaline, very faintly tinged, the venation yellowish. Legs dark honey-yellow, the coxæ dusky. Abdomen as long as the thorax, polished, terminating in a stylus-like point, with a few long hairs.

The ♂ differs only in its antennal characters and in having a broader head. The antennae are nearly thrice as long as the head, filiform, tapering at tips, the scape shorter, the joints of the flagellum a little longer.

HABITAT.—Indiana, Florida, and Virginia.

Specimens in Coll. Ashmead.

I have had no difficulty in recognizing this and the following species of Say's from his remarks at the end of his description, as he says: "This species and the following (*B. centratus*) are remarkable for the brevity of the inflected tip of the radial nervure, which is not at all arcuated, but points obliquely inwards."

Bethylus centratus Say.

Lec. Ed. Say's Works, II. p. 727; Ashm. Ent. Am., III. p. 97; Cress. Syn. Hym., p. 247.

♀. Length 3ᵐᵐ. Polished black, with some scattered hairs. Head across the eyes fully as wide as long. Antennae 12-jointed, brownish-piceous, the scape black, the pedicel yellowish, the joints of the flagellum a little longer than thick. Mesonotum with 2 more or less distinct furrows. Metathorax closely, finely punctured, with raised longitudinal lines. Wings subhyaline, the venation brown. Legs black, the tibiae and tarsi brown or piceous. Abdomen ovate, scarcely as long as the thorax, polished black, with a slight projecting point at apex.

HABITAT.—Indiana, Colorado, and Florida.

Specimens in Coll. Ashmead.

EUPSENELLA Westwood.

Thes. Ent. Oxon. p. 168 (1874).

(Type E. *agilis* Westw.)

(Pl. IV, Fig. 1, ♂.)

Head oblong, subconvex, broader than the thorax; clypeus angulated at the middle; eyes smooth; ocelli 3, distinct.

Antennae 13-jointed, rather short, the scape thickened, the 12 following joints short nearly equal.

Maxillary palpi 6-jointed; labial palpi 3-jointed.

Mandibles stout, 5- or 6-dentate, the teeth obtuse.

Thorax: prothorax large, semicircular; mesonotum with 2 furrows; metathorax transverse, the posterior angles rounded, between punctured with a semicircular impression at base and a slender median carina.

Front wings with a large stigma, a complete marginal, one submarginal, two discoidal and two basal cells; the radial vein is recurved at apex and the second discoidal cell is very long, extending beyond the apex of the marginal cell.

Abdomen oblong ovate, smooth.

Legs short, the anterior femora especially stout; claws strongly curved, dilated at base.

A genus described from Australia and unknown to me. Prof. Westwood further describes the "maxillæ as minute, terminating in 3 ciliated lobes; the mentum minute; the labium minute, membranaceous, hairy." The closed marginal cell and the two discoidal cells readily distinguish the genus from all others.

SIEROLA Cameron.

Trans. Lond. Ent. Soc., 1881, p. 556

(Type *S. testaceipes* Cam.)

Head oblong.

Antennæ 13-jointed, not much longer than the head; the scape thicker than the other joints and as long as the following, the third and fourth slightly longer than the succeeding and produced beneath on the longer side; remaining joints not much longer than broad, distinctly separated from each other.

Palpi (?).

Mandibles (?).

Thorax (?).

Front wings with a stigma and a prostigma, the two being separated by a hyaline space, costa thickened in front of the stigma; radial cell completely closed; two humeral cells unequal and closed; from the upper end of the lower (and smaller) cellule there proceeds a small oval cellule, which is united to the prostigma by a short thick nervure, so that the upper humeral cellule is thus completely closed. From the end of the radial cellule runs to the edge of the wing a white spurious vein; another runs in the same direction from the small oval cellule above mentioned, the two being united by a cross nervure halfway between the radial cellule and the apex of the wing; another spurious vein runs from the lower humeral cellule to the bottom of the wing.

Abdomen longer than the head and thorax, the third and fourth segments contracted in the middle at the junction; in length subequal; the last is acuminated.

Legs as in *Perisemus*.

A genus unknown to me and not yet recognized in North America, the type, *S. testaceipes* Cameron, coming from the Sandwich Islands. The species described by me as such, *S. maculipennis*, Entomologica Americana, vol. III, p. 75, is a peculiar little braconid belonging to the subfamily *Euphorinæ*, and will probably form the type of a new genus.

According to Cameron: "This genus differs from all the genera of the Bethylloidæ in having the radial cellule completely closed and in the presence of the small oval cellule uniting the humeral cellules. It comes nearest to *Goniozus* Förster. In the shape of the prostigma it more resembles *Perisemus*."

Cameron does not appear to be acquainted with *Eupsenella* Westw., which also has a closed radial cell and next to which I have placed it, believing it closely allied.

Since the above was written I have received from Mr. G. C. Davis, of Agricultural College, Michigan, a remarkable male Bethylid, taken in South Dakota, that agrees with *Sierola* in having a closed marginal cell, but in other particulars differs so widely as to lead me to believe it will yet prove quite distinct and form the type of a new genus.

As Mr. Cameron, however, in his diagnosis, fails to define the palpal, mandibular, and thoracic characteristics of *Sierola*, I am unable to decide the question at present, and rather than run the risk of creating a synonymn prefer to describe it doubtfully under this genus. The front wing is represented on Pl. IV, Fig. 2.

(?) Sierola ambigua sp. nov.

(Pl. IV, Fig. 2, ♂.)

♂ Length about 4.5ᵐᵐ. Black, polished, finely sericeous; sutures of trochanters, extreme tips of all femora and tibiæ and tarsi, except last joint, honey-yellow; tibial spurs 1, 2, 2; claws with a small tooth at base; wings subfuscous; palpi fuscous.

The head is transverse, not quite as broad as the mesothorax between the wings; ocelli 3, triangularly arranged; eyes oblong-oval, bare, reaching not quite to the base of the mandibles; antennæ 13-jointed, filiform, extending to base of metathorax; scape subglobose, obliquely truncate at tip; pedicel small, rounded; first flagellar joint scarcely two-thirds the length of second; the second joint longer than any other except the last, the third and following to last joint very gradually shortened the last a little longer than the second. Thorax shaped much as in *Epyris*, except that the metanotum is smooth and polished instead of rugose; the pronotum is large, trapezoidal, the collar anteriorly transversely impressed; the mesoscutum with two deep parapsidal furrows, and with a longitudinal furrow on the parapsides; scutellum with a transverse furrow at base; metathorax subquadrate, the posterior angles slightly rounded, margined at sides. Wings ample, subfuscous, the venation as in figure, brown-black or fuscous; the marginal cell closed and not longer than the stigma; the second recurrent and second transverse cubital veins partially visible as hyaline lines. Abdomen lost.

HABITAT.—Brookings, S. Dak.

Type in coll. Ashmead.

Described from a single specimen kindly given me by Mr. G. C. Davis.

CALYOZA Westwood.

(Pl. IV, Fig. 3, ♂.)

Thes. Ent. Oxon.. p. 156.   1874.

(Type, *C. staphylinoides* Westw.)

Head large, subconvex; eyes placed at the side of the head before the middle; ocelli distinct.

Antennæ 13-jointed, joints 2 and 3 minute, joints 4 to 12 in ♂ ramose, in ♀ simple, cylindrical.

Maxillary palpi moderately elongated, 6-jointed, joints 1 and 2 short, the 4 following longer, nearly equal and gradually more slender; maxilla small, terminating at apex in three flat, membranaceous, ciliated lobes; labial palpi short, 3-jointed; mentum subobovate.

Mandibles elongate, in ♂ broadly obliquely truncate at apex, with a large apical tooth, followed by small obtuse teeth; in ♀ with large irregular teeth, internal tooth obtuse; beneath armed with rigid bristles; the apical middle tooth large, triangular; external small, obtuse.

Thorax: Prothorax large, subtriangular, anteriorly rounded; mesothorax without furrows; metathorax longitudinally sulcate and transversely striolated.

Front wings with a large stigma, an incomplete marginal and two basal cells.

Legs of the usual form; claws acute, broad at base, and armed with a tooth at the middle.

A genus not found in America, north of Mexico, but readily known by the branched antennæ in the male.

EPYRIS Westwood.

Phil. Mag., 1832, p. 129; Hal. Ent. Mag., IV, p. 432; Först. Hym. Stud., II, p. 96; Westw. Thes. Ent. Oxon., p. 157.

(Type, *E. niger* Westw.)

Head oblong, not much longer than wide, subconvex, wider than thorax; eyes hairy; ocelli distinct.

Antennæ 13-jointed, long, slender, cylindrical, much shorter in ♀, the scape thickened and several times longer than the pedicel, the following joints short, submoniliform, nearly equal in length.

Maxillary palpi long, 6-jointed, the three basal joints thickened, the first minute, the second and third gradually longer, the last three long, slender; maxilla terminating in two oval membranaceous lobes, ciliated at apex; labial palpi 3-jointed, the joints gradually becoming longer; mentum small, somewhat broad at apex.

Mandibles falcate, the apex oblique, 5-dentate, the outer tooth large, acute.

Thorax: Prothorax triangular, the apex rounded and with a deep transverse furrow above; mesothorax trapezoidal, the mesonotum with two distinct furrows: scutellum with a transverse furrow or fovea at base; metathorax transverse or subquadrate, longitudinally carinated.

Front wings with a moderate sized stigma, an incomplete marginal and two basal cells, the discoidal cells obliterated; the basal cells are about equal in length, the first oblique at apex, the second rounded or somewhat angulate.

Abdomen ovate, subpetiolate, segments 2 and 3 the longest; the apical margins of the segments are entire.

Legs stout, the femora much swollen, especially the anterior pair, claws often with a tooth towards the base.

This genus might easily be mistaken for forms in *Calyoza*, *Mesitius*, and *Isobrachium*. The much swollen fore femora separate it from *Isobrachium*; the transverse fovea or grooved line at the base of the scutullum from *Mesitius*; while the bilobed maxilla separates it from *Calyoza*; but when one is in doubt, or until one becomes familiar with the habitus of the genus, the full generic description must be consulted.

It is quite well represented in our fauna and no doubt many new species will be discovered when the *Proctotrypidæ* are more carefully collected.

The species at present known to me may be thus tabulated:

TABLE OF SPECIES.

FEMALES.

Coxæ and legs rufous or pale rufous 2
Coxæ black.
  Legs rufous, the femora more or less fuscous or dusky.
    Wings subhyaline.
      Head and thorax æneous-black; metathorax with 5 longitudinal raised lines
        with transverse lines between............E. OCCIDENTALIS, sp. nov.
  Legs and antennæ rufous.
    Wings subfuscous.
      Head and thorax opaque black, very finely, closely punctulate; metathorax
        with a central raised line, the surface on each side being reticulate
        with raised lines, ♂ .... E. CARBONARIUS, sp. nov.
2. Head black, not at all metallic.. 3
Head æneous, abdomen bluish.
  Wings fuscous..E. ÆNEICEPS, sp. nov.
3. Front coxæ black; antennæ and legs rufous.
  Wings subfuscous.
    Metathorax with many longitudinal raised lines and irregular crosslines be-
      tween; apex of abdomen rufous.......................E. ANALIS Cr.
All coxæ and legs rufous or honey-yellow.
  Wings subhyaline.
    Head not unusually large or broad.
      Metathorax with 6 or 8 raised longitudinal lines.
        Head and thorax black.
          Mesopleura foveated; *mandibles 5-dentate*..............E. RUFIPES, Say.
          Mesopleura areolated; *mandibles 6-dentate* ....E. COLUMBIANUS, sp. nov.
        Head and thorax with a bluish tinge, or submetallic....E. GRANDIS, Ashm.
      Head very large and broad.............. ....E. MEGACEPHALUS, sp. nov.

Epyris occidentalis, sp. nov.

♀. Length 5ᵐᵐ. Æneous black, shining; head with sparse, distinct punctures; mandibles large, 5-dentate, rufous; palpi and antennæ rufous, the latter brownish toward tips. Metathorax with 5 longitudinal carinæ, the two lateral abbreviated before attaining the posterior margin, the surface between the carinæ transversely rugulose, sides finely longitudinally striate. Wings yellowish-hyaline. Legs rufous, the coxæ blackish, the anterior pair always black, anterior and posterior femora blackish or rufo-piceous.

♂. Length 4.3ᵐᵐ. Agrees with female in color except the posterior coxæ are blackish and the antennæ are brown; structurally it differs in the long, acuminate antennæ, the joints being at least 3 times as long as thick; and in having a smaller less pointed abdomen.

HABITAT.—California.

Types in Coll. Ashmead.

Described from 1 ♂ and 1 ♀ specimen.

Epyris carbonarius, sp. nov.

(Pl. iv, Fig. 4, ♂.)

♂. Length 3.5ᵐᵐ. Black, opaque, closely, finely punctate, sparsely covered with a glittering white pubescence. Head very large, one-fourth longer than wide. Antennæ, mandibles, palpi and legs rufous, the coxæ black. Antennæ 13-jointed, extending beyond the tegulæ; the scape curved, narrowed basally, three times as long as thick at the tip; pedicel wider than long, hardly half as long as the first flagellar joint; the first and second flagellar joints about equal, much longer than thick; remaining to the ultimate, shorter but all of about an equal length, the ultimate longer and thinner, 2½ times as long as thick. Prothorax longer than wide, laterally impressed; mesonotum broader than long, without distinct furrows, although with a strong lens faint traces of them can be discerned anteriorly; scutellum with a transverse impressed line at base; metathorax longer than broad, finely punctate, slightly rugose basally, with a trace of a median longitudinal carina. Wings fusco-hyaline, the venation brown, the stigma subquadrate, the transverse median nervure oblique, with a spurious nervure almost interstitial with the median nervure.

HABITAT.—District of Columbia.

Type in Coll. Ashmead.

Epyris æneiceps, sp. nov.

♀. Length 4ᵐᵐ. Bluish æneous; the head with a slight greenish metallic luster, sparsely punctate: mandibles, palpi, antennæ, and tegulæ, rufous; wings subfuscous; metathorax with 5 close, longitudinal carinæ on disk, interstices and laterally, transversely rugulose, the abrupt sides longitudinally striated.

HABITAT.—Fort Capron, Fla.

Type in Coll. Ashmead.

Described from a single specimen taken by Mr. E. A. Schwarz.

Epyris analis, Cress.

Trans. Am. Ent. Soc. IV. p. 193, ♀ ; Ashm. Ent. Am., III, p. 76; Cress. Syn. Hym., p. 247.

♀. Length 6ᵐᵐ. Head and thorax dark metallic green, shining, sparsely clothed with pale pubescence; head subquadrate, flattened, rather closely and finely punctured; mandibles dark ferruginous; palpi whitish; antennæ short, robust, ferruginous; pro and mesothorax sparsely and finely punctured; metathorax black, quadrate, abrupt laterally and abruptly truncate at tip, above minutely and transversely aciculated, the disk with 4 or 5 approximate longitudinal carinæ, posterior face transversely aciculated, the sides longitudinally so; tegulæ ferruginous, wings tinged with yellowish-fuscous; legs short, robust, pale ferruginous, anterior coxæ black, all the femora incrassated; abdomen elongate ovate, smooth and polished, black, immaculate, apex ferruginous.

HABITAT.—Texas.

Type in Coll. American Entomological Society.

Epyris rufipes, Say.

Bethylus rufipes Say, Lec. Ed. Say's Works I, p. 221.
Epyris rufipes Cr. Trans. Am. Ent. Soc. IV. p. 193; Ashm. Ent. Am. III, p. 76; Cr. Syn. Hym., p. 247.

♀. Length 4ᵐᵐ. Black; head large, with sparse punctures; eyes hairy; mandibles large, broad, rufous, 6-dentate; legs and antennæ rufous, the latter a little dusky at tips.

Antennæ 13-jointed, long, the scape as long as the first three flagellar joints together; flagellum tapering off at tip, the first joint the smallest, the others all much longer than thick. Thorax coriaceous, the metathorax with longitudinal raised lines, the surface between transversely rugulose. Wings subhyaline, the nervures brownish, the radius very long. Abdomen ovate, pointed at tip, sparsely hairy, black and polished, the extreme apical edges of segments 3 to 6 rufous.

HABITAT.—Indiana and Southern States.

Specimens in Coll. American Entomological Society and Coll. Ashmead.

Epyris columbianus, sp. nov.

♀. Length 5ᵐᵐ. Black, shining, coriaceous; head sparsely punctate, mandibles rufous, large, 6-dentate; antennæ, except flagellum, legs and tegulæ rufous; flagellum dusky; wings subfuscous. Metathorax with 5 longitudinal carinæ on the disk, the two between the central and the outer abbreviated, the outer curving outwardly towards apex, the surface transversely rugulose.

The ♂ is only 4ᵐᵐ in length, and agrees with the female, except that the antennæ are longer, the flagellar joints being 2½ times as long as thick, the thorax with some sparse punctures, the abdomen elongate, the second, third, and fifth joints about equal, the fourth being only half the length of the third.

HABITAT.—Washington, D. C., and Jacksonville, Fla.

Types in Coll. Ashmead.

Distinct from *E. rufipes* Say in its larger size, darker wings and the 6-dentate mandibles.

Epyris grandis Ashm.

Goniozus grandis Ashm. Ent. Am. III, p. 76; Cress. Syn. Hym., p. 247.

♂ . Length 6.25^{mm}. Polished black, with a few faint scattered punctures. Head one-fourth longer than wide, carinated between the bases of antennæ. Mandibles rufous, piceous at base. Palpi pale. Antennæ 13-jointed, long, filiform, brown, tapering at tips and extending to the apex of the metathorax; the scape is only twice as long as thick at tip; the pedicel a little shorter than the first flagellar joint, not longer than thick; the first flagellar joint slightly longer than thick and about one-fourth shorter than the second; the remaining joints longer, about equal in length but becoming slenderer towards the apex. Prothorax as long as the head, trapezoidal; mesonotum with two abbreviated furrows in ♂, entire in ♀ ; scutellum with a furrow across the base; metathorax quadrate, with numerous faint raised lines at base; mesopleura smooth with a round fovea just beneath the tegulæ, and divided into two parts by a longitudinal furrow which extends from tegulæ to middle coxæ. Wings hyaline, the venation brown, the transverse median nervure arcuate outwardly with a spurious vein directed towards the disk of the wing.

HABITAT.—Jacksonville, Fla.

Types ♂ and ♀ in Coll. Ashmead.

The spurious backward directed vein from the transverse median nervure caused me to confound it with the genus *Goniozus* in my original description of the species.

Epyris megacephalus, sp. nov.

♀ . Length, 6^{mm}. Æneous black; head large, broader than the thorax, rather closely, strongly punctate; mandibles large, the fourth tooth very broad; antennæ and legs rufous, the anterior coxæ black; metathorax with several longitudinal carinæ on the disk, the surface transversely rugulose, the surface at posterior angle smooth, polished. Abdomen pointed ovate, shorter than the thorax, smooth, polished; the apical margins of the segments, after the second, narrowly rufous; segments 2, 3, and 4 nearly equal, longer than the following.

HABITAT.—California.

Type in Coll. Ashmead.

Described from a single specimen.

MESITIUS Spinola.

Compte. rendu Hym. de Para (1846); Mém. Acad. Turin, Sér. II, tom 13 (1853); Westw., Thes. Ent. Oxon., p. 222.

(Type *M. ghilianii* Spin.)

Head oblong, subconvex, not much longer than wide; eyes oval; ocelli distinct.

Antennae 13-jointed, in ♂ long, in ♀ much shorter, the scape thickened, about four times as long as the pedicel, the following joints short.

Maxillary palpi rather long, 6-jointed, the three basal joints gradually increasing slightly in length, the three terminal ones longer and subequal; maxilla terminates in three membraneous ciliated lobes; labial palpi 3-jointed, the joints nearly equal.

Mandibles oblong, with the apex oblique and but slightly sinuated; in the ♀ the outer tooth is small, acute, followed by a very small tooth, the rest of the surface scarcely denticulate; in ♂ 4- or 5-dentate.

Thorax: Prothorax long, triangular or trapezoidal, the apex at the junction with the head contracted, with a deep transverse furrow above; mesonotum usually with two distinct furrows, often abbreviated posteriorly; scutellum with two foveae at base; metathorax with prominent posterior angles, the dorsum with many longitudinal carinae.

Front wings with a moderate sized stigma, a long, incomplete marginal cell and two basal cells, the apices of both being more or less oblique.

Abdomen ovate or oblong-ovate, smooth, the second segment the longest, the apical margins sinuate or emarginate.

Legs as in *Epyris*, the claws slender, nearly straight, with a tooth at the middle.

This genus closely resembles *Epyris* and great care is necessary to distinguish it from that genus. As far as the North American species are concerned I have had no difficulty in separating them by the two foveae at the base of the scutellum.

Westwood, in Thesaurus Entomologicus Oxoniensis, p. 222, and in the Transactions of the Entomological Society of London, 1881, p. 125, states that the genera *Isobrachium* Förster (Hym. Stud., II, p. 96, 1856) and *Heterocoelia* Dahlbom (Hym. Europ., II, p. 21, 1854) are synonymous with *Mesitius*, an opinion in which I can not concur. The apical segments of the abdomen of *Heterocoelia nigriventris* Dahl., the type of the genus, is figured by Dahlbom, loc. cit., p. 23, and it, as well as the description, plainly point to a chrysidid. Dahlbom also figures it on Pl. I, Fig. 15. A careful comparison of this figure with Westwood's (Thes., Pl. 31, Fig. 10) plainly shows that Dahlbom has a genuine chrysidid and Westwood a genuine proctotrypid.

It is inexplicable to me how so careful a worker as Westwood could have made so grave an error. Förster, in his definition of the genus *Isobrachium*, evidently confused and correlated as sexes two distinct insects. His *Isobrachium dichotomus* is a ♀, and evidently a genuine

Mesitius, while its supposed ♂, *Omalus fuscicornis* Jurine, is a genuine *Isobrachium*. An error of this kind does not necessarily invalidate the genus, the male still holding good as the type of the genus. A male specimen of *Omalus fuscicornis* is in my collection, and although it closely resembles males of *Epyris* and *Mesitius* it is quite distinct, as I have pointed out, and it is here made the type of the genus *Isobrachium* Förster.

The following table will be found useful in separating the North American species:

TABLE OF SPECIES.

Coxæ and legs pale rufous or honey-yellow 2
Coxæ and femora black; wings subhyaline or fuscous.
  Antennæ and legs black, the tarsi fuscous ♂ ....M. MONTICOLA Ashm.
  Antennæ and legs dark rufous ♀ ....M. VANCOUVERENSIS, sp. nov.
Front coxæ and femora black, middle and posterior femora rufous, or only slightly dusky.
  Trochanters, tibiæ, and tarsi rufous.
    Wings subfuscous ♀ ....M. NEVADENSIS, sp. nov.
    Wings hyaline.
      Anterior tibiæ and tarsi and middle and posterior legs fuscous; antennæ dark fuscous ♂ ....M. MINUTUS, sp. nov.
      Anterior tibiæ and tarsi and middle and posterior legs, brownish-yellow; their femora more or less dusky; antennæ rufous ♀.
        M. CALIFORNICUS, sp. nov.
2. Wings subhyaline.
  Legs brownish-yellow or pale rufous.
    Tip of abdomen not rufous; metathorax with about 7 raised lines, the lateral abbreviated; ♂ antennæ fuscous: ♀ antennæ yellow.
      M. BIFOVEOLATUS, sp. nov.
    Tip of abdomen rufous; metathorax with about 12 raised lines ♀.
      M. TEXANUS, sp. nov.
Wings not fully developed, reaching only slightly beyond base of abdomen.
  Legs rufous..................................M. BRACHYPTERUS, sp. nov.

Mesitius monticola Ashm.

Epyris monticola Ashm. Bull. No. 1, Col. Biol. Assoc., p. 8, 1890, ♂.

♂. Length 4ᵐᵐ. Black, smooth, shining. Head sparsely punctate, each ocellus surrounded in front by a depression. Antennæ black, covered with a fuscous pubescence, 13-jointed, reaching to the middle of the metathorax, the first joint the thickest, the length of the fourth, the second joint the shortest, the third slightly longer, the fourth and following joints slightly longer than the third and about of an equal length or very nearly so; terminally the antennæ are a little thinner than at base. Thorax above with some rather long, fuscous hairs, trapezoidal, the mesonotum with two grooves, slightly converging toward each other posteriorly but still widely separated. Scutellum flattened, with two small, widely-separated foveæ at base, each fovea being exactly opposite the termination of the mesonotal grooves. Mesopleura finely, microscopically sculptured, with a rounded impression or fovea at the middle. Metathorax quadrate, truncate behind, the margins delicately keeled and a delicate longitudinal keel down the center, the disk toward the base delicately longitudinally wrinkled; on each side, between the wrinkled portion and the lateral margins, is a smoother space that exhibits a minutely transverse strigose sculpture; the truncature is a little depressed in the middle and with a delicate median keel. Legs black; the tibiæ and tarsi are rather densely covered with a brownish pubescence and in consequence appear brown. Abdomen black, polished, the second segment the longest. Wings hyaline, veins brown; the second basal cell is half the length of the basal nervure longer than the first, the radius very long, as long as the second basal cell.

HABITAT.—Wales Canyon, Colorado.

Type ♂ in Coll. Ashmead.

Described from a single specimen, received from T. D. A. Cockerell.

Mesitius vancouverensis sp. nov.

(Pl. IV, Fig. 5, ♀.)

♀. Length 6.4^{mm}. Black, shining, impunctured, except a few small scattered punctures on the head and thorax. Head oval, 1½ times as long as wide. Mandibles and palpi rufous; the large outer tooth black. Antennæ 13-jointed, acuminate at tips, extending to base of metathorax, the flagellar joints after the first all longer than thick, the first about equal with the pedicel, not, or scarcely, longer than thick. Prothorax very long, three times the length of the mesonotum, the latter with 2 parallel furrows, faint anteriorly, more deeply impressed posteriorly; scutellum with 2 foveæ at base directly opposite the mesonotal lines; metathorax almost smooth, quadrate; there is a median longitudinal carina extending to the apex and the surface on each side is finely sculptured. Tegulæ pale rufous. Wings subfuscous, the venation brown; the stigmal vein is very long, almost forming a closed radial cell; the transverse medial nervure is strongly angularly curved, so that its apex is parallel with the apex of the basal nervure. Legs black, the tibiæ and tarsi dark rufous. Abdomen ovate, shorter but wider than the thorax, black and shining, with the extreme apical edges of the segments tinged with rufous, sparsely pilose.

HABITAT.—Vancouver Island.

Type ♀ in Coll. Ashmead.

Described from a single specimen received from W. Hague Harrington.

Mesitius nevadensis sp. nov.

♀. Length 6^{mm}. Very close to *M. vancourerensis*, but differs as follows: The legs, except the front coxæ and femora, are rufous, the middle and posterior femora rufo-piceous; mandibles terminating in a long blunt tooth, the surface within obtuse, scarcely denticulate, while the flagellar joints, after the first, are twice as long as thick.

HABITAT.—Nevada.

Type ♀ in Coll. Ashmead.

Described from a single specimen.

Mesitius minutus, sp. nov.

♂. Length, 2.2^{mm}. Black, shining, the surface minutely coriaceous, very finely sparsely pubescent. Head a little longer than broad across the eyes. Mandibles rufous, not broadened at tips, the outer tooth acute, followed by about 4 very minute denticulations. Antennæ 13-jointed, long, brown; scape stout, not more than twice as long as thick and very little longer than the first funiclar joint; pedicel and first funiclar joint about equal; the remaining joints longer, fully thrice as long as thick. Prothorax twice as long as the mesonotum, rounded anteriorly; mesonotum with 2 furrows more deeply impressed posteriorly; scutellum with two oblique foveæ opposite the mesonotal furrows; metathorax subquadrate, the posterior angles a little rounded, minutely coriaceous, with a median longitudinal carina and some very short raised lines at base. Tegulæ brownish-yellow. Wings hyaline, the venation yellowish or pale; the transverse medial nervure curved outwardly. Legs rufous, coxæ and anterior femora black, the middle and posterior femora fuscous. Abdomen polished black, shorter than the thorax, the third segment one-third longer than the fourth.

HABITAT.—Arlington, Va.

Type ♂ in Coll. Ashmead.

Described from a single specimen.

Mesitius californicus, sp. nov.

♀. Length, 3.1^{mm}. Polished black, the head alutaceous. Head oblong, 1½ times as long as broad, the eyes whitish. Mandibles brownish-yellow, truncate but not broadened at apex; the outer tooth acute, followed by about 4 minute teeth. Antennæ 13-jointed, extending to the tegulæ; scape stout, piceous, as long as the three following joints united, a little curved and narrowed at base; flagellum brown, tapering towards tip; pedicel and first flagellar joint about equal, the following joints 1½ times as long as thick, the last longer. Prothorax fully thrice as long as the metanotum, rounded before; mesonotum with 2 nearly parallel furrows and a delicate short line on the scapulæ; scutellum with 2 oblique foveæ at base; metathorax about 1½ times as long as wide, minutely sculptured with a median carina and some faint, short, irregular raised lines at base. Wings hyaline, very faintly tinged, the nervures brown, the venation as in *vancouverensis*. Legs rufous, coxæ and anterior femora black, middle and posterior femora more or less piceous. Abdomen oblong ovate, nearly as long as the thorax, black,

sparsely pilose, especially at tip; the extreme apical margins of segments very slightly rufous.

HABITAT.—California.

Types ♀ in Coll. Ashmead and National Museum.

Described from several specimens.

Mesitius bifoveolatus, sp. nov.

♀. Length, 5.5ᵐᵐ. Polished black, impunctured, the surface microscopically alutaceous; antennæ, mandibles, and legs entirely rufous.
Head about one and a half times as long as wide. The outer tooth of
mandibles large, acute. Antennæ 13-jointed, reaching to the tegulæ,
twisted; the scape swollen, as long as the 4 following joints; pedicel
longer but slightly narrower than the first flagellar joint; the flagellar
joints very slightly but gradually increasing in length, the first wider
than long, the last very slightly more than twice as long as thick.
Prothorax three and a half times as long as the mesonotum, rounded
before, the superior edge of collar rufous; mesonotum with 2 distinct
furrows on a delicate line on the scapulæ; scutellum with 2 oblique
foveæ at base; metathorax quadrate, with numerous raised lineations
at the middle and base, almost smooth towards the lateral margins, the
truncate sides and apex microscopically striated. Tegulæ yellow.
Wings subhyaline, the venation yellow and as in *vancouverensis*. Abdomen pointed ovate, much shorter than the thorax, polished black,
pilose towards apex and with the extreme edges of apical margins of
the last three or four segments rufous.

The ♂ is from 3.5 to 4ᵐᵐ long and is recognized at once by the long
fuscous antennæ, as long, or nearly as long, as the body, the first flagellar
joint being half again as long as the pedicel, the joints beyond fully
four times as long as thick, the last 5 or 6 times as long as thick, and
by the more fuscous wings; otherwise it is quite similar.

HABITAT.—Georgia, Florida, Canada.

Types in National Museum and Coll. Ashmead.

Described from several specimens.

Mesitius brachypterus, sp. nov.

♀. Length, 3.6ᵐᵐ. Polished black, with only a few scattered punctures on the head and prothorax. Head oblong, one-fourth longer than
wide. Antennæ 13-jointed, pale ferruginous, about twice as long as
the head; scape as long as the first three joints of the flagellum united;
pedicel a little longer than wide; first flagellar joint not longer than
wide and narrower than the following joints, which are slightly longer
than thick. Prothorax long, as long as the head; mesonotum very
short, scarcely as long as the scutellum, with 2 delicate furrows converging posteriorly and terminating in a rounded puncture before attaining the posterior margin; scutellum with 2 oblique foveæ at base;
metathorax quadrate, finely shagreened at base, with a faint median

longitudinal carina. Wings not fully developed, reaching only to the tip of the first abdominal segment. Legs rufous, the anterior femora above slightly fuscous. Abdomen pointed ovate, black, shining, sparsely pilose, the apical margins of the segments tinged with rufous.

HABITAT.—Carolina.

Type ♀ in Berlin Museum.

Described from a single specimen, labeled "Carolina, Zimmermann."

Mesitius texanus, sp. nov.

♀. Length 3.2ᵐᵐ. Apterous; black, shining; head oblong, feebly sparsely punctate; thorax coriaceous; antennæ, mandibles, and legs brownish-yellow; metathorax with a single central longitudinal carina, and three or four very short raised lines on either side at base. Abdomen ovate, as long as the thorax and much wider, polished black, the apical margin of the third and following segments narrowly testaceous. The pedicel and the first flagellar joint are about equal, smaller than the following, the following joints being a little longer than thick.

HABITAT.—Texas.

Type in Coll. Ashmead.

Described from a single specimen, distinguished by the absence of wings.

ANOXUS Thomson.

Öfvers. af K. Vet.—Akad., 1861, p. 452.

(Type *A. boops* Thoms.)

Head oblong; the ocelli 3, in a triangle on a slight prominence, in ♀ subobsolete; eyes in ♂ oblong, in ♀ more rounded, slightly hairy.

Antennæ 12-jointed, filiform, submoniliform in both sexes, in the ♂ pilose.

Maxillary palpi short, 4-jointed; labial palpi 3-jointed.

Mandibles short, tridentate at apex, the outer tooth acute.

Thorax smooth, without furrows, the prothorax in ♀ much elongated, in ♂ short, rounded anteriorly; mesonotum very short; scutellum bifoveated at base; metathorax subquadrate, truncate posteriorly.

Front wings with two basal cells of nearly an equal length, and a long radial or stigmal vein; the stigma is minute, quadrate, smaller than the parastigma.

Abdomen ovate subpetiolate, the third segment the longest.

Legs as in *Bethylus*.

A genus allied to *Epyris* and *Mesitius*, but readily distinguished by the 12-jointed antennæ, mesonotal, mandibular and palpal characters. The venation also sufficiently separates it from *Bethylus* and *Cephalonomia*.

Two species have been discovered as follows:

Anoxus Chittendenii, sp. nov.

(Pl. IV, Fig. 6, ♂.)

♂. Length, 1.5ᵐᵐ. Black, shining, impunctured, except the pleura and metathorax. The head is very slightly longer than wide, with a prominent clypeal carina. Mandibles short, black or piceous, the outer tooth conic acute. Eyes oval, slightly hairy. Antennæ 12-jointed, piceous, extending beyond the base of the metathorax, sparsely covered with white hairs; the scape is about the length of the eye or twice as long as the pedicel; the first flagellar joint is slightly smaller than the second; remaining joints very slightly larger, the last the longest, fusiform. Prothorax as long as the mesonotum and scutellum together, rounded before; mesonotum broader than long with a slight impressed scapular line near the tegulæ; scutellum small, rounded posteriorly, with an impressed line at base; mesopleura with a small round fovea on the disk; metathorax quadrate, truncate behind and a little contracted at sides towards the insertion of hind wings, the angles slightly rounded, the dorsum closely punctulate. Wings hyaline, fringed, the venation pale, the anal nervure in consequence almost obliterated. Legs piceous black, the tarsi white. Abdomen shorter than the thorax, depressed, black, smooth, and shining.

HABITAT.—Ithaca, New York.

Types in National Museum and Coll. Ashmead.

Described from 3 specimens, received from Mr. F. H. Chittenden, and in honor of whom the species is named. Mr. Chittenden informs me he reared them from a coleopteron, *Cis* sp.

Anoxus lævis, sp. nov.

♀. Length, 2.2ᵐᵐ. Polished black, impunctured. Head a little longer than wide; ocelli subobsolete, partially hidden; eyes large, oblong-oval, with a wide space between them and the mandibles, the cheeks separated from the face by a grooved line. Antennæ 12-jointed, twice the length of the head, piceous; scape clavate, about one-third the length of the head, pedicel larger than the first flagellar joint; remaining joints, except the last, submoniliform, the last twice as long as the penultimate. Prothorax long, narrowed in front, about four times as long as the mesonotum; scutellum with 2 oblique foveæ at base; metathorax much longer than wide, with a median carina, towards the base minutely shagreened. Wings hyaline, ciliated, the venation yellowish; the radial vein is very long, nearly one-half longer than the first basal cell. Legs piceous black, the trochanters, a dot at base of tibiæ and tarsi, rufous. Abdomen pointed-ovate, highly polished black, a little shorter than the thorax, the third segment the longest.

HABITAT.—Washington, D. C.

Type ♀ in Coll. Ashmead.

Described from a single specimen taken on a window pane.

Anoxus musculus Say.

Bethylus musculus Say, Bost. Jour., 1, 280; Lec. Ed. Say, II, p. 726; Cress. Syn.
Hym., p. 217.

Black; antennæ and feet yellowish; abdomen depressed.
Inhabits Indiana.

Body somewhat polished, impunctured, black; antennæ dusky, honey-yellow
towards the base; mandibles honey-yellow; thorax with the anterior segment not
much elongated; dorsal impressed lines very obvious; wings hyaline; radial nervure extended, equally distinctly near to the tip of the wing; discoidal cellule none;
metathorax minutely and distinctly punctured or granulated above, and minutely
lineated each side; abdomen depressed, polished, piceous black, distinctly petiolated; feet honey-yellow; thighs a little dusky in the middle; length over one-
twentieth of an inch. (Say.)

Unknown to me. The long radial nervure will exclude the species
from *Bethylus* as now restricted, and it is placed here temporarily or
until it is rediscovered and its true position ascertained.

PERISEMUS Förster.

Hym. Stud., II, p. 95, 1856.
Episemus Thoms. Öfv., 1861, p. 452.

(Type *P. triareolatus* Först.)

Head large, oblong, much wider than the thorax; eyes prominent,
smooth; ocelli small, but distinct.

Antennæ 12-jointed, subsetaceous, somewhat distant at base, separated by a frontal lamina.

Maxillary palpi 6-jointed; labial palpi 4-jointed.

Mandibles 4- or 5-dentate.

Thorax elongate, the prothorax rounded anteriorly; mesonotum
short, smooth, without furrows; metathorax subquadrate, obliquely
rounded off posteriorly.

Front wings with a subquadrate stigma, a parastigma (rarely wanting), an incomplete radial cell, and two basal cells, the first of which
is longer than the second; the basal vein with a branch or spurious
vein directed backwards, or at least broken by a stump of a vein.
Apterous forms rare.

Abdomen ovate, subsessile, the apical margins of some of the segments sinuate or emarginate.

Legs short, stout, the femora much swollen, the tarsi slender, the basal
joint of hind tarsi twice as long as the second, claws simple.

This genus could only be confused with *Goniozus*, with which it
agrees, except in having 12-jointed antennæ, and in having a slightly
narrower head.

The species known to me may be separated by the aid of the following table:

TABLE OF SPECIES.

FEMALES.

Perisemus oregonensis, sp. nov.

♀. Length, 5^{mm}. Black, shining; the head and thorax closely, microscopically punctate, the former with some larger scattered thimble-like punctures; metathorax subopaque, without carinæ, finely punctate; legs black; all trochanters, tibiæ, and tarsi pale brown, the middle tibiæ with only one spur. Head large, almost twice as wide as the thorax across the eyes, and a long, prominent, frontal carina. Antennæ 12-jointed, one-fourth longer than the head, honey-yellow, tapering towards tips; the first flagellar joint is about as long as the pedicel, which is a little more than twice as long as thick; remaining joints to the last gradually subequal, the last longer than the penultimate. Wings hyaline, with a slight fuscous tinge at the middle, the veins brown; stigma small, quadrate, black; parastigma wanting; the submedian cell is one-third shorter than the median; the basal vein broken by a stump of a vein beyond the middle; stigmal vein very long, abruptly curved upwards at tip and almost forming a complete marginal cell. Abdomen ovate, not quite as long as the head and thorax together, black and shining, the second, third, fourth, and fifth segments emarginate at the middle.

HABITAT.—Portland, Oregon.

Described from a single specimen obtained from H. F. Wickham.

This is the only species known to me in this genus without a parastigma, and this character, in connection with the single tibial spur on the middle legs, may be sufficient to found a new genus.

Perisemus formicoides Prov.

Bethylus formicoides Prov., Add. et Corr., p. 179; Cress. Syn. Hym., p. 247.

♀. Length, 2.5 to 3^{mm}. Black, polished; legs and antennæ honey-yellow, the latter brownish toward tips; the anterior and posterior femora blackish toward base. Prothorax very long, flattened, narrowed before, shining, with a fine and somewhat dense punctuation; mesonotum short; metathorax elongate, the sides polished, the disk punctured.

Wings hyaline, the parastigma and stigma brown.

Abdomen elongate oval, polished, black, the apex with sparse hairs.

HABITAT.—Ottawa, Canada.

Type in Coll. Provancher.

Perisemus floridanus Ashm.

(Pl. IV, Fig. 7, ♀.)

Ent. Am., III, p. 76. ♂ ♀ ; Cress. Syn. Hym., p. 247.

♂ ♀ . Length, 2.5 to 3ᵐᵐ. Black, shining, finely, closely punctulate, the head with some scattered, thimble-like punctures. Head a little longer than wide, the frontal carina not extending behind the insertion of antennæ. Mandibles and palpi honey-yellow. Antennæ 12-jointed, honey-yellow, the two or three apical joints dusky; the scape is thick, less than thrice as long as thick; flagellar joints submoniliform, scarcely longer than thick, the last twice as long as thick. Prothorax trapezoidal, longer than wide across the base; mesonotum about as long as the scutellum, the latter with a faint impressed line across the base; metathorax quadrate, smooth, with a faint trace of a median longitudinal carina. Wings hyaline, the costa, stigma, and parastigma dark brown, the other veins honey yellow; the backward-directed branch of the basal vein is not as long as the first branch of the basal. Legs, except coxæ and femora, honey-yellow.

HABITAT.—Jacksonville, Fla.

Types in Coll. Ashmead.

Perisemus minimus, sp. nov.

♀ . Length, 1.8 to 2ᵐᵐ. Black, shining; legs brown; anterior tibiæ and all tarsi honey-yellow; mandibles black; antennæ honey-yellow, the scape thick, 2½ times as long as thick; pedicel longer than the second flagellar joint, first flagellar joint minute; flagellar joints, after the second, not longer than thick; wings hyaline, the venation pale hyaline, the stigma and parastigma brown. Abdomen pointed-ovate, not longer than the head and thorax together, polished black.

HABITAT.—Arlington, Va.

Type in Coll. Ashmead.

The small size, shining surface, black mandibles, and color of the legs sufficiently distinguish the species. The head is more like that in *Epyris*, being longer and more narrowed than is usual in this genus.

Perisemus mellipes Ashm.

Ent. Am., III, p. 76, ♀ ; Cress. Syn. Hym., p. 247.

♀ . Length, 3.2ᵐᵐ. Black; the head opaque, finely, densely punctulate; thorax smooth, shining, microscopically punctulate; legs entirely honey-yellow. The head is slightly longer than wide and wider than the thorax, with a frontal carina.

Antennæ 12-jointed, honey-yellow; the scape less than thrice as long as thick; the joints of the flagellum submoniliform, scarcely longer than thick. Thorax as in *floridanus*, the metathorax smooth with a ridge in the middle at base. Wings hyaline, the parastigma and stigma black, the other veins pale yellow; the second basal cell is much shorter than the first; the backward directed branch of the basal vein is curved, but not as long as the first branch of the basal.

HABITAT.—Jacksonville, Fla.

Type in Coll. Ashmead.

Perisemus prolongus Prov.

Bethylus prolongus Prov., Can. Nat., XII, p. 265, ♀ ; Ashm. Ent. Am., III, p. 97; Cress. Syn. Hym., p. 248.

♀. Length, 4.2ᵐᵐ. Black, shining; the head very large, much broader than the thorax, about twice as long as wide at base and a little broader in front than behind, densely, minutely punctulate, with some feeble thimble-like punctures. Mandibles black. Palpi pale. Antennæ 12-jointed, honey-yellow; the scape stout, more than thrice as long as thick; the flagellum much thinner, tapering towards tip; pedicel as long as the first flagellar joint; the second and third a little stouter than the first and of an equal length; the following a little shorter, all longer than thick. Prothorax subquadrate; mesonotum very short; scutellum with a short, faint transverse line at base; metathorax longer, obliquely rounded off posteriorly, shining, at base finely rugose. Wings aborted, not extending to tip of the metathorax. Legs black, trochanters, tibiæ, and tarsi honey-yellow. Abdomen pointed ovate, much wider than the thorax, highly polished, black, segments 2, 3, and 4 emarginate at apex.

HABITAT.—Ottawa, Canada, and Lafayette, Indiana.

Type ♀ in Coll. Ashmead.

Described from the type specimen, kindly given me by Mr. Harrington. In the National Museum is a specimen, reared August 28, 1888, by Mr. F. M. Webster, from *Crambus caliginosellus.*

GONIOZUS Förster.

Hym. Stud., II, p. 96 (1856); Parasierola Cam., Trans. Ent. Soc. Lond., 1883, p. 197.

(Type *G. claripennis* Först.)

Head much as in *Perisemus;* in ♂ much broader than the thorax, with a prominent clypeal carina; in ♀ longer and less broad.

Antennæ 13-jointed, subsetaceous, not inserted so far apart as in *Perisemus,* the scape swollen, twice as long as thick, the first flagellar joint minute, smaller than the pedicel, the following joints moniliform, except in the males, rarely much longer than thick.

Maxillary palpi 6-jointed; labial palpi 4-jointed.

Mandibles oblong, truncate at tips, with 4 to 5 minute teeth.   •
Thorax as in *Perisemus*.

Front wings as in *Perisemus*, except that sometimes the backward-directed branch of the basal nervure is often bent backwards and joins the basal nervure near its origin, forming a small triangular closed discoidal cell (= *Parasierola* Cameron).

Abdomen in ♀ pointed ovate or long conical, in ♂ oblong-oval and more depressed.

Legs stout, the femora very much swollen.

Distinguished from *Perisemus* by having 13, not 12-jointed antennæ. *Parasierola* Cameron is apparently only a section of this genus, having the backward directed branch of the basal nervure a little more elongated and forming a distinct cellule; otherwise it is identical, and is here conjoined to *Goniozus*.

TABLE OF SPECIES.

FEMALES.   •

Species with a small triangular discoidal cell.
  Coxæ and femora black; anterior tibiæ and all tarsi yellow...G. CELLULARIS Say.
Species without a closed discoidal cell.
  Head unusually large 2
  Head normal.
    Backward-directed branch of the basal nervure as long as the first branch
      of the basal nervure; legs black.
      Anterior tibiæ and all tarsi yellowish.
        Wings subhyaline...............................G. FOVEOLATUS Ashm.
      All tibiæ and tarsi yellowish; ♀ with middle and hind tibiæ dusky.
        Wings clear hyaline...........................G. PLATYNOTÆ, sp. nov.
    Backward-directed branch of the basal nervure only one-third the length of
      the first branch of the basal nervure.
      Abdomen very long and cylindrical, pointed at tip, much longer than the
        head and thorax united.
        Wings subfuscous.............................G. POLITUS, sp. nov.
      Abdomen very little longer than the thorax, ovate.
        Legs brown, the tarsi pale................G. COLUMBIANUS, sp. nov.
      Abdomen not longer than the head and thorax united.
        All legs, including coxæ, yellow ....G. HUBBARDII How.
2. Legs piceous, tips of tibiæ and tarsi pale.
    Wings hyaline...................................G. MEGACEPHALUS, sp. nov.

MALES.

Wings without a discoidal cellule.
  Legs honey-yellow; wings hyaline ....G. HUBBARDII How.
  Legs brownish or piceous, the coxæ black.
    Wings fuscous or subfuscous.
      Anterior tibiæ, trochanters and all tarsi brownish-yellow.
                                 G. FOVEOLATUS Ashm.
      All tibiæ and tarsi honey-yellow ....G. COLUMBIANUS
    Wings hyaline.
      All tibiæ and tarsi brownish-yellow.....................G. PLATYNOTÆ
Wings with a discoidal cellule.
    Legs piceous, the tibiæ and tarsi pale ....G. CELLULARIS Say.

Goniozus cellularis Say.

Bethylus cellularis Say, Lec. Ed. Say, ii, p. 276; Ashm. Ent. Am., iii, p. 97; Cress. Syn. Hym., p. 247.

♂ ♀. Length, 2.2 to 3ᵐᵐ. Black, shining, the head with some fine, scattered punctures. Antennæ varying from a honey-yellow to fuscous. Legs piceous, the tibiæ and tarsi often honey-yellow, especially the anterior pair; femora often black. In the male the wings are clear hyaline, in the female fusco-hyaline; the parastigma brow nor black; the branch from the basal nervure curves backwards and joins the median nervure near the tip, forming a complete cell which readily distinguishes the species.

HABITAT.—United States.

Specimens in National Museum and Coll. Ashmead.

Found in various parts of the United States. A single specimen is in the National Museum reared by Prof. F. M. Webster, September 17, 1884, from a geometrid larva in wheat stubble; while my collection contains a specimen, reared June 5, 1885, from wheat stalks infested with *Isosoma tritici*.

The species is, without doubt, parasitic on various microlepidoptera.

Goniozus megacephalus, sp. nov.

♀. Length, 2.8ᵐᵐ. Black, shining, with a fine, microscopic punctuation and a few larger punctures scattered over the surface. The head is large and long, a little more than one and a half times as long as wide. Mandibles large, black. Antennæ 13-jointed, inserted wide apart, and not much longer than the head, moniliform, yellow; the scape swollen, twice as long as wide. Metathorax smooth without carinæ. Wings hyaline, the parastigma and stigma piceous. The other veins yellowish; the branch of the basal vein is reduced to a mere stump.

Legs brownish piceous, the tips of tibiæ and the tarsi yellow.

HABITAT.—Key West, Fla.

Types in National Museum.

Described from a single ♀ taken by E. A. Schwarz.

Goniozus foveolatus Ashm.

Ent. Am., iii, p. 76, ♀ ; Cress. Syn. Hym., p. 247.

♀. Length, 2.5 to 3ᵐᵐ. Black, shining, but finely, delicately punctate, the head with a few coarse, scattered punctures; transverse furrow at base of scutellum terminates in small, oblique foveæ.

Antennæ 13-jointed, honey-yellow, the scape short, thick, the following joints all small, moniliform. Legs black or piceous; anterior tibiæ and tarsi and the middle and posterior tarsi, honey-yellow. Wings subhyaline; stigma and parastigma black, the other nervures honey-yellow; the basal nervure is broken by a stump of a vein which is less than half the length of the first branch of the basal nervure. Abdomen cylindric-ovate, a little longer than the head and thorax united.

In the ♂ the middle and posterior tibiæ are pale brown, the abdomen being shorter than the head and thorax united.

HABITAT.—Florida and District of Columbia.

Types in National Museum and Coll. Ashmead.

The species is closely allied to *G. platynotæ*, but its slightly smaller and more slender form, the subhyaline wings, and the black stigma and parastigma, readily distinguish it.

The National Museum contains a specimen, reared April 14, 1882, at Georgiana, Fla., from a tineid larva in dry fungus.

Goniozus platynotæ, sp. nov.

(Pl. iv, Fig. 8, ♀.)

♂ ♀. Length, 3^{mm}. Black, shining, very sparsely pilose; head with a prominent keel between the antennae, microscopically closely punctulate, with a few larger punctures scattered irregularly over the surface, subopaque. Mandibles, antennæ, and legs pale honey-yellow; the anterior femora vary from a dark brown to black, the middle and posterior femora in ♂ brownish only in the middle; in the female the mandibles are black, and all the femora are black or brownish-piceous, their tibiæ brownish.

Antennæ 13-jointed, extending to the tegulæ (in ♀ a little shorter), moniliform; pedicel small, rounded, much smaller than the first flagellar joint; all the flagellar joints but the last moniliform, the last oval. Metathorax smooth, with a delicate median carina, and carinated along the lateral margins.

Wings hyaline, the parastigma and stigma large, brown, the latter twice as long as wide, rest of the nervures pale yellow; the first basal cell is longer than the second, the basal vein broken by a backward directed branch.

Abdomen as long as the head and thorax united, polished black, the sutures piceous.

HABITAT.—Virginia, Maryland, District of Columbia, and Florida.

Types in National Museum and Coll. Ashmead.

The species is not rare. Dr. Riley reared a single male July 30, 1886, from *Platynota sentana*.

Goniozus politus, sp. nov.

♀. Length, 2.4^{mm}. Black, highly polished, impunctured. Head not wider than the thorax, nearly twice as long as wide, smooth, impunctured.

Antennæ 13-jointed, short, scarcely longer than the head, moniliform, brownish-yellow, dusky or black at apex; scape short, stout, about twice as long as thick; pedicel as long as the last joint; first flagellar joint very minute, half the length of the pedicel, the following joints rounded.

Prothorax very long, rounded before; mesonotum very short, not longer than the scutellum; metathorax subquadrate, longer than wide, truncate behind, the angles a little rounded, smooth; mesopleura polished with a small round fovea at the middle.

Wings fusco-hyaline, the venation brown, the parastigma and stigma black, the latter quadrate, but slightly longer than wide, the stigmal vein long, scarcely curved; the second basal cell is much shorter than the first, the basal vein broken only by a short stump of a vein.

Legs short, stout, black, the tips of tibiæ and the tarsi yellowish.

Abdomen very long, cylindrical, conic, longer than the head and thorax together, highly polished, black.

HABITAT.—Virginia Beach, Md.

Type ♀ in Coll. Ashmead.

Taken by Mr. E. A. Schwarz July 17, 1890.

Goniozus columbianus, sp. nov.

♂ ♀. Length, 1.5 to 2ᵐᵐ. Black, shining, the head very feebly microscopically punctate, in ♀ 1¼ times as long as wide, in ♂ not longer than wide across the eyes.

Antennæ 13-jointed, a little longer than the head, the scape dusky or piceous towards base, the flagellum yellow; the pedicel is very small, very little longer than thick, the two following joints subtriangular, those beyond moniliform.

Wings hyaline, the parastigma and stigma brown, the other veins pale or tinged with yellow; the stigma is oblong, 2¼ times as long as wide; the stigmal vein long, bent upwards at tip; the branch of the basal nervure short.

Legs piceous brown, in ♂ brown, with the trochanter, all tibiæ and tarsi yellow, in the ♀ the middle and posterior tibiæ yellow except at the middle above.

Abdomen not longer than the thorax, black polished.

HABITAT.—Washington, D. C.

Types ♂ and ♀ in Coll. Ashmead.

Comes nearest to *G. Hubbardi* How., but quite distinct by the color of the legs.

Goniozus Hubbardi Howard.

Hubbard's Orange Ins. app., p. 217 ♀ ; Ashm. Ent. Am., III, p. 119; Cress. Syn. Hym., p. 247..

♂ ♀. Length, 1.6 to 2ᵐᵐ. Black, shining, alutaceous; head feebly, sparsely punctate, with a distinct carina anteriorly between the antennæ.

Antennæ 13-jointed, the scape and pedicel yellow, the flagellum brownish-yellow, the joints after the second submoniliform, the first minute.

Legs, including all coxæ, yellow, the anterior femora much swollen; wings hyaline, the parastigma and stigma piceous.

Abdomen longer than the thorax, pointed at apex, smooth and polished, the venter sometimes piceous towards base.

The male is the smaller and agrees perfectly with the female, the only noticeable difference being in the shape of the abdomen, which is oblong-oval, slightly depressed, not pointed at apex, and scarcely as long as the thorax. The flagellum is more distinctly brown, the joints very slightly longer than in the female.

Types in National Museum.

Bred by H. G. Hubbard at Crescent City, Fla., from *Platynota rostrana*, and by myself at Jacksonville, Fla., from the orange case-worm, *Platœceticus Glorerii* Packard.

NOTE.—The following genus has not been recognized:

Notwithstanding the fact that the antennæ are described as 26-jointed, the genus evidently belongs in this group. I merely copy Stefano's description with the hope that it may some day aid in its identification. Were it not for the 26-jointed antennæ I should say the genus was identical with either *Pristocera* or *Scleroderma*. It is possible, though, the antennæ will be found to be only 13-jointed, each joint being constricted and having the appearance of forming two joints. Such mistakes have frequently occurred in systematic entomology.

SCLEROGIBBA Stefano.

Nat. Siciliano, 1887-'88, p. 145.

Scler. caput depressum, ocelli nulli. Antennæ crassæ cum 26 articulis. Os inferior. Thorax oblongus distincte divisus; prothorax cylindricus antice rotundatus posticeque emarginatus; mesothorax antice attenuatus postice incrassatus; metathorax supra planus, dilatatus. Femora et tibiæ antice crassissima. Abdomen sessile. Ovipositor brevissimus.

Scler. crassifemorata, Tav. 1, Figs. 3c, 3a, b c, p. 146, Hab. Ustica.

Subfamily II—EMBOLEMINÆ.

Head globose. Ocelli 3 in a triangle, close together; in apterous females very minute. Mandibles oblong-quadrate, 3-dentate. Antennæ porrect, filiform, in ♂ 10-, in ♀ 13-jointed, inserted on the middle of the face. Maxillary palpi 3- to 5-jointed; labial palpi 2- or 3-jointed. Pronotum usually well developed, triangular or quadrate; mesonotum transverse, or at the furthest not longer than wide, the scapulæ not separated; scutellum transverse, quadrate, or subtriangular; metathorax large, quadrate. Front wings with a lanceolate stigma, a closed costal cell, two basal, two discoidal cells, and an open radial cell; hind wings distinctly lobed, veinless, except along the costa to two-thirds its length, the tip of this vein being furnished with 6 hooklets. Abdomen subpetiolate, ovate or oblong oval, depressed, composed of 7 segments, the second and third being very large. Legs long, the anterior femora much swollen, the middle and posterior pairs obclavate, the tibiæ subclavate, their spurs 1, 0, 2, tarsi 5 jointed, claws simple.

The shape of the head, the difference in the numbers of joints in the antennæ, and non-chelate anterior tarsi in the females, sufficiently differentiate this group from the Bethylinae and the Dryininae, to which it is most closely allied.

The group is rare and only a few species in it are known, their habits still remaining unknown. From their close structural affinity with the Dryininae it is not unlikely they have similar habits.

The genera may be distinguished by the aid of the following table:

TABLE OF GENERA.

FEMALES.

Antennæ 13-jointed.
 Wings rudimentary; eyes flat; ocelli very small; scape much longer than the
  first funiclar joint ....PEDINOMMA Först.
 Wings fully developed.
  Eyes arched; ocelli large; scape shorter than the first funiclar joint.
                                                            EMBOLEMUS Westw.

MALES.

Antennæ 10-jointed.
 Prothorax as long as the mesonotum, with a deep median sulcus; scape much longer
  than the first flagellar joint ....AMPULICOMORPHA Ashm., gen. nov.
 Prothorax shorter than the mesonotum, without a median sulcus; scape much
  shorter than the first flagellar joint...........EMBOLEMUS Westw.

PEDINOMMA Förster.

Hym. Stud., II, p. 91 (1856).

Myrmecomorphus Westw., Lond. Mag., 1833, p. 496.

(Type *M. rufescens* Westw.)

Head rounded, a little oblong, with a frontal tubercle; eyes rounded; ocelli subobsolete.

Antennæ 10-jointed, as long as the body, cylindric and filiform, the scape as long as the head and longer than the first flagellar joint.

Maxillary palpi 3-jointed; labial palpi 2-jointed.

Wings in ♀ rudimentary or wanting, the ♂ unknown but probably winged.

Abdomen ovate, much longer that the thorax.

Legs thick, the posterior pair long.

A very rare genus, so far only known from Europe, and I have been unable to obtain specimens for examination. There is a slight discrepancy between Westwood and Förster in the description of the ocelli. The former says "without ocelli," the latter that "die Nebenaugen sind sehr klein." Nothing is known of the habits of the genus and the male is still to be discovered.

The name given to it by Westwood was changed by Förster on account of its being preoccupied in Diptera.

EMBOLEMUS Westw.

(Phil. Mag., 1832, p. 411, ♂.)

Polyplanus Nees, Mon., II, p. 349. ♀.

(Type *E. Ruddii* Westw.)

(Pl. v, Fig. 2, ♂.)

Head small, subglobose, pubescent; eyes small, rounded; ocelli 3, distinct.

Antennæ in ♂ 10-jointed, setaceous, much longer than the body, inserted on a frontal tubercle; the scape short, half the length of the head; pedicel small, rounded; flagellar joints all elongate, longer than the scape; in ♀ 13-jointed, subclavate, shorter than the body; scape short; pedicel small, flagellar joints to the twelfth subequal; thirteenth fusiform, longer and stouter than the twelfth.

Maxillary palpi 5-jointed, setaceous, slender, the first joint slender, slightly arcuated, the second dilated, the third slender, the fourth shorter than the third, the fifth linear, much longer than the fourth; maxilla small, short, subovate; labium small, narrow, sublinear; ligula transverse, short.

Mandibles oblong quadrate, straight, 3-dentate, the teeth acute, subequal.

Thorax ovoid, the prothorax distinct, triangular, impressed laterally, rounded anteriorly; mesonotum as broad as long, without furrows; metathorax slightly longer than high, rounded posteriorly.

Front wings with a lanceolate stigma, a marginal, two submarginal, and three discoidal cells; the third discoidal is nearly divided into two others, the transverse median not interstitial with the basal nervure; in the ♀ the nervures are not well developed.

Legs rather long, the femora obclavate, the coxæ large, the tarsi slender, claws and pulvilli small.

Abdomen long ovate.

This genus still remains to be discovered in our fauna, the species described by me as such, *E. nasutus*, being a dryinid, belonging to the genus *Labeo* Hal.

E. Ruddii Westw., the type of the genus, I had the pleasure of seeing in the Berlin Museum.

AMPULICOMORPHA Ashmead, gen. nov.

(Type *A. confusa.*)

Head globose, with a frontal protuberance for the insertion of the antennæ; eyes rather small, oval, placed at the sides, a little before the middle of the head; ocelli very small, arranged in a triangle on the vertex; clypeus convex, slightly projecting, the anterior margin regularly rounded; superal clypeal piece distinctly separated by deep grooved lines, extending from the base of each antenna.

Antennæ 10-jointed, longer than the body, setaceous, the scape very long, longer than the head, slender and curved at base, the pedicel small, the remaining joints lengthened.

Maxillary palpi long, 5-jointed, the terminal joint the longest, the third the shortest; labial palpi very short; apparently 3-jointed (?).

Mandibles small, truncate at tips, 3-dentate.

Thorax elongate, abruptly truncate behind, the prothorax quadrate, as long as the mesonotum, but narrower, the sides compressed and with a deep median channel above, as in *Ampulex*; mesonotum without furrows, except two very short oblique grooves anteriorly; scutellum triangular, with three confluent foveæ at base; metathorax quadrate, rugose, the posterior truncature abrupt, the angles a little rounded.

Front wings smoky, the stigma moderate, lanceolate, the marginal cell almost closed, two basal and two distinct discoidal cells; besides there is an indistinct submarginal cell.

Hind wings lobed at base, with a distinct venation.

Abdomen ovate, subpetiolate, the second and third segments occupying most of its surface.

Legs rather long, the femora obclavate, the anterior pair the shortest and stoutest, the tibial spurs on the hind legs long.

In venation this remarkable genus resembles *Embolemus* Westwood, but otherwise it is quite distinct, and recalls *Rhinopsis* and *Ampulex*. The long, quadrate, sulcate pronotum at once distinguishes the genus. I know of but a single species, as follows:

Ampulicomorpha confusa, sp. nov.

(Pl. v, Fig. 1, ♂.)

♂. Length, 4^{mm}. Polished black; metathorax above and on the truncature rugose; antennæ brown; legs black, the trochanters, tibiæ, and tarsi piceous or dark rufous.

Wings fuliginous, the venation dark brown, the second basal cell shorter than the first; all the nervures are well developed.

HABITAT.—California.

Type ♂ in Coll. Ashmead.

Described from a single specimen. Since the above was written I have seen another specimen in the collection of the American Entomological Society, taken in Nevada. It differs in being slightly smaller and with paler colored wings.

Subfamily III.—DRYININÆ.

Head transverse or subquadrate; when viewed from in front often triangular. Ocelli 3, in a triangle. Mandibles stout, 3- or 4-dentate. Antennæ porrect, filiform or subclavate, pubescent or pilose, 10-jointed in both sexes, inserted just above the clypeus. Maxillary palpi 3- to 5-jointed; labial palpi 2- to 4-jointed. Pronotum variable, very short, scarcely visible from above, or very long, and separated from the mesonotum by a strong constriction; mesonotum not longer than wide, with or without distinct parapsides; scutellum generally large, rounded or truncate posteriorly, entirely wanting only in *Gonatopus;* metathorax quadrate or subquadrate, truncate or rounded off behind. Front wings with a lanceolate or ovate stigma, a closed costal cell, two basal cells, and an open radial cell; rarely with a distinct discoidal cell; hind wings distinctly lobed, veinless except along the costa to two-thirds its length, the tip of this vein being furnished with hooklets. Abdomen subpetiolate or petiolate, ovate or oblong-oval, depressed, or occasionally compressed, composed of 8 segments, the ovipositor not exserted. Legs long, the anterior femora much swollen, the middle and posterior pairs obclavate, the tibiæ subclavate, their spurs 1, 1, 2; tarsi 5-jointed; the anterior tarsi in the females in all the genera, except *Aphelopus* and *Mystrophorus*, are chelate.

A very interesting group, and evidently an ancient phylogenetic type of the order, the chelate anterior tarsi in the females being found in no other group among the Hymenoptera.

The *Dryininæ* confine their attacks to homopterous insects belonging principally to the families Fulgoridæ, Membracidæ and Jassidæ, living in felt-like sacks protruding from the abdominal spiracles.

Prof. Joseph Mik has given a most interesting account of the biology of *Gonatopus pilosus* Thoms., living on *Deltocephalus xanthoneurus,* while *G. pedestris* Dalm. has been reared from *Athysanus maritimus* and *Thamnotettix sulphurella. Aphelopus melaleucus* Dalm. has been discovered by Giard, in France, to prey upon *Typhlocyba hippocastani* and *T. douglasi.* The genus *Labeo* Haliday, known only in the male sex, bears a superficial resemblance to *Aphelopus* and has similar habits, my *Labeo typhlocybæ* having been reared by Dr. Riley from a *Typhlocyba* living on the sycamore. I believe this genus will yet prove to be the opposite sex of *Gonatopus.*

The following table will be found all that is necessary to recognize the genera:

TABLE OF GENERA.

FEMALES.

1. Vertex convex, not impressed 2
   Vertex deeply impressed; prothorax very long; anterior tarsi chelate.
      Wingless, without a scutellum...Gonatopus Ljungh.
      Winged, with a scutellum...Dryinus Latr.
2. Stigma oval or ovate... 3
   Stigma lanceolate; anterior tarsi chelate.
      Prothorax not quite as long as the mesonotum, much contracted; fourth joint of anterior tarsi not much longer than the third, the first twice as long as the three following united; maxillary palpi 4-jointed.
                                                                Bocchus Ashm. gen. nov.

3. Anterior tarsi not chelate 4
Anterior tarsi chelate.
> Prothorax almost as long as the mesonotum; fourth joint of anterior tarsi much
> longer than the third, the first joint not, or scarcely, longer than
> the three following united; maxillary palpi 5-jointed.
>> CHELOGYNUS Hal.
> Prothorax much shorter than the mesonotum; fourth joint of anterior tarsi
> scarcely longer than the third, the first not, or scarcely, longer than
> the three following united; maxillary palpi 4-jointed.
>> ANTEON Jurine.
4. Prothorax much longer than the mesonotum, the latter without a trace of a furrow; head large, broad...................MYSTROPHORUS Förster.
Prothorax above not or only slightly visible; mesonotum strongly developed, with
   furrows...APHELOPUS Dalm.

MALES.

1. Stigma oval or ovate... 2
Stigma lanceolate.
> Occiput deeply concave; vertex and neck separated by a sharp angle; mesono-
> tum with distinct furrows; a discoidal cell; maxillary palpi 4-
> jointed..LABEO Hal.
> Occiput not concave, straight and broad; mesonotum without furrows; no dis-
> coidal cell; maxillary palpi 5-jointed ....PHORBAS Ashm. gen. nov.
2. Prothorax much longer than the mesonotum.
> Mesonotum with furrows; maxillary palpi 5-jointed ....CHELOGYNUS Hal.
> Mesonotum without furrows; maxillary palpi 4-jointed..MYSTROPHORUS Först.
Prothorax much shorter than the mesonotum.
> Mesonotum with or without traces of furrows; maxillary palpi 4-jointed.
>> ANTEON Jurine.
Prothorax not, or scarcely, visible from above.
> Mesonotum strongly developed, with furrows; maxillary palpi 5-jointed.
>> APHELOPUS Dalm.

GONATOPUS Ljungh.

Web. und Mohr. Beitr., 1810, p. 161; Dicondylus Hal., Ent. Mag., IV, p. 410 (1837).

(Type *G. pedestris* Dalm.)

Head large, transverse; when viewed from in front very short, triangular, the vertex deeply impressed and sloping off towards the neck;
the occiput convex, not margined; eyes very large, prominent, occupying the whole side of the head; ocelli small.

Antennæ 10-jointed, filiform or subclavate; scape about one-third
the length of the head, thicker but not much longer than the second.

Maxillary palpi 5-jointed.

Mandibles 3-dentate, the outer tooth long and acute.

Thorax greatly elongated and of a peculiar shape; the prothorax
much wider than the mesothorax, the latter elongate and humped at
the middle, and separated from the metathorax by a strong constriction; the metathorax clavate.

Wings always absent.

Abdomen oblong-ovate, subpetiolate, the second segment the longest.

Legs very long, the posterior pair greatly lengthened; coxæ large,

the anterior pair greatly lengthened; femora obclavate, the front pair
the stoutest; tibiæ very long and slender, very little thicker at tips
than at the base; anterior tarsi chelate, middle and posterior tarsi,
5-jointed, shorter than their tibiæ, the claws and pulvilli small.

This genus is readily distinguished from all the other genera in the
group by its peculiar shape, the strong constriction between the meso-
and meta- thorax, the absence of wings and scutellum, and the much
longer and more slender tibiæ. In its cephalic and antennal characters it approaches nearest to *Dryinus*, but otherwise it is quite different, that genus having wings, a distinct large scutellum, and is without
the strong constriction between the meso- and meta- thorax.

The male is unknown, but I have reasons for believing that the genus
Labeo Haliday, known only in the male sex, will yet prove to be the
opposite sex of this peculiar genus.

Mr. Cameron, in Biologia Centrali-Americana, has described two or
three species that he supposed were wingless males, but as these so-
called males have chelate claws, I suspect he has mistaken females for
males, as all the males in this group known to me have simple, not
chelate claws, the chelate claws being a character peculiar to the females.

Our species may be tabulated as follows:

TABLE OF SPECIES.

FEMALES.

Species more or less brown or pale brownish-yellow.......................... 2
  Species, excepting a portion of the head, black.
    Thorax punctulate.
      Head not twice as broad as long; metathorax, anteriorly and posteriorly,
        transversely striated.
      Occiput, face, and two basal joints of antennæ, yellow, rest of antennæ
        fuscous ....G. CONTORTULUS, Patton.
      Head twice as broad as long; metathorax, anteriorly and posteriorly, trans-
        versely striated.
      Occiput, face, and antennæ, except the three terminal joints, yellow.
                       G. FLAVIFRONS, sp. nov.
    Thorax polished, shining.
      Head entirely black ....G. DECIPIENS, Prov.
2. Abdomen black; head and thorax dark reddish-brown.
    Metathorax coarsely transversely striated; antennæ yellow; legs piceous or
      brown, the coxæ and tarsi yellow........G. CALIFORNICUS, sp. nov.
  Abdomen piceous; head, thorax, and legs pale or yellowish.
    Metathorax smooth, polished, with only a few faint, transverse striæ.
                    G. BICOLOR, sp. nov.

Gonatopus contortulus Patton.

Can. Ent. XI, p. 65 (1879), ♀; Ashm. Ent. Am. III, p. 74; Cress. Syn. Hym., p. 246.

"♀. Length 3.5ᵐᵐ. Head testaceous, mandibles and scape of antennæ white; the teeth of the mandibles, second joint of the antennæ,
and a line on the scape posteriorly, pale testaceous, remainder of antennæ fuscous. A large fuscous spot on the under side of head, and

another above in front of and including the ocelli: a raised line extending forwards from anterior ocellus to the face. Head transverse, broader than thorax or abdomen; convex beneath, concave behind, above and in front; the mouth prominent; the eyes longitudinally ovate, prominent, not reaching the posterior border of the head. Antennae 10-jointed, the basal joint stout, the second joint more slender and one-half as long as the first, the third very slender and equal in length to the first and second together, the fourth and following joints slender but gradually becoming thicker, the fourth one-half as long as the third, the fifth a little shorter than the fourth and a little longer than each of the following joints. Thorax and abdomen piceous black. The thorax slender, binodose. The trochanters formed of only one joint. Anterior coxae long and robust, pale testaceous with a darker stripe above; anterior trochanters whitish, more slender, clavate; femora large, obclavate, dark testaceous, paler at tip; tibiae as long as the femora and, together with the first tarsal joint, pale testaceous; terminal joint of the tarsi and the chelae whitish. The chelae at rest extending back to the tip of the first joint of the tarsus, the outer claw pointed and slightly curved at the extremity, the inner claw more robust, ciliated internally and with a wrench-shaped curve at the extremity; pulvillus tipped with fuscous. The other legs slender, the coxae and the base of femora dilated, testaceous, the coxae, base of femora, tibiae above and claw-joint of tarsi, darker. Abdomen ovate, pointed at tip and with a short petiole." (*Patton.*)

HABITAT.—Waterbury, Conn.

Type in Coll. American Entomological Society, at Philadelphia.

The type is in poor condition and I have copied Mr. Patton's description. It was captured at Waterbury, Conn., on herbage a few inches above the ground, August 18, 1879.

Gonatopus flavifrons, sp. nov.

(Pl. v, Fig. 4, ♀.)

♀. Length, 4.4^{mm}. Black, shining, with a fine shagreened punctuation. Occiput, face, mandibles, except teeth, palpi, antennae, except the three terminal joints, which are fuscous, and legs, yellow; the long large anterior coxae have two black spots beneath; the greatly swollen anterior femora above are almost entirely black, while their tibiae have a black streak above; the middle and posterior coxae and femora basally are also more or less black above. The anterior or constricted part of the metathorax and its posterior face are transversely striated.

The abdomen, except blotches at the sides of the first segment, the apical edge of the third and fourth segments, especially laterally, and the terminal segment, which are rufous, is highly polished, black.

HABITAT.—Albany, N. Y.

Type ♀ in Coll. Ashmead.

Described from a single specimen received from Mr. W. H. Harrington, who informs me that it was captured by Mr. Van Duzee at the above place.

The species comes nearest to *G. contortulus* Patton; but it is larger, the head broader, and it differs also in colorational detail.

Gonatopus decipiens Prov.

Add. et Corr., p. 179, ♀ ; Ashm. Ent. Am., III, p. 71; Cress. Syn. Hym., p. 216.

♀. Length, 2.4ᵐᵐ. Black, with the antennæ and legs, in part, testaceous. Head large, flat, very finely punctured, with a large ocellus on the vertex. Antennæ inserted near the mouth, 10-jointed, the first the longest and thickest, the following elongated, slender, the terminal joints dusky. Thorax polished, shining, narrow, elongated with respect to the constricted part of the scutellum. Abdomen subsessile, oval. Anterior femora strongly swollen, black, their tibiæ testaceous, their tarsi brownish, with the claws long, in the form of pincers; the four posterior femora with their tibiæ black, their tarsi testaceous.

HABITAT.—Cap Rouge, Canada.
Type ♀ in Coll. Abbé Provancher.
Unknown to me.

Gonatopus californicus, sp. nov.

♀. Length, 3ᵐᵐ. Uniformly piceous-brown, except the apex of the metathorax and the abdomen, which are more or less black. The antennæ, the tarsi, and the slender part of the posterior femora, and the middle of their tibiæ, honey-yellow. The head is very closely, finely punctulate, pubescent, while the metathorax is transversely rugulose.

HABITAT.—California.
Type ♀ in National Museum.
Described from a single specimen received from A. Koebele.

Gonatopus bicolor, sp. nov.

♀. Length, 3ᵐᵐ. Head, thorax and legs yellow or reddish-yellow; middle and posterior knees and tips of posterior tibiæ black. Abdomen piceous black. Mandibles 4-dentate. Eyes brown. Metathorax smooth, polished, the posterior face feebly transversely aciculated.

Antennæ subclavate, brown, the 4 or 5 basal joints yellow, the first flagellar joint about twice as long as the second.

HABITAT.—Selma, Ala.
Type ♀ in National Museum.
Described from a single specimen collected by Mr. W. H. Patton.

DRYINUS Latreille.

Hist. Nat., XIII, p. 228 (1805); Först. Hym. Stud., II. p. 90.
Chelothelius Reinh., Berl. Ent. Zeits, 1863, p. 409.

(Type *D. formicarius* Latr.)

Head transverse, the vertex impressed, when viewed from in front triangular; eyes large, prominent; ocelli 3 in a triangle.

Antennæ 10-jointed, filiform or slightly subclavate, inserted low down on the face, just above the clypeus; the scape about half the length of the head, slightly thickened and more than twice as long as the pedicel; the first flagellar joint usually greatly lengthened.

Maxillary palpi long, 6-jointed; labial palpi very short, 2-jointed.

Mandibles 3-dentate.

Thorax elongated; the prothorax very long, separated from the mesothorax by a strong constriction, above convex, anteriorly rounded; mesothorax broader than long, shorter than the prothorax, and usually without parapsidal furrows; the scutellum distinct; metathorax subquadrate, more or less rounded off posteriorly, rugose.

Front wings with a large stigma, two basal cells and an incomplete marginal cell.

Abdomen oblong oval or oblong ovate, petiolate, the second and third segments the largest, of which the second is the longer.

Legs long, the femora obclavate, the tibiæ long, slender, the anterior tarsi chelate; in ♂ anterior tarsi long, simple.

Only three species in this genus are known in our fauna, which may be recognized by the aid of the following table:

TABLE OF SPECIES.

FEMALES.

Not entirely brownish-yellow.. 2
  Entirely brownish-yellow.
    Wings with two fuscous bands.........................D. BIFASCIATUS, Say.
    Wings fuscous, with a large rounded hyaline spot beneath the stigma.
                                                D. ALATUS, Cr.
2. Abdomen shining black.
    Wings hyaline with two fuscous bands.................D. AMERICANUS Ashm.

D. bifasciatus Say.

Lec, Ed. Say's Works, I, p.384, ♀ ; Ashm. Ent. Am., III, p. 74; Cress. Syn. Hym., p. 246.

Yellowish; wings bifasciate.
Inhabits Indiana.
Body honey-yellow, varied with blackish: anterior thighs dilated; wings with two fuscous bands, the apical one broader.
Length rather more than one-fifth of an inch. (Say.)
Unknown to me.

Dryinus alatus Cr.

Gonatopus? alatus Cr., Trans. Am. Ent. Soc., IV, p. 193, ♀. Dryinus alatus Patton, Can. Ent., XI, p. 65; Ashm. Ent. Am., p. 246; Cress. Syn. Hym., p. 246.

♀. Length 5.75^{mm}. Pale ferruginous, head shining, much broader than thorax, transversely compressed; front flat; eyes large, prominent, ovate; clypeus transverse, subconvex, bituberculate at tips; mandibles and scape beneath whitish; antennæ slender, as long as head and thorax, 10-jointed, first joint short and robust, second small, scarcely half the length of first, third very long and slender, more than twice the length of first and second together, black; scape and two or three apical joints yellow; prothorax elongate, as long as metathorax, convex, subtruncate anteriorly, narrowed and somewhat contracted posteriorly; metathorax short, convex, fusco-ferruginous; metathorax elongate, rounded behind, longitudinally rugose at base above; wings narrow, scarcely reaching tips of posterior femora, fuscous, with a large rounded hyaline spot beneath base of stigma nearly as broad as the wing; legs rather paler than body, anterior pair very long; coxæ and trochanters as long as femora, which are longer than tibiæ, tarsi chelate; four posterior legs much shorter than anterior pair, all the femora much thickened towards base; abdomen elongate, smooth and shining. (*Cresson*.)

HABITAT.—Texas.

Types in Coll. American Ent. Soc. and National Museum.

The type specimen in the National Museum came with the Belfrage collection.

Dryinus americanus Ashm.

(Pl. v, Fig. 3, ♀.)

Mystrophorus americanus Ashm., Ent. Am., III, p. 128.

♀. Length 6.3ᵐᵐ. Head, thorax, and legs brownish yellow; head above dusky; eyes brown-black; antennæ dusky at tips; abdomen smooth, polished, black. The head when viewed from in front is triangular, finely and regularly punctate, impressed above. Antennæ 10-jointed, the third joint as long as the fourth, fifth, and sixth joints united. Prothorax much elongated, narrower, and more than thrice as long as the mesonotum, finely microscopically sculptured. The short spoon-shaped wings are hyaline, except a smoky transverse band across the front wing, broad enough to inclose the stigma and stigmal vein.

HABITAT.—Jacksonville, Fla.

Types in Coll. Ashmead.

Two specimens. The first specimen captured was in poor condition, and when my original description was drawn up the anterior chelate tarsus was not observed, and it was consequently placed in the genus *Mystrophorus*, which is distinguished by having the anterior tarsi simple, not chelate.

LABEO Haliday.

Ent. Mag., I, p. 273 (1833).

(Type *L. excisus* Westw.)

Head transverse, the vertex broad, subconvex, the occiput deeply concave; eyes large, oval, prominent, but much smaller than in either *Gonatopus* or *Dryinus*, and occupying only the anterior half of the sides; the cheeks oblique; ocelli 3, small, triangularly arranged; a delicate keel extends forward from the front ocellus.

Antennæ 10-jointed, filiform, inserted just above the clypeus, slightly thickened towards the tips; scape thickened, a little curved, about twice as long as the pedicel, or a little shorter, or as long as the first flagellar joint.

Maxillary palpi rather long, 4-jointed; labial palpi very short, 2-jointed.

Mandibles truncate at tips and 3-dentate, the outer two conical, equal; the inner smaller, blunt.

Thorax subovate; prothorax narrowed before and impressed at sides, scarcely as long as the mesonotum, the anterior margin ridged; mesonotum broader than long, with 2 distinct furrows converging posteriorly; metathorax short, rounded behind, scarcely sculptured or indistinctly areolated.

Front wings with a moderate sized lanceolate stigma, two basal cells, and an angulated stigmal vein; sometimes one or two discoidal cells are more or less visible, but usually they are wanting or subobsolete.

Abdomen subovate or oblong oval, petiolate, the second and third segments nearly equal.

Legs moderate; the hind pair the longest, femora obclavate, the tibiae subclavate, the tarsi usually long and slender.

I am strongly of the opinion that this genus, known only in the male sex, will yet prove to be the opposite sex of *Gonatopus*. It is reared from *Typhlocyba* and other jassids.

The four species in our fauna may be distinguished by the aid of the following table:

TABLE OF SPECIES.

```
Hind legs very long, the tibiae very long.
  Hind tarsi longer than their tibiae.
    Antennae very long, the flagellar joints nearly 5 times as long as thick.
      Marginal cell open at tip..........................L. LONGITARSIS, sp. nov.
      Marginal cell completely closed....................L. TEXANUS, sp. nov.
  Hind legs not especially long, the tibiae not long.
    Antennae not longer than the thorax, the flagellar joints never more than 3 times
        as long as thick.
      Pedicel and scape equal in length ..................L. NASUTUS Ashm.
      Pedicel twice as long as the scape..................L. TYPHLOCYBAE, sp. nov.
```

Labeo longitarsis sp. nov.

♂. Length 2.6^{mm}. Black, shining; head finely closely punctate, the occiput deeply emarginate; eyes large, rounded, pubescent. Mandibles and palpi rufous. Antennae 10-jointed, long, setaceous, pubescent; first and last flagellar joints about of an equal length, longer than the scape, the remaining slightly shorter, four times as long as thick. Prothorax scarcely visible from above; mesonotum with two distinct furrows, converging posteriorly; scutellum with a transverse impressed line at base; metathorax longer than high, roundedly truncate behind, finely closely punctate, and with a median carina. Wings hyaline, pubescent, the venation distinct, brown; the stigma long lanceolate, the stigmal vein very long, forming nearly a closed marginal cell. Legs fuscous, the posterior pair unusually long.

HABITAT.—Jacksonville, Fla.

Type ♂ in Coll. Ashmead.

Labeo texanus sp. nov.

♂. Length 2.5ᵐᵐ. Allied to *L. longitarsis*, but differs as follows: The head is more finely punctate; mandibles black or piceous; marginal cell in anterior wing completely closed; legs black, the anterior knees, tibiæ, and tarsi pale brownish; the middle and hind knees brownish; tibiæ and tarsi fuscous. The metathorax is more rugosely punctate.

HABITAT.—Texas.

Type in Coll. Ashmead.

Described from a single specimen.

Labeo nasutus Ashm.

Embolemus nasutus Ashm., Ent. Am., III, p. 75 ♂.

♂. Length 2ᵐᵐ. Black, subopaque. Head vertically short, the frons punctate; ocelli prominent; eyes large, convex, hairy; mandibles rufous; 3-dentate, the outer tooth a little the longest. Antennæ 10-jointed, brown-black, densely pubescent; the scape and the pedicel very short, together scarcely longer than the first flagellar joint; first flagellar joint the longest, about thrice as long as thick, the following nearly equal, about twice as long as thick. Prothorax scarcely visible from above, narrowed into a little neck anteriorly; mesonotum without distinct furrows, smooth; metathorax short, smooth, with a slight carina.

Abdomen subsessile, small, black, compressed along the venter.

Legs brown-black, the knees, tips of tibiæ, and the tarsi pale or yellowish.

Wings hyaline, the veins of the basal cells hyaline, scarcely distinguishable.

HABITAT.—Jacksonville, Fla.

Type ♂ in Coll. Ashmead.

Labeo typhlocybæ, sp. nov.

(Pl. v, Fig. 5, ♂.)

♂. Length 2.5ᵐᵐ. Black, subopaque, shagreened, covered with a sparse whitish pubescence. Eyes densely pubescent. Mandibles and palpi yellowish. Antennæ and legs brown, except the anterior tibiæ and all the tarsi, which are yellowish. Antennæ 10-jointed, extending to the middle of the metathorax, very pubescent; the pedicel is more than twice as long as the scape; the first flagellar joint two-thirds the length of the pedicel; the fourth and fifth equal and a little longer than the first; remaining joints, except the last, subequal, the last a little longer, pointed. Mesonotum with 2 furrows, converging and almost meeting at the base of the scutellum. Wings hyaline, the costa and stigma brown, the other veins pale.

HABITAT.—Washington, D. C.

Type ♂ in National Museum.

Reared by Dr. C. V. Riley, at Washington, D. C., July 23, 1883, from *Typhlocyba* sp. occurring on *Celtis* and elm.

PHORBAS Ashm., gen. nov.

(Type *P. laticeps* Ashm., ♂.)

Head broadly transverse, wider than the thorax, but not especially thick through antero-posteriorly, and when viewed from in front rounded; eyes oblong oval, pubescent; ocelli 3, triangularly arranged, neither especially small nor very close together.

Antennæ 10-jointed, thick, filiform, not longer than the thorax, pubescent, the scape only a little longer than the first flagellar joint.

Maxillary palpi 5-jointed.

Mandibles 3-dentate.

Thorax as in *Aphelopus*, but without distinct parapsidal furrows, the prothorax only slightly visible from above.

Front wings with a lanceolate stigma, two basal cells, and an open marginal cell.

Abdomen subpetiolate, oval.

Legs as in *Aphelopus*.

A genus described from a single male specimen, but the lanceolate stigma and 5-jointed maxillary palpi will, however, distinguish it from all other males, in the genera in which the males are known. There is a possibility that it may be the opposite sex of my genus *Bocchus*, as it agrees with it in maxillary, mandibular, and wing characters, but otherwise, in the shape of the head, thoracic, and abdominal characters, it is quite distinct. Of the other genera, it comes nearest to *Aphelopus*, but the widely transverse head, absence of parapsidal furrows, and the lanceolate stigma, sufficiently separate the two.

The single specimen, described below, was captured in Florida by sweeping.

Phorbas laticeps, sp. nov.

(Pl. v, Fig. 7, ♂.)

♂. Length 2^{mm}. Robust, black, subopaque, finely shagreened and sparsely pubescent. Head transverse, much wider than the thorax, the occiput very wide, not at all emarginate. Eyes prominent, oval, pubescent. Mandibles 3-dentate, rufous. Antennæ 10-jointed, extending to the tegulæ, pubescent, the scape longer than the first flagellar joint, the flagellar joints, except the penultimate, about of an equal length, twice as long as thick, the last pointed, the penultimate only 1½ times as long as thick. Mesonotum with two faint furrows; scutellum smooth, impunctured, with a transverse furrow at base; metathorax very short, obliquely truncate behind. Wings hyaline, without pubescence, the stigma and veins white or hyaline. Legs fuscous, a dot on knees, anterior tibiæ and tarsi and middle and posterior tarsi, yellowish-white. Abdomen small, black, subcompressed.

HABITAT.—Jacksonville, Fla.

Type ♂ in Coll. Ashmead.

BOCCHUS Ashm., gen.nov.

(Type *B. flavicollis* Ashm., ♀.)

Head large, broad, shaped much as in *Dryinus*, but the vertex not impressed, the occiput very slightly concave; eyes large oval, prominent; ocelli small, close together in a triangle. Antennæ 10-jointed, subfiliform, very slightly thickened toward tips, inserted just above the clypeus, the scape longer than the first flagellar joint.

Maxillary palpi 4-jointed.

Mandibles 3-dentate, nearly equal, the inner tooth a little the smallest.

Thorax not much lengthened, much narrowed anteriorly and truncate posteriorly, the angles of the truncature rounded; the pronotum is almost as long as the mesonotum but much narrower, a little wider anteriorly than at base, only about half the width of the very broad head; mesonotum wider than long, with furrows; scutellum semicircular with a transverse furrow all across the base; metathorax much shorter than high, abruptly truncate.

Front wings with a lanceolate stigma, two basal cells, and an open marginal cell, the radius being long and curved.

Abdomen globose, distinctly petiolate, the petiole slender, cylindrical, as long as the hind coxæ.

Legs as in *Dryinus*, the anterior tarsi chelate, the fourth joint not much larger than the third.

This genus comes nearest to *Dryinus*, but is at once separated by the 4-jointed maxillary palpi, the difference in the shape of the prothorax, the globose abdomen, the relative length of the anterior tarsal joints, and by the vertex of the head not being impressed.

Only a single species is known, as follows:

Bocchus flavicollis, sp. nov.

(Pl. v. Fig. 6, ♀.)

♀. Length 3^{mm}. Black, closely, finely punctulate, the head and mesonotum tinged with brown. Head large, broad, subquadrate, not impressed on the vertex, a little narrowed behind the eyes; eyes large. Prothorax strongly contracted, yellow; two basal joints of antennæ, the legs and petiole pale rufous; flagellum subclavate, brown black. Antennæ 10-jointed, the pedicel about half the length of the first flagellar joint, the first being about as long as the scape, the remaining joints shorter but thicker. The pronotum is much narrower than the mesonotum, although about as long; mesonotum with two furrows; scutellum bifoveated, the foveae widely separated but connected by a transverse line; metathorax short, coarsely rugose, with some dorsal raised lines. Wings hyaline, with a fuscous band across the marginal cell two-thirds the width of wing, almost devoid of pubescence; stigma lanceolate; stigmal vein long curved; abdomen globose, black, petiolated, the petiole slender, rufous, as long as the hind coxae, the second and third segments very large, equal, occupying nearly the whole surface, the following segments retracted.

HABITAT.—Marquette, Mich.

Type ♀ in Coll. Ashmead.

Described from a single specimen captured by Mr. E. A. Schwarz.

CHELOGYNUS Haliday.

. Ent. Mag., I, p. 273 (1833).

(Type *C. fascicornis* Dalm.)

Head transverse, the vertex broad, convex; the cheeks oblique, the occiput not deeply concave, margined; ocelli 3 in a triangle, but widely separated; eyes oval, subprominent.

Antennae inserted just above the clypeus, 10-jointed, the first flagellar joint scarcely as long as the scape, usually shorter, the following joints short, in ♂ the following joints lengthened.

Maxillary palpi 5-jointed.

Mandibles truncate at tips, 3- or 4-dentate; the teeth acute, slightly subequal, the inner tooth the smallest.

Thorax elongate; prothorax in ♀ as long, or nearly as long, as the mesonotum. subquadrate, obliquely impressed at sides, in ♂ very short; mesonotum usually with two distinct furrows or indicated anteriorly; metathorax long, rounded posteriorly.

Front wings with a large ovate or subovate stigma, two basal cells and a stigmal vein.

Abdomen subpetiolate, ovate, the second segment the largest, the others gradually subequal.

Legs of moderate length, the femora obclavate, stout, the anterior tarsi in the ♀ chelate.

The ovate stigma and the 5-jointed labial palpi sufficiently separate this genus from *Bocchus;* and these characters, in connection with the difference in the shape of the head and pronotum, and the differences to be pointed out in the chelae distinguish it from *Dryinus;* while the longer pronotum, the 5-jointed maxillary palpi, and the differences in the relative length of the anterior tarsal joints and the chelae can be depended upon to separate it from *Anteon.*

Chelogynus atriventris Cress.

Dryinus atriventris Cr., Trans. Am. Ent. Soc., IV, p. 193, ♀ ; Ashm., Ent. Am.,
    III, p. 74; Cr., Syn. Hym., p. 246.

♀. Length, 4.5^{mm}. Ferruginous, with pale glittering pubescence; head much broader than thorax; face, mandibles, and base of scape beneath, pale yellowish; tips of antennæ blackish; pleura with dense silvery pubescence; metathorax rugulose, posterior face depressed and transversely aciculated; wings hyaline, with a dusky band beneath stigma; posterior tibiæ blackish at tips, their tarsi pale; abdomen black, smooth, and polished.

HABITAT.—Texas.

Type ♀ in Coll. Amer. Ent. Soc.

Chelogynus Henshawi sp. nov.

♀. Length, 5^{mm}. Black, shining; head and prothorax finely rugose; mesonotum and scutellum smooth, polished; metathorax coarsely rugose. Antennæ and legs pale rufous; the 6 terminal joints of the antennæ fuscous, and the posterior femora towards apex above have a large black spot. The antennæ are 10-jointed, filiform, the first flagellar joint the longest, about one-third longer than the scape. Clypeus and mandibles rufous, the latter 4-dentate. Mesonotum with two sharply defined furrows. Tegulæ rufous. Wings hyaline, with two fuscous bands, one at the basal nervure, the other below the upper half of the stigma; veins pale brown. Anterior femora much swollen, the tarsi with large pincher-like claws, as long as the tibiæ, and fringed with stiff bristles on the inside.

HABITAT.—Milton, Mass.

Type ♀ in Coll. Ashmead.

Described from a single specimen received from Mr. Saml. Henshaw. It is the only species I have seen with 4-dentate mandibles.

Chelogynus canadensis sp. nov.

(Pl. VI, Fig. 1, ♀.)

♀. Length, 2.5^{mm}. Black; head and collar very finely punctate, the mesonotum and scutellum smooth, the former with two furrows; metathorax rugose. Face sparsely covered with glittering silvery hairs. Mandibles and palpi white. Antennæ 10-jointed, brown, the first flagellar joint about half the length of the scape, the remaining

joints very little longer than thick.   Wings clear hyaline, stigma
brown, veins pale or hyaline.   Legs pale rufous, the posterior coxæ
blackish basally; the pinchers of the anterior feet very small.

HABITAT.—Ottawa, Canada.

Type in Coll. Ashmead.

This species is described from a single specimen received from W.
H. Harrington, and comes nearest to the European *C. fascicornis* Dahn.

ANTEON Jurine.

Hymn., p. 302 (1807); Först., Hymn. Stud., II, p. 93.

Heterolepis Nees, Mon., II, p. 271.

(Type *A. Jurineanus* Latr.)

Head transverse, the vertex wide, convex or subconvex, the occiput
slightly concave, margined; ocelli in a triangle; eyes ovate.

Antennæ inserted just above the clypeus, 10-jointed; in the ♀ subclavate, the first flagellar joint hardly one-third the length of the scape,
the following joints short, the terminal joints thickened gradually;
in ♂ filiform, pilose, or pubescent, the first flagellar as long or slightly
longer than the scape, the following joints all lengthened, cylindrical.

Maxillary palpi 4-jointed; labial palpi very short, 2-jointed.

Mandibles truncate at tips, 3-dentate, the teeth acute, the outer
slightly longer than the others.

Thorax subovate, the prothorax short, narrowed and margined anteriorly, the sides oblique, impressed; mesonotum usually smooth and
without distinct furrows; metathorax at least as long as wide, rugose,
rounded posteriorly.

Front wings with a large ovate stigma, two basal cells and a stigmal
vein; other cells obliterated.

Abdomen ovate or oblong-oval, subpetiolate, the second segment the
largest, the others gradually subequal.

Legs moderate, the femora swollen or obclavate, the anterior tarsi
in ♀ chelate.

A genus usually confused with *Chelogynus* and *Dryinus*, but quite
distinct.   It has not a particle of resemblance to *Dryinus*, except in the
anterior tarsi being chelate.   The shape of the head, the short collar,
and ovate stigma in front wing, readily separate it from that genus.
With *Chelogynus*, however, there is a closer resemblance, but the pronotum is always much shorter, often only visible from above as a slight
collar, while the maxillary palpi are 4-jointed.   In *Chelogynus* the pronotum is always as long, or nearly as long, as the mesonotum, while
the maxillary palpi are 5-jointed.

The species are numerous, although nothing positively is known of
their habits.

The following table will afford assistance in distinguishing our
species:

TABLE OF SPECIES.

FEMALES.

Wings hyaline, not banded.. 2
    Wings with a wide dusky band below the stigma.
        Head closely, densely punctate, opaque; mesonotum sparsely punctate; legs reddish-yellow, tarsi white..............A. UNIFASCIATUS, sp. nov.
2. All coxæ pale, except sometimes the hind coxæ.................................. 3
    All coxæ and femora black.
        Posterior tibiæ black, rest of the legs rufous.
            Head coarsely rugose; antennæ long, dull rufous, densely pilose; mesonotum with furrows ♂ .... A. RUGOSUS, sp. nov.
        Tibiæ and tarsi dull honey-yellow, the tibiæ more or less fuscous at the middle.
            Head finely, closely punctulate; antennæ dark brown, the scape and pedicel rufous; mesonotum without furrows ♀ ....A. TIBIALIS Say.
3. Hind coxæ black or at least basally.
    Legs rufous, mesonotal furrows indicated only anteriorly.
        Head and thorax shining, sparsely irregularly shagreened, the collar transversely rugulose; antennæ brown-black, the scape rufous. ♀
                                      A. POLITUS, sp. nov.
    Posterior femora black above, legs yellowish.
        Head and thorax shining, the head wrinkled, collar rugoso-punctate anteriorly; antennæ honey-yellow ....A. PALLIDICORNIS, sp. nov.
    Posterior femora and tibiæ embrowned.
        Head closely punctate, opaque, the thorax sparsely punctate, shining; antennæ long, brown, pilose, the scape rufous ♀ ....A. PUNCTICEPS, sp. nov.
    Hind coxæ black above toward base, rest of the legs rufous; mesonotal furrows not at all indicated.
        Head closely, finely punctate, opaque, collar rugulose anteriorly, mesonotum closely microscopically punctate ♀ .... A. MINUTUS, sp. nov.
    All coxæ pale, legs reddish-yellow.
        Mesonotum polished, the furrows slightly indicated anteriorly.
            Head finely closely punctate, subopaque, collar very short, rugulose at sides; antennæ as long as the body, pilose, the second and third joints shorter than the scape, the fourth and beyond as long as the scape. ♂
                                    A. POPENOEI Ashm.

Anteon unifasciatus, sp. nov.

♀. Length 1.8ᵐᵐ. Black, shining; the head opaque, closely, densely punctured; the meso- and pronotum more sparsely punctate, sparsely pubescent, the latter somewhat transversely rugulose anteriorly; metathorax rugose. Antennæ, mandibles, palpi and tegulæ reddish-yellow. Antennæ 10-jointed, a little thickened at tips, the scape 3½ times as long as the pedicel, the first flagellar joint a little shorter than the pedicel, the second and third nearly equal, much shorter than the first, the remaining joints all longer and stouter. Wings hyaline, with a broad dusky band across the stigmal region of the wing, the venation pale or hyaline, the large stigma brown.

HABITAT.—Biscayne Bay, Florida.

Type ♀ in National Museum.

Described from a single specimen taken by E. A. Schwarz.

Anteon rugosus, sp. nov.

♂. Length, 4.3ᵐᵐ. Head and thorax, except the disk of mesonotum and scutellum, very coarsely rugose, the mesonotum and scutellum sparsely but coarsely punctate; head very large and thick, the face with silvery white hairs; mandibles, scape, tips of anterior femora, their tibiae and tarsi, ferruginous; rest of legs, except the tarsi, black; flagellum brown-black, very pilose; wings hyaline, the stigma large, brown-black, with a pale spot at base; nervures distinct but pale. Abdomen very narrow, small, polished, black, subcompressed. Antennae 10-jointed, as long as the head and thorax united, tapering toward tips, the flagellar joints nearly equal in length, four or more times longer than thick, the scape thicker and a little longer than the first flagellar joint.

HABITAT.—Illinois.

Type in Coll. Ashmead.

Described from a single specimen.

Anteon tibialis Say.

Bost. Jour., I, p. 284; Lec, Ed. Say's Works, II, p. 730; Ashm. Ent. Am., III, p. 71; Cress. Syn. Hym. p. 216.

♂. Length, 3ᵐᵐ. Black, shining; head and prothorax finely punctured; mesonotum smooth, without furrows; metathorax rugoso-punctate with some elevated lines toward base. Antennae 10-jointed, fuscous, covered with a short pubescence, the scape and pedicel dull yellow; flagellum subclavate, brown, the joints after the first less than twice as long as thick. Wings hyaline, the stigma brown, the nervures hyaline, subobsolete, the stigmal vein very short, scarcely half the length of the stigma. Legs black, extreme tips of the femora and tibiae and tarsi dull honey-yellow.

HABITAT.—Indiana and District of Columbia.

The specimen from which this description is drawn up was taken by Mr. Schwarz, and agrees exactly with Say's description.

Anteon politus, sp. nov.

(Pl. VI, Fig. 2, ♀.)

♀. Length, 2.6ᵐᵐ. Polished black; head, collar, and mesopleura irregularly microscopically shagreened; metathorax rather coarsely rugose; collar distinct, less than half the length of the mesonotum; mesonotum with furrows indicated only anteriorly; legs (except the posterior coxae, which are black basally, and the posterior tibiae and tarsi, which are fuscous) rufous. Antennae 10-jointed, reaching only a little beyond the tegulae, brown-black, the scape rufous; flagellar joints only a little longer than thick. Wings hyaline, the stigma large, brown, the nervures distinct pale yellow, the stigmal vein about as long as the stigma. Abdomen ovate, subpetiolate, polished black, depressed, not quite as long as the thorax.

HABITAT.—Toronto, Canada.

Type in Coll Ashmead.

Described from a single specimen received from Mr. W. Hague Harrington.

Anteon pallidicornis sp. nov.

♀. Length, 1.5ᵐᵐ. Black, shining; head large, broad, finely shagreened; collar, mesopleura, and metathorax rugose; legs yellowish, posterior coxae, their femora, above and beneath, and tips of their tibiae black or fuscous. Antennæ 10-jointed, rather short, entirely honey-yellow, the flagellum a little thickened toward tip; first flagellar joint about one-third longer than the pedicel, the two following subequal, joints 4 to 7 stouter, scarcely longer than thick, the last ovate, nearly twice as long as the penultimate. Mesonotum and scutellum polished, impunctured, the mesonotal furrows not extended through to the scutellum. Wings hyaline, the stigma brown, the nervures yellowish, distinct. Abdomen ovate, depressed, polished black, much narrower, but as long as the thorax.

HABITAT.—Utah Lake, Utah.

Type in Coll. Ashmead.

Described from a single specimen given to me by Mr. E. A. Schwarz.

Anteon puncticeps sp. nov.

♂. Length, 1.75ᵐᵐ. Black, closely punctate, the disk of the mesonotum and the scutellum polished, impunctate; head rather coarsely punctate; mandibles, palpi, scape, and anterior and middle legs, pale or brownish-yellow; posterior legs dark brown, the trochanters, knees, and tarsi pale; middle tibiae a little dusky towards tip above. Antennae 10-jointed, the flagellum brown, pilose; the scape is three times as long as the pedicel; first flagellar joint very slightly shorter than the following, the following joints being 3 times as long as thick. Parapsidal furrows only indicated anteriorly. Mesopleura and metathorax rugose. Wings hyaline, the stigma pale brownish-yellow, the nervures distinct but pale; the stigmal nervure is a little longer than the stigma, angularly bent before the tip.

HABITAT.—Arlington, Va., and Vancouver Island.

Types in Coll. Ashmead.

Two specimens. The specimen from Vancouver was obtained through the collector, Mr. H. F. Wickham, the other was taken by myself at Arlington, Va.

Anteon minutus sp. nov.

♀. Length, 1.2ᵐᵐ. Black, shining; the head and thorax with a close, delicate punctuation; the scutellum smooth, highly polished. Antennae and legs brownish-yellow, the former dusky toward tips;

posterior coxæ above, the middle tibiæ above, and the posterior femora and tibiæ above, more or less piceous. Antennæ 10-jointed, reaching to the tip of metathorax, incrassated toward the tips, the terminal joint twice as long as the penultimate; the pedicel stouter but not longer than the first flagellar joint, the joints beyond submoniliform. The collar is about one-third the length of the mesonotum, above finely, transversely rugulose; mesonotum without furrows; scutellum smooth, polished, with a row of punctures surrounding the posterior margin; metathorax finely rugose. Wings hyaline, the veins pale, the stigma large, brown, with a pale spot at base, the stigmal vein short. Abdomen small, ovate, subpetiolate, smooth and shining, about as long as the thorax, excluding the collar.

HABITAT.—Washington, D. C.

Type in Coll. Ashmead.

Described from a single specimen given to me by E. A. Schwarz. The species is the smallest known, and could easily be mistaken for a small *Aphelopus*.

Anteon Popenoei Ashm.

Dryinus Popenoei Ashm., Bull. No. 3, Kans. Exper. Sta., App., p. 1. June, 1888.

♂. Length, 2ᵐᵐ. Black: head subopaque, microscopically punctuate; thorax and abdomen polished, black; metathorax and mesopleura finely rugose; the superior edge of the truncature in the former margined. Clypeus anteriorly arcuate, smooth. Mandibles 4-dentate, pale brown. Antennæ 10-jointed, brown, with sparse, long, white hairs; pedicel small, half the length of the scape; first flagellar joint longer, but shorter than the fourth; fourth joint and those beyond, except the last, nearly equal, about as long as the scape. Wings hyaline, the stigma large, pale-brown, the stigmal vein short. Legs, including coxæ, reddish-yellow.

HABITAT.—Riley County, Kansas.

Type in Kansas State Agricultural College.

MYSTROPHORUS Förster.

Hym. Stud., II, p. 91 (1856); Ruthe, Berl. Ent. Zeits., 1859, p. 120.

(Type *M. formicæformis* Ruthe.)

Head large, broad, subquadrate, nearly twice broader than the thorax, the occiput not emarginate; ocelli 3, triangularly arranged; eyes oval.

Antennæ in both sexes 10-jointed, in ♀ subclavate, the pedicel small, the first flagellar joint a little longer and slenderer than the scape, the following shorter.

Maxillary palpi 4-jointed; labial palpi 2-jointed.

Mandibles truncate at tips, with 3 nearly equal, acute teeth.

Thorax elongate; the prothorax nearly twice as long as the mesothorax; mesonotum short, semicircular, without furrows, the scutellum small; metathorax large, longer than the prothorax, not areolated.

Front wings in ♂ short, spoon-shaped, with two basal cells, a stigma, and a stigmal vein; ♀ apterous.

Abdomen short, ovate.

Legs as in *Aphelopus*.

A peculiar little genus, closely allied to *Aphelopus*, but readily distinguished by the long and broad prothorax and the absence of mesonotal furrows. The genus is unknown in our fauna.

APHELOPUS Dalman.

Anal. Ent., pp. 8–14 (1823); Förster Hym. Stud., ii, p. 91.

(Type *A. melaleucus* Dalm.)

Head transverse, wider than the thorax, the vertex broad, convex, the occiput slightly emarginate; eyes large, oval, hairy; ocelli 3 in a triangle, but widely separated, the lateral ocelli nearer the margin of the eye than to each other.

Antennæ inserted just above the clypeus, 10-jointed, in ♀ shorter than in the ♂, subclavate, the scape very short, scarcely longer than the second, the others variable in length, in ♂ filiform, hairy, the scape usually longer than the third, the last joint sometimes thickened.

Maxillary palpi long, 5-jointed; labial palpi short, 2-jointed.

Mandibles truncate at tips, with 3 small teeth.

Thorax ovoid; the prothorax not or scarcely visible from above; mesonotum as broad as long, convex, without furrows; metathorax not longer than high, rounded posteriorly.

Front wings with a large oval stigma, a short stigmal vein and two basal cells, which are often subobliterated on account of the paleness of the nervures.

Legs moderate, the femora not obclavate, the tarsi simple in both sexes.

TABLE OF SPECIES.

Head, thorax, and abdomen black, the head sometimes maculate 2
Head and thorax black, the abdomen rufous.
    Antennæ and legs yellow ♀ .... A. RUFIVENTRIS sp. nov.
2. Head with white or yellow marks anteriorly................................. 3
    Head entirely black.
        Clypeus subtriangular.
            Legs honey-yellow; antennæ nearly as long as the body, brown, pilose, the
                  scape yellow ♂ .... A. AMERICANUS Ashm.
        Clypeus truncate.
            Legs dull rufous, the posterior femora and tibæ fuscous or black; antennæ
                much shorter than the body, black, pilose ♂ . A. MELALEUCUS Dalm.

3. Head anteriorly from a little above the insertion of the antennae and including
    the lower part of cheeks, the clypeus, mandibles, and palpi white.
   Clypeus anteriorly truncate.
      Anterior and middle legs white, sometimes the middle femora fuscous, pos-
         terior legs fuscous, with sometimes most of the coxae, trochanters
         and base of the femora white ♀ .... A. MELALEUCUS Dalm.
   Clypeus anteriorly slightly emarginate at the middle.
      Pronotum and all legs, except the apical half of posterior femora and their
         tibiae, white ♀ .... A. ALBOPICTUS sp. nov.
   Head anteriorly from a little above the insertion of the antennae and including
      the lower part of the cheeks, clypeus, and mandibles, pale brownish
      yellow; palpi white.
   Clypeus anteriorly arcuate.
      Legs honey-yellow, the posterior tibiae fuscous ♀ .... A. AFFINIS sp. nov.

Aphelopus rufiventris sp. nov.

♀. Length, 2 mm. Head and thorax black, minutely punctate; abdomen rufous; antennae and legs honey-yellow. Antennae 10-jointed,
short, the flagellum subclavate; scape stout, as long as the pedicel and
first flagellar joint united; first flagellar joint scarcely as long as the
pedicel, joints 2 and 3 subequal, those beyond scarcely as long as thick.
Mesonotum with furrows indicated only anteriorly. Wings hyaline,
the nervures, excepting the stigmal, obsolete; stigma very large, pale
brown, the stigmal vein very short, oblique, only about one-third as long
as the stigma. Metathorax rounded off posteriorly, rather coarsely
rugose. Abdomen oblong-oval, much narrower than the thorax, but
fully as long, rufous.

HABITAT.—Jacksonville, Fla.

Type in Coll. Ashmead.

A species readily distinguished by the color of the abdomen, the antennae, and the venation of anterior wings.

Aphelopus americanus Ashm.

Ent. Am., III, p. 74, ♂; Cress., Syn. Hym., p. 247.

♂. Length, 1.5 mm. Black, subopaque, finely punctate, and covered
with a fine whitish pubescence; mesonotum and scutellum smooth,
polished, with only a few faint widely separated punctures; clypeus
subtriangular. Antennae 10-jointed, long, reaching to the middle of the
abdomen, the scape brownish-yellow, the flagellum brown, covered with
a short, dense pubescence, the joints of the flagellum three times as
long as thick. Mesonotum with 2 faint furrows. Wings hyaline, the
stigma large brown, the other nervures very pale, or hyaline, the stigmal vein being as long as the stigma. Legs honey-yellow, the posterior
femora and tibiae fuscous or black. Abdomen very small, black, shining.

HABITAT.—Jacksonville, Fla.

Type ♂ in Coll. Ashmead.

Aphelopus melaleucus Dalman.

(Pl. VI, Fig. 3, ♀.)

Gonatopus melaleucus Dalm., Sv. Ak. Handl., 1818, p. 82 ♀ ; Dalm., An. Ent., p. 14.
Dryinus atratus Dalm. *loc. cit.*, p. 15.
Aphelopus melaleucus Nees., Mon., II, p. 388, ♀ ; Walk., Ent. Mag., IV, p. 127, pl. XVI, fig. 3 ♀.
Aphelopus atratus Nees., Mon., II, p. 389 ♂.
Aphelopus melaleucus Thoms., Öfv., 1860, p. 179, ♂ ♀ ; Marsh., Cat. Brit. Hym., p. 7.

♀. Length, 1.8ᵐᵐ. Black; anterior orbits, face below the eyes, clypeus, except sometimes a central line, mandibles and anterior coxae and legs, white; middle legs, honey-yellow; posterior legs, except a spot beneath the coxae and the trochanters, usually brown or piceous; antennæ brown, the flagellum a little thickened towards the tip, the first four joints nearly equal, the three following stouter, but much shorter, the last joint nearly twice as long as the preceding.

♂. Length, 1.5ᵐᵐ. In this sex the clypeus and the mandibles are white, the legs brown or fuscous, the apical half of anterior femora, their tibiæ and tarsi, honey-yellow or whitish; the other legs are also sometimes varied with yellow.

HABITAT.—Europe, Canada.

This European species is recognized in our fauna from a single female, taken at Ottawa, Canada, by Mr. W. H. Harrington. Monsieur Alfred Giard has bred it in France, from *Typhlocyba hippocastani* and *T. douglasi*, and has given a most interesting account of its habits, in the Comptes Rendus des Séances de l'Académie des Sciences for 1889.

Aphelopus albopictus sp. nov.

♀. Length, 1.5ᵐᵐ. Black, shining, the head alone finely, closely punctulate, subopaque; face below the front ocellus, clypeus, mandibles, lower part of cheeks, palpi, anterior and middle legs, pronotum to the tegulæ, and the propectus, white or yellowish-white; posterior legs embrowned, the coxæ, trochanters and base of femora and tibiæ, white; clypeus anteriorly slightly emarginate at the middle. Head large, subquadrate, with a slight carina between the antennæ. Antennæ 10-jointed, dark brown, slightly thickened toward tips, densely pubescent; the first flagellar joint is a little longer than the short scape, the former the stouter; joints 4, 5, and 6 slightly longer than the first, the seventh, eighth, and ninth shorter, only a little longer than wide, the tenth about twice as long as the penultimate. Mesonotum with the furrows indicated only anteriorly. Metathorax rugose, obliquely rounded off posteriorly. Abdomen very small, compressed, about half the length of the thorax, polished, black. Wings hyaline, the nervures, except the large brown stigma, very pale and indistinct, or subobsolete; stigmal nervure as long as the stigma, slightly arcuate.

HABITAT.—Washington, D. C., and Bladensburg, Md.

Types in Coll. Ashmead.

Two specimens, both captured by Mr. E. A. Schwarz. The species comes nearest to *A. melaleucus*, but shows more white, has a different shaped clypeus and with the joints of the antennæ relatively different.

Aphelopus affinis, sp. nov.

♀. Length, 2.2ᵐᵐ. Black, shining; head and thorax very minutely punctate, metathorax rugulose; face from the frons, including the clypeus and mandibles, ferruginous; clypeus anteriorly rounded; antennæ brown; legs (except the posterior tibiæ, which are fuscous), entirely honey-yellow. Antennæ 10-jointed, a little thickened toward tips, the scape not longer than the first flagellar joint, flagellar joints 1 to 3 about three times as long as thick, the following to the last about twice as long as thick, the last much longer and thicker than any of the others. Wings hyaline, the stigma and stigmal vein brown, the other nervures hyaline, subobsolete; the stigmal vein is arcuate and a little longer than the stigma. Abdomen as long as the thorax, strongly compressed.

HABITAT.—Canada.

Type ♀ in Coll. Ashmead.

Described from a single specimen.

Subfamily IV.—CERAPHRONINÆ.

Head transverse, when viewed from in front oval, rounded, or oblong. Ocelli 3 in a triangle, rarely wanting. Mandibles oblong, bidentate at tips. Antennæ elbowed, inserted at base of the clypeus; in males 11-jointed (in a single case 10-jointed), filiform, dentate or subramose; in females 9-, 10-, or 11-jointed, filiform or clavate. Maxillary palpi 4- or 5-jointed; labial palpi 2- or 3-jointed. Pronotum short, visible from above only as a transverse line; mesonotum large, transverse, rarely without grooved lines, usually with 1, 2, or 3 grooved lines; scutellum large, convex or subconvex, the axillæ distinct; metathorax very short, rounded behind, the angles slightly prominent. Front wings with a large stigmated, or linear, marginal vein, a radial nervure, a more or less distinct basal cell, and always without a postmarginal nervure; hind wings simple, rather broad at base, and entirely veinless. Apterous forms frequent. Abdomen subsessile, ovate, composed of 8 segments, the second occupying about half its surface, striated at base. Legs moderate, the posterior femora the stoutest, the tibial spurs 1, 1, 1, the anterior spur large, divided into two prongs, the middle and posterior spurs weak, short; tarsi 5-jointed, the claws small, simple.

A somewhat extensive and widely distributed group, more closely allied to the three preceding subfamilies than to those that follow, except possibly the tribe Telenomini, in the Scelioninæ, but with quite a different habitus; the hind wings are never lobed, the anterior wings with a wholly different venation, while the structure of the thorax and abdomen is quite distinct. The non-lobed hind wings, venation, different antennæ, and shape of mandibles, at once separate the species in this group from the Bethylinæ, Emboleminæ, and the Dryininæ; while from the only two groups that follow, with which they would likely be confused—the Scelioninæ and the Platygasterinæ, which like themselves have non-lobed hind wings and the antennæ inserted close to the mouth—they may be readily distinguished by the widely different antennæ, the shape of the head, venation, and non-carinated abdomen.

If we accept the published records as accurate, this group has great diversity of habits.

Ratzeburg has recorded *Megaspilus* sp. from a bombycid, a syrphid, two cecidomyiids, a *Chermes*, a scolytid and a tortricid; *Ceraphron* species from a tineid, a *Tomicus*, a *Curculio* and a *Brachonyx;* and a *Lygocerus* from a cynipid. Unquestionably, so far as the hosts are concerned, some of these records are inaccurate. The records in this country and others published abroad, show the group attacks almost exclusively the homopterous family *Aphididæ*, and the dipterous family *Cecidomyiidæ*. All reared by me, as well as those reared by Dr. Riley, except in a single instance (the rearing of a *Lygocerus*, by Dr. Riley, from a tortricid, *Sarrothripa rawayana*, which is of questionable accuracy) have been from Aphids and Cecidomyiids. It is not improbable, therefore, that a parasitized Aphid was on the same leaf when the Lepidopteron was placed in the breeding jar.

Some such explanation may also account for the rearing of these insects by Ratzeburg from Lepidoptera, and his rearings from Coleoptera may be accounted for by the supposition that they were accompanied by inquilinous Dipterous larvæ.

To accept without question such diversity of habits in the genera of this group it seems to me would be unscientific.

The group seems to divide naturally into two tribes distinguished as follows:

Marginal vein stigmated; antennæ, with the same number of joints in both sexes, 11-jointed ....Tribe I.—MEGASPILINI
Marginal vein linear, never stigmated; antennæ with a less number of joints in the females than in the males; males with 10- or 11-jointed antennæ; females 9- or 10-jointed ....Tribe II.—CERAPHRONINI

Tribe I.—MEGASPILINI.

The genera in this tribe are not numerous and may be tabulated as follows:

TABLE OF GENERA.

FEMALES.

1. Mesonotum without or with 1 or 2 impressed lines............................2
   Mesonotum with 3 impressed lines.
      Metathorax with a forked spine at base ....HABROPELTE Thoms.

Metathorax not spined at base.
  With wings.
    Wings bare, without cilia ........................TRICHOSTERESIS Förster
    Wings pubescent, with cilia.
      Eyes usually bare; mesonotum not narrowed anteriorly.
                                              LYGOCERUS Förster
      Eyes pubescent; mesonotum narrowed anteriorly...MEGASPILUS Westw.
  Wingless.
    Maxillary palpi 5-jointed; labial 3-jointed; thorax not much narrowed.
                                          MEGASPILUS Westw.
    Maxillary palpi 4-jointed; labial 2-jointed; thorax much narrowed.
                                          EUMEGASPILUS Ashm.
2. Mesonotum with 2 impressed lines; apterous................DICHOGMUS Thoms.
  Mesonotum with 1 impressed line; apterous.
    Eyes small; ocelli wanting ....LAGYNODES Förster
    Eyes large; ocelli present; winged..................ATRITOMUS Förster
  Mesonotum without furrows; subapterous.
    Eyes large, bare; ocelli distinct ....ATRITOMUS Förster

MALES.

1. Mesonotum without impressed lines... 2
  Mesonotum with 3 impressed lines.
    Metathorax with a forked spine at base ....HABROPELTE Thoms.
    Metathorax without a forked spine at base.
      Wings bare, without cilia ........................TRICHOSTERESIS Förster
      Wings pubescent, with cilia.
        Antennæ dentate or ramose ....LYGOCERUS Förster
        Antennæ simple, filiform.
          Maxillary palpi 5-jointed; labial palpi 3-jointed....MEGASPILUS Westw.
          Wingless.
            Maxillary palpi 4-jointed; labial palpi 2-jointed...EUMEGASPILUS Ashm.
  Mesonotum with 1 impressed line (Atritomus).
2. Antennæ toothed or serrate.................................ATRITOMUS Förster
  Antennæ with 5 long branches, a branch on each of the first five flagellar joints,
                                     DENDROCERUS[1] Ratzb.

HABROPELTE Thomson.

Öfv., 1858, p. 288.

Megaspilodes Ashm., Can. Ent., xx, p. 48 (1888).

(Type *H. tibialis* Boh.)

Head transverse, the face not greatly lengthened, occiput margined; ocelli 3, triangularly arranged; eyes oval, hairy.

Antennæ inserted just above the clypeus, long, filiform, 11-jointed in both sexes, the scape scarcely as long as the first flagellar joint, pedicel very small, hardly longer than wide, the flagellar joints lengthened, equal or very gradually subequal.

---

[1] This genus is unknown to me, and in the preparation of this work was at first overlooked. It was described and figured by Ratz. in Die Ichn. d. Forstins., Band 3, p. 181 (1852), where a full description of the male, the only sex known, is given. While apparently closely allied to *Lygocerus*, I believe it to be distinct.

Maxillary palpi 5-jointed; labial palpi 3-jointed.

Mandibles bidentate at tips.

Thorax subovate, the prothorax not visible from above, contracted into a little neck at the junction with the head; mesonotum subdepressed, with 3 furrows; scutellum large, subconvex, longer than wide, rounded posteriorly and with a distinct frenum; metathorax short, slightly emarginate, with 2 teeth at base.

Front wings with a large, oval stigma, a parastigma and a curved stigmal vein.

Abdomen subsessile, ovate, subconvex above, convex beneath, the second segment occupying fully two-thirds of the whole surface, longitudinally striated, the following segments very short, the last pointed.

Legs moderate, pilose, the femora slightly swollen, the posterior pair the thickest, tibiæ subclavate, the tibial spurs not very well developed, tarsi shorter than their tibiæ.

The genus is distinguished from other genera with 3 mesonotal furrows, by the 2 erect teeth or spines at the base of the metathorax, the larger flattened scutellum, and the longer, denser pilosity of the legs.

Two closely allied species are known in our fauna, separable as follows:

FEMALES.

Legs black, base and tips of tibiæ and the tarsi, dark rufous.

Antennæ, except sometimes the first 2 flagellar joints, black; the first flagellar joint thrice as long as the pedicel; axillæ separated from the scutellum by 5 large punctures...H. ARMATUS Say

Legs black, trochanters, bases, and tips of femora, tibiæ except at the middle and tarsi, rufous.

Antennæ black, the scape, pedicel and first 2 flagellar joints, rufous; the first flagellar joint less than thrice as long as the pedicel; axillæ separated from the scutellum by 8 punctures........................H. FUSCIPENNIS Ashm.

MALES.

Axillæ separated from the scutellum by 5 large punctures.

First flagellar joint a little longer than the scape; last joint of palpi nearly twice as long as the fourth...H. ARMATUS Say

Axillæ separated from the scutellum by 8 punctures.

First flagellar joint not longer than the scape; last joint of palpi not one-half longer than the fourth................................H. FUSCIPENNIS Ashm.

H. fuscipennis Ashm.

Megaspilodes fuscipennis Ashm., Bull. 3, Kans. Exp. Sta. App., p. ii, 1888.

♂ ♀. Length, 3.5ᵐᵐ. Shining black, pilose; head and thorax anteriorly rugose, rest of the thorax smooth, the face with a narrow, smooth space in front of the ocelli, antennal depression absent. Maxillary palpi long, the last joint 1½ times longer than the fourth. Antennæ 11-jointed, in ♀ incrassated toward tips, the scape long, obclavate; the scape, pedicel, and first two flagellar joints are brownish-yellow, the first flagellar joint more than twice as long as the pedicel; in ♂ filiform; the scape is equal in length with the first flagellar joint, the following joints slightly shorter, all black or dark fuscous, about 5 times as long as thick; last joint of palpi one-half longer than the fourth. Prothorax very short, narrower than the mesothorax and below the dorsal line depressed and produced into a slight neck. Mesothorax truncate before, with 3 coarsely punctate furrows. Scutellum with a punctate frenum. Axillæ separated from the scutellum by about 8 punctures. Metathorax armed in the middle, behind the scutellum, with two blunt teeth or spines. Wings fusco-hyaline, darker beneath the stigma, the stigmal vein about twice as long as the stigma; both brownish-black. Legs dark rufous or reddish-brown, the coxæ and femora black, with sometimes the tibiæ black or dusky. Abdomen with coarse longitudinal striæ to near the apex of the second segment.

HABITAT.—Washington, D. C.; Arlington, Va; Manhattan, Kans., and Wyoming.

Types in Coll. Ashmead and Kansas State Agricultural College.

Described from several specimens.

Habropelte armatus Say.

(Pl. vi, Fig. 4. ♂.)

Ceraphron armatus Say, Bost. Jour., i, p. 276; Lec. Ed., Say, ii, p. 724.
Lygocerus armatus Ashm., Ent. Am., iii. p. 98.
Ceraphron armatus Cress. Syn., Hym., p. 248.
Megaspilodes armatus Ashm., Bull. 3, Kans. Exp. Sta., ii, 1888.
Telenomus stygicus Prov. Add. et. Corr., p. 189.

♂ ♀. Length, 3.4 to 3.8mm. Much like fuscipennis, but may readily be separated by the following differences: The ♀ antennæ are usually black, except sometimes the first two flagellar joints, which are pale brownish, or at least beneath, the first flagellar joint being thrice as long as the pedicel; in the ♂ the first flagellar joint is a little longer than the scape; while the axillæ are separated from the scutellum by five large punctures, which is a constant character in both sexes. In the ♂ the last joint of the palpi is nearly twice as long as the fourth. The legs in both species are variable in color, exhibiting more red in some specimens than in others.

HABITAT.—Indiana and Arlington, Va.

Types in Coll. Ashmead.

Either one of these species could be used as typical of Say's species; but as Say's type is no longer in existence it devolves upon me to designate which should be known as the type, his description not being sufficiently definite to decide, and I have, therefore, separated them as above.

TRICHOSTERESIS Förster.

Hym. Stud., ii, p. 97 (1856).

Thliboneura Thoms., Öfv., 1858.

(Type *T. glabra* Boh.)

Head transverse, the face, seen from before, not longer than wide, the occiput margined; ocelii 3 in a curved line; eyes broadly oval, not hairy.

Antennæ inserted just above the clypeus, 11-jointed, subfiliform, the scape 3 or 4 times as long as the first funiclar joint, the pedicel longer than thick, the joints beyond the third short.

Maxillary palpi 5-jointed; labial palpi 3-jointed.

Mandibles bidentate.

Thorax ovoid, the prothorax not visible from above; mesonotum not or scarcely narrowed anteriorly, with 3 distinct impressed lines; scutellum large, subconvex, longer than wide, rounded posteriorly; metathorax short, rounded posteriorly, unarmed, the spiracles large, orbicular.

Front wings entirely bare, without pubescence, with a large oval stigma and a short, almost straight, stigmal vein, which is usually shorter than the stigma; no parastigma.

Abdomen ovate, smooth, the second segment the largest, occupying about half the length of the abdomen.

Legs not pilose.

The non-pubescent wings will distinguish the genus from *Lygocerus* and *Megaspilus*, with which it is most closely allied.

Only one species, closely resembling the European *T. glaber* Boheman, is known in our fauna.

Trichosteresis floridanus Ashm.

(Pl. vi, Fig. 5, ♀.)

Ent. Am. iii. p. 98; Cress. Syn. Hym, p. 313.

♀. Length 2.5ᵐᵐ. Black, shining, finely alutaceous. Antennæ 11-jointed, dark-brown, the first flagellar joint a little longer than the pedicel, the following joints, except the last, scarcely longer than thick, the last conic. Legs piceous black, the trochanters, knees, tips of tibiæ, and the tarsi dark honey-yellow or reddish. Wings clear hyaline, without pubescence, the stigma large, semirotund, brown, the stigmal vein not quite as long as the stigma.

HABITAT.—Jacksonville, Fla.

Type in Coll. Ashmead.

LYGOCERUS Förster.

Hym. Stud., ii, p. 97 (1856).

Ceraphron Thoms., Öfv. 1858, p. 287.

(Type *L. ramicornis* Boh.)

Head transverse, seen from before, the face broader than long, the occiput faintly margined; ocelli 3, arranged nearly in a straight line; eyes large, oval, smooth, or but faintly pubescent.

Antennae inserted just above the clypeus, 11-jointed in both sexes, the scape long, extending above the ocelli; in ♀ subfiliform, the joints of flagellum only slightly longer than thick; in ♂ some of the flagellar joints always dentate or with short hairy branches.

Maxillary palpi 5-jointed; labial palpi 3-jointed.

Mandibles bidentate.

Thorax ovoid, the prothorax not visible from above; mesonotum not narrowed anteriorly, with 3 impressed lines; scutellum large, longer than wide, subconvex, without a distinct frenum; metathorax very short, rounded posteriorly.

Front wings pubescent, with a large oval or semicircular stigma, and a slightly arcuate stigmal vein, distinctly longer than the stigma.

Abdomen ovate, smooth, subpetiolate, the second segment the longest, not striate at base.

Legs finely pubescent, but not pilose, the last joint of posterior tarsi not longer than the second.

A genus parasitic principally on aphidsid, and probably universally distributed.

The males are readily separated from those in *Megaspilus*, and in the other genera in the tribe having 3 mesonotal furrows, by the ramose or serrate antennae, while the females are separated with difficulty; the non-pubescent or but slightly pubescent eyes, the mesonotum being as wide anteriorly as posteriorly, and the lateral mesonotal furrows, before reaching the anterior margin, curving somewhat obliquely toward the anterior angles, will, however, at once distinguish them.

Our species are not numerous, and may be separated by the aid of the following table, except possibly *L. triticum* Taylor, which at one time I considered to be identical with *L. (Megaspilus) niger* Howard, but which I now consider distinct, Miss Taylor's figure of the male antennae of her species being wholly different from that of Mr. Howard's species:

TABLE OF SPECIES.

FEMALES.

Legs, including coxae, uniformly reddish-yellow.............. L. FLORIDANUS Ashm.
Posterior coxae black; legs pale brownish.
    Antennae black, scape pale brownish beneath; first flagellar joint as long as
        the pedicel ....L. PICIPES, sp. nov.
All coxae black.
    Legs black, tibiae piceous, knees, anterior tibiae, and all tarsi, honey-yellow.
        Antennae wholly black; first flagellar joint not quite as long as the pedicel.
                                        L. STIGMATUS Say.
Legs black or piceous, the tibiae, and tarsi brownish-yellow.
    Mesopleura almost smooth; antennae black, the first flagellar joint much
        longer than the pedicel......................L. NIGER How.

Legs brownish-yellow.

Mesopleura smooth polished; antennæ black, the first flagellar joint not
    longer than pedicel.........................L. 6-DENTATUS, sp. nov.

Mesopleura shagreened; first flagellar joint much longer than pedicel.
                     L. PACIFICUS, sp. nov.

MALES.

Antennæ ramose.

Flagellar joints 1 to 7 ramose; legs uniformly reddish-yellow.
                  L. FLORIDANUS Ashm.

Flagellar joints 1 to 4 ramose...........................L. TRITICUM Taylor.

Antennæ serrate, black.

Flagellar joints 1 to 5 dentate; first flagellar joint twice as long as thick.
                  L. STIGMATUS Say.

Flagellar joints 1 to 5 dentate; first flagellar joint scarcely twice as long as thick.

Coxæ black, the legs brown, with the anterior tibiæ and all tarsi honey-yellow.
    Stigmal vein 1½ times as long as the stigma...........L. 6-DENTATUS, sp. nov.
    Stigmal vein not longer than the stigma...........L. CALIFORNICUS, sp. nov.

Flagellar joints 1 to 7 dentate; first flagellar joint stout, 2½ times as long as thick.
    Mesopleura scaly-punctate.......................................L. NIGER How.

Lygocerus floridanus Ashm.

Chirocerus floridanus Ashm., Trans. Am. Ent. Soc. Mo. Proc., 1881, p. 34.
Lygocerus floridanus Ashm., Ent. Am., III, p. 98.

♂ ♀. Length, 2 to 2.2ᵐᵐ. Black, closely, finely punctate, subopaque; abdomen black, highly polished, pointed at tip, the petiole striate; legs, including coxæ, reddish-yellow. Antennæ in ♀ brown, acuminate at tips, the flagellar joints a little longer than thick; in ♂ flagellar joints 1 to 7 ramose, covered with sparse long hairs. Wings hyaline, the stigma large, semicircular, brown, the stigmal vein about one and one-third times as long as the stigma.

HABITAT.—Jacksonville, Fla.

Types, ♂ and ♀, in Coll. Ashmead.

All my specimens were reared from a large aphis on pine, *Lachnus australis* Ashm.

Lygocerus picipes, sp. nov.

♀. Length, 1.8ᵐᵐ. Black, subopaque, finely shagreened; abdomen black, polished; legs brownish, piceous, the trochanters, knees, and tarsi yellowish, hind coxæ black. Antennæ 11-jointed, acuminate at tips, brown-black, the scape yellowish beneath; the pedicel is a little longer than the first flagellar joint; the second flagellar joint a little shorter than the first; the third and following slightly longer. Wings hyaline, the large stigma dark brown, the stigmal vein less than twice as long as the stigma.

HABITAT.—Ottawa, Canada.

Type in Coll. Ashmead.

Described from a single specimen received from W. Hague Harrington.

The color of the anterior and middle coxæ and the scape beneath will at once separate the species from those that follow.

Lygocerus triticum Taylor.

Ceraphron triticum Taylor, Am. Agric. 1860, p. 300, f. 1; Cress. Syn. Hym., p. 248.

This fly does not correspond with the above (*Ceraphron destructor* Say). therefore I have named it *triticum*, from the botanical name of wheat. It is not of such a shining black as Mr. Say's fly, but is rather rusty in appearance from a few hairs scattered over its body. In some specimens, when very fresh, the legs have a bright tinge of yellow. The antennæ (*b*, Fig. 2) are termed setigerous (having the basal joint large) and the last four globular, the intermediate one furnished with four long bristles resembling plumes. This is a very sure mark for distinguishing this family according to European classification. The eyes are large in proportion, the palpi 3-jointed. The fore wings have submarginal cells, with a faint nervure running to apex. The under wings have a long nervure running through and two smaller ones descending to the inferior region; these are so very slight, that you can only see their existence by a deep shade of the wings in a strong light, but are evidently nervures, indistinct as they are. The ovipositor is retractile and tubular. The fly deposits her eggs in the pupa of the Hessian fly. (*Taylor.*)

Unknown to me, and the above description is copied from the American Agriculturist. Miss Taylor further informs us that "this fly can be found in every wheat field throughout the country, from spring to autumn." Her description is very imperfect, and her figure of the male antenna strongly recalls the branched antenna of an *Eulophus*.

Lygocerus stigmatus Say.

Ceraphron stigmatus Say, Boston Journ., I. p. 277; Lec. ed. Say's Works, II, p. 724.
Lygocerus stigmatus Ashm., Ent. Am., III, p. 98.
Ceraphron stigmatus Cr., Syn. Hym. p. 248.

♂ ♀. Length, 1.4ᵐᵐ. Black, pubescent, closely, minutely punctulate, subopaque; abdomen ovate, pointed at tip, highly polished, the petiole striate; legs piceous black, the anterior tibiæ and tarsi and knees on middle and hind legs and their tarsi, pale or honey yellow. Antennæ in ♀ slightly thickened toward tips, the flagellar joints beyond the third, scarcely longer than wide; in the ♂ the first five flagellar joints dentate, pilose, the first joint being about twice as long as thick, excluding the short pedicel.

Wings hyaline, the stigma semicircular, brown, the stigmal vein about 1¾ the length of the stigma.

HABITAT.—Indiana; Canada.

Specimens in Coll. Ashmead and National Museum.

Recognized from specimens bred by Mr. James Fletcher, at Ottawa, Canada, from the raspberry aphis.

Lygocerus niger Howard.

Megaspilus niger How., Ins. Life, vol. II, p. 247., ♀, f. 52.
Ceraphron triticum Smith. Rep. N. J. Exp. Sta., 1890, p. 502, f. 18.

♀. "Length, 1.6ᵐᵐ; expanse, 3.33ᵐᵐ; greatest width of fore wing, 0.62ᵐᵐ. Scape of antennæ very long, somewhat swollen beyond middle; funicle long, curved, all joints increasing gradually in width from pedicel to club; joint 1 of funicle somewhat longer than pedicel, joint 3 shorter, joints 4 to 8 increasing in length very slightly. Head and mesonotum very faintly shagreened, but still glistening: lower portion of mesopleura and all of abdomen perfectly smooth. Abdomen sub-ovoid in shape, acutely pointed at tip. Radial vein only slightly curved, extending a little more than half way from stigma to tip of wing. General color jet black; all trochanters, femora, and wing veins dark-brown; all tibiæ and tarsi lighter brown." (*Howard.*)

The ♂ is slightly smaller, with the flagellar joints 1 to 7 dentate, the first being 2½ times as long as thick, excluding the pedicel, while the mesopleura are scaly-punctate.

HABITAT.—United States.

Types in National Museum.

Parasitic on the wheat aphis, *Siphonophora avenæ* Fabr.

Allied to *L. stigmatus* Say, but relatively larger, and at once distinguished by the difference in the length of the flagellar joints.

Lygocerus 6-dentatus, sp. nov.

(Pl. vi, Fig. 8, ♂.)

♀ . Length, 1.2ᵐᵐ. Black, coriaceous, faintly sericeous; mesopleura, coxæ, and abdomen highly polished, shining, impunctured; legs brownish yellow, the coxæ polished black; wings hyaline, the stigma semicircular, brown, the stigmal vein 1½ times as long as the stigma; antennæ brown-black, the pedicel and first flagellar joint equal, the last joint longer than the pedicel.

♂ . Length, 1.1ᵐᵐ. Agrees well with the ♀, except that all the femora and the posterior tibiæ are darker, the antennæ black, the flagellar joints 1 to 6 serrate, with long hairs, the first being scarcely twice as long as thick.

HABITAT.—District of Columbia.

Types, ♂ and ♀, in National Museum.

This species is labeled as having been bred July 22, 1886, from *Sarrothripa rawayana*, on willow; but evidently the record is unreliable, and in all probability it came from some Aphidid or Cecidomyiid overlooked by the recorder.

Lygocerus californicus, sp. nov.

♂ . Length, 1.1ᵐᵐ. Black, shining, alutaceous; abdomen piceous or obscure, rufous basally; legs brown, the anterior tibiæ and tarsi honey-yellow, the coxæ black. Antennæ black, flagellar joints 1 to 6 serrate, pubescent. Wings hyaline, the stigma semicircular, the stigmal nervure nearly straight, not, or scarcely, longer than the stigma.

HABITAT.—Los Angeles, Cal.

Type in National Museum.

Described from a single ♂ specimen reared by Mr. A. Koebele, from a Cecidomyiid gall on *Larrea mexicana*,

This species comes quite close to *L. 6-dentatus*, but is distinguished at once from that species and all the others by the brevity of the stigmal nervure and in having the abdomen piceous basally.

Lygocerus pacificus, sp. nov.

♀. Length 2.5ᵐᵐ. Black, opaque, shagreened or closely punctate; antennæ long, black, the tip of pedicel reddish-yellow; first flagellar joint long, distinctly longer than the pedicel, the following joints a little shorter, about equal in length; mesopleura shagreened; wings subhyaline, the stigma large, oblong, brown, the stigmal vein much longer than the stigma; legs brownish-yellow, the coxæ black; abdomen emarginate at apex, the ventral valve and ovipositor slightly prominent.

HABITAT.—Placer County, Cal.

Type in National Museum.

Described from a single specimen. It is the largest species I have as yet seen in this genus; and its size, opaque surface, color of the legs, and the shape of the stigma will at once separate it from the others in our fauna.

MEGASPILUS, Westw.

Phil. Mag., I, p. 128 (1832); Först., Hym. Stud., II, p. 97 (1856); Thoms., öfv., 1858, p. 287.

Eumegaspilus Ashm. (pars) Can. Ent., XX, p. 48, 1888.

(Type *M. abdominalis*, Boh.)

Head transverse, the occiput margined; ocelli small, in a triangle; eyes ovate or long-oval, pubescent or hairy.

Antennæ inserted just above the clypeus, 11-jointed in both sexes, filiform or subfiliform, the scape long, the first flagellar joint usually much lengthened, the male with simple joints, never dentate nor ramose.

Maxillary palpi 5-jointed; labial palpi 3-jointed.

Mandibles bifid at tips.

Thorax ovoid, the prothorax slightly visible from above and produced into a slight collar anteriorly; mesonotum slightly narrowed anteriorly, with three impressed lines; scutellum longer than wide, subconvex, rounded behind, with a distinct frenum; metathorax short, with acute angles.

Front wings pubescent, with a large subovate stigma, truncate at apex, the stigmal vein usually long and curved, the parastigma slightly developed. Wingless forms not uncommon.

Abdomen ovate, subpetiolate, the petiole short, stout, and strongly channeled, the second segment very large, striated at base.

Legs pilose, the posterior femora somewhat swollen, the last joint of the posterior tarsi longer than the second.

This extensive genus was very properly separated from *Ceraphron* Jurine, by Prof. Westwood, as early as 1832. In 1856 Dr. Arnold Förster again separated from it the genus *Lygocerus*, to contain those species having bare eyes and serrate antennæ in the males. The habits of the two genera are identical, and their structural differences have been sufficiently pointed out in my remarks under the genus *Lygocerus*.

In 1888, I erected the genus *Enmegaspilus* for some wingless forms which I now find are true *Megaspili*. The name, however, is retained for a closely allied apterous form, differing in several particulars from *Megaspilus*.

Our species are numerous, and may be distinguished by the following synoptical table:

TABLE OF SPECIES.

FEMALES.

1. Face very finely closely punctate or shagreened 2
   Face smooth, polished, impunctured.
     Apex of abdomen compressed, truncate, and gaping open, so that the valves of the ovipositor project.
       Legs brownish-yellow; antennæ brown-black.
         Pedicel shorter than the first flagellar joint ..M. ANOMALIVENTRIS, sp. nov.
         Pedicel as long as the first flagellar joint; scape and pedicel pale.
                                                    M. POPENOEI, sp. nov.
     Apex of abdomen normal.
       Legs brownish-yellow; antennæ brown-black.
         Pedicel as long as the first flagellar joint ....M. AMBIGUUS, sp. nov.
         Pedicel much shorter than the first flagellar joint.
           Posterior tibiæ with a brown streak above.......M. STRIATIPES, sp. nov.
2. Wings wanting or not fully developed....................................... 3
   Wings fully developed.
     Species shining.
       Face finely shagreened or rugose.
         Scape and legs pale brownish-yellow.
           Pedicel concolorous with flagellum ....M. SCHWARZI, sp. nov.
           Pedicel brownish-yellow.
         Face rather coarsely rugose.
           Posterior coxæ not black.......................M. MARYLANDICUS, sp. nov.
           Posterior coxæ and femora dusky.................M. HARRINGTONI Ashm.
             Stigma piceous-black, wings subfuscous.
             Stigma brown, wings clear hyaline...........M. HYALINIPENNIS, Ashm.
     Species opaque, the thorax sometimes subopaque.
       Face closely finely punctulate.
         Hind coxæ black.
           Scape, pedicel, and legs rufous, the posterior femora and tibiæ fuscous;
             pleura scaly-punctate ....M. PENMARICUS, sp. nov.
           Scape, pedicel, and legs brownish-yellow ....M. VIRGINICUS, sp. nov.
3. Wings abbreviated and narrow, reaching not quite to the middle of the abdomen.
   Scape, pedicel, first three flagellar joints, and legs, pale yellow.
                                                    M. OTTAWENSIS Ashm.
   Wings reaching only to the tip of the petiole.
     Antennæ brown-black, scape beneath and legs brownish-yellow.
                                                    M. CANADENSIS Ashm.

MALES.

Species smooth, shining.
First flagellar joint much longer than the second, the second and the following
at least thrice as long as thick.
  Scape and legs brownish-yellow.
    Wings clear hyaline; length 1.6^{mm} .... M. LÆVICEPS, sp. nov.
    Wings fusco-hyaline; length 2^{mm} .... M. AMBIGUUS, sp. nov.
    Antennæ wholly brown-black, legs rufous .... M. CALIFORNICUS, sp. nov.
First flagellar joint not much longer than the second, the second and the fol-
lowing scarcely twice as long as thick.
  Antennæ brown; legs rufous, abdomen piceous or rufous toward base.
                                      M. PERGANDEI.

Megaspilus anomaliventris, sp. nov.

♀. Length, 2.4^{mm}. Black, shining; head and thorax finely alutaceous; antennæ brown-black; legs reddish-yellow, the posterior coxæ
a little dusky basally; abdomen polished black, becoming piceous at
apex. Eyes very large, occupying the greater portion of the sides of
the head, pubescent. Antennæ 11-jointed, the flagellum flagellate, the
scape long, projecting beyond the ocelli; pedicel small, about half the
length of the first flagellar joint, the latter the longest joint, the second
shorter than the first, the third and following joints to the last stouter
and nearly of an equal length, very slightly longer than the second,
the last more slender and a little longer than the penultimate. Wings
hyaline, pubescent, with a faint dusky blotch below the stigmal vein,
the stigma large oval, truncate behind, brown, the stigmal vein less than
one and a half times as long as the stigma.

HABITAT.—Marquette, Mich.

Type ♀ in Coll. Ashmead.

Described from a single specimen collected by E. A. Schwarz.

Megaspilus Popenoei sp. nov.

♀. Length, 1.8^{mm}. Black, shining, impunctured, with a sparse greyish
pubescence; scape, pedicel, collar, and legs, including the coxæ, brownish-yellow; flagellum black, abdomen beneath and towards base piceous.
Antennæ 11-jointed, subclavate, the flagellum slightly more than twice
as long as the scape; the first flagellar joint is equal to the pedicel, the
second, third, and fourth shorter than the first but stouter, the following longer than thick, the last conic, nearly twice as long as the penultimate. Wings subfuscous, hyaline at base, the parastigma slightly
developed, the stigma ovate, the stigmal vein about one and a half
times as long as the stigma; all the nervures dark-brown. Abdomen
longer than the head and thorax together, compressed and gaping open
at apex.

HABITAT.—Manhattan, Kansas.

Type ♀ in Kansas State Agricultural College.

Described from a single specimen received from Prof. E. A. Popenoe
and in honor of whom the species is named.

Megaspilus ambiguus sp. nov.

♂ ♀. Length, 1.2 to 1.6ᵐᵐ. Black, shining, finely alutaceous; scape and legs brownish-yellow. The posterior coxæ black basally. In the ♂ the prosternum and the collar laterally are yellow. Mandibles and palpi pale. Antennæ 11-jointed, in ♀ the flagellum flagellate, the pedicel and first flagellar joint of an equal length, the following to the last only slightly longer than thick, the last joint two and a half times as long as thick; in the ♂ filiform, the flagellum black, pubescent, the pedicel and the first flagellar joint together distinctly shorter than the scape; the first joint of the flagellum is not more than 4 times as long as thick, the second shorter, the following to the last nearly equal, a little less than thrice as long as thick, the last joint about one-fourth longer than the penultimate. Wings hyaline, very faintly tinged with fuscous, the stigma brown, the stigmal vein two and a half times the length of the stigma.

HABITAT.—Arlington, Va., and Washington, D. C.

Types ♂ ♀ in Coll. Ashmead.

Described from several specimens taken by Schwarz, Pergande and myself.

The species is closely allied to *M. læviceps*, but it is slightly larger, with a fine alutaceous sculpture, and with a difference in the relative length of the antennal joints.

Megaspilus striatipes sp. nov.

(Pl. VI, Fig. 7, ♀.)

♀. Length, 1.4ᵐᵐ. Black, shining; vertex of head and thorax very faintly shagreened, the face smooth and highly polished. Scape and legs brownish-yellow, the posterior femora and tibiæ with a fuscous streak above. Antennæ 11-jointed, the flagellum flagellate, stout, black, a little more than twice the length of the scape; the scape is long, extending beyond the ocelli; the pedicel is much shorter than the first flagellar joint; the latter is much longer and slenderer than the second; remaining joints widening, but still all a little longer than thick, the last very slightly longer than the penultimate. Wings fuscous, the stigma large, brown, the stigmal vein less than twice the length of the stigma.

HABITAT.—Ottawa, Canada.

Type ♀ in Coll. Ashmead.

Described from a single specimen received from Mr. W. Hague Harrington.

Megaspilus Schwarzii sp. nov.

♀. Length, 1.8ᵐᵐ. Black, shining, very finely shagreened, the scutellum smooth; scape and legs brownish-yellow; the propectus piceous. Antennæ 11-jointed, the flagellum brown-black, about two and a half times as long as the scape, the pedicel as long as the first flagellar joint, the latter the narrower, the following joints to the last, a little shorter, but gradually thickened, the last joint twice as long as the penultimate. Wings hyaline, the stigma brown, the stigmal vein one and a half times as long as the stigma. Abdomen very slightly longer than the thorax, smooth and polished, pubescent toward tip, the venter flatter than usually, the petiole and second segment at base, striate.

HABITAT.—Washington, D. C.

Type ♀ in Coll. Ashmead.

Described from a single specimen received from Mr. E. A. Schwarz.

Megaspilus marylandicus sp. nov.

♀. Length, 2.4ᵐᵐ. Black, subopaque, finely shagreened and sparsely pubescent; the scape, pedicel, and legs reddish-yellow. Antennæ 11-jointed, the flagellum brown, slightly paler basally, not longer than twice the length of the scape; the pedicel yellow, distinctly shorter and thicker than the first flagellar joint, the second shorter than the first, third and fourth about equal, shorter than the fifth, the fifth to the last about equal, very slightly longer than thick, the last twice as long as thick. Wings fuscous, the stigma brown, the stigmal vein slightly more than twice the length of the stigma. Abdomen a little longer than the head and thorax together, black, polished, boat-shaped below, somewhat flat above.

HABITAT.—Oakland, Md.

Type ♀ in Coll. Ashmead.

Described from specimens received from Mr. E. A. Schwarz.

Megaspilus Harringtoni Ashm.

Can. Ent. vol. XX, p. 48.

♂ ♀. Length, 2 to 2.5ᵐᵐ. Black; head and thorax with a fine reticulate punctuation; abdomen polished black. Antennæ 11-jointed, the scape and pedicel dull yellow, the flagellum brown black; the pedicel is not quite as long as the first flagellar joint. Legs dull yellow, the posterior coxæ black, the anterior and middle coxæ dusky basally, the posterior femora fuscous or dusky. Wings subhyaline, pubescent, the large stigma and stigmal vein brown.

Abdomen the length of the thorax, the petiole yellowish.

The ♂ differs from ♀ only in its slightly smaller size and in the long filiform antennæ, the joints of the flagellum being about four times as long as thick, the scape yellowish toward the base.

HABITAT.—Ottawa, Canada.

Types ♂ ♀ in Coll. Ashmead.

The types were taken by Mr. W. H. Harrington.

Megaspilus hyalinipennis Ashm.

Ent. Am., III, p. 98; Cress. Syn. Hym., p. 313.

♀. Length, 2ᵐᵐ. Robust, black, alutaceous, with a very sparse, pale pubescence. Antennæ 11-jointed, filiform, dark brown; the pedicel is a little shorter than the first flagellar joint, the second joint a little shorter than the first, the remaining, except the last, which is longer, subequal with second. Legs pale brown, the femora and tibiæ fuscous. Wings clear hyaline, the stigma large, brown, the stigmal vein about twice the length of the stigma. Abdomen polished black, pointed and sparsely pubescent at tip, the petiole and second segment, at base, striolate.

HABITAT.—Jacksonville, Fla.

Type ♀ in Coll. Ashmead.

Megaspilus virginicus, sp. nov.

♀. Length, 2.2mm. Robust, black, opaque, shagreened, with a sparse whitish pubescence; the head more coarsely shagreened than the thorax; the scutellum finely shagreened and shining; scape and legs brownish yellow. Antennæ 11-jointed, the flagellum brown-black, twice the length of the scape, the latter not extending beyond the ocelli; pedicel not quite as long as the first flagellar joint; the second and third flagellar joints equal, about half the length of the first, but stouter, the joints beyond a little longer and stouter, the last not much longer than the penultimate. Wings fusco-hyaline, the stigma brown-black, the stigmal vein about one and a half times as long as the stigma. Abdomen as long as the head and thorax together, black, shining, with a rather dense, fine pubescence toward apex.

HABITAT.—Arlington, Va.

Type ♀ in Coll. Ashmead.

Described from a single specimen.

Megaspilus canadensis Ashm.

Eumegaspilus canadensis Ashm., Can. Ent., xx, p. 49.

♀. Length, 2.2mm. Black, shining; the head and thorax alutaceous, with two small foveæ on each side of the front ocellus; occiput distinctly margined; eyes pubescent. Antennæ 11-jointed, subclavate, brown-black, the scape pale beneath, the first flagellar joint longer than the pedicel, the following very slightly subequal to the last, the last one-third longer than the penultimate. Wings extending only to tip of the petiole. Legs, including coxæ, brownish-yellow. Abdomen highly polished, the petiole striate.

HABITAT.—Canada.

Type ♀ in Coll. Ashmead.

Described from a single specimen received from W. Hague Harrington.

Megaspilus ottawensis Ashm.

Eumegaspilus ottawensis Ashm., Can. Ent. xx, p. 49.

♀. Length, 2mm. Black, shining, impunctured; eyes large, pubescent. Antennæ 11-jointed, subclavate, the scape, pedicel, first three flagellar joints, and the legs honey-yellow. Wings linear, extending not quite to the middle of the abdomen. Abdomen black, polished, with striæ at base.

HABITAT.—Ottawa, Canada.

Type ♀ in Coll. Ashmead.

Described from a single specimen received from W. Hague Harrington.

Megaspilus læviceps, sp. nov.

♂. Length, 1ᵐᵐ. Black, highly polished; scape and legs pale brownish yellow; flagellum brown-black; mandibles black: the short petiole and a streak at base of abdomen yellowish. Head transverse, sparsely hairy, the occiput with a distinct margin, the face convex, smooth, shining. Palpi pale. Antennæ 11-jointed, longer than the body; the pedicel and the first flagellar joint together are as long as the scape, the pedicel very small, rounded; the flagellum is cylindrical, covered with a short, rather dense pubescence, the first joint about 5 times as long as thick, the second one-third shorter, the following very slightly shorter. Wings hyaline, the stigma semicircular, brown; the stigmal vein nearly twice as long as the stigma.

HABITAT.—Arlington, Va.: Washington, D. C., and Bladensburg, Md.

Types, all ♂ s, in Coll. Ashmead.

A specimen was taken by E. A. Schwarz at Bladensburg, Md.; the others were captured by myself on the outskirts of Washington and along the banks of the Potomac River, in Virginia.

Megaspilus californicus, sp. nov.

♂. Length, 2.2ᵐᵐ. Black, smooth, shining; the head transverse, obliquely narrowed behind the eyes. Antennæ 11-jointed, filiform, longer than the body, brown-black, the scape long, reaching much beyond the ocelli, the pedicel very small, rounded, the first flagellar joint more than six times as long as thick, the following joints shorter, subequal. Legs pale rufous. Wings subhyaline, the stigma large, brown, the stigmal vein two and a half times as long as the stigma.

HABITAT.—California.

Type ♂ in Coll. Ashmead.

Megaspilus Pergandei, sp. nov.

♂. Length, 1.2ᵐᵐ. Black, smooth, shining; the vertex of head and thorax faintly alutaceous. Antennæ 11-jointed, slightly longer than the body, brown, the scape not reaching beyond the ocelli, the pedicel small, oval, the first flagellar joint scarcely thrice as long as thick, the following shorter, about twice as long as thick, except the last, which is almost as long as the first. Legs rufous or reddish yellow. Wings subhyaline, pubescent. Abdomen a little longer than the thorax, polished black, a little rufous or piceous toward base.

HABITAT.—District of Columbia.

Type in Coll. Ashmead.

Described from a single specimen.

DICHOGMUS Thomson.

Öfv. 1859, p. 301.

(Type *D. dimidiatus* Thoms.)

Head large, wider than the thorax, the front convex, the occiput narrowly margined; ocelli 3 triangularly arranged, the eyes rather large.

Antennæ inserted just above the clypeus, 11-jointed in both sexes, in ♂ longer than the body, filiform, densely pubescent.

Maxillary palpi 5-jointed; labial palpi 3-jointed..

Mandibles bidentate.

Thorax subovoid, with a slight collar, the mesonotum with 2 impressed lines, the metathorax very short.

Wings entirely wanting.

Abdomen ovate, the apex slightly pointed; the second segment very large, very faintly stiolate at base.

Legs pubescent.

A European genus not yet recognized in America. It resembles *Lagynodes* Förster and *Eumegaspilus* Ashm., but is easily separated from both by the two distinct mesonotal furrows. I have seen specimens of the genus in Europe.

EUMEGASPILUS Ashm.

Can. Ent., Vol. xx, p. 49, 1888.

(Type, *E. erythrothorax*.)

Head broadly transverse, across the eyes twice as wide as the thorax, the occiput distinctly margined; ocelli 3, small, triangularly arranged; eyes large, oval, faintly pubescent.

Antennæ inserted just above the clypeus, 11-jointed in both sexes; in ♀ subclavate, the first flagellar joint longer than the pedicel; in ♂ filiform, the first flagellar joint two-thirds the length of the scape, the following joints lengthened, cylindrical.

Maxillary palpi 4-jointed; labial palpi 2-jointed.

Mandibles bifid.

Thorax narrowed, the prothorax visible from above only as a slight neck; mesonotum trilobed; scutellum semicircular, margined behind, the axillæ scarcely traceable; metathorax short, truncate behind, the upper edge of truncature margined,

Wings entirely wanting.

Abdomen as in *Megaspilus*, the petiole short, striate, the second segment occupying a little more than half the whole surface, the following segments short.

Legs pubescent, the posterior tarsi with the first joint as long as all the remaining joints united, the last joint not longer than the second.

Allied to the wingless forms in *Megaspilus*, but with a much larger head, narrower thorax, and with 4-jointed, not 5-jointed, maxillary palpi. It has also a superficial resemblance to the genus *Lagynodes* Förster, but the thorax in that genus is not trilobed and the male has an acute spine between the base of the antennæ.

Two species, *E. canadensis* and *ottawensis*, described in Canadian Entomologist, Vol. xx, p. 49, under this genus I find are nothing but wingless species belonging to the genus *Megaspilus*. The genus as now restricted will contain but a single species, as follows:

Eumegaspilus erythrothorax sp. nov.

(Pl. vi, Fig. 6 ♀.)

♀. Length, 1.9^{mm}. Head and abdomen black; face, scape, thorax, and petiole pale rufous or brownish-yellow; legs pallid yellow. Head and thorax closely, minutely punctulate; petiole striate; pedicel and flagellum black, the first flagellar joint longer than the pedicel. Abdomen smooth, shining, much longer than the head and thorax united.

♂. Length, 1.5^{mm}. Agrees with the ♀, except that the head, the pedicel, and the base of the abdomen as well as the thorax are brownish yellow, while the flagellar joints are long and cylindrical, the first being 4 times as long as the pedicel.

HABITAT.—Jacksonville, Fla.

Types ♂ and ♀ in Coll. Ashmead.

A pretty and easily recognized species, apt to be mistaken for a *Lagynodes*, but readily distinguished from the characters pointed out in the generic description.

LAGYNODES Förster.

Beitr. 1841, p. 46.

Microps Hal., Ent. Mag. i (1833).

(Type *L. pallidus* Boh.).

Head deflexed, when viewed from in front, longer than wide, convex, in ♀ with an impression above the antennæ, in the ♂ with a spine between the antennæ; occiput delicately margined; ocelli wanting; eyes very small, rounded.

Antennæ inserted close to the clypeus, 11-jointed in both sexes, the scape long, longer than the head, the pedicel more than twice as long as thick, the flagellum subclavate, the terminal joint not longer than thick, submoniliform.

Maxillary palpi short, 4-jointed; labial palpi 2-jointed.

Mandibles bifid.

Thorax oblong, compressed at sides, subconvex above, the collar not apparent, the mesonotum with a delicate impressed median line, scutellum small, metathorax very short.

Wings wanting.

Abdomen pointed ovate, subpetiolate, twice the length of the thorax and much wider, beneath strongly convex, above subconvex, the second segment very large, occupying about two-thirds of the whole surface.

Legs pilose, the last joint of posterior tarsi twice as long as the second, claws small, simple.

The small eyes, absence of ocelli, and the form of the thorax, sufficiently separate this genus from all the others in the tribe *Megaspilini*. The male is readily distinguished by having an acute spine between the antennæ.

Only a single species has been recognized in our fauna, as follows:

Lagynodes minutus, sp. nov.

(Pl. vi, Fig. 10, ♀.)

♀. Length, 1 to 1.5ᵐᵐ. Honey-yellow to reddish-yellow, polished, impunctured, sparsely pilose; the antennæ basally and the legs pale or whitish. Antennæ 11-jointed, thickened toward tips, the apical joints brown; the flagellar joints, except the last, which is oblong, are not longer than thick. The mesonotal line is only indicated posteriorly, or entirely wanting. Abdomen with a few raised lines at base, the second segment at apex tinged with fuscous or brown.

HABITAT.—Washington, D. C., and Arlington, Va.

Types in Coll. Ashmead and National Museum.

Described from many specimens taken by myself and Schwarz.

The species is much smaller, more slender, and paler colored than the European *L. rufus* Förster (= *L. pallidus* Boh.).

ATRITOMUS Förster.

Kl. Mon., p. 56 (1878).

(Type *A. coccophagus* Först.)

Head transverse, stout, wider than the thorax, when viewed from in front, wider than long; eyes large, rounded, prominent, bare; ocelli 3, triangularly arranged.

Antennæ inserted at the clypeus, 11-jointed, in ♂ serrate, in ♀ filiform, the first three flagellar joints small.

Maxillary palpi 4-jointed; labial palpi 2-jointed.

Mandibles bidentate.

Thorax subovoid, the prothorax not visible from above; mesonotum with a single central impressed line or without impressed lines; the scutellum large, longer than wide, the axillæ usually, but not always, separated; metathorax very short, the angles rounded.

Front wings as in *Lygocerus*, with a large semicircular stigma.

Abdomen ovate, subpetiolate, the second and third segments the longest, subequal, the following short.

Legs as in Ceraphron.

This genus, as here recognized, differs from *Ceraphron* in having a stigmated marginal nervure and serrate antennæ in the male; and from *Lygocerus* and *Megaspilus* in having, at the most, only a single mesonotal furrow.

Förster characterized the genus as being without trace of a mesonotal furrow; but, as I have found to be the case in *Aphanogmus*, species probably occur with and without a furrow.

A single species is known in America.

Atritomus americanus, sp. nov.

(Pl. vi, Fig. 9, ♂.)

♂. Length, 1.3ᵐᵐ. Black, shining, feebly microscopically punctate; antennæ brown; legs reddish-brown, the coxæ black; abdomen piceous; mandibles black or piceous. Head transverse, a little wider than the thorax. Eyes large, oval, bare. Antennæ as long as the body, the pedicel small, triangular, the first, second, third, and fourth funiclar joints subserrate, the first the longest, a little more than thrice as long as thick, the following a little shorter. Thorax scarcely narrowed before, with a delicate, central grooved line. Metathorax rounded behind. Wing hyaline, pubescent, the stigma large semicircular, reddish brown, the stigmal vein a little longer than the stigma. Abdomen ovate, as long as the thorax, the petiole very short, rugose, the second segment occupying about half of the remaining surface, the third and fourth equal, longer than the following, the following very short; claspers flat, distinctly extended.

HABITAT.—Odenton, Md.

Type ♂ in Coll. Ashmead.

Described from a single specimen taken by Mr. E. A. Schwarz.

Tribe II.—CERAPHRONINI.

The species in this tribe can always be distinguished from those in the *Megaspilini* by the linear marginal nervure and the paucity of joints in the female antennæ, the mesonotum, at the most, having but a single impressed line.

The genera may be recognized by the aid of the following table:

TABLE OF GENERA.

FEMALES.

Antennæ 9-jointed 3
Antennæ 10-jointed.
    Wingless forms 2

Winged.
  Mesonotum with a median impressed line.
    Scutellum flat or subconvex, with a marginal frenum....CERAPHRON Jurine.
    Scutellum convex, acuminate, without a frenum ....APHANOGMUS Thoms.
    Mesonotum without a furrow (= Synarsis Först.)........APHANOGMUS Thoms.
2. Mesonotum with a median impressed line.........................CERAPHRON.
    Scutellum flat or subconvex with a frenum......................CERAPHRON.
    Scutellum convex, without a frenum............................APHANOGMUS.
  Mesonotum without a median furrow...........................APHANOGMUS.
3. Mesonotum with a median impressed line............NEOCERAPHRON, gen. nov.

MALES.

Antennæ 10-jointed... 3
Antennæ 11-jointed.
  Mesonotum without a median impressed line............................. 2
  Mesonotum with a median impressed line.
    Scutellum depressed with a marginal frenum.
      Antennæ simple......................................CERAPHRON.
    Scutellum convex, acuminate, without a frenum.
      Antennæ serrate................................APHANOGMUS Thoms.
2. Scutellum convex, acuminate, without a frenum.
    Antennæ serrate ....APHANOGMUS Thoms.
3. Mesonotum with a median impressed line.
    Antennæ filiform...............................NEOCERAPHRON, gen. nov.

CERAPHRON Jurine.

Hym., p. 303 (1807).

Calliceras Nees, Mon., II, p. 280, 18; Thoms. Öfv., 1858, p. 302.
Hadroceras Först., Beitr., p. 46 (1841).
Ceraphron Först., Hym. Stud., II, p. 97, 1856.
Megaspilidea Ashm., Can. Ent., XX, p. 49 (apterous forms).

(Type *C. sulcatus* Jurine.)

Head transverse, when viewed from in front oblong or oval, convex, impressed above the antennæ; ocelli 3, triangularly arranged; eyes oval.

Antennæ inserted just above the clypeus, in ♀ 10-jointed, subclavate, the scape long, obclavate; in ♂, 11-jointed, filiform, pubescent, or pilose.

Maxillary palpi 4-jointed; labial palpi 2-jointed.

Mandibles bidentate.

Thorax subovoid, the prothorax not visible from above, mesonotum with a median impressed line, rarely entirely wanting, the scutellum elongate, flattened or subconvex, with a distinct frenum; metathorax short, the angles acute or prominent, and with 1 or 2 teeth at base.

Front wings pubescent, with a short linear marginal vein and a rather long curved stigmal vein; apterous individuals of frequent occurrence.

Abdomen subsessile, ovate, compressed beneath, subconvex above and longer than the thorax, the second segment large, striolate at base.

Legs pubescent, the posterior tarsi with the first joint elongate, joints 2-4 subequal.

The majority of the species in this genus are minute and closely resemble those in *Aphanogmus* Thoms., although they are readily separated by the shape of the scutellum, which is broader, more flattened, or subconvex, without a frenum, the antennæ in the males being filiform, not dentate or serrate, as in that genus. Apterous forms are quite common and will probably prove to be dimorphic forms of the winged species.

The records of rearings, with but few exceptions, show the group is almost exclusively parasitic on the dipterous family Cecidomyidæ and the homopterous family Aphididæ.

The species known to me in our fauna may be thus tabulated:

TABLE OF SPECIES.

FEMALES.

Winged forms 2
Wingless forms.
  Brownish yellow, the vertex of head fuscous ....C. FUSCICEPS, sp. nov.
  Black, the base of abdomen pale or yellowish.
    Antennæ, except scape at base, black......................C. MINUTUS Ashm.
  Wholly black, with wing pads.
    Legs golden yellow......................................C. AURIPES, sp. nov.
2. Wholly black 3
  Thorax and abdomen brownish yellow, the head black.
                      C. MELANOCEPHALUS Ashm.
  Head and thorax black.
    Abdomen obscure rufous or piceous, yellowish at base and beneath.
      Shining, but distinctly punctulate; scape rufous, the flagellum black.
                      C. PALLIDIVENTRIS, sp. nov.
      Polished, impunctured, angles of metathorax prominent; antennæ brownish-yellow, 4 or 5 terminal joints black ....C. BASALIS, sp. nov.
  Head and thorax brownish yellow; abdomen black; base and apex of front wings hyaline ....C. CALIFORNICUS, sp. nov.
3. Head and thorax distinctly, closely punctate.
  Pleura aciculated; antennæ brown; legs brownish yellow.
                      C. PUNCTATUS, sp. nov.
  Head and thorax smooth, or the punctuation exceedingly delicate.
    Antennæ, tegulæ, and legs black; pleura not aciculated; flagellum slender, the joints all longer than thick..............C. UNICOLOR sp. nov.
    Antennæ, tegulæ, and legs brownish piceous; trochanters, tips of tibiæ, and the tarsi yellowish white ....C. SALICICOLA, sp. nov.
    Scape, tegulæ, trochanters, knees, tibiæ, and tarsi yellow; flagellum subclavate ....C. AMPLUS, sp. nov.
    Scape and legs brownish yellow.
      Tegulæ black; flagellum black.
        Wings fuscous......................C. MELANOCERUS, sp. nov.
        Wings subhyaline ....C. PEDALIS, sp. nov.
      Tegulæ pale; antennæ brown, the scape and pedicel yellow; pedicel longer than the first flagellar joint, smooth, highly polished.
    Wings hyaline; flagellar joints 1 to 5 transverse, 6 and 7 quadrate.
                      C. FLAVISCAPUS, sp. nov.

Wings subhyaline; flagellar joints 1 to 5 transverse, 6 and 7 longer than wide ....C. GLABER, sp. nov.

Tegulæ piceous; antennæ fuscous; first flagellar joint twice as long as the second; wings subhyaline ....C. CARINATUS, sp. nov.

Scape, tegulæ, and legs, including the coxæ, honey-yellow.

First flagellar joint only a little longer than the second; wings clear hyaline ....C. MELLIPES, sp. nov.

MALES.

Wholly black 2

Not wholly black.

Thorax piceous, collar, base of abdomen, and legs bright yellow.

C. LONGICORNIS, sp. nov.

Thorax black, base of abdomen pale ....C. BASALIS, sp. nov.

2. Distinctly punctate.

Scape and legs yellow ....C. PUNCTATUS, sp. nov.

Faintly punctate or nearly smooth.

Antennæ black ....C. CARINATUS, sp. nov.

Smooth, polished, impunctate.

Head and abdomen black, the thorax piceous.

Antennæ dark brown, the scape beneath and legs, pale yellow.

C. GLABER, sp. nov.

Ceraphron fusciceps sp. nov.

♀. Length, 1 ^{mm}. Brownish-yellow or honey-yellow, the head more or less fuscous on vertex: eyes and flagellum black or brown-black; mesopleura sometimes tinged with fuscous. Head and thorax microscopically punctulate: scutellum finely striate at base; flagellum subclavate, the pedicel more than twice as long as the first flagellar joint, flagellar joints 2 to 6 transverse, the 7th quadrate, the last fusiform, thrice as long as the 7th.

HABITAT.—District of Columbia and Oakland, Md.

Types in Coll. Ashmead.

Described from 3 specimens.

Ceraphron minutus Ashm.

Megaspilidea minuta Ashm., Can. Ent., xx, p. 49.

♀. Length, 1.2 ^{mm}. Black; abdomen fuscous, yellowish at base and beneath; scape and legs yellow. The head and thorax are closely, minutely punctulate; abdomen at extreme base striate; pedicel and flagellum black, subclavate; pedicel twice the length of first flagellar joint: flagellar joints 2 to 5 transverse, the 7th quadrate; 8th or last fusiform, three times as long as the 7th.

HABITAT.—Ottawa, Canada.

Type in Coll. Ashmead.

One specimen; received from Mr. W. Hague Harrington.

Ceraphron auripes, sp. nov.

♀. Length, 1 ^{mm}. Black, shining; scape brownish-yellow: legs bright yellow. Head and thorax minutely, closely punctulate; wings but slightly developed, extending to the base of the abdomen; abdomen highly polished black, more than twice the length of the thorax; flagellum subclavate, joints 2 to 4 transverse, the 5th larger, transverse, the 6th and 7th quadrate, the last fusiform, thrice as long as the 7th.

HABITAT.—Ottawa, Canada.

Type in Coll. Ashmead.

Described from a single specimen received from Mr. W. Hague Harrington.

Ceraphron melanocephalus Ashm.

Copidosoma melanocephalum Ashm. Trans. Am. Ent. Soc., Vol. XIII, p. 131.

♀. Length, 1.2ᵐᵐ. Head black; thorax abdomen and legs, brownish-yellow; flagellum black, paler or brownish at base. Head and thorax closely minutely punctate, the occiput with a delicate median carina from the ocelli; flagellum clavate, the 3 terminal joints being very large and thick and as long as all the others (excepting the scape) united, the funicle joints 2 to 5 transverse; first club joint quadrate, the second, a little longer than wide, the last not quite thrice as long as the second. Wings subhyaline.

HABITAT.—Jacksonville, Fla.

Type in Coll. Ashmead.

Described from one specimen reared May, 1885, from the Cynipid gall *Belonocnema Treatæ* Mayr.

Ceraphron pallidiventris sp. nov.

♀. Length, 1.6ᵐᵐ. Head and thorax black, shining, faintly shagreened; abdomen obscure rufous or piceous above, at base and beneath yellow; scape, pedicel beneath and at tip, and legs, pallid-yellow; tegulæ yellowish; wings hyaline, the stigmal vein yellowish, long and curved, forming almost a closed marginal cell; marginal vein brownish.

Antennæ 10-jointed, the flagellum subclavate, brown-black; scape long, half the length of the flagellum; pedicel a little longer than the first flagellar joint; second flagellar joint scarcely two-thirds the length of the first; the third, fourth and fifth, transverse, but increasing in length and width, the sixth, quadrate, the seventh, oblong-quadrate, the last fusiform, as long as the two preceding joints united. Angles of metathorax tubercular, the metapleura finely rugose, bounded by a carina above. Abdomen a little longer than the head and thorax together, pointed at tip and subcompressed at apex beneath.

HABITAT.—Fort Pendleton, Md.

Type in Coll. Ashmead.

Described from a single specimen taken by Mr. E. A. Schwarz.

The species comes nearest to *C. basalis*, but it is slightly larger, the relative length of the flagellar joints different, the legs pallid yellow, the wings hyaline, while the sculpture is wholly different.

Ceraphron basalis, sp. nov.

♂ ♀. Length, 0.8 to 1.2^{mm}. Head and thorax polished black; abdomen rufous or rufo-piceous, yellowish at base; scape, and sometimes the pedicel, and two or three of the basal flagellar joints, and the legs yellow or brownish-yellow. Wings subhyaline, slightly tinged, the stigmal vein long, curved. In the ♀ the flagellum is twice the length of scape, subclavate, the joints 2 to 5 transverse, the 5th longer and wider than the 4th, 6th and 7th quadrate, the last fusiform, as long as the 6th and 7th united. In the ♂ the flagellum is filiform, 1½ times as long as the body, or about 5 times as long as the scape, pale brown, the first joint the longest.

HABITAT.—District of Columbia and Arlington, Va.

Types in Coll. Ashmead.

Described from 8 specimens (2 ♂ 6 ♀), collected by Mr. E. A. Schwarz and myself.

Its smaller size, very highly polished, impunctate surface, and the difference in the antennae and color of wings at once distinguish it from *C. pallidiventris*.

Ceraphron californicus, sp. nov.

♀. Length, 0.8^{mm}. Head and thorax brownish-yellow; abdomen black; eyes brown. Antennae 10-jointed, two-thirds the length of the body, the flagellum brownish beyond the first joint; scape a little longer than half the length of the flagellum; pedicel and first flagellar joint equal, nearly twice as long as the second; last joint as long as the pedicel. Thorax shining, but feebly, minutely punctulate. Wings hyaline at base and apex, fuscous from the basal third to near the apex.

HABITAT.—Folsom, Cal.

Type in National Museum.

One specimen; taken by Mr. A. Koebele, July 12, 1885.

Ceraphron punctatus, sp., nov.

(Pl. VII, Fig. 2, ♀.)

♂ ♀. Length, 1.1 to 1.6^{mm}. Black, finely, closely punctate, subopaque; antennae 10-jointed, brownish-yellow, the apical half fuscous or black; scape obclavate, nearly two-thirds the length of the flagellum; pedicel longer than the first flagellar joint; flagellum slightly incrassated toward tip; first joint twice as long as thick; joints 2 to 5 transverse, sixth and seventh very little longer than wide, the last joint fusiform, as long as the two preceding united. Vertex with a slight grooved line between the lateral ocelli. Eyes pubescent. Mandibles pale rufous. Mesonotum with a central impressed line. Pleura shining, but more or less aciculated. Metathorax rugose, very short, the posterior angles toothed. Tegulæ dull rufous or piceous. Wings subfuscous, the venation piceous, the stigmal nervure long, slightly curved, about three times as long as the linear marginal vein, and forming almost a closed marginal cell. Legs, including coxæ, brownish-yellow. Abdomen stout, longer than the head and thorax united, polished black, with a slight striated elevation at base of the second segment above.

The ♂ agrees with the ♀, except that the abdomen is shorter and slightly piceous at base, the antennæ filiform, 11-jointed, the scape and legs brownish-yellow, while the flagellum is brown. The antennæ are as long as the body, the scape as long as the first three flagellar joints united; pedicel very small; first flagellar joint slightly more than twice as long as thick, a little stouter than the others, and subequal with the last joint, the latter the longer joint; the three following joints are subequal, the four following gradually become a little longer, the last very slightly longer than the first.

HABITAT.—Virginia, Maryland, and District of Columbia.

Types in Coll. Ashmead and National Museum.

Described from many specimens.

Ceraphron unicolor, sp. nov.

♀. Length, 2 to 2.1ᵐᵐ. Wholly black, shining, the articulations of the legs and the tarsi alone pale brown; head and thorax faintly sculptured; face deeply impressed above the insertion of antennæ, with a grooved line extending forward from the front ocellus; eyes large, pubescent. Antennæ 10-jointed, slender, the scape long, obclavate, a little longer than half the length of the flagellum; first flagellar joint longer than the pedicel, about four times as long as thick, the second two-thirds the length of the first, the two following subequal, those beyond longer, at least 2¼ times as long as thick, the last joint being the longest, and longer than the first. Thorax with a central grooved line, which is subobsolete anteriorly. Post-scutellum toothed. Metathorax very short, with the posterior angles acute. Wings subfuscous; the stigmal vein long. Abdomen stout, highly polished, black, as long as the head and thorax together, with a striate space at base.

HABITAT.—Cheyenne, Wyo., and Alta, Utah.

Types in Coll. Ashmead.

A single specimen of this species was taken by Schwarz in Utah, and another by Wickham in Wyoming.

Ceraphron salicicola, sp. nov.

♀. Length, 0.9ᵐᵐ. Black, shining, feebly punctate; face emarginate: antennæ dark brown; pedicel yellow at apex, as long as the first and second flagellar joints together; flagellum subclavate, the joints gradually widening after the second, the last ovate, about as long as the two preceding joints united. Legs brown; the trochanters, spot on knees, tip of tibiæ, and the tarsi yellowish-white. Wings hyaline, iridescent, pubescent. Abdomen piceous-black, shining.

HABITAT.—Los Angeles, Cal.

Type in National Museum.

A single specimen, labeled as having been "bred from old willow wood partly covered with fungus and infested with Coleopterous larvæ."

In all probability, with the coleopterous larvæ were associated Dipterous larvæ from which the Ceraphron came.

Ceraphron amplus, sp. nov.

♀. Length, 2.2ᵐᵐ. Polished black, but showing some faint microscopic punctures; face deeply emarginated; vertex with a grooved line between the lateral ocelli; eyes pubescent; mandibles brownish; legs brownish-yellow; the coxæ black; the femora and tibiæ dark-brown or fuscous at the middle. Antennæ 10-jointed, black, with a yellow ring between the pedicel and the first funicle joint. Scape long, obclavate, as long as the pedicel and the first three funicle joints united; flagellum subclavate, the first joint being as long as the pedicel, the second, about two-thirds the length of the first, the following to the penultimate subequal, the penultimate longer than preceding joint, or about two-thirds the length of the last, which is long, very slightly longer than the first. Thorax with a central grooved line. Metathorax with the angles but slightly prominent, the post-scutellum toothed. Wings subfuscous, the stigmal vein long and curved. Abdomen long and stout, longer than the head and thorax together, pointed at tip, the second segment occupying fully half the whole length of the abdomen, with striæ at base.

HABITAT.—Washington, D. C.

Type in Coll. Ashmead.

Ceraphron melanocerus, sp. nov.

♀. Length, 1.6ᵐᵐ. Polished black; legs brownish-yellow, the coxæ and femora obfuscated or fuscous; face with an antennal impression and a deep grooved line in front of the front ocellus; eyes pubescent. Antennæ 10-jointed, black, with a pale ring at apex of pedicel; scape long, obclavate; flagellum subclavate, the first funicle joint long, but scarcely as long as the pedicel, the second, only half the length of the first, the joints from the third to the last gradually increasing in size, the third and fourth quadrate, the fifth and sixth a little longer than thick, the last fusiform, twice as long as the penultimate. Thorax with a central grooved line; pronotum deeply impressed at the sides; tegulæ black; metathorax rugose, the posterior angles not acute. Wings subfuscous, the stigmal vein long, curved. Abdomen polished black, as long as the head and thorax together, the second segment with striæ at base, a little longer than all the following segments united.

HABITAT.—Ottawa, Canada.

Types in Coll. Ashmead.

Two specimens received from Mr. W. H. Harrington.

Ceraphron pedalis, sp. nov.

♀. Length, 1.5ᵐᵐ. Polished black, with some sparse, minute punctures; legs entirely brownish-yellow; antennæ black; wings subhyaline, or with a yellowish tinge. Differs principally from *C. melanocerus* in having the axillæ united before the base of the scutellum, the metapleura bounded above by a high carina, the post-scutellum distinctly toothed, the posterior angles of metathorax produced into a tooth, the paler, more uniformly colored legs, while the abdomen is longer, the second segment being fully twice as long as all the following joints united.

HABITAT.—Arlington, Va.

Type in Coll. Ashmead.

Ceraphron flaviscapus, sp. nov.

♀. Length, 0.9 to 1ᵐᵐ. Polished black, impunctured; scape, pedicel, and legs, yellow or brownish-yellow; flagellum subclavate, brown or fuscous; face impressed with a grooved line extending from the front ocellus; eyes pubescent; thorax with a central grooved line; scutellum a little longer than wide; post-scutellum with a small tooth; posterior angles of metathorax acute; wings hyaline; tegulæ yellowish.

The antennæ are 10-jointed, the flagellum subclavate; first flagellar joint much shorter than the pedicel, obconic; the joints 1 to 5 transverse, 6 and 7 quadrate, not longer than wide, the last fusiform, as long as 6 and 7 united.

HABITAT.—Arlington, Va., and District of Columbia.

Types in Coll. Ashmead.

Three specimens.

Ceraphron glaber, sp. nov.

♂ ♀. Length, 1.1 to 1.2ᵐᵐ. Agrees very closely with *C. flaviscapus*, but it is slightly larger, the pedicel brown or fuscous, the sixth and seventh flagellar joints distinctly longer than wide, the abdomen longer, being twice as long as the thorax, while the wings are subhyaline or subfuscous.

In the ♂ the antennæ are long, filiform, 11-jointed, the scape yellow, the flagellum brown, the joints loosely joined or slightly pedicellate, rounded at ends, and sparsely pubescent; pedicel small; first funicle joint about thrice as long as thick, the second shorter, the following to the last about twice as long as thick; collar and petiole yellowish; while the basal joint of hind tarsi is as long as joints 2 to 4 united.

HABITAT.—District of Columbia, and Bladensburg, Md.

Types in Coll. Ashmead.

Three specimens, all taken by Mr. E. A. Schwarz.

Ceraphron carinatus, sp. nov.

♂ ♀. Length, 1.5 to 2ᵐᵐ. Polished black, feebly punctate; antennæ brown-black, except a yellowish ring between the pedicel and first flagellar joint; flagellum very slightly thickened toward tip; pedicel and first flagellar joint very long, the latter slightly the longer, joints 2 to 4 subequal, the second two-thirds the length of the first; joints 5, 6, and 7 about equal, longer than the second; last joint not quite as long as the sixth and seventh united. Thorax with a central grooved line; scutellum a little longer than wide; post-scutellum with a tooth; posterior angles of metathorax acutely prominent; metapleura bounded above by a high carina. Wings subfuscous, the venation brown, the stigmal vein long, curved. Abdomen polished black, 1½ times as long as the thorax, the second segment more than twice as long as the following segments united, striate at base. Basal joint of hind tarsi as long as the following joints united.

In the ♂ the 11-jointed antennæ are long, filiform, the flagellum alone being longer than the body; scape and legs brownish-yellow; flagellum brown-black; first flagellar joint about 5 times as long as thick, the following joints to the last shorter, nearly equal, the last as long as the first; collar and base of abdomen yellowish, the latter not much longer than the thorax.

HABITAT.—Oakland and Bladensburg, Md.; District of Columbia, and Virginia.

Types in Coll. Ashmead and National Museum.

Several specimens.

Ceraphron mellipes, sp. nov.

♂ ♀. Length, 0.8 to 1ᵐᵐ. Polished black, impunctured; scape, pedicel, and legs, bright honey-yellow; tegulæ yellowish; wings clear hyaline. Antennæ in ♀ 10-jointed, the flagellum brown, subclavate, the first funicle joint obconic, two-thirds the length of the scape; joints 2, 3, and 4 as wide as long, fifth subquadrate, sixth and seventh a little longer than wide, the last a little longer than the two preceding united. In the ♂ the antennæ are 11-jointed, filiform, not longer than the body, the first flagellar joint 1½ times as long as thick, longer than the pedicel, the following joints, except the last, scarcely longer than thick, pubescent, the last joint a little longer than the first. In both sexes the face is impressed above the insertion of the antennæ with a grooved line in front of the front ocellus; thorax with a central impressed line;

post-scutellum with a minute tooth, while the posterior angles of meta-thorax are slightly acute.

HABITAT.—Jacksonville, Fla.

Type in Coll. Ashmead.

Described from 1 ♂ and 2 ♀ specimens. It might be mistaken for *C. flaviscapus*, but it is a smaller and more slender form; the antennæ are not so stout, the joints relatively different in length, while the wings are clearer.

Ceraphron longicornis, sp. nov.

♂. Length, 1.2ᵐᵐ. Highly polished, impunctured; the head and api-cal two-thirds of abdomen, black; thorax piceous; collar, scape, legs, and basal one-third of abdomen, yellow or flavo-testaceous. The antennæ are long, filiform, 11-jointed, the flagellum brown; first flagellar joint nearly three times as long as the pedicel and as long as the terminal joint; the other joints are slightly shorter, about four times as long as thick, and all with sparse long hairs. Wings hyaline, strongly fringed, the stigmal vein long, curved. The face is impressed above the an-tennæ; there is a fovea in front of the front ocellus, the mesonotum with a central grooved line, the post-scutellum toothed, while the posterior angles are scarcely prominent, reduced to a minute tubercle.

HABITAT.—Fort George Island, Florida.

Type in Coll. Ashmead.

Described from a single specimen collected by Dr. R. S. Turner.

APHANOGMUS Thomson.

Synarsis Förster, Kleine Monog., p. 57 (1878).

(Type *A. fumipennis* Thoms.)

Head transverse, the frons convex, the occiput slightly excavated and delicately margined; ocelli 3, close together, triangularly arranged; eyes oval or rounded, usually pubescent.

Antennæ inserted just above the clypeus; in ♀ 10-jointed, clavate, in ♂ 11-jointed, subserrate, pilose.

Maxillary palpi 4-jointed, labial palpi 2-jointed.

Mandibles bifid.

Thorax subovoid, compressed at sides, convex above, the collar small; mesonotum usually with a delicate median impressed line, which is often subobsolete or entirely wanting; scutellum conical, convex, at least twice as long as wide, the frenum usually wanting; metathorax very short, abrupt, the angles not prominent.

Front wings pubescent, with a short, linear marginal vein and a short, slightly curved, stigmal vein.

Abdomen ovate, subsessile, compressed beneath, convex above, and a little pointed at tip; the petiole very small and short, the second segment very large.

Legs pubescent, the posterior coxæ pilose behind; the last joint of posterior tarsi about twice as long as the second.

This genus is very closely allied to *Ceraphron* Jurine and it requires great care to separate it from that genus. The frons is more convex, the facial impression less distinct, the scutellum longer and convex, without teeth at base of metathorax: while, as a rule, the metathoracic angles are less distinct.

Synarsis Förster, seems to be without doubt identical, his type *S. pulla* being evidently a small species of *Aphanogmus*, without the central mesonotal furrow and with the head held horizontally, a position often assumed by many species in the group.

All the species known are minute and could easily be mistaken for species in the tribe *Telenomini* in the subfamily *Scelioninæ*.

The species known to me in our fauna may be separated by the aid of the following table:

TABLE OF SPECIES.

Winged 2
Wingless.
    ♂. Black, shining; antennæ dark brown, the third joint the longest, all the
        flagellar joints with long hairs .... A. NIGER, sp. nov.
    ♀. Black; abdomen, legs, and antennæ, except club, bright honey-yellow.
                                     A. BICOLOR, sp. nov.
2. Black.
    Wings not banded 3
    Wings with a fuscous band; no mesonotal furrow.
      Antennæ and legs brown; tarsi pale; stigmal vein oblique, not curved ♀.
                                 A. FLORIDANUS, sp. nov.
      Antennæ and legs dark fuscous, trochanters and tarsi pale; stigmal vein
        curved.
      Scape pale ♀ .... A. VIRGINIENSIS, sp. nov.
      Scape and legs pale yellowish; base of abdomen piceous, ♂.
3. With a mesonotal furrow.
    Antennæ clavate, brown-black; coxæ black; legs dark brown, ♀.
                           A. MARYLANDICUS, sp. nov.
    Without a mesonotal furrow.
      Antennæ subclavate, yellow.
      Legs pale yellow, ♀ .... A. PALLIDIPES, sp. nov.
      Antennæ black or brown-black, clavate, the last joint 4 times as long as the
        penultimate.
      Legs black, coxæ and tarsi brown .. ⎱
                           ⎰ .... A. VARIPES, sp. nov.
      Legs rufopiceous (var.) .... ⎰

Aphanogmus niger, sp. nov.

♂. Length, 1ᵐᵐ. Black, shining; antennæ and legs dark brown, the tarsi whitish. Antennæ 11-jointed, very slightly thickened towards tips; scape obclavate, not reaching beyond the ocelli; first flagellar joint long, about four times as long as the pedicel, the following joints of nearly an equal length, all emarginate at base and covered with long, sparse white hairs. The thorax is compressed, much narrower than the head, convex above, with a single median impressed line; mesopleura smooth, polished; scutellum fully twice as long as wide, extending to the apex of the metathorax, its tip slightly projecting. Wings aborted. Abdomen small, not as long as the thorax, subpetiolated, highly polished and subcompressed.

HABITAT.—Washington, D. C.

Type ♂ in Coll. Ashmead.

Described from a single specimen, given me by Mr. E. A. Schwarz.

Aphanogmus bicolor, sp. nov.

♀. Length, 0.8ᵐᵐ. Apterous; head and thorax black, microscopically punctulate; antennae, except the large terminal joint, the legs and the abdomen, bright yellow, the latter with an obscure spot above at tip. The head is large, quadrate; eyes very large, rounded, occupying the whole side of the head, subpubescent. Antennae 10-jointed, clavate; the scape long, obclavate; the flagellum less than twice as long as the scape; pedicel oval, much stouter and larger than the first flagellar joint; flagellar joints, except the last, submoniliform, increasing in size from the first, which is very small; the last joint is very large, stout, oblong, and as long as the four preceding joints united. Mesonotum with a delicate central grooved line; scutellum extending to the tip of the very short metathorax, convex, longer than wide; angles of metathorax subtubercular, the pleura slightly wrinkled and bounded by a carina above. Abdomen subsessile, as long as the head and thorax together, bright brownish-yellow and highly polished.

HABITAT.—Ottawa, Canada.

Type in Coll. Ashmead.

Taken by Mr. W. H. Harrington.

Aphanogmus floridanus, sp. nov.

♀. Length, 0.75ᵐᵐ. Polished black; antennae and legs brown, the scape, trochanters and tarsi pale or yellowish. Thorax compressed, without a median impressed line. Antennae 10-jointed, clavate, brown, pubescent, the flagellar joints, except the last, transverse, the last ovate; ♂ antennae 11-jointed, slightly thickened at tips, the first flagellar joint longer than the scape, the following shorter. Wings hyaline, fringed, with a fuscous band across the disk before the marginal vein; stigmal vein oblique, not curved, one and a half times as long as the marginal. Legs brown, the tarsi white. Abdomen polished black; not longer than the thorax.

HABITAT.—Jacksonville, Fla.

Type ♂ ♀ in Coll. Ashmead.

Described from two specimens, taken on the edge of a swamp by sweeping.

Aphanogmus virginiensis, sp. nov.

♀. Length, 1mm. Polished black; antennæ and legs fuscous, the scape, trochanters and tarsi pale or whitish. Thorax compressed, convex, without a median impressed line. Antennæ 10-jointed, subclavate, pubescent, the flagellar joints, except the first and the last, transverse quadrate, the last conic. Wings hyaline, fringed, with a fuscous band below the region of the parastigma, the stigmal vein curved. Abdomen conic-ovate, black and polished, as long as the head and thorax together.

HABITAT.—Arlington, Va.

Type ♀ in Coll. Ashmead.

Described from several specimens. The species comes close to *A. floridanus*, but it is slightly larger, the antennæ slenderer, the abdomen longer, more pointed, while there is a distinct curve to the stigmal vein.

Aphanogmus marylandicus, sp. nov.

♀. Length, 1mm. Polished black: antennæ wholly brown-black; legs dark-brown, the coxæ black, knees and tarsi pale or yellowish. Thorax compressed, highly convex, shining, with a distinct median impressed line. Antennæ 10-jointed, clavate, the flagellar joints, except the last, transverse; the last very large, ovate. Wings clear hyaline, fringed, the stigmal vein long and curved. Abdomen pointed, ovate, subcompressed, as long as the head and thorax together, black and polished.

HABITAT.—Oakland, Md.

Type ♀ in Coll. Ashmead.

Described from one specimen received from Mr. E. A. Schwarz. This may be the ♀ of *A. niger*.

Aphanogmus pallidipes, sp. nov.

♀. Length, 0.8mm. Black, shining; antennæ brownish-yellow; legs pale yellow. Antennæ 10-jointed, subclavate, the flagellar joints all a little longer than thick, the last conic. Thorax compressed, convex, without a median impressed line. Wings clear hyaline, the nervures brown, the stigmal vein a little longer than the marginal, slightly curved. Abdomen a little longer than the thorax, ovate, pointed at tip, subcompressed, black and shining.

HABITAT.—Arlington, Va.

Type ♀ in Coll. Ashmead.

Described from a single specimen.

Aphanogmus varipes, sp. nov.

(Pl. VII, Fig. 1, ♀.)

♀. Length, 0.8mm. Black, shining; antennæ and legs vary from brown or rufous to black; the trochanter and tarsi pale brownish-yellow. Antennæ 10-jointed, clavate, the last joint very large, about four times as large as the penultimate, the two preceding joints transverse, the others a little longer than wide, the first flagellar joint being scarcely as long as the pedicel. Thorax convex, without a trace of a furrow. Wings hyaline, pubescent, the parastigma as long and thick as the marginal vein, the stigmal vein very little longer than the marginal. Abdomen not longer than the thorax.

HABITAT.—Manhattan, Kans.

Types in Kansas State Agricultural College and Coll. Ashmead.

Described from two specimens received from Prof. Popenoe.

NEOCERAPHRON Ashm., gen. nov.

Agrees in all particulars with *Ceraphron*, except that the antennæ in the male are 10-jointed, not 11-jointed, and in female 9-jointed, not 10-jointed, the flagellum being strongly clavate.

Neoceraphon macroneurus, Ashm.

(Pl. VII, Fig. 3, ♂.)

Ceraphron macroneurus Ashm., Ent. Am., III, p. 97, ♂ ; Cress. Syn. Hym., p. 312.

♂. Length, 0.8^{mm}. Polished black, impunctured; abdomen yellow, blackish above towards apex; legs yellowish-white; antennæ 10-jointed, filiform, reaching to the base of the abdomen, dark-brown, the scape brownish-yellow, the flagellar joints, except the first and last, not longer than wide, loosely joined, moniliform, pubescent, the last joint twice as long as the first. Wings hyaline, the marginal vein linear, the stigmal vein very long, curved, almost attaining the apex of wing, forming a large marginal cell.

♀. Length, 0.75 ^{mm}. Black, polished, impunctured, the abdomen honey-yellow, blackish above; legs whitish; antennæ 9-jointed, brownish-yellow, fuscous toward tips, the flagellum much incrassated towards apex, all the joints, except the first, which is as long as thick, are wider than long, the last large, fusiform.

HABITAT.—Jacksonville, Fla., and Virginia.

Types in Coll. Ashmead.

Subfamily V.—SCELIONINÆ.

Head transverse or quadrate, often very broad and large. Ocelli 3, always present. Mandibles most frequently bidentate, although occasionally 3-dentate. Antennæ elbowed, inserted on a clypeal prominence or at the base of the clypeus, usually clavate, 11- or 12-jointed in the females, or if the club is unjointed, but 7-jointed; in the males filiform or setaceous, 12-jointed except in *Scelio*, where they are but 10-jointed. Maxillary palpi 2, 3, 4, or 5-jointed, labial palpi 2- or 3- jointed. The pronotum is often not visible from above, or it is large, transverse, or quadrate; mesonotum generally short, transverse, with or without grooved furrows; scutellum generally semicircular, the axillæ not distinctly separated; it is rarely spined or wanting, although the postscutellum is frequently spined; metathorax short, frequently with acute angles or spines. Front wings most frequently with submarginal, marginal, post-marginal, and stigmal veins; the post-marginal and marginal veins are rarely absent, except in the tribes *Baeini* and *Teleasini;* if absent, in the tribe *Scelionini,* the submarginal vein terminates in a stigma or knob. Abdomen sessile, or subsessile, inserted above the coxæ, depressed and sharp-edged or strongly carinated along the sides, where the tergites join the urites; in shape it is variable: it may be broadly oval, oblong, ovate, fusiform, or linear, and often greatly elongated, composed of several segments; the second and third segments are usually much the largest, but occasionally the segments are nearly of an equal length. Legs moderate, the femora clavate, the tibiae subclavate or slender, the tibial spurs usually 1, 1, 1, the middle and posterior spurs generally weak or poorly developed; the tarsi long, slender, 5-jointed.

This group is probably the most extensive one in the whole family and of the greatest economic importance, all the species comprising it being strictly egg parasites, scarcely a single order of insects being free from their attack.

It may be subdivided into four natural tribes distinguished as follows:

TABLE OF TRIBES.

Abdomen without distinct lateral carinæ, most frequently broadly oval, rarely pointed ovate, depressed, the second segment always the largest and longest; post-marginal and stigmal veins long; ♀ with 11-jointed antennæ, rarely with 12 joints, clavate; ♂ antennæ 12-jointed.

Tribe I.—TELENOMINI.

Abdomen always with distinct lateral carinæ.

Abdomen broadly oval or long oval, the third segment much the longest; post-marginal vein not developed.

Marginal vein very short, punctiform or thickened, not or scarcely as long as the stigmal vein; stigmal vein short, thickened at base and ending in a rounded stigma; ♀ antennæ 7-jointed with an unjointed club; ♀ usually apterous; ♂ antennæ 12-jointed, filiform-moniliform; lateral ocelli usually close to the inner margin of the eye..Tribe II.—BAEINI.

Marginal vein very long, 5 or 6 times as long as the exceedingly short stigmal vein; stigmal vein not thickened at base; ♀ antennæ 12-jointed, clavate, the club 5 or 6 jointed; ♂ antennæ 12-jointed, filiform, the funicle joints long; lateral ocelli far away from the inner margin of the eye, never very close ....Tribe III.—TELEASINI.

Abdomen sessile, most frequently long, fusiform, or linear, extending beyond the tips of the wings when folded, rarely broadly oval, the segments more nearly equal, or the third segment the longest, but rarely much longer than some one of the others; post-marginal vein present, rarely wanting, if wanting the submarginal vein terminating in a stigma; marginal vein seldom twice as long as the stigmal; the stigmal not especially short, oblique, rarely entirely absent; ♀ antennæ 12-jointed, clavate; ♂ antennæ 12-jointed, usually filiform (in a single case 10-jointed) ....Tribe IV.—SCELIONINI.

Tribe I.—TELENOMINI.

This tribe was first separated by Thomson, Öfvers. af K. Vet.-Akad.
Förh., 1860, p. 169, with two genera, *Telenomus* and *Phanurus*, to which
I here add *Trimorus* Förster, and three new genera, viz., *Trissolcus*,
Dissolcus, and *Aradophagus*, all distinguished from other genera in the
Scelioninæ by the absence of the lateral carinæ on the abdomen.

The species are all minute or microscopic in size, and are parasitic in
the eggs of various insects, those of Lepidoptera and Hemiptera being
particularly subject to their attack.

The six genera included in the tribe may be thus distinguished:

TABLE OF GENERA.

Females, with 11-jointed, clavate antennæ...................................... 2
Females, with 12-jointed antennæ 4
2. Lateral ocelli not touching the margin of the eye........................ 3
    Lateral ocelli touching the margin of the eye.
    Mesonotum with 2 furrows.
        Post-scutellum spined...TRIMORUS Förster.
    Mesonotum without furrows.
        Post-scutellum not spined.
        Head quadrate; abdomen pointed ovate, the ovipositor usually ex-
          serted ....PHANURUS Thom.
        Head transverse, often very broad, abdomen broadly oval, usually
          truncate at tip...TELENOMUS Hal.
3. Mesonotum with 3 furrows, abbreviated anteriorly.
    Frons very broad; a short but distinct groove extends from the eye back
      of the lateral ocellus to the occiput ..TRISSOLCUS Ashm., gen. nov.
    Mesonotum with 2 furrows, abbreviated anteriorly.
    Frons not very broad...........................DISSOLCUS Ashm., gen. nov.
4. Mesonotum without furrows.
    Head large, flat, the ocelli in a triangle, the laterals nearer to the front
      ocellus than to margin of eye; wings banded.
                    ARADOPHAGUS Ashm., gen. nov.

TRIMORUS Förster.

Hym. Stud., II, p. 101 (1856).

(Type *P. nauno*, Walk., ♂ ; *P. philias*, Walk., ♀.)

Head transverse, a little wider than the thorax, the occiput delicately
margined; ocelli 3, the lateral close to the eye; eyes ovate; subpubescent.

Antennæ in both sexes 12-jointed; in ♀ clavate, the club 5-jointed;
in ♂ filiform-moniliform or submoniliform.

Maxillary palpi 3-jointed.

Mandibles bifid.

Thorax ovoid, the prothorax depressed above, produced into a little
neck anteriorly; mesonotum with 2 furrows; scutellum convex, semi-

circular; post-scutellum armed with a short, stout spine or tooth; metathorax short.

Front wings fringed, with a marginal vein about half the length of the stigmal, the latter oblique, ending in a small knob.

Abdomen oval, depressed, somewhat broader than the thorax; the first segment short, striate, the second very large, the following short.

Legs rather long, slender, the femora subclavate, tarsi 5-jointed, the basal joint of hind tarsi more than three times as long as the second.

Distinguished by the spined scutellum.

Trimorus americanus, sp. nov

(Pl. vii, Fig. 4, ♂.)

♂. Length, 1.2ᵐᵐ. Polished black, impunctured, sparsely pubescent. Head transverse, not wider than the thorax; lateral ocelli on the margin of the eye; eyes oval, pubescent. Antennæ 12-jointed, filiform-moniliform, extending to the middle of the abdomen; the scape scarcely reaches to the middle ocellus, brownish-yellow; the flagellum rust-brown, the pedicel very small, the first and third funicle joints equal, twice as long as the pedicel, the second slightly smaller than the first, the fourth and the following to the last about equal, moniliform, the last oval.

Thorax ovoid, the mesonotum with 2 furrows abbreviated anteriorly; scutellum semicircular; the post-scutellum produced into a small triangular tooth; metathorax very short. Abdomen oval, black, not longer than the thorax, the first segment striated. Wings hyaline, pubescent, the venation brown; the marginal vein is rather thick, the length of the shaft of the stigmal, the latter terminating in a small knob.

HABITAT.—Arlington, Va.

Described from a single specimen.

PHANURUS Thomson.

Öfv., 1860, p. 169.

(*P. angustatus* Thom.)

Head quadrate or subquadrate, the frons smooth, shining, the occiput usually but slightly emarginate, not margined. Ocelli 3, triangularly arranged, the lateral touching the margin of the eyes. Eyes large, oval, sometimes slightly pubescent.

Antennæ in ♀ 11-jointed, subclavate, rarely distinctly clavate, from the last two funicle joints being widened; in ♂ 12-jointed, submoniliform, shorter than the body; the pedicel longer than the first funicle joint, the second funicle joint about twice as long as the first, the third shorter, joints beyond moniliform or submoniliform, the last ovate.

Maxillary palpi 2-jointed.

Mandibles bifid.

Thorax ovate or long ovate, the mesonotum longer than wide, without furrows, the metathorax not especially shortened.

Front wings rather narrowed, longly fringed, the marginal vein very short, the stigmal vein rather long, very oblique, the post-marginal long.

Abdomen long, pointed ovate and at least as long as the head and thorax together, and narrower, the ovipositor often exserted, the second segment very long, occupying two-thirds the whole surface.

Legs rather long and but slightly thickened.

Although this genus closely resembles *Telenomus*, it is readily distinguished by the shape of the head, the subclavate antennæ, narrower wings, and the longer, narrower, pointed abdomen, the second segment always being two or three times longer than wide.

The habits of the genus are identical with *Telenomus*.

Our species may be tabulated as follows:

TABLE OF SPECIES.

```
Head and thorax dark brown.......................................................2
Black.
  Antennæ black or brown black.
    Legs brownish-piceous, the knees and tarsi pale........P. OVIVORUS, sp. nov.
    Legs rufous, the tarsi white.........................P. FLORIDANUS, sp. nov.
  Scape pale rufous, the flagellum brown.
    Legs yellow.......................................................P. FLAVIPES, sp. nov.
2. Abdomen pale brown..................................................P. OPACUS How.
```

Phanurus ovivorus, sp. nov.

♀. Length 0.6ᵐᵐ. Polished black, impunctured. Head transverse-quadrate, scarcely wider than the thorax. Eyes oval, with a few hairs. Antennæ 11-jointed, subclavate, brown-black or black, the flagellum nearly thrice as long as the scape, the pedicel longer than the first funicular joint, the joints of funicle all longer than thick, the club scarcely separable from the 4-jointed funicle, the joints a little longer than wide. Thorax ovoid, the disk somewhat flattened and highly polished, the mesonotum longer than wide, without furrows. Legs piceous, the knees and tarsi paler. Abdomen pointed ovate, depressed, flat above and longer than the head and thorax together, the ovipositor slightly projecting. Wings rather narrow, hyaline, iridescent, and ciliated, the nervures yellow, the marginal vein shorter than the stigmal.

HABITAT.—Washington, D. C.

Types in National Museum and Coll. Ashmead.

Bred by Dr. Riley at Washington, D. C., September 10, 1885, from Heteropterous eggs; also from Curculionid in catkins of Black Birch, June 19, 1889. The last record I consider unreliable. In all probability there were insect eggs in the catkins overlooked by Dr. Riley. My collection contains specimens captured at large.

Phanurus floridanus, sp. nov.
(Pl. VII, Fig. 5, ♀.)

♀. Length 0.6ᵐᵐ. Polished black, impunctured. Head subquadrate, as wide as the thorax. Antennæ 11-jointed, clavate, black, the club stouter than in *P. orivorus;* pedicel much longer and thicker than the first funiclar joint, the funiclar joints after the second, transverse, the club 4-jointed, the joints excepting the last transverse, the last ovate. Legs pale brownish-yellow, the coxæ black or dark fuscous. Abdomen pointed ovate, about as long as the head and thorax together, the first segment short, narrowed, the second occupying half the rest of the surface, the segments beyond short, about equal. Wings hyaline, ciliated, the venation yellowish, the marginal vein about one-third the length of the stigmal.

HABITAT.—Jacksonville, Fla.

Type in Coll. Ashmead.

Phanurus flavipes, sp. nov.

♀. Length, 0.65ᵐᵐ. Polished black. Head quadrate, a little wider than the thorax across from tegula to tegula, and with the occiput slightly emarginated. Eyes large, oval, subpubescent. Mandibles pale or yellowish. Antennæ 11-jointed; the scape is half the length of the flagellum, pale rufous or yellowish; the flagellum subclavate, brown, the pedicel about 2½ times as long as thick or nearly twice as long as the first funiclar joint, the two last funiclar joints a little shorter and stouter than the preceding joints, club joints stouter, a little longer than thick. Legs, including coxæ, bright yellow. Abdomen as long as the head and thorax together, polished, impunctured, the second segment nearly twice as long as wide at apex, the first short and narrow. Wings hyaline, somewhat narrowed, with long cilia.

HABITAT.—Arlington, Va.

Type in Coll. Ashmead.

Described from a single specimen.

The color of the antennæ and legs readily separate the species from all the others.

Phanurus opacus How.

Thoron opacus How., Ins. Life, I, p. 268, f. 64.

" ♂. Length, 0.84ᵐᵐ; expanse, 1.2ᵐᵐ; greatest width of forewing, 0.163ᵐᵐ; length of antennæ, 0.6ᵐᵐ. Joint 1 of funicle rather shorter and slightly narrower than pedicel; funicle joints distinctly separated, subequal in length, increasing very slightly in width from 4 to 9, joints 2 and 3 equal in width and slightly slenderer than either 1 or 4; club one-third longer than joint 9 of funicle, ovate, at base of same width as joint 9 of funicle, without a trace of dividing sutures. Metanotal spiracles large, oval; metascutellum with a straight median longitudinal furrow. Abdomen flattened, ovate, rather longer than thorax. General surface of the body with no visible punctation, opaque. Head, antennæ, and thorax dark brown; abdomen rather lighter; all legs brown; tarsi nearly white; base of all tibiæ nearly white. Wings hyaline; veins slightly dusky." (*Howard.*)

HABITAT.—Los Angeles, Cal.

According to Mr. Howard this insect was reared by Mr. D. W. Coquillett, July 27, 1887, from the adult female of *Icerya purchasi*. I am not willing to accept this statement as accurate, the species being an egg-parasite. As with the European *Telenomus coccidivorus*, additional evidence is needed to substantiate its divergent habit.

TELENOMUS Haliday.

Ent. Mag. 1, p. 271 (1833).

Syn. (?) *Hemisius* Westwood, Phil. Mag., 11, p. 445 (1832).

(Type *T. brachialis* Hal.)

Head large, transverse, usually very wide, very rarely quadrate, the occiput concave, not margined. Ocelli 3, triangularly arranged and widely separated, lateral contiguous to the margin of the eye. Eyes oval, often pubescent.

Antennae inserted close to the clypeus, in ♀ 11-jointed, clavate, the club 4- or 5-jointed, the pedicel usually larger than the first funicle joint, the first funicle joint longer than wide, the last two funicle joints minute, transverse; in ♂ filiform, pubescent, moniliform or submoniliform, the joints after the fifth rarely elongate-cylindric.

Maxillary palpi 2-jointed.

Thorax ovoid, the mesonotum wider than long, without furrows; the metathorax short, rounded, unarmed.

Front wings pubescent, ciliated, the submarginal vein joining the marginal at about ⅓ the length of the wing, the marginal vein linear, rarely punctiform, and usually shorter than the stigmal, the latter oblique and rather long, the post-marginal vein long.

Abdomen subsessile, broadly oval, depressed, not or rarely longer than the thorax, the apex usually truncate, the first segment wider than long, the second always the largest and longest segment, although often wider than long; the following segments all short.

Legs moderate, the femora subclavate, the tibial spurs weak, the tarsi 5-jointed, not as long as the tibiae.

A well known and widely distributed genus found in all parts of the world. Haliday and Walker have described in it species that should now be relegated to *Phanurus*. The species are numerous and probably susceptible of a still further generic subdivision.

The description of Westwood's genus *Hemisius*, with one species, *minutus*, appeared in the Phil. Mag, Vol. 11, p. 445, 1832, and consequently antedates that of Haliday (which did not appear until 1833) just one year. From Westwood's brief description I am unable to separate it from *Telenomus*, and if *Telenomus* is not again subdivided it must replace that now well-known genus.

Species of the genus *Telenomus*, according to Dr. Gustav Mayr, have been reared in Europe from the following insect eggs:

Lepidoptera..13 species.
Hemiptera..3 species.
Diptera...1 species.

He also records the rearing of *Telenomus coccidivorus* Mayr from a coccid; but in all probability this is a mistake, and it came from the egg of some Lepidopteron, Syrphid, or other predaceous insect associated with the coccid.

In America *Telenomi* have been reared from the following insect eggs:

Species.

Lepidoptera 14

Hemiptera 3

Neuroptera 1

Our species are quite numerous and separated with difficulty, but it is hoped that the student will find assistance in the following dichotomous table:

TABLE OF SPECIES.

FEMALES.

Second abdominal segment not, or very little, longer than wide............... 2
Second abdominal segment about twice as long as wide.
  Head transverse-quadrate.
    Legs, including coxæ and scape, reddish-yellow; thorax flattened.
      Moderate.......................................T. HUBBARDI, sp. nov.
    Legs, including coxæ and scape, pallid yellow; thorax convex.
      Minute..T. PUSILLUS, sp. nov.
  Head transverse.
    Legs and antennæ black, knees and tarsi honey-yellow..T. KOEBELEI, sp. nov.
2. Pedicel distinctly longer and thicker than the first funicle joint............... 3
Pedicel not longer than the first funicle joint.
  Scape black, flagellum brown-black, tip of pedicel yellow, funicle joints 2, 3, and 4 equal, moniliform.
    Legs black or piceous, trochanters, knees, anterior tibiæ, and tarsi, yellow.
      T. NIGRISCAPUS, sp. nov.
  Antennæ brown-black, pedicel yellow at tip, shorter than first funicle joint, funicle joints 2 and 3 subequal, the third transverse.
    Legs piceous, trochanters, anterior tibiæ beneath and tarsi, brown.
      T. UTAHENSIS, sp. nov.
3. Scape of antennæ pale or pale at base or tip 4
Scape of antennæ black or brown-black.
  Coxæ pale.
    Marginal vein punctiform.
      Claval joints 2, 3, and 4 about equal, quadrate ....T. NOCTUÆ, sp. nov.
  Coxæ black.
    Funicle joints 2 and 3 longer than thick.
      Legs black or piceous, trochanters, knees, tips of tibiæ and tarsi, yellow.
      First abdomal segment and the second, at base, striate.
        Marginal vein ⅓ the length of the stigmal.
          Second abdominal segment wider than long...T. GNOPHLELÆ, sp. nov.
          Second abdominal segment longer than wide.
            T. CALIFORNICUS, sp. nov.
        Marginal vein punctiform...................T. GRACILICORNIS, sp. nov.
  Legs pale rufous.
    Thorax and scutellum closely punctate; head large, polished.
      T. PERSIMILIS, sp. nov.

Funicle joints 2 and 3 rounded, not longer than thick.
    Legs piceous or brown-black, trochanters, knees, tips of tibiæ and tarsi pale
      or yellowish.
      First abdominal segment striate.
        Marginal vein $\frac{1}{4}$ the length of the stigmal; head $3\frac{1}{2}$ times as wide as
          long..T. GRAPTÆ How.
        Marginal vein nearly $\frac{1}{2}$ the length of the stigmal; head 3 times as wide
          as long..............................T. SPILOSOMATIS, sp. nov.
        Marginal vein as long as the stigmal................T. MINIMUS, sp. nov.
      Legs dark-brown, the tarsi whitish.
        Marginal vein about $\frac{1}{4}$ the length of the stigmal..T. HELIOTHIDIS, sp. nov.
    Legs piceous or brown-black, trochanters, knees, tips of tibiæ and tarsi,
      pale or yellowish.
      First abdominal segment not striate.
        Eyes pubescent ....T. ORGYLÆ, Fitch
        Eyes bare or only faintly pubescent.
          Marginal vein punctiform...................T. ICHTHYURÆ, sp. nov.
          Marginal vein long...........................T. INFUSCATIPES Ashm.
      Legs yellow or brownish-yellow.
        Marginal vein $\frac{1}{4}$ the length of the stigmal; head about $3\frac{1}{2}$ times as long
          as wide...T. BIFIDUS, Riley
4. Scape wholly pale.. 5
  Scape not wholly pale, pale beneath or at base or apex.
    Coxæ black or piceous.
      Wings with a dusky submarginal blotch ....T. MACULIPENNIS, sp. nov.
      Wings with no submarginal blotch, hyaline or subhyaline.
        Funicle joints 2 and 3 not longer than thick.
          Coxæ black.
            Legs dark brown or fuscous, trochanters, knees and tarsi, pale yellow.
              Second abdominal segment $1\frac{1}{2}$ times as long as wide
                              T. SPHINGIS Ashm.
           Second abdominal segment not or scarcely longer than wide.
             Head very wide.
               Funicle joints 2 and 3 subequal...................T. RILEYI How.
               Funicle joints 2 and 3 equal...........T. GOSSYPICOLA, sp. nov.
          Coxæ pale.
             Legs brown, the trochanters, knees, tips of tibiæ and tarsi, whitish or
            pale. Pedicel longer and stouter than the first funicular joint, the
            second very little longer than thick, the third small, transverse.
                          T. GEOMETRÆ, sp. nov.
5. Pedicel scarcely longer than the first funicular joint.
    Second abdominal segment not striate at base.
      Legs rufous or reddish-yellow, the femora and tibiæ often obfuscated.
                             T. ARZAMÆ Riley.
      Legs brownish-yellow.....................................T. LAVERNÆ, sp. nov.
  Pedicel much longer than the first funicular joint.
    Closely minutely punctate.
      Second abdominal segment striate at base.
        Legs, including coxæ, honey-yellow ....T. PODISI, sp. nov.
    Coarsely cribrato punctate.
      Mesopleura tinged with rufous.
        Legs and antennæ, except club, brownish-yellow.
                      T. PENNSYLVANICUS, sp. nov.

MALES.

Antennæ not, or scarcely, longer than the body 2
Antennæ much longer than the body.
   Legs pale brownish-yellow; joints of flagellum oval...T. DOLICHOCERUS Ashm.
2. Flagellar joints 1 and 2 equal, or nearly equal 4
  Flagellar joints 2 and 3 about equal, longer than the first 3
  Flagellar joints 2 and 3 about equal, shorter than the first.
    Coxæ black.
      Legs fuscous or brown, knees, tibiæ, and tarsi yellow; marginal vein half the
          length of the stigma ....T. CHRYSOPÆ, sp. nov.
      Legs honey-yellow, the femora, except tips, brown; marginal vein one-fourth
          the length of the stigma ....T. UTAHENSIS, sp. nov.
3. Coxæ black.
   Pedicel longer than the first funicle joint.
     Legs pale brownish-yellow, the femora usually more or less dusky.
       Flagellar joints, after the third, moniliform........T. SPILOSOMATIS Ashm.
       Flagellar joints, after the third, long oval.............T. SPHINGIS Ashm.
   Pedicel shorter than the first funicle joint.
     Legs black or piceous, trochanters, knees, tips of tibiæ, and tarsi honey-
       yellow.
       Scape black, flagellum dark brown ....T. GRAPTÆ How.
       Scape and flagellum pale brown ....T. CŒLODASIDIS sp. nov.
   Coxæ pale.
   Pedicel shorter than the first funicle joint.
     Legs honey-yellow; flagellar joints longer than thick ......T. ARZAMÆ Riley.
     Legs reddish-yellow; flagellar joints round, moniliform.T. NIGRISCAPUS Ashm.
   Pedicel a little longer than the first funicle joint.
     Legs whitish, the femora and tibiæ tinged with brown; flagellar joints trans-
      verse...T. GEOMETRÆ Ashm.
4. Flagellar joints 1 and 2 about equal, the third shorter.
   Coxæ black.
     Legs black or piceous, the trochanters, tips of femora, and tibiæ and the tarsi
      honey-yellow ....T. CLISIOCAMPÆ Riley
   Coxæ pale.
     Legs brownish-yellow; flagellar joints 1 to 3 not especially elongate, the joints
      beyond transverse.............................T. BIFIDUS Riley
     Legs yellow; flagellar joints 1 to 3 stout and elongate, the joints beyond
      moniliform ....T. PODISI Ashm.
  Flagellar joints 1 and 2 about equal, the third longer.
     Legs, including coxæ, pale brownish-yellow.......................T. NOCTUÆ
  Flagellar joints 1 and 2 equal, the joints beyond oval. Coxæ black; legs reddish-
     yellow ....T. GNOPHELÆ

T. dolichocerus Ashm.

Teleas dolichocerus Ashm., Ent. Am., III, p. 100, ♂ ; Cress. Syn. Hym., p. 313.

♂ . Length, 0.8^{mm}. Black, shining; the thorax microscopically punctate and pubescent; antennæ black, the scape beneath brownish-yellow; legs yellow. Antennæ 12-jointed, very long, filiform, much longer
than the body, pubescent; the pedicel is hardly half the length of the
first funicle joint; the latter is stouter than the pedicel or any of the
following joints, about twice as long as thick; the second funicle joint is
greatly elongated, longer than the pedicel and the first joint united;

the 3 following joints subequal; those beyond, to the last, elliptic, oval; the last fusiform, as long as the third; all the joints from the third briefly pedicellated. Wings hyaline, with long cilia; the venation pale brown, the marginal vein half the length of the shaft of the stigmal, the latter ending in a knob.

HABITAT.—Jacksonville, Fla.

Type, ♂ in Coll. Ashmead.

Described from a single specimen. The greatly elongated antennae and the shape of the joints render the species easy of recognition.

Telenomus Hubbardi, sp. nov.

♀. Length, 1^{mm}. Black, smooth, shining; head quadrate, not wider than the thorax; mandibles piceous. Antennae 11-jointed, the scape reddish-yellow, the flagellum dark brown; the pedicel is much longer than the first funiclar joint; the second, third, and fourth funiclar joints nearly equal, the second being very slightly the longest; the club is stout, the joints broadly transverse. Thorax subconvex, much longer than wide. Legs, including coxae, uniformly reddish-yellow. Abdomen as long as the head and thorax together, the apex subtruncate, highly polished; the first segment transverse, with some coarse striae at base; the second segment long, twice as long as wide; the following segments very short, equal, but all distinctly visible. Wings hyaline, fringed, the venation yellowish, the marginal vein punctiform.

HABITAT.—Centerville, Fla.   .

Type in National Museum.

Described from a single specimen received from Mr. H. G. Hubbard, reared September 4, 1880, from the eggs of a Reduviid.

Telenomus pusillus, sp. nov.

♀. Length, 0.6^{mm}. Black, shining; head transverse quadrate, polished; eyes pubescent; thorax microscopically punctate; scape and legs pallid yellow or whitish. Antennae 11-jointed, the flagellum brownish; pedicel nearly twice as long as the first funiclar joint; first and second funiclar joints equal; the third and fourth shorter, subequal, the fourth being transverse. Wings hyaline, fringed; the nervures pale yellowish, the marginal nervure being one-half the length of the shaft of the stigmal. Abdomen not quite as long as the head and thorax united, triangularly pointed at apex, when viewed from above, polished; the first segment narrow, striated; the second, widened at apex, but still twice as long as wide at apex.

HABITAT.—Arlington, Va.

Type in Coll. Ashmead.

Its minute size and pallid legs distinguish it from *T. Hubbardi* with which it agrees in the shape of the head and the abdomen.

Telenomus Koebelei, sp. nov.

♀. Length, 1ᵐᵐ. Black, shining, the thorax finely punctate, devoid of pubescence, scutellum smooth, polished. Head transverse-quadrate, as wide as the thorax, the face rather flat, smooth. Eyes with a fine pubescence. Mandibles black. Antennæ 11-jointed, rather short, black, the pedicel longer than the first funiclar joint, the second subequal with the first, the third much shorter than the second; the club is gradually fusiform from the fourth funiclar joint, the first and second joints transverse-quadrate, the third quadrate, the fourth slightly smaller than the third quadrate, the last cone-shaped. Legs black; trochanters, a small spot on knees, and the tarsi, dull honey-yellow. Abdomen a little longer than the thorax, smooth, shining, the first segment, and the second at base, striated; the second segment is at least 1½ times longer than wide; all the following segments are exceedingly short, but distinct, in this respect differing from the typical forms in the genus. Wings hyaline, with a short fringe; the venation yellowish, the marginal vein half the length of the stigmal.

HABITAT.—Alameda, Cal.

Types in National Museum.

Described from three specimens reared by Mr. A. Koebele from an egg of some large unknown bombycid, probably an *Attacus*.

Telenomus nigriscapus, sp. nov.

♂ ♀. Length, 0.8ᵐᵐ. Black, shining, the thorax with a fine white pubescence. Head very broad, the face polished. Mandibles black. Antennæ in ♀ 11-jointed, the scape black, the flagellum brown-black, tip of pedicel yellowish; the pedicel is not longer than the first funiclar joint; the second, third, and fourth funiclar joints are about equal, moniliform, the first four joints of club transverse-quadrate. Thorax convex. Legs black, the trochanters, anterior tibiæ, and all knees and tarsi, honey-yellow, the middle and hind tibiæ piceous. Abdomen fully as long as the thorax, truncate at tip, polished, the second segment about one-fourth longer than wide. Wings hyaline, fringed, the venation pale yellowish; the marginal vein is at least one-third the length of the stigmal.

In the ♂ the legs, including the coxæ, are uniformly reddish-yellow; antennæ 12-jointed, pale brown, the scape paler; the pedicel is distinctly smaller than the first funiclar joint, the second and third a little longer and stouter than the first, the joints beyond to the last round-moniliform, the last conic, twice as long as the penultimate.

HABITAT.—Agricultural College P. O., Mich.

Types in National Museum.

Described from 1 ♂ and 1 ♀ received from Prof. A. J. Cook.

Telenomus utahensis, sp. nov.

♂ ♀. Length, 0.8ᵐᵐ. Black, shining; the thorax microscopically punctate, covered with a fine pubescence; the head very little more than thrice as wide as long, finely shagreened; antennæ black, the pedicel at tip tinged with yellow; legs black, trochanters, knees, anterior tibiæ, except sometimes a blotch above, and the tarsi, honey-yellow. Antennæ in ♀ 11-jointed, the first funiclar joint longer than the pedicel, the second half the length of the first, the third and fourth short, transverse; the fourth the broader, nearly as wide as the club; the club joints quadrate, the first the largest and a little the broadest, last joint conic. Wings subhyaline, pubescent, iridescent, with short cilia, the nervures brown, the marginal about one-third the length of the stigmal.

Abdomen broadly oval, truncate at apex, the petiole striate, the second segment one-half broader than long.

In the ♂ the antennæ are 12-jointed, filiform, the first funiclar joint about twice as long as the pedicel, the second and third funiclar joints a little shorter than the first, subequal; the joints beyond, to the ninth, moniliform; the ninth, a little longer than thick, and two-thirds the length of the last joint, which is pointed, conic.

HABITAT.—Wasatch and Salt Lake, Utah.

Types, ♂ and ♀, in Coll. Ashmead.

Described from specimens received from Mr. E. A. Schwarz; collected June 13 and 27, 1891.

Telenomus noctuæ, sp. nov.

♂ ♀. Length, 0.6ᵐᵐ. Black, polished, pubescent, the thorax highly convex, microscopically punctate; antennæ in ♀ brown-black, in ♂ yellow; legs yellow to brownish-yellow, the femora and tibiæ in the ♀ dusky. Head about thrice as wide as long antero-posteriorly. Eyes pubescent. Antennæ ♀ 11-jointed, less than twice as long as the scape; the pedicel is stouter, and nearly twice as long as the first funiclar joint, the latter only a little longer than thick, the second moniliform, the third and fourth equal, transverse-moniliform; the club rather stout, the first joint transverse, twice as wide as long, the second, third, and fourth joints about equal, transverse-quadrate, a little broader than long, the last conic. Wings hyaline, ciliated, the nervures brownish-yellow, the marginal vein punctiform, about twice as long as thick. Abdomen not quite as long as the thorax, truncate behind, the second segment wider than long.

In the ♂ the antennæ are 12-jointed, filiform-moniliform, yellow, the joints loosely articulated, with short bristly hairs; the pedicel is smaller than the first funiclar joint; the first three funiclar joints are elongate, the second being a little the largest, the third the smallest, the following joints to the last moniliform, briefly pedicellate, the last cone-shaped, as long as the first; legs yellow.

HABITAT.—Washington, D. C.

Types in Coll. Ashmead.

Described from many specimens reared in June from the eggs of an unknown Noctuid moth.

Telenomus gnophælæ, sp. nov.

♂ ♀. Length, 0.8ᵐᵐ. Black, shining, the thorax microscopically punctate. Head transverse, wider than the thorax, the face polished, impunctured, the vertex faintly shagreened. Eyes slightly bristly. Mandibles rufous. Antennæ ♀ 11-jointed, black, the pedicel much longer than the first funiclar joint, the second funiclar joint very slightly shorter than the first, the third not more than half the length of the second, the club much more slender than usual. Legs black; trochanters, knees, base and apex of tibiæ, and the tarsi, honey-yellow. Abdomen shorter than the thorax, broadly truncate behind, the first and second segments at base, striate, the second being broader than long. Wings hyaline, fringed, the venation brown, the marginal vein half the length of the stigmal.

In the ♂ the antennæ are 12-jointed, filiform, brown, the pedicel smaller than the first funiclar joint, the first and third funiclar joints about equal, the second very slightly longer than either of the others, the joints beyond to the last, moniliform, the last conic, twice as long as the penultimate; legs reddish-yellow, the coxæ dusky or black.

HABITAT.—Sisson, California.

Types in National Museum.

Described from many specimens reared August 24, 1890, by Mr. A. Koebele, from the eggs of *Gnophaela hopferi* Grote.

Telenomus gracilicornis, sp. nov.

♀. Length, 0.8ᵐᵐ. Head, scutellum, and abdomen polished black, shining; thorax microscopically punctate, subopaque, with a fine pubescence; scape black, the flagellum brown black, the pedicel yellowish at tip; legs black; the trochanters, knees, anterior tibiæ, except a blotch above, and the tarsi, honey-yellow. Antennæ 11-jointed, the club rather slender; the flagellum twice as long as the scape; the first funiclar joint is about two-thirds the length of the pedicel, the second a little shorter than the first but still a little longer than thick; the third and fourth moniliform, about equal in size; the first joint of club is submoniliform, transverse, a little larger than the last funiclar joint, the second, transverse-quadrate, the third and fourth quadrate, slightly longer than the second, the last pointed, conic. Wings subfuscous, the venation brown, the marginal vein punctiform, hardly twice as long as thick. Abdomen not longer than the thorax, truncate at apex, spatulate, not striate basally, the second segment scarcely longer than wide at apex.

HABITAT.—Ottawa, Canada.

Type, ♀, in Coll. Ashmead.

Described from a single specimen.

Telenomus californicus, sp. nov.

♀. Length, 0.8ᵐᵐ. Black, shining, the thorax microscopically punctate and covered with a fine pubescence. Head transverse, as wide as the thorax, the face smooth, polished. Eyes pubescent. Mandibles black. Antennæ 11-jointed, black, the pedicel distinctly longer than the first funiclar joint, the first three funiclar joints subequal, all longer than thick, the fourth moniliform, the first joint of the club small, transverse, the second, third, and fourth nearly equal, transverse-quadrate, the last short, cone-shaped. Legs black, the trochanters, knees, extreme tips of the tibiæ and the tarsi, honey-yellow. Abdomen very slightly longer than the thorax, polished, the first segment striated, the second about as long as wide. Wings hyaline, fringed, the venation pale brownish-yellow, the marginal vein one-third the length of the stigmal.

HABITAT.—Los Angeles, Cal.

Types in National Museum.

Described from 4 ♀ specimens, reared by Mr. D. W. Coquillett from the eggs of an unknown *Orgyia*.

Telenomus persimilis, sp. nov.

♀. Length, 1.5ᵐᵐ. Black, shining, the thorax distinctly punctulate, subopaque, pubescent; the head 2½ times as wide as long, polished, the vertex toward the eyes alutaceous; legs rufous, the coxæ black. Antennæ 11-jointed, black, the pedicel shorter than the first funiclar joint, its apical margin yellow, the second funiclar joint is two-thirds the length of the first, the third and fourth shorter than the second, subequal, the fourth being rounded; the club is rather slender, the first joint transverse and shorter than the following, the second, third, and fourth equal, quadrate, the last conic. Wings hyaline, fringed, the venation pale brown, the marginal vein one-third the length of the shaft of the stigmal. Abdomen oval-subtruncate at apex, a little longer than the thorax, smooth, polished, the first segment wider than long, striate, the second not longer than its width at apex, the following short, the third being twice as long as the fourth; the suture between the first and second segments is striated.

HABITAT.—Arlington, Va.

Type in Coll. Ashmead.

Described from a single specimen.

Since this was written I have seen specimens of this species reared from unknown hemipterous eggs in Michigan.

Telenomus graptæ How.

Scudd. But. New Eng., p. 1896.

♂ ♀. Length, 0.6ᵐᵐ. Black, shining. Head transverse, wider than the thorax, the face polished, the vertex subopaque. Eyes slightly bristly. Mandibles piceous-black. Antennæ in ♀ 11-jointed, black, very gradually clavate, the club less distinctly defined than usual; the pedicel a little longer and stouter than the first funiclar joint, the second and third funiclar joints moniliform, the second very slightly longer than the third, the latter a little transverse; first and second joints of club transverse, the third and fourth quadrate, the last conic. Legs piceous-brown, almost black, the trochanters, knees, tips of tibiæ, and tarsi, pale. Abdomen a little shorter than the thorax, truncate posteriorly, the first segment transverse, finely striated, the second wider than long, smooth and polished, the suture between it and the first with some striæ. Wings hyaline, fringed, the venation yellowish, the marginal vein one-third the length of the shaft of the stigma.

In the ♂ the antennæ are 12-jointed, filiform, the scape black, the flagellum brown; the pedicel is shorter than the first funiclar joint, the second and third funiclar joints about equal, and longer than the first, the joints beyond to the last, round-moniliform, slightly pedicellate, the last conic, twice as long as the penultimate. Legs brown, the anterior tibiæ, trochanters, base and apex of the tibiæ, and the tarsi, honey-yellow.

HABITAT.—District of Columbia.

Types in National Museum.

Redescribed from many specimens reared by Dr. Riley, September 16, 1886, from the eggs of *Grapta interrogationis* Fabr.

Telenomus spilosomatis, sp. nov.

♂ ♀. Length, 0.6ᵐᵐ. Polished black, impunctured, the thorax with a fine microscopic pubescence. Eyes bare. Mandibles black. Antennæ ♀ 11-jointed, the scape black, the flagellum dark brown; the pedicel is longer than the first funiclar joint; the second and third funiclar joints are equal, moniliform; the first joint of the club is transverse, the second, third, and fourth about equal, quadrate, the last cone-shaped. Legs piceous-brown, the trochanters, knees, apices of tibiæ, and the tarsi, pale brownish-yellow. Abdomen not longer than the thorax, smooth and shining, the first segment transverse, the second wider than long, the terminal segments retracted within the second. Wings hyaline, fringed, the venation pale yellowish, the marginal vein very slightly longer than half the length of the stigmal.

In the ♂ the antennæ are 12-jointed, filiform, pale brown, the pedicel distinctly longer than the first funiclar joint, the second and third funiclar joints about equal, longer than the first, the joints beyond round,

moniliform, the last cone-shaped, twice as long as the penultimate; legs
pale brownish-yellow, the coxæ black.

HABITAT.—District of Columbia.

Types in National Museum.

Described from several specimens reared by Dr. Riley, from the eggs
of *Spilosoma virginica* Fabr.

Telenomus minimus, sp. nov.

♀. Length, 0.6ᵐᵐ. Black, shining, pubescent, the dorsum of thorax
flattened, the pedicel at apex and the two last funicler joints, trochanters, knees, and tarsi, yellowish, rest of the legs pale brownish. The head
is twice as wide as long, or very slightly wider, highly polished, face
convex, eyes pubescent. Mandibles and throat brownish-yellow. Antennæ 11-jointed, the pedicel more than twice as long as the first funicler
joint, the latter not longer than thick, the three following joints exceedingly short, transverse, the last the smallest, the first joint of club short,
crescent-shaped, the second semicircular, the third and fourth quadrate,
the last ovate. Wings hyaline, pubescent, the venation pale brown, the
marginal vein about half the length of the stigmal. Abdomen scarcely
larger than the thorax, polished black, without striæ at base, the
second segment not as long as wide, the third about one-third the length
of the second, the following very short.

HABITAT.—Arlington, Va.

Type in Coll. Ashmead.

Telenomus heliothidis, sp. nov.

♀. Length, 0.6ᵐᵐ. Black, smooth, impunctured. Head large, much
wider than the thorax. Eyes not pubescent. Antennæ 11-jointed, dark
brown, the flagellum twice as long as the scape, the pedicel stout and
as long as the first and second funicler joints together, the first funicler
joint scarcely longer than thick, the second, third and fourth about
equal, not longer than thick, the fifth larger, moniliform, club 4-jointed,
the second and third joints quadrate, the last conic. Thorax very
faintly pubescent, almost bare. Wings hyaline, with a long fringe.
Legs dark brown, the tarsi pale. Abdomen not longer than the thorax,
broadly truncate behind, the first segment exceedinly short.

HABITAT.—Shreveport, La.

Type in National Museum.

Described from one ♀ specimen, reared in January, 1891, by Mr. F. W.
Mally, from the eggs of *Heliothis armigera*.

Telenomus orgyiæ Fitch.

Telenomus orgyiæ Fitch, Eighth N. Y. Rep., p. 197.
Teleas orgyiæ Ashm., Ent. Am., III, p. 100.

♂ ♀. Length, 0.8ᵐᵐ. Black, shining; the thorax microscopically
punctate, with a fine sericeous down; the head a little more than 3 times

as wide as thick antero-posteriorly, the face highly polished, the eyes pubescent; thorax highly convex; legs black or piceous-brown, the trochanters, knees, tips of tibiæ, and tarsi, pale or yellowish; in male always pale.

Antennæ in ♀ 11-jointed, black; pedicel longer than the first funiclar joint, pale at tip; second and third funiclar joints, moniliform, the second slightly the larger, third very small; in ♂ 12-jointed, filiform, hairy, brown; first three funiclar joints almost equal in length, about twice as long as the pedicel, the first the stoutest, the third a little curved; remaining joints, except the last, moniliform, loosely joined, the last conic. Wings hyaline, fringed, the venation pale brown, the marginal vein about one-third as long as the stigmal.

Abdomen oval, about as long as the thorax, black and highly polished, the first and second segments without striæ.

HABITAT.—New York and Ottawa, Canada.

Specimens in Coll. Ashmead.

Described from several specimens, in both sexes, bred by Mr. W. H. Harrington from eggs *Orgyia* sp. at Ottawa.

Telenomus ichthyuræ, sp. nov.

♂ ♀. Length, 5.6^{mm}. Black, shining, impunctured, the thorax covered with a fine microscopic pubescence. Head very wide, wider than the widest part of the thorax, the face convex, polished. Mandibles piceous. Antennæ in ♀ 11-jointed, black, the flagellum one and a half times as long as the scape, the pedicel much longer than the first funiclar joint, the latter only a little longer than thick, the second and third joints equal, moniliform, the fourth still smaller, the joints 2, 3 and 4 of club, quadrate, the last short, conic. Thorax high, convex. Legs piceous-brown, coxæ black, trochanters, knees, base, and apex of tibiæ and the tarsi honey-yellow. Abdomen not as long as the thorax, broadly truncate posteriorly, polished, the first segment transverse, thrice as wide as long, the second much wider than long and occupying most of the surface, the remaining segments scarcely visible, more or less retracted within the second. Wings hyaline, fringed, the venation pale; the marginal vein is very short, punctiform, the stigmal oblique, nearly four times the length of marginal.

In the ♂ the antennæ are filiform, 12-jointed, pale brown, the pedicel slightly longer than the first funiclar joint, the second and third stouter and longer than the first, about equal in length, the joints beyond to the last distinctly transverse, the last conic; the legs, except the black coxæ, are pale brownish-yellow.

HABITAT.—Washington, D. C.

Types in National Museum.

Described from many specimens, in both sexes, reared by Dr. Riley from the eggs of *Ichthyura inclusa* Hübn.

Telenomus infuscatipes Ashm.

Teleas infuscatipes Ashm., Ent. Am., III, p. 100; ♀ ♂.

♂ ♀. Length, 0.8ᵐᵐ. Polished black; the thorax finely pubescent; the head thrice as wide as long, the face very convex; eyes bare; antennæ brown-black, the scape pale at base; legs brown, the coxæ and middle of femora and tibiæ dusky; mandibles rufous. Antennæ ♀ 11-jointed, the flagellum one and a half times as long as the scape, the pedicel twice as large as the first funiclar joint, the second and third funiclar joints moniliform, a little transverse, the fourth very small, transverse; the first joint of club very short, crescent-shaped, thrice as wide as the last joint of the funicle; the second, third, and fourth joints very wide, transverse-quadrate, twice as wide as long, the last ovate. Wings hyaline, pubescent, the venation yellowish-brown, the marginal vein about half the length of the stigmal. Abdomen not longer or wider than the thorax, the first segment striate, the second a little wider than long.

In the ♂ the antennæ are 12-jointed, long, filiform, the first funiclar joint incrassated.

HABITAT.—Jacksonville, Fla.

Types in Coll. Ashmead.

Described from one ♂ and two ♀ specimens. The ♂ of this species is now destroyed, but it should be recognized again if rediscovered by the incrassated first joint of the funicle.

Telenomus bifidus Riley.

Rep. U. S. Dept. Agric. 1886, p. 531, ♂ ♀; Cress. Syn. Hym., p. 248.

♂ ♀. Length, 0.6ᵐᵐ. Black, shining, the thorax microscopically punctulate and covered with a fine white pubescence. Head as broad as the widest part of the thorax, highly polished, impunctured. Eyes almost bare. Mandibles brown. Antennæ ♀ 11-jointed, the scape black, the flagellum brown-black, the pedicel brownish-yellow, distinctly longer and stouter than the first funiclar joint, the second and third funiclar joints equal, moniliform, the first joint of the club small, transverse, the second longer, the third and fourth transverse-quadrate, the last cone-shaped. Legs honey-yellow, the femora slightly obfuscated, the coxæ black. Wings hyaline, fringed, the venation yellowish, the marginal vein nearly half the length of the stigmal. Abdomen slightly shorter than the thorax, polished, the first segment slightly wider than long.

In the ♂ the antennæ are 12-jointed, filiform, and of a uniform pale brownish-yellow, the pedicel scarcely longer than the first funiclar joint, the first and second funiclar joints nearly equal in length, the third slightly shorter, the joints beyond round, or moniliform, the last conic a little less than twice as long as the penultimate; legs pale yellow.

HABITAT.—District of Columbia.

Types in National Museum.

The species was reared by Dr. Riley, July 27, 1886, from the eggs of *Hyphantria textor* Harris.

Telenomus maculipennis, sp. nov.

♀ . Length, 0.6^{mm}. Polished black, impunctured, the first segment and the second, for half its length, striate; tarsi whitish.

Head transverse, very little more than twice as wide as thick antero-posteriorly. Eyes bare. Antennæ 11-jointed, the scape short, less than half the length of the flagellum, the pedicel large and much stouter than the funiclar joints, as long as the first and second funiclar joints together, pale at apex, first and second funiclar joints equal, scarcely longer than thick, the third and fourth still smaller, minute, moniliform; club fusiform, longer than the scape, the first joint transverse, the second a little wider and longer, the third still wider, quadrate, the fourth a little narrower, quadrate, the last conic. Thorax subconvex. Wings hyaline, ciliated, with a dusky band across the middle below the marginal vein; the stigmal vein short, with a spurious vein extending into the middle of the wing from the knob; marginal vein about as long as the stigmal. Abdomen about as long as the head and thorax together, the second segment a little longer than wide, with the basal half very finely, longitudinally striated.

HABITAT.—Jacksonville, Fla.

Type ♀ in Coll. Ashmead.

Described from a single specimen captured while sweeping.

The banded wings and the short stigmal vein, with a branch or uncus, from its tip, readily distinguish the species.

Telenomus sphingis Ashm.

(Plate VII, Fig. 7, ♀.)

Teleas sphingis Ashm., Bull. No. 14, Div. Ent., U. S. Dept. Agric., p. 18; Ent. Am. III, p. 100; Cress. Syn. Hym.. p. 313.

♂ ♀ . Length, 0.85 to 1^{mm}. Black, shining, the thorax very faintly microscopically punctate, finely pubescent; antennæ brown to dark brown, the scape rarely entirely black, usually pale beneath or at base and apex; legs pale brown, or brownish-yellow, the coxæ black, the femora and tibiæ more or less embrowned. Head thrice as wide as long; the eyes pubescent, the mandibles piceous or brown. Antennæ ♀ 11-jointed, the pedicel longer and stouter than the first funiclar joint, yellow at tip, the latter scarcely longer than thick, second and third joints not longer than thick, the fourth short, transverse, a little wider than the third, the club about as long as the funicle and pedicel united, the first joint transverse not so wide as the second, the second, third, and fourth, quadrate, the last conic. Wings hyaline, ciliated, the venation pale brownish or yellowish, the marginal vein a little longer than half the length of the stigmal. Abdomen black, polished, not longer than the thorax, truncate at apex, the first segment and the suture between the first and second striate; the second segment is half again as long as wide.

In the ♂ the antennæ and legs are yellow, the coxæ black, or dusky, the pedicel slightly shorter than the first funiclar joint, the first three funiclar joints equal, one and a half times as long as thick, the following to the last, moniliform, subpedicellate.

HABITAT.—Jacksonville, Fla.

Types in Coll. Ashmead.

Described from many specimens, reared from the eggs of *Sphinx carolina* Linn.

Telenomus Rileyi How.

Scudder, Butterflies New Eng., p. 1896.

♀. Length, 0.6ᵐᵐ. Black, shining, the thorax with a fine microscopic punctation and down. Head broadly transverse, slightly more than thrice as wide as long. Eyes slightly pubescent. Mandibles brown. Antennæ 11-jointed, dark brown, the scape pale at extreme base; the pedicel is a little longer than the first funiclar joint; the second, third, and fourth funiclar joints moniliform; the first joint of the club transverse, the second, third, and fourth joints about equal, transverse-quadrate, the last cone-shaped. Legs dark brown or piceous, the trochanters, knees, and tarsi, honey-yellow. Abdomen very slightly longer than the thorax, smooth, polished, the first segment striated, the second a little wider than long. Wings hyaline, fringed, the venation pale, the marginal vein punctiform.

HABITAT.—Fairbury, Ill.

Types in National Museum.

Described from 3 ♀ specimens, reared March 22, 1884, from the eggs of *Apatura clyton* by A. H. Mundt.

Telenomus gossypiicola, sp. nov.

♀. Length, 0.6ᵐᵐ. Black, shining, impunctured. Head broadly transverse, not quite thrice as wide as long antero-posteriorly. Eyes bare. Mandibles pale rufous. Antennæ 11-jointed, brown, the scape blackish above; the pedicel is slightly longer than the first funiclar joint; the second and third funiclar joints equal, moniliform; the fourth smaller, transverse; the first joint of the club is transverse, the second, third, and fourth transverse-quadrate, the fourth slightly the largest, the last cone-shaped. Legs dark brown, the trochanters, knees, tips of tibiæ and the tarsi, yellow. Abdomen slightly shorter than the thorax, smooth, polished, the first segment short, transverse, the second wider than long. Wings hyaline, fringed, the venation pale, the marginal vein about half the length of the stigmal.

HABITAT.—Bougère P. O., Concordia Parish, La.
Types in National Museum.

Described from 6 ♀ specimens, reared August 16, 1880, from lepidopterous eggs found on cotton.

Telenomus geometræ, sp. nov.

♂ ♀. Length, 0.45 ᵐᵐ. Black, shining, impunctured. Head transverse, thrice as wide as long, the face convex, highly polished. Eyes bare. Mandibles piceous-brown. Antennæ ♀ 11-jointed, dark brown; the pedicel is stout and at least twice as long as the first funiclar joint; the first funiclar joint only a little longer than thick, the second subequal with it, the third moniliform, the fourth very small, transverse; the first joint of the club is transverse, the second larger, transverse, the third and fourth larger, quadrate, the last conic. Legs brown, the posterior coxæ blackish, the trochanters, knees, tips of tibiæ and tarsi, pale. Abdomen not longer than the thorax, smooth, polished, the second segment a little wider than long, apex truncate. Wings hyaline, fringed, the venation pale yellowish, the marginal vein one-third the length of the stigmal.

In the ♂ the antennæ are 12-jointed, filiform, pale brown, the pedicel about as long as the first funiclar joint, the first three funiclar joints nearly equal, the first being slightly the shortest, the joints beyond round, moniliform, slightly pedicellate, the terminal joint conic, slightly more than twice as long as the preceding. Legs pale or yellowish, the femora and tibiæ dusky.

HABITAT.—(?) District of Columbia.
Types in National Museum.

Described from many specimens reared from the eggs of an unknown geometrid moth found on wild cherry; no date of rearing is given.

Telenomus arzamæ, sp. nov., Riley.

"♀. Length, 0.8 ᵐᵐ. Black, shining, the head and thorax with a faint microscopic punctuation and finely pubescent. Head transverse, slightly more than thrice as wide as long, polished; eyes slightly pubescent; mandibles brown; antennæ 11-jointed, brown, the flagellum darker above than beneath; the pedicel is scarcely longer than the first funicular joint, the funicular joints subequal, the first joint of club transverse, the second, third, and fourth about equal, transverse-quadrate, the last cone-shaped: legs rufous, or reddish-yellow, the trochanters and tarsi paler; sometimes the femora and tibiæ are more or less obfuscated: wings hyaline, fringed, the venation pale brownish, the marginal vein about half the length of the stigmal. Abdomen as long as the thorax, polished, the first segment striated, the second about as long as its width at apex.

"In the ♂ the antennæ are 12-jointed, filiform, pale brownish yellow, the flagellum dusky brown; the pedicel shorter than the first funicular joint; the second and third equal, a little longer than the first, the joints beyond oval-moniliform, the last conic, much longer than the preceding; legs reddish yellow or brownish yellow, the posterior coxæ dusky.

"Described from many specimens reared June 4 to 23, 1884, from eggs of *Arzama densa* Walk."—[*From Riley's MS.*]

HABITAT.—District of Columbia.

Types in National Museum.

Telenomus lavernæ, sp. nov.

♀. Length. 0.6ᵐᵐ. Black, shining, the thorax closely, microscopically punctulate. Head thrice as wide as long. Eyes bare. Mandibles piceous. Antennæ 11-jointed, brown, the scape pale rufous, the pedicel slightly longer than the first funicular joint, the funicular joints subequal, the joints of the club only a little wider than long. Wings hyaline, fringed, the venation yellowish; the marginal vein about one-third the length of the stigmal. Legs brownish-yellow. Abdomen as long as the thorax, polished, the first segment striate, the second not longer than wide.

HABITAT.—District of Columbia.

Types in National Museum.

Described from specimens reared by Dr. Riley, June 21, 1884, from from eggs of *Laverna luciferella* Clem.

Telenomus podisi, sp. nov.

♂ ♀. Length, 1ᵐᵐ. Black, shining, the thorax very finely but distinctly punctate, and with a white pubescence. Head broadly transverse, the face smooth, the vertex under a strong lens exhibiting a faintly shagreened or alutaceous surface. Eyes a little bristly. Mandibles brownish-yellow. Antennæ ♀ 11-jointed, brown, the scape and pedicel yellow, or brownish-yellow; the pedicel is distinctly longer than the first funicular joint, the second subequal with the first, the third and fourth small, moniliform; the first joint of the club is transverse, about half the length of the second, but not so wide, the second and third transverse-quadrate, the fourth quadrate, the last cone-shaped, not longer than the preceding. Legs, including coxæ, honey-yellow. Abdomen as long as the thorax, polished, the first segment striate, the second as wide as long, striate at basal suture. Wings hyaline, fringed, the venation pale yellow, the marginal vein less than half the length of the stigmal.

In the ♂ the antennæ are 12-jointed, filiform, the first three funicular joints lengthened and thickened, the first and second, about equal, the third a little shorter; the pedicel is only half the length of the first funiclar joint, the joints after the third moniliform, the last conic; legs, including coxæ, honey-yellow.

HABITAT.—St. Louis, Mo.

Types in National Museum.

Described from specimens reared from eggs of *Podisus spinosus* Dall., June 9, 1879.

Telenomus chrysopæ. sp. nov.

♂. Length, 0.6ᵐᵐ. Black, shining, the thorax microscopically punctate, with a fine down. Head transverse, thrice as wide as long. Eyes with a few hairs. Mandibles brown. Antennæ 11-jointed, brown, the pedicel hardly as long as the first funiclar joint, the second and third funiclar joints about equal, longer than the first, the joints beyond to last oval-moniliform, slightly pedicellate, the last conic, about twice as long as the penultimate. Legs dark fuscous or brown, the coxæ black or blackish, the trochanters, knees and tarsi, pale. Abdomen not as long as the thorax, spatulate, smooth, polished, without striæ, the second segment not longer than its width at apex. Wings hyaline, fringed, the venation pale brown, the marginal vein about half the length of the stigmal.

HABITAT.—District of Columbia.

Types in National Museum.

Described from 4 ♂ specimens, reared from the eggs of *Chrysopa* sp. in July.

Telenomus cœlodasidis, sp. nov.

♂. Length, 0.8ᵐᵐ. Black, shining, the thorax very faintly, microscopically punctate, with a fine, white pubescence. Head transverse, a little broader than the thorax, the vertex exhibiting a faint, shagreened punctuation, the face smooth, highly polished. Eyes pubescent. Mandibles pale brown, or yellowish. Antennæ 12-jointed, brown, the flagellum, fully as long as the body, very bristly, the pedicel very small, rounded, not half the length of the first funiclar joint, the first three funiclar joints stout and long, the first, much shorter than the second, the third, a little shorter than the second, the joints beyond oval-moniliform, the last cone-shaped, twice as long as the penultimate. Legs piceous, the coxæ black, trochanters, knees, anterior tibiæ, and tips of the others and all the tarsi, honey-yellow. Wings hyaline, fringed, the venation pale yellowish, the marginal vein about half the length of the shaft of the stigmal. Abdomen scarcely as long as the thorax, polished, the first segment and the second at base, striate, the latter wider at apex than long.

HABITAT.—Washington, D. C.

Types in National Museum.

Described from 2 specimens reared August 31, 1882, from the eggs of *Cœlodasys leptinoides* Grote.

Telenomus clisiocampæ sp. nov., Riley.

"♀. Length, 0.6ᵐᵐ. Black, shining, the thorax alone exhibiting a faint microscopic punctation. Head transverse, about thrice as wide as long; eyes faintly bristly; mandibles brown; antennæ 11-jointed, black, the pedicel much longer than the first funicular joint, the second funicular joint subequal with the first, the third very short, moniliform; the first joint of club transverse, small, the second, third, and fourth equal, quadrate, the last bluntly cone-shaped, scarcely longer than the preceding and narrower; legs dark brown, almost black, the trochanters, a small spot on knees, and the tarsi, pale; wings hyaline, fringed, the venation pale brown, the marginal vein about one-third the length of the stigmal. Abdomen longer than the thorax, pointed at apex, smooth and polished, the second segment longer than wide.

"In the ♂ of what I take to be this species, the antennæ are 12-jointed, filiform, brown, the pedicel not as long as the first funicular joint, the second and third about equal, very slightly longer than the first, the joints beyond ovoid-moniliform. The legs show much more yellow than in the ♀, the knees, broadly, base and apex of tibiæ and tarsi, yellow. The first and second abdominal segments are striated at base; while the second is almost twice as long as wide.

"Described from one ♀ reared by Albert Koebele from the eggs of a *Clisiocampa* sp. in California, and one ♂ reared by C. F. Waters in Nebraska, March 23, 1889, from the eggs of *Clisiocampa americana* Harris." [*From Riley's MS.*]

HABITAT.—Placer County, Cal., and Westerville, Nebr.

Types in National Museum.

? Telenomus pennsylvanicus, sp. nov.

♀. Length, 1.2ᵐᵐ. Brown-black, opaque, coarsely cribrately punctate; the face with a deep impression just above the antennæ and with a carina between the antennæ. Antennæ 12-jointed, brownish-yellow, the club fuscous; the pedicel is longer than the first funicular joint, the second subequal with the first, joints 3 and 4, scarcely longer than thick, the club 5-jointed. Mesopleura cribrate punctate, with a rufous margin. Wings hyaline, pubescent, the marginal vein a little longer than half the length of the stigmal.

Abdomen broadly oval, sessile, the first segment transverse with longitudinal raised lines, the following segments coarsely shagreened.

HABITAT.—Pennsylvania.

Type, ♀, in Berlin Museum.

Described from a single specimen labeled "Penn. Zimmermann." This species is at once distinguished from all others by the cribrate punctuation, and it may ultimately form the type of a new genus. My recollection of it is that it closely resembled an *Hadronotus*, and it is placed here doubtfully.

Telenomus rufoniger Prov.

Add. et corr., p. 103.

♂. Long. 10 pce. D'un noir brillant en dessus, le scape des antennes, la poitrine, les pattes avec la base de l'abdomen d'un jaune plus ou moins roux. Ailes avec l'humérus se confondant avec la nervure costale, ulna élargie, ayant l'apparence d'un second stigma; cellule radiale ouverte, la partie antérieure de l'aile plus ou moins obscurcie de roussâtre. Abdomen assez court, claviforme, subsessile. Cap Rouge.

Unknown to me. The species is certainly not a true *Telenomus*, and appears to me, judging from his description, a Ceraphronid.

TRISSOLCUS Ashmead, gen. nov.

(Type *T. brochymena*, Ashm.)

Head very large, transverse, much broader than the thorax, the frons convex, the occiput deeply concave, the upper edge of which is sharp. Ocelli 3, triangularly arranged, widely separated, the lateral a little distant from the margin of the eye, and connected with it by an oblique grooved line. Eyes large, subovate.

Antennæ inserted close to the mouth, in ♀ 11-jointed, clavate, the funicle 3-jointed, the club 6-jointed, in ♂ 12-jointed, filiform, with the flagellar joints moniliform or submoniliform, pubescent.

Maxillary palpi 3-jointed.

Mandibles bifid.

Thorax short, ovoid, convex, the collar not visible from above, the mesonotum with 3 furrows abbreviated anteriorly, the scutellum semicircular, the metathorax very short.

Front wings pubescent, the marginal vein usually short, the stigmal vein rather long, oblique; postmarginal long.

Abdomen broadly oval, depressed, subsessile, the second segment the longest, but always broader than long.

Legs as in *Telenomus*.

This genus is closely allied to *Telenomus*, but is readily distinguished by the three abbreviated mesonotal grooves, broader head, and the wide second abdominal segment.

It seems to be parasitic only on the eggs of plant bugs belonging to the family *Pentatomidæ*.

Our species may be thus tabulated:

TABLE OF SPECIES.

Coxæ and femora black.
  Trochanters, tips of femora, tibiæ, and tarsi honey-yellow.
    Antennæ wholly black, rarely with the distal ends of scape pale.
      First funicular joint as long as, or a little longer than, the pedicel.
        Second abdominal segment above longitudinally aciculated; scutellum finely punctate ....T. EUSCHISTI Ashm.
        Second abdominal segment above smooth, not aciculated; scutellum smooth, polished...T. PODISI, sp. nov.

First funiclar joint shorter than the pedicel.
　Thorax finely and closely punctate.
　　Second abdominal segment longitudinally aciculated.
　　　Scutellum smooth, polished. . . . . .T. THYANTÆ, sp. nov.
　Thorax rugose.
　　Second abdominal segment perfectly smooth.
　　　Scutellum punctate or rugose. . . . . .T. MURGANTIÆ, sp. nov.
Legs and scape yellow, or pale ferruginous; coxæ black.
　Pedicel longer than the first funiclar joint.
　　Scutellum rugoso-punctate, subopaque. . . . . .T. RUFISCAPUS, sp. nov.
　　Scutellum smooth, polished. . . . . .T. BROCHYMENÆ Ashm.

Trissolcus euschisti Ashm.

Telenomus euschistus Ashm., Bull. No. 3, Kans. Ex. Sta., App., p. ii (1888).

♀. Length, 1.5ᵐᵐ. Black, shining, very finely closely punctate, the thorax with a white pubescence. Head very large and broad, the face with a median furrow. Mandibles piceous-black. Palpi pale. Antennæ 11-jointed, dark brown, the scape at base and tip, the pedicel and one or two funiclar joints more or less pale brown or yellowish; the pedicel is as thick and as long as the first funiclar joint, the latter thrice as long as thick, the two following joints small, the club slender, fusiform, 6-jointed. Mesonotum with 3 abbreviated furrows posteriorly; scutellum smooth, shining, impunctured. Wings hyaline, the venation pale brown or yellowish, the marginal vein about one-third the length of the stigmal. Legs black; the trochanters, tips of femora, tibiæ, and tarsi, honey-yellow or brownish-yellow, the tibiæ sometimes dusky or fuscous at the middle. Abdomen broadly oval, polished, as long as the thorax, the first segment striate.

HABITAT.—Manhattan, Kans.

Types in Kansas State Agricultural College and Coll. Ashmead.

Described from many specimens reared by Prof. E. A. Popenoe, from the eggs of *Euschistus servus* Say.

Trissolcus podisi, sp. nov.

♀. Length, 1 to 1.2ᵐᵐ. Black, subopaque, the thorax microscopically punctate; head broad, smooth, and polished, with a few punctures surrounding the orbits: mandibles black. Antennæ 11-jointed, black, the pedicel not longer than the first funiclar joint, the second and third funiclar joints transverse-quadrate: the club large, pointed at tip, the joints transverse. Thorax with 3 abbreviated furrows posteriorly; scutellum smooth, polished, impunctured. Wings hyaline, finely pubescent; the venation fuscous, the marginal vein about half the length of the shaft of the stigmal. Legs black, trochanters, knees, tips of tibiæ and tarsi, pale or yellowish. Abdomen broadly oval, about as large as the thorax, polished, the first segment striate, the second about thrice as wide as long, with only faint traces of aciculations at base.

HABITAT.—Philadelphia, Pa.

Types in National Museum.

Described from four specimens received from Mr. E. T. Cresson, reared from the eggs of *Podisus spinosus* Dallas.

Trissolcus thyantæ, sp. nov.

♀. Length, 0.8 to 1ᵐᵐ. Black, subopaque, closely, finely, minutely punctulate; mandibles piceous; legs black, tips of all the femora and the tibiæ and tarsi, honey-yellow. Antennæ 11-jointed, black, the pedicel shorter than the first funiclar joint, the second quadrate, the third very small, transverse.

Thorax with 3 abbreviated impressed lines posteriorly; scutellum smooth, polished. Wings hyaline, pubescent, the venation pale brownish-yellow, the marginal vein about one-third the length of the shaft of the stigmal. Abdomen broadly oval, truncate at tip, the first segment, and the second at base, coarsely striate.

HABITAT.—Selma, Ala.

Types in National Museum.

Described from specimens reared by Mr. E. A. Schwarz from the eggs of *Thyanta custator*, but which were wrongly determined as those of *Podisus spinosus*. The species comes near *T. podisi*, but differs in size, color of the legs, punctuation, and in venation.

Trissolcus murgantiæ, sp. nov.

♂♀. Length, 1 to 1.4ᵐᵐ. Black, rugose, the abdomen smooth, polished; first abdominal segment striate; trochanters, knees, distal ends of tibiæ, and tarsi, dark honey-yellow; wings hyaline, the venation pale brownish, the marginal vein about as long as the stigmal. The female is the larger, with the face rugose or closely punctate; antennæ 11-jointed, wholly black or brown-black, the pedicel a little longer than the first flagellar joint; funiclar joints 2, 3, and 4 transverse, the third the largest; club 5-jointed, slightly wider than the last funiclar joint, the last joint minute; mandibles black.

The male averages only 1ᵐᵐ in length, with the face almost smooth or only faintly punctate; antennæ 12-jointed, filiform-moniliform, the scape brownish-yellow, flagellum brown-black, with all the joints, except the first and last, moniliform; the last fusiform, twice as long as the penultimate; mandibles rufo-piceous.

HABITAT.—Baton Rouge, La.

Types in National Museum.

Described from several specimens reared by Mr. H. A. Morgan, from eggs of the cabbage bug, *Murgantia histrionica* Hahn. Comes nearest to *T. thyantæ* and *T. rufiscapus*, but is distinguished from them by sculpture, color, and differences in the antennæ.

Trissolcus rufiscapus, sp. nov.

♀. Length, 1.4ᵐᵐ. Black, subopaque, very finely punctate; front shining, finely, feebly punctate: a large fovea above the clypeus; legs rufous; the coxæ black.   Antennæ 11-jointed, black, the scape rufous, slightly dusky at tip above, the pedicel nearly as long as the first and second funiclar joints together, the second and third funiclar joints transverse; the club 5-jointed, not quite as long as the scape.   Thorax minutely, closely punctulate, slightly pubescent, with three abbreviated furrows posteriorly; the scutellum rugoso-punctate, slightly lustrous, pubescent.   Wings subhyaline, pubescent, the nervures pale brownish, the marginal vein very short.   Abdomen broadly oval, depressed, polished, the first segment striate, the second, twice as wide as long.

HABITAT.—Washington, D. C.

Type in National Museum.

Trissolcus brochymenæ Ashm.

(Pl. VII, Fig. 6, ♀.)

Telenomus brochymenæ Ashm., Fla. Agric., IV, 1881, p. 193; Ent. Am., III, p. 118; Cress. Syn. Hym., p. 344.

♀.   Length, 0.8 to 0.9^{mm}.   Black, shining, the thorax feebly, microscopically punctulate, with a fine pubescence; head smooth, fully four times as wide as thick antero-posteriorly, a fine punctulate line surrounding the orbits.   Ocelli very widely separated, the lateral close to the eye margin on an oblique grooved line extending from the eye to the occiput.   The occiput concave, the upper edge sharp; cheek flat. Eyes pubescent.   Antennæ 11-jointed, the scape, pedicel, and first funiclar joint, brownish-yellow, the following joints brown-black; the scape does not extend above the front ocellus; the pedicel is longer and thicker than the first funiclar joint, the latter about twice as long as thick, joints 2 and 3 transverse, the club large, 6-jointed, all the joints, except the last, transverse.   Thorax rounded, the mesonotum convex, wider than long, with 3 abbreviated grooved lines posteriorly.   Legs honey-yellow, the coxæ black.   Abdomen small, flattened, not longer than the thorax, truncate behind, black, shining, the first segment transverse, striated, separated from the second by a deep suture, the second occupying most of the remaining surface.   Wings hyaline, the venation yellow, the marginal vein punctiform, not longer than thick.

HABITAT.—Jacksonville, Fla.

Types in Coll. Ashmead.

Described from several specimens reared from the eggs of *Brochymena arborea* Say.   Specimens are also in the National Museum, reared from the eggs of an unknown Hemipteron by Mr. H. G. Hubbard.

DISSOLCUS Ashmead, gen. nov.

(Type *D. nigricornis* Ashm.)

Head transverse or subquadrate, not wider than the thorax, the occiput flat but not concave or emarginate.   Ocelli 3, in a triangle, the lateral a little away from the margin of the eye.   Eyes ovate, pubescent.

Antennæ inserted close to the clypeus, in ♀ 11-jointed, the flagellum very gradually increasing in thickness toward the tip, or subclavate; the pedicel is larger than the first funiclar joint, which is a little longer than thick, the second shorter, the third and following joints, transverse, or wider than long. ♂ unknown.

Thorax ovoid, the mesonotum scarcely longer than wide, subdepressed, with 2 short impressed lines posteriorly, abbreviated anteriorly, the scutellum flattened, the metathorax short, rounded behind.

Front wings as in *Telenomus* except the marginal nervure is as long as the stigmal.

Abdomen subsessile, depressed, ovate, about as long as the head and thorax together, subacute at tip, the second segment nearly twice as long as wide.

Legs as in *Telenomus* except the tarsi are distinctly longer than the tibiæ.

In the shape of the abdomen and in antennal and thoracic characters this genus approaches nearest to *Phanurus*, but otherwise it is like *Telenomus*.

The broader wings and the longer marginal nervure will distinguish it from the former, while its cephalic, antennal, and thoracic characters readily separate it from the latter, the two abbreviated mesonotal furrows being found in no other genus. It bears not the slightest resemblance to *Trissolcus*, which has a very large, broad head, convex thorax, a broadly oval abdomen, and three abbreviated mesonotal furrows.

Only a single species is known.

Dissolcus nigricornis, sp. nov.

(Pl. VII. Fig. 8, ♀.)

♀. Length, 0.8ᵐᵐ. Black, subopaque, the surface finely alutaceous, subpubescent; head hardly twice as wide as thick antero-posteriorly, the face convex, smooth, shining. Eyes pubescent. Antennæ 11-jointed, very gradually subclavate, the club not distinctly separable; the flagellum is twice the length of the scape; the pedicel stouter than the first three or four funiclar joints, twice as long as thick; the first funiclar joint longer than thick, the second, not longer than thick, the following transverse and very gradually widening to the tip, the last conic or ovate. Thorax ovoid, the mesonotum scarcely longer than wide, a little depressed, with 2 delicate abbreviated impressed lines posteriorly. Wings hyaline, iridescent, the nervures pale brown, the marginal as long as the oblique stigmal. Legs black, the knees and tarsi, brown. Abdomen black, polished, as long as the thorax, the first segment striate, the second longer than wide.

HABITAT—Jacksonville, Fla.

Type in Coll. Ashmead.

Described from a single specimen taken while sweeping.

ARADOPHAGUS Ashm., gen. nov.

Head oblong, very flat, attached to the thorax high up on the occiput, the space between the eyes very wide. Ocelli 3, subtriangularly arranged, the lateral far from the eye. Eyes oblong-oval, bare.

Antennæ inserted at the mouth, subclavate, 12-jointed in both sexes; the scape very long, cylindrical; the pedicel long, one-third longer than the first flagellar joint, the following joints shorter but stouter, subequal.

Maxillary and labial palpi very short, inconspicuous, 2-jointed.

Mandibles small, bifid.

Thorax ovoid, very flat, the prothorax not at all developed or visible from above; mesonotum smooth, without furrows, rounded before; scutellum short, semicircular, with a very delicate cross-line before the tip; metathorax short, with delicate lateral keels.

Front wings fringed as in *Cerocephala*, the submarginal vein reaching the costa before attaining half the length of the wing, the marginal vein longer than the oblique stigmal vein, while the post-marginal vein is well developed.

Abdomen very flat, ovate, with a short petiole, the second segment the largest, about twice the length of the first, the third and fourth about equal, two-thirds the length of the second; ovipositor exserted. Legs slender.

This genus in its thoracic and antennal characters is quite distinct from all others in the group, and requires no special comment. It shows, strongly, affinities with the *Cerapbroninæ* and was originally placed there in my collection; it also resembles a "Spalangiid," and caused me, at one time, to contemplate removing the *Spalangiinæ* from the *Chalcididæ* to the *Proctotrypidæ*.

Aradophagus fasciatus, sp. nov.

(Pl. VII, Fig. 9, ♀.)

♂ ♀. Length, 1.5ᵐᵐ. Honey-yellow, smooth, polished, impunctured. Scape, pedicel, and basal half of first funicular joint pale or whitish, the rest of the antennæ, brown-black. Front wings fuscous, the base, a transverse band at the middle, and the apical margin white; hind wings hyaline. Abdominal sutures narrowly banded with fuscous.

HABITAT.—Jacksonville, Fla.

Types in National Museum and coll. Ashmead.

The two specimens in my collection were taken by myself in April, 1887, under live oak bark badly infested with *Brachyrhynchus granulatus* Say and *Pityophthorus querciperda* Swz., the old egg masses of the Aradid being quite plentiful. The single specimen in the National Museum was taken by Mr. E. A. Schwarz in Florida, who thinks he reared it from *Pityophthorus consimilis* Lec., together with several Cerocephalæ. I believe its habits will prove identical with other Scelionids, and it will be found to be a parasite in the eggs of the *Aradidæ*.

Tribe II.—BÆINI.

This tribe is of small extent, apparently, and for the most part, comprises species of the smallest size, the majority of which rarely attain one millimeter in length.

The habitus of the species more closely resembles the Telenomini, but the solid antennal club of the female, the difference in venation of the winged forms, the marginal vein being punctiform, or short and stout, the stigmal vein short, thickened at base and terminating in a small rounded stigma, while the basal nervure is usually present, and the third abdominal segment, not the second, being the largest, readily distinguish them. These differences and the position of the lateral ocelli will also separate them at once from the other tribes.

The tribe *Bæini*, as at present defined, will contain five genera: *Thoron* Hal., *Acolus* Först., *Acoloides* How., *Ceratobæus* Ashm., and *Bæus* Hal., the parasitism of three of which is known.

Species in the genera *Acolus, Acoloides,* and *Bæus* have been reared in America from spider eggs; and I should not be in the least surprised to learn, from future observations, that the whole group confine their attacks to Arachnid eggs.

The genera may be distinguished by the aid of the following table:

TABLE OF GENERA.

FEMALES.

Apterous forms... 2
Winged.
  First abdominal segment without a horn at base.
    Mesonotum with 2 furrows; first abdominal segment petioliform; eyes bare; lateral ocelli distant from the margin of the eye......THORON Hal.
  First abdominal segment with a horn at base.
    Mesonotum without furrows..................CERATOBÆUS Ashm., gen. nov.
2. Scutellum distinct.
  Mesonotum without furrows; lateral ocelli close to the eye.
    First abdominal segment with a horn at base ....CERATOBÆUS.
  First abdominal segment without a horn at base.
    First abdominal segment as broad as the metathorax and only visible as a transverse line; face with an antennal impression, the occiput concave, the superior margin sharp.....................ACOLUS Först.
    First abdominal segment subpetiolate, much narrower than the metathorax; face not or but slightly impressed, the superior margin of the occiput rounded ....ACOLOIDES How.
  Scutellum wanting ....BÆUS Hal.

MALES.

Winged.
  Mesonotum with two furrows; lateral ocelli distant from the margin of the eye.
    Antennae filiform, the flagellar joints about three times as long as thick.
                                                        THORON Haliday.

Mesonotum without furrows; lateral ocelli close to the margin of the eye.
  Antennæ filiform, moniliform, or submoniliform.
    Basal nervure distinct; mandibles bifid.
      Head much broader than the thorax; eyes bare; antennæ tapering toward
        apex; basal abdominal segment as wide as the metathorax.
                                            ACOLUS Förster.
      Head subquadrate, but slightly broader than the thorax; antennæ slightly
        thickened toward apex; basal abdominal segment petioliform, much
        narrower than the metathorax...................BAEUS Haliday.
    Basal nervure wanting; mandibles tridentate.
      Head transverse, scarcely broader than the thorax; eyes hairy; antennæ
        slightly thickened at tip; basal abdominal segment transverse, not
        quite as wide as the apex of metathorax.......ACOLOIDES Howard.

THORON Haliday.

Ent. Mag., I, p. 271 (1833); Först. Hym. Stud., II, p. 100 (1856).

(Type *T. formicatus* Nees. = *T. metallicus* Hal.)

Head rather large, transverse, the occiput and cheeks margined;
ocelli 3, triangularly arranged, and wider apart than in *Prosacantha;*
eyes large, oval.

Antennæ inserted close to the clypeus, 12-jointed in both sexes; in ♀
clavate, the club solid, although showing traces of sutures; in ♂ long,
filiform, the pedicel small, oval, the flagellar joints nearly equal, about
4 times as long as thick, pubescent.

Maxillary palpi 4-jointed; labial palpi 2-jointed.

Mandibles tridentate.

Thorax ovoid, the prothorax visible from above as an arcuate line,
depressed anteriorly; mesonotum with 2 impressed lines; metathorax
very short, rounded at the sides.

Front wings with the submarginal vein joining the costa behind the
middle; the marginal vein very short, thickened; postmarginal vein
not, or scarcely, developed; the stigmal vein longer than marginal, thickened at base, and terminating in a little knob.

Abdomen oval or oblong-oval, strongly narrowed at base, the first
segment longer than wide, the third the longest.

Legs long, slender, pilose, the femora clavate, the tibiæ subclavate,
with spurs 1, 1, 1; the tarsi very long and slender.

Thoron pallipes Ashm.

(Pl. XVI, Fig. 3, ♀.)

Ent. Am., III, p. 99.

♀. Length, 2ᵐᵐ. Black, polished; head subquadrate, smooth, the
face convex; mandibles bluntly 3-dentate at tips, pale rufous. Mesonotum sparsely punctate, with 2 furrows posteriorly; metathorax and
pleura rugose. Antennæ 12-jointed, dark brown, the scape and pedicel brownish-yellow. Legs, including coxæ, pale or yellowish. Abdomen oblong-oval, a little longer than the head and thorax together,
attached high up on the metathorax, the first segment petioliform,

scarcely as long as the second, a little humped basally; first and second segments striated; rest of the abdomen smooth, polished, sparsely hairy toward the apex. Wings hyaline, the venation brown, the marginal vein very short, thick, the stigmal vein short, stout at base.

HABITAT.—Jacksonville, Fla.

Type in Coll. Ashmead.

This is the only true *Thoron* in our fauna, the species described as such, *Thoron opacus* How., Ins. Life, Vol. II, p. 268, supposed to have been reared from the Fluted Scale, *Icerya purchasi*, being a ♂ Telenomid and belonging to the genus *Phanurus*.

ACOLOIDES Howard.

Ins. Life, II. p. 269 (1890).

(Type *A. saitidis* How.)

Head transverse, wide, the frons convex; ocelli 3, triangularly arranged, but widely separated, the lateral being close to the margin of the eye; eyes large, oval, hairy.

Antennæ inserted just above the clypeus; in ♀ apparently but 7-jointed, the club being large and inarticulate, the pedicel lengthened, the last three funiclar joints, small, transverse; in ♂ 12-jointed, filiform, submoniliform.

Maxillary palpi 4-jointed; labial palpi 2-jointed.

Mandibles 3-dentate.

Thorax oval, the prothorax not visible from above; mesonotum without furrows; metathorax with the posterior angles subacute.

Front wings with the marginal vein punctiform, the postmarginal not, or scarcely, developed, the stigmal vein long, oblique, thickened at base.

Abdomen short, oval, the first and second segments short, usually striated, the first much narrower than the metathorax or subpetiolate, the third very large, occupying half, or a little more than half, the whole surface.

Legs moderate, the tibial spurs very weak, scarcely developed, the tarsi 5-jointed, slender, the basal joint of posterior tarsi twice the length of the second.

This genus may be identical with *Acolus* Förster, as species occur in it with and without wings, the apterous species fitting exactly into the brief diagnosis of the genus by Förster; but as Förster's type *Acolus picivcntris*, so far as I know, was never described, and as I have discovered another wingless form closely allied to *Acoloides*, which is evidently quite distinct, that will also fit into Förster's brief description, I have here, in my perplexity, made the latter the type of Förster's genus, so as to enable me to retain the well-characterized Howardian genus. Kirchner's description of *Acolus luteus*, which might assist me, I have not been able to see.

The difference between the two genera, as here understood, is sufficiently brought out in my table, and I must leave to others the thankless task, by the inspection of the type, if it is still in existence, of the proper identification of Förster's genus.

Mr. Howard's genus is parasitic on spiders. The species may be synoptically represented as follows:

TABLE OF SPECIES.

Wingless or subapterous species 3
Winged.
  Species not entirely black...2
  Species, except sometimes the base of the abdomen, entirely black.
    Opaque, closely, minutely punctate.
      Thorax and scutellum flat.
        Antennae and legs, except coxae, yellow ....A. SAITIDIS How.
    Lustrous, but still microscopically punctulate.
      Thorax and scutellum more convex.
        First and second abdominal segments striate.
        Petiole pale rufous or yellowish ....A. HOWARDII, sp. nov.
        Petiole and base of abdomen yellowish ....A. EMERTONII How.
2. Head and thorax black, microscopically punctate.
  Abdomen at base and disks of two or three of the following segments, yellow.
                      A. EMERTONII How.
  Abdomen yellow, the third segment across the base and the two apical segments
    fuscous ....A. BICOLOR, sp. nov.
  Wholly honey-yellow, eyes and ocelli, black or brown....A. MELLEUS, sp. nov.
3. Head and thorax black, subopaque.
  With wing scales; abdomen black, sericeous, the petiole yellow.
                      A. SUBAPTERUS, sp. nov.
  Without wing scales; abdomen brownish-yellow ....A. SEMINIGER, sp. nov.

Acoloides saitidis How.

Ins. Life, II, p. 270, fig. 58; loc. cit., p. 359.

" ♀. Length, 1.4^{mm}; expanse, 3.6^{mm}; greatest width of fore wing, 0.46^{mm}. Antennae short; pedicel long, nearly one-half the length of scape; joint 1 of funicle one-half as long as pedicel; joints 2, 3, and 4 very short; club very large, oval, and one-third longer than four preceding joints together; no articulations can be distinguished, but it is homologically composed of six joints. Eyes hairy; lateral ocelli touching the eye margin. Head, face, and mesonotum densely and finely punctate; parapsidal furrows not present; first and second abdominal segments with fine, close, longitudinal striae, wanting at smooth posterior border; the very large third segment and short fourth, densely and finely punctate, and clothed irregularly with short, whitish pile, which is also present, although sparser, upon the mesonotum, and is quite thick on the vertex; mesopleura finely punctate below; metapleura smooth. The marginal vein is very short, and not quite coincident with costa; the post marginal is extremely short; the stigmal is long and slender, and terminated by a small rounded knob. General color deep black; all legs and antennæ honey-yellow; all coxæ black, litter at tips; scape brownish and pedicel darker than club.

"♂. Differs from female only in antennæ, which are plainly 12-jointed; joint 1 of funicle as long as pedicel, joints 2 to 7 subequal in length and width, and each as broad as long, and well separated; club oval, nearly as long as three preceding joints together. Antennæ uniformly honey-yellow." (*Howard.*)

HABITAT.—Lincoln, Nebr., and Oxford, Ind.

Types, 9 ♂ and 1 ♀ in National Museum.

The types of this species, as recorded by Mr. Howard, were reared by Mr. L. Bruner, at Lincoln, Nebr., from the eggs of the Araneid *Saitis pulex.* "The eggs of this spider are a little more than a millimeter in circumference, and each egg harbors but one parasite, which issues by splitting the eggs open rather than by gnawing a regular hole." The same species was also bred by Mr. F. M. Webster, at Oxford, Ind., from a spider egg-sac found under the bark of a log, in October, 1884.

Acoloides Howardii, sp. nov.

♀. Length, 1^{mm}. Black, shining, with a microscopic white pile, and feebly microscopically punctate; petiole and legs, brownish-yellow, the coxæ black. Antennæ 7-jointed, brown, the pedicel large, the first funiclar joint very little longer than thick, scarcely one-third the length of the pedicel, the three following joints transverse, the last the widest, club large, not jointed. Scutellum semicircular, subconvex. Mesopleura with a long, femoral furrow, crenate at bottom. Metapleura divided into two parts by the spiracular furrow, the upper portion smooth, impunctate, the lower portion punctate. Wings subhyaline, the nervures brown, the stigmal vein long, thickened at base. Abdomen broadly oval, shining, but microscopically punctate, and finely pubescent, the first segment twice wider than long, striated and yellow; the second and following segments black, the second with striæ at base.

HABITAT.—Washington, D. C.

Types in Coll. Ashmead.

Described from two specimens.

This species is closely allied to *A. saitidis* How., but it is slightly smaller, and not so densely punctate, the scutellum shorter, subconvex, the petiole yellow, while the first flagellar joint is scarcely one-third the length of the pedicel.

Acoloides Emertonii How.

Ins. Life, IV, p. 202.

♀. Length, 1.4^{mm}; expanse, 1.5^{mm}. Black, shining, but closely microscopically punctulate; antennæ brown-black, the scape pale at extreme base; legs, including coxæ, brownish-yellow; abdomen mostly yellow, the second segment, lateral and apical margins of third, and the following segments, fuscous or black. Head very wide, more than three times as wide as thick antero-posteriorly; eyes large, rounded, whitish (after death) and pubescent; mandibles pale rufous, the tips black. Antennal club large, fusiform, as long as the pedicel and funicle united; first funicle joint less than half the length of the pedicel, the other joints transverse. Thorax convex, with two punctate lines in front of the scutellum, the latter semicircular, convex; metanotum very short, striated, bounded by a carina posteriorly, the angles produced into a minute tooth. Wings subhyaline, pubescent, extending beyond tip of abdomen, the venation dark brown, the marginal vein short, stout, very little longer than thick, the stigmal long, slender, ending in a small knob. Abdomen oblong-oval, one-third longer than the head and thorax together, microscopically sculptured, but lustrous, the first and second segments about equal in length, striated. (*Howard.*)

HABITAT.—Massachusetts.

Type in National Museum.

Described from seven ♀ specimens, reared by Mr. J. H. Emerton, from a spider's cocoon.

Acoloides bicolor, sp. nov.

♀. Length, 0.8^{mm}. Black, shining, sparsely, faintly microscopically punctate and pubescent; the abdomen, antennæ, and legs, yellow, the third abdominal segment across the base, two apical segments and the femora slightly fuscous or brownish; eyes large, rounded, whitish. Antennæ 7-jointed, very short; the club large, conic-ovate, as long as the pedicel and funicle together; the first funicular joint not longer than thick, about one-fourth the length of the pedicel, the three following joints minute, transverse. Wings hyaline, with a faint yellowish tinge, the nervures dark brown, the stigmal vein thickened at base. Abdomen rotund, the petiole short, striated, rest of the segments smooth, shining.

HABITAT.—Ottawa, Canada.

Type in Coll. Ashmead.

Described from two specimens received from Mr. W. Hague Harrington. This species is much smaller than the others, the surface smoother, less distinctly punctate, and the abdomen more rotund; while the color of the eyes and abdomen, and the shortness and relative length of the joints of the antennæ, render it easy of recognition.

Acoloides melleus, sp. nov.

♀. Length, 1.4^{mm}. Uniformly honey-yellow, the eyes and ocelli brown, the wings subfuliginous. The head is very large and wide, the lower part of face being longitudinally striated. Antennæ short, 7-jointed, the club large fusiform, inarticulated; the first funicular joint is twice as long as thick and fully half as long as the pedicel; the second and third are as long as wide; the fourth, transverse.

Abdomen oval, the first segment and the second, at the suture, striated, the following segments feebly, microscopically punctate.

HABITAT.—Arlington, Va.

Type in Coll. Ashmead.

Described from a single specimen.

In color, size, width of head, non-pubescent eyes, and the relative length of the antennal joints, this species is very distinct from any other species in our fauna.

Acoloides subapterus, sp. nov.

♀. Length, 0.8ᵐᵐ. Head and thorax black, minutely punctulate; abdomen fuscous or dark brown, the petiole yellow; antennæ and legs brownish-yellow, the club fuscous.

Antennæ 7-jointed; the club large, inarticulate, as long as the pedicel and funicle united; pedicel stout, 2½ times as long as thick, fuscous above; first funiclar joint very little longer than thick; second and following joints scarcely as long as thick. Thorax with slight indication of parapsidal furrows posteriorly in the form of two very short grooved lines. Wings present in the form of wing pads which do not extend beyond the tip of the short metathorax.

Abdomen oblong-oval, much longer than the head and thorax together, shining, sericeous; the petiole yellow and separated from the second segment by a constriction; the petiole and second segment are of an equal length, but the latter is fully twice as wide as the former; third segment nearly thrice as long as the second.

Habitat.—Ottawa, Canada.

Type ♀ in Coll. Ashmead.

Described from a single specimen received from Mr. W. H. Harrington.

Acoloides seminiger, sp. nov.

♀. Length, 0.8ᵐᵐ. Head and thorax black, closely finely punctate, with a sparse, sericeous pubescence; abdomen and legs, brownish yellow, pubescent.

Antennæ 7-jointed, yellowish, the club very large, inarticulate, fuscous, pedicel stout, very little more than twice as long as thick; first funiclar joint scarcely longer than thick; second and third, not longer than thick; fourth, transverse. Thorax with slight indications of grooves posteriorly. Metathorax with subacute angles. Abdomen oblong-oval, a little longer than the head and thorax together, the second segment slightly longer than the petiole, the suture between, striate, the third segment 2½ times as long as the second.

Habitat.—Ottawa, Canada.

Types in Coll. Ashmead.

Described from two specimens received from W. Hague Harrington. This species comes very close to A. subapterus, but the abdomen is yellow, the relative length of the segments is different, and there are no wing pads.

ACOLUS Förster.

Hym. Stud., II, pp. 100 and 102 (1856).

(Type *A. piceiventris* Först. MS.)

Head large, transverse, the frons very convex, the face with an impression just above the antennae, the occiput concave, emarginated, and faintly margined; ocelli 3, very minute, the lateral close to the margin of the eye, but scarcely discernible; eyes large, oval, bare, or hairy.

Antennae inserted just above the clypeus, in ♀ 7-jointed, clavate, the club large, inarticulated, the pedicel stout, the first funiclar joint a little longer than thick, the three following joints small; in ♂ 12-jointed, filiform, pilose.

Maxillary palpi 3-jointed; labial palpi (?) 2-jointed.

Mandibles bifid.

Thorax subovoid, not wider than long, and narrower than the head; pronotum not visible from above; mesonotum transverse without furrows; scutellum semicircular; metathorax extremely short, abrupt.

Wings in ♀ wanting; in ♂ present, the front wings with a basal nervure, a short marginal and short stigmal nervure; the latter clavate.

Abdomen broadly oval, sessile, the first segment the width of the metathorax and only visible as a transverse line, the second about one-third the length of the third, which is the largest, the following segments short.

Legs as in *Bæus*, the basal joint of hind tarsi one-third longer than the second, the tibial spurs weak.

Differs from *Acoloides* Howard, in the concave emarginated occiput, in the deep facial impression, bare eyes, 3-jointed maxillary palpi, and in the broadly oval sessile abdomen, the base being as wide as the metathorax, the first segment-visible only as a transverse line.

Acolus xanthogaster, sp. nov.

♀. Length, 0.75ᵐᵐ. Head and thorax black, shining, faintly microscopically punctate; antennae, abdomen, and legs, yellow. Head a little more than twice as wide as thick antero-posteriorly, the frons convex, the face with an antennal impression, the eyes large, rounded, bare. Antennae with an unjointed club; the pedicel longer than the first funiclar joint; first and second funiclar joints about equal, longer than thick; third and fourth, minute. Abdomen oval, sessile, the first segment only visible as a transverse line, the second about two-thirds the length of the third.

HABITAT.—District of Columbia.

Type in Coll. Ashmead.

Described from a single specimen collected by Mr. E. A. Schwarz, June 29, 1891.

Acolus Zabriskiei, sp. nov.

(Pl. XV, Fig. 7. ♂.)

♂. Length, 1ᵐᵐ. Black, polished, finely sericeous; mandibles reddish; scape honey-yellow; flagellum pale brown; legs, including coxæ, reddish-yellow. Head very large and broad, a little more than three times as wide as thick antero-posteriorly, and much wider than the thorax, the space between the eyes alone fully as wide as the thorax; occiput concave, with the superior margin sharp; eyes broadly oval, convex, bare; ocelli triangularly arranged, the laterals situated in the posterior angle of the vertex, close to the eye. Antennæ 12-jointed, filiform; the scape clavate, about one-third as long as the flagellum; pedicel cyathiform, longer than any of the funiclar joints; funiclar joints all short, scarcely, if at all, longer than thick, pubescent. Thorax short, ovate, rounded before, narrowed and truncate behind, the collar not visible from above; mesonotum a little broader than long, without furrows; scutellum convexly elevated posteriorly, margined, or with a delicate rim, laterally; metathorax short, with short striæ above. Wings hyaline, with long cilia, a basal nervure, a short marginal and a short stigmal vein ending in a small rounded stigma, the stigmal vein thickened at base.

Abdomen oval, narrowed and truncate at base, rounded and broadened posteriorly, the first two segments transversely linear, the third very large, occupying most of its surface, the fourth about as long as the first two united, the following exceedingly short, scarcely visible, except as transverse lines.

HABITAT.—Flatbush, Long Island.

Types, two ♂ specimens, in National Museum.

Bred October 4. 1892, by Rev. J. L. Zabriskie, from the nest of an unknown spider.

CERATOBÆUS Ashm., gen. nov.

(Type *C. cornutus* Ashm.)

Head very large and broad, much broader than the thorax, the occiput margined; ocelli 3 in a triangle, wide apart, the lateral touching the eye; eyes broadly oval.

Antennæ inserted at the base of the clypeus, in ♀ short, clavate. 7-jointed, the club very large, inarticulate; ♂ unknown.

Maxillary palpi (?) 3-jointed; labial palpi 2-jointed.

Mandibles bifid.

Thorax oval, the prothorax not visible from above; mesonotum without furrows, or with but a slight trace of them posteriorly; scutellum semicircular; metathorax very short, abrupt.

Front wings, when present, with a distinct but short marginal vein, an oblique stigmal, and no post marginal, the stigmal vein terminating in a slight knob.

Abdomen oval, or pointed-ovate, depressed, subsessile, composed of

6 or 7 visible tergites, the third the largest, the first, furnished with a horn at base extending over the metathorax.

Legs as in *Acolus*.

Male unknown. Allied to *Baeus*, *Acolus*, and *Acoloides*, but readily separated by the projecting horn at the base of the abdomen, as in *Baryconus*, *Calotelcia*, etc.

The two species known to me may be thus tabulated:

TABLE OF SPECIES.

Wingless 2
Winged.
  Black, finely, closely punctulate; legs and antennae, yellow.
    Abdomen pointed ovate, longer than the head and the thorax together ......
                                               C. CORNUTUS, sp. nov.
2. Head and thorax black, finely, closely punctulate.
    Abdomen, legs, and antennae bright yellow, the former with two fuscous spots
        toward base .... C. BINOTATUS, sp. nov.

Ceratobaeus cornutus sp. nov.

(Pl. VIII, Fig. 8, ♀.)

♀. Length, 1.5ᵐᵐ. Black, finely, closely punctulate; the head with some faint, thimble-like punctures; antennae and legs, yellow. Head about 3½ times as wide as thick antero-posteriorly; the lower part of face smooth, shining, with a slight grooved line from front ocellus, the rest of the head closely punctulate. Antennae 7-jointed, the club inarticulate; the first funicular joint is half the length of the pedicel, or twice as long as thick; the three following, transverse. The thorax exhibits two very short grooved lines posteriorly, just in front of the scutellum, that are evidently the beginning of the parapsidal furrows. Mesopleura with a crenulate femoral furrow. Metapleura punctate, divided into two parts by a fovea and a grooved line. Wings subhyaline, the nervures brown; the marginal vein is about half the length of the stigmal. Abdomen pointed-ovate, depressed, longer than the head and thorax together, minutely punctate, with a microscopic pubescence; the horn on the first segment reaches to the apex of the scutellum and is wholly longitudinally striated; the second segment is much wider but no longer than the first; the third segment is the longest and widest, being 2½ times as long as the second; the fourth, is about half the length of the second; the fifth still shorter; the sixth, very short, scarcely discernible; the seventh, pointed or conical, and longer than the fourth.

HABITAT.—Washington, D. C.

Type in Coll. Ashmead.

Described from a single specimen taken by Mr. E. A. Schwarz.

Ceratobaeus binotatus sp. nov.

♀. Length, 0.8ᵐᵐ. Head and thorax black, closely, minutely punctulate; antennae, legs, and abdomen bright yellow, the latter with two fuscous spots at base of second segment, but toward the lateral corners. Head very wide and thick antero-posteriorly. Eyes large, rounded, coarsely faceted. Antennæ 7-jointed, the club very large, inarticulated, and longer than the pedicel and funicle together; first funiclar joint scarcely longer than thick, the other three exceedingly short, transverse. Thorax entirely without trace of furrows. Wings entirely wanting. Abdomen oval, not longer than the head and thorax together, faintly, microscopically punctate and shining; the first segment with a horn-like prominence, striated posteriorly; the second segment is a little longer than the first; the third, as long as the first and second together; the following segments short.

HABITAT.—Washington, D. C.

Type in Coll. Ashmead.

Described from 3 specimens, taken by Mr. E. A. Schwarz and myself.

This wingless species, but for the horn-like structure at base of abdomen, could easily be mistaken for an *Acolus*.

BÆUS Haliday.

Ent. Mag., ι, p. 270 ♀.

Hyperbæus Förster Hym. Stud., II, p. 111, 1856.
Trichasius Prov. Add. et Corr., p. 209, 1887.

Head very large, transverse, about twice as wide as the thorax, the frons broad, convex; ocelli 3, triangularly arranged but widely separated, the lateral being close to the margin of the eye; eyes very large, oval.

Antennæ inserted close to the clypeus; in ♀ 7-jointed, clavate, the club large, inarticulate; in ♂ 12-jointed, subclavate, the funiclar joints after the second scarcely longer than wide, very slightly widened toward tips.

Maxillary palpi 2-jointed.

Thorax scarcely as long as wide, the pronotum not visible from above, the scutellum in ♀ not differentiated, while the metathorax is extremely short; in the ♂ the scutellum is distinct, semicircular, subconvex.

Wings in the ♂ with a distinct basal nervure, a short marginal, a long knobbed stigmal nervure and no post marginal; the ♀ always apterous.

Abdomen in ♀ oval, sessile, convex above, flat beneath, the third segment occupying nearly the whole surface, the first and second segments being visible only as transverse lines or wrinkles; in ♂ smaller, rounded and more flattened, the first and second segments very short, transverse, the third the largest segment.

Legs rather long, the femora clavate, the tibial spurs weak, the tarsi 5-jointed, the hind tarsi not longer than their femora, the basal joint being one-third longer than the second.

The absence of a scutellum readily distinguishes the females of this genus. The male was unknown to Haliday and Förster and is here

described for the first time. A single ♂ specimen (together with several
females) was reared from spider eggs in an orange-colored cocoon by
Mr. J. H. Emerton, in 1871, and is now in the Department of Agriculture. This single specimen has enabled me to perfect the diagnosis
of the genus. It bears a close resemblance to the male in the genus
Acoloides, but differs in having subclavate antennæ, a little longer
marginal vein, longer stigmal vein, and a distinct basal nervure.

The habits of the genus were unknown to the European authorities,
and we are indebted to an American, Mr. L. O. Howard, for first making
us acquainted with them in this country. All that have been reared
are from spider eggs.

The species known to me may be tabulated as follows:

TABLE OF SPECIES.

Body not entirely black .... 2.
Body entirely black.
   Head and thorax faintly microscopically punctulate; abdomen smooth, shining.
     Legs and antennæ fuscous.............................B. MINUTUS, sp. nov.
     Legs, scape and funicle, brownish-yellow ....B. NIGER, sp. nov.
2. Head black.
   Thorax and abdomen fuscous or piceous; legs and antennæ, brownish-yellow, the
     scape and pedicel at base, fuscous ....B. PICEUS, sp. nov.
   Thorax and abdomen brownish-yellow; legs and antennæ yellow, the club
     black..B. CLAVATUS, Prov.
   Wholly dark honey-yellow........................B. AMERICANUS, How.

Bæus minutus, sp. nov.

♀. Length, 0.5ᵐᵐ. Black, shining, faintly microscopically punctulate, with a microscopic sericeous pubescence; antennæ and legs fuscous, or dark brown, the tarsi paler. Antennæ 7-jointed, very short,
the club large, pointed-ovate; pedicel large, pale at tip; first funiclar
joint scarcely longer than thick; the remaining joints of funicle very
minute, transverse. Abdomen broadly oval, convex, shining, impunctate, wider than the head and as long as the head and thorax united.

HABITAT.—Ottawa, Canada.

Types in Coll. Ashmead.

Described from 2 ♀ specimens received from W. Hague Harrington.

Bæus niger, sp. nov.

♀. Length, 0.7ᵐᵐ. Black, shining, with a fine sericeous pubescence; head impunctate; eyes very large, nearly round, occupying the
whole side of the head, pubescent; thorax not longer than the length
of the head, microscopically punctate. Antennæ very short, the scape
and funicle pale rufous or brownish-yellow, the club large, black. Legs
brownish-yellow, the tibiæ with a slight dusky streak at the middle.

Abdomen long-oval, twice as long as the head and thorax together the widest part as broad as the head.

HABITAT.—Washington, D. C.

Type in National Museum.

Described from a single specimen, taken September 21, 1884.

Bæus piceus, sp. nov.

♀. Length 0.6ᵐᵐ. Brownish-piceous, smooth, shining, impunctured; antennæ and legs brownish-yellow, the scape and pedicel above fuscous, pale at tips; the three last funicle joints very minute transverse, the first not longer than thick, less than one-third the length of the pedicel. Abdomen oval, convex, polished and shining, a little longer than the head and thorax together.

HABITAT.—Ottawa, Canada.

Type in Coll. Ashmead.

Described from a single specimen received from W. Hague Harrington.

Bæus clavatus Prov.

Trichasius clavatus Prov., Add. et Corr., p. 209,
Bæus clavatus Hargtn., Ins. Life, II, p. 359.

♀. Length 0.05 inch. Of a uniform reddish-brown, with the legs yellow. The club of the antennæ black. Thorax densely punctured, metathorax rugose. Legs pale yellow, the tarsi with the last joint brown. Abdomen browner than the rest, polished but not metallic. (*Translated from Provancher.*)

HABITAT.—Ottawa, Canada.

Type in Coll. Harrington.

Bæus americanus How.

(Pl. VIII, Fig. 9, ♂ ♀.)

Ins. Life, II, p. 270, Fig. 59.

♀. Length 0.65ᵐᵐ. Length of antennal club 0.185ᵐᵐ, or in other words the entire body is only 3½ times as long as the antennal club. Width of antennal club 0.082ᵐᵐ. General color dark honey-yellow; scape and funicle of antennæ brownish, club lighter, dark at tip; vertex and face light honey-yellow; dorsum of thorax and abdomen dark honey-yellow, almost approaching mahogany; legs throughout concolorous with the head; middle and hind tibiæ a little darker near base. Surface of abdomen smooth, shiny; mesonotum very faintly punctate. Thorax and abdomen with extremely fine, sparse, whitish pile; tip of abdomen with a short and contracted fringe of white pile. Antennal club very large, longer than rest of funicle and pedicel together; funicle joints very narrow and short, subequal; pedicel wider and as long as entire funicle, except club. (*Howard.*)

♂. Length 0.8ᵐᵐ. Dark honey-yellow; the head piceous, with a median carina anteriorly; antennæ and legs pale yellowish; abdomen small, rounded, fuscous, the basal segment short, transverse, striate. Wings hyaline, fringed, the venation pale brown, the basal nervure distinct, the marginal nervure short, about twice as long as thick, the post-marginal nervure wanting, the stigmal long, oblique.

HABITAT.—Brooklyn, New York.

Types in National Museum.

This species, as we are informed by Mr. Howard, was reared by Col. Nicholas Pike from spider eggs in an orange cocoon belonging to the family *Epeiridæ*.

The male, here described, was reared by Mr. J. H. Emerton, in 1871, together with several females, from an orange-colored spider's cocoon.

Tribe III.—TELEASINI.

An extensive tribe with few genera, but numerous in species, distinguished by the antennæ being inserted close together on a clypeal prominence; by the venation, the postmarginal vein never being developed, the marginal vein always greatly lengthened, and the stigmal vein always minute, scarcely developed; by the abdomen always being distinctly carinated along the sides, with the third segment the longest; and by the weak tibial spurs.

Under this tribe six genera are brought together, *Teleas*, *Prosacantha*, and allied genera.

Except in a single instance, the rearing of *Prosacantha caraborum*, by Dr. Riley, from the eggs of a beetle, *Chlænius impunctifrons*, nothing positively is known of the parasitism of the group. It is possible they confine their attacks to Coleopterous eggs, although *Prosacantha basalis* Förster is recorded as having been reared from Dipterous larvæ. Since this record is totally at variance with the habits of the whole subfamily *Scelioninæ*, I question its accuracy, as I believe the whole group are egg parasites.

The genera may be tabulated as follows:

TABLE OF GENERA.

FEMALES.

Abdomen broadly oval, the first segment wider than long......................3
Abdomen long-oval, or long ovate, the first segment petioliform (longer than wide).
  First abdominal segment bearing a horn; postscutellum with three spines.
                                          PENTACANTHA Ashm.
  First abdominal segment without a horn.
    Mesonotum without furrows................................2
    Mesonotum with 2 furrows.
      Metascutellum with 3 spines......................TRISSACANTHA Ashm.
      Metascutellum with 1 spine.........................XENOMERUS Walk.
2. Postscutellum with a single large spine, mandibles bifid, the outer tooth the longer.
  Posterior femora, tibiæ, and tarsi slender; tibial spurs weak. PROSACANTHA Nees.
  Posterior femora swollen, tibiæ dilated at apex, the basal tarsal joint short, stout; tibial spurs not weak...............................TELEAS Latr.

3. Winged; mandibles bifid, the teeth equal.
    Metascutellum with a spine, or tuberculate.....HOPLOGRYON Ashm., gen. nov.
    Metascutellum simple...GRYON Haliday.
Apterous.
    Metascutellum with a small spine or tubercle.................(HOPLOGRYON).
    Metascutellum without a spine, simple...........................(GRYON).

MALES.

Abdomen broadly oval, the first segment wider than long.....................3
Abdomen long-oval, the first segment petioliform; the marginal vein very long.
    Mesonotum without furrows; postscutellum spined.........................2
    Mesonotum with 2 distinct furrows; posterior angles of metathorax usually
        spined or toothed.
    Metascutellum with 3 spines; antennæ very long, filiform, pubescent.
                          TRISSACANTHA Ashm.
    Metascutellum with 1 spine; antennæ with whorls of hair.
                          XENOMERUS Walk.
2. Antennæ long, filiform, the flagellar joints at least four times as long as thick, the
        third joint excised at base or angulated; posterior femora not
        swollen, the tibial spurs not developed, the basal tarsal joint long,
        slender....................................PROSACANTHA Nees.
    Antennæ filiform, the flagellar joints usually less than thrice as long as thick;
        posterior femora swollen, the tibial spurs developed, the basal
        joint short, stout ....TELEAS Latreille.
3. Marginal vein long, at least four or five times as long as the stigma.
    Metascutellum with a small spine or tubercle; antennæ filiform, the flagellar
        joints elongate....................HOPLOGRYON Ashm., gen. nov.
    Metascutellum without a spine, simple; antennæ filiform, the flagellar joints
        scarcely longer than thick....................GRYON Haliday.

PENTACANTHA Ashm.

Can. Ent., xx, p. 51 (1888).

(Type, P. canadensis Ashm).

Head transverse, the occiput straight, not emarginate, with a delicate margin extending on the cheeks, which are flattened; ocelli 3, in a triangle; eyes long-oval, hairy.

Antennæ inserted at the base of the clypeus, 12-jointed in both sexes, in the ♀ clavate, the club 6-jointed, the first funicular joint twice as long as the pedicel.

Maxillary palpi very short, 3-jointed.

Mandibles bidentate at apex, sinuate, and with a lobe at base within.

Thorax ovoid, the prothorax not visible from above; mesonotum wider than long, the scapulae not separated; scutellum semicircular, rugose; postscutellum with 3 erect, conic teeth; metathorax very short, with prominent acute posterior angles that are connected with a carina.

Front wings as in <i>Prosacantha.</i>

Abdomen oblong-oval, narrowed at base, attached to metathorax far above the hind coxæ, and composed of six visible dorsal segments, the basal segment having a short, smooth, blunt horn, partly prolonged over the metathorax, the third segment the longest.

Legs rather long, slender, pilose, the tibiae clavate, their spurs weak, the tarsi much longer than the tibiae, slender, the claws simple.

Closely allied to *Prosacantha*, but readily distinguished by the postscutellum being armed with 3 spines, and the first abdominal segment bearing a horn, as in *Caloteleia* Westwood, in the tribe *Scelionini*. The male is unknown.

Pentacantha canadensis Ashm.

(Pl. VIII, Fig. 1, ♀.)

loc. cit. p. 51.

♀. Length, 2ᵐᵐ. Black; thorax and scutellum coarsely shagreened, opaque; the head smooth, shining, with some grooved lines back of the eyes and on the occiput, the vertex rather acute. Antennæ 12-jointed, brown, the first funiclar joint twice as long as the pedicel, the second one-third shorter than the first, the third about as long as wide, the fourth wider than long; club 6-jointed. Thorax a little wider than long, without furrows, the postscutellum with 3 erect spines, the lateral posterior angles of the metathorax acutely spined. Legs, including anterior coxæ, honey-yellow or brownish-yellow; middle of the femora and tibiæ and the posterior coxæ, dark brown or fuscous. Abdomen polished black, the third segment the longest and widest, the first with a horn at base, this as well as the second and the third, excepting on its disk, longitudinally striated. Wings subhyaline, the venation rufo-piceous.

HABITAT.—Ottawa, Canada, and Kansas.

Type, ♀ in Coll. Ashmead.

The type was described from a single specimen received from W. H. Harrington, but I have since seen a specimen in Prof. Popenoe's collection, taken in Riley County, Kans.

TRISSACANTHA Ashm.

Ent. Am., III, p. 101 (1887).

(Type, *T. americana* Ashm.)

Head transverse, the occiput almost straight, margined; ocelli 3, close together in a triangle; eyes oval.

Antennæ inserted on a clypeal prominence, 12-jointed, in ♂ very long, filiform, cylindrical, the scape extending to the ocelli, the pedicel very small, annular, the third funiclar joint bent and angulated at the middle.

Maxillary palpi 3-jointed.

Mandibles bidentate at tips, the teeth equal.

Thorax ovoid, the prothorax slightly visible from above, the mesonotum smooth, wider than long, with two delicate but distinct furrows, scutellum semicircular, convex, metascutellum armed with 3 erect spines, metathorax short, the pleura more or less covered with a grayish woolly pubescence, the angles acute.

Front wings pubescent, ciliated, the cilia on hind margin very long, the marginal vein linear, about five times as long as the short stigmal, the postmarginal wanting.

Abdomen long-oval, depressed, sparsely pilose, the first segment longer than wide, fluted, the third the largest; the first three segments are longitudinally striated.

Legs rather long, slender, pilose, the tibial spurs minute, the tarsi slender but spinulose.

This genus, known only in the male sex, is distinguished from *Prosacantha* by the 3-spined postscutellum, and in having distinct mesonotal furrows; from *Xenomerus*, which also has mesonotal furrows, it is separated by the postscutellum and the filiform antennæ.

It may be the opposite sex of *Pentacantha*, agreeing with it in its postscutellar character, but *the mesonotum has two distinct furrows*, and as I know of no genus in the *Proctotrypidæ* having the mesonotum grooved in one sex and plain in the opposite. I believe it to be distinct. This character, used in the table of the males, is therefore repeated in the generic table of the females.

Only two species are known to me, which may be separated as follows:

Black; legs rufous, coxæ black.
    Scutellum smooth, polished...T. AMERICANA Ashm.
    Scutellum rugose...T. RUGOSA, sp. nov.

Trissacantha americana Ashm.

(Pl. VIII, Fig. 2. ♂.)

Ent. Am., III, p. 117; Cress. Syn. Hym., p. 313.

♂. Length, 2.5ᵐᵐ. Polished black, pubescent; head and occiput with some striæ; mandibles piceous; legs rufous, the coxæ black. Antennæ 12-jointed, filiform, black, longer than the body, the scape not much longer than the first flagellar joint, the pedicel annular, the flagellar joints all long, cylindrical, the first about five times as long as thick, the second shorter, the third angulated at the middle and slightly excised at base. Mesonotum with two distinct furrows; scutellum semicircular, smooth, polished; postscutellum with three erect spines, the middle the longest. Abdomen longer than the head and thorax together, longitudinally striated. Wings fusco-hyaline, pubescent.

HABITAT.—Florida, District of Columbia and Virginia.

Type in Coll. Ashmead; specimens in National Museum.

This species was originally described from a specimen taken by myself in Florida, but I have since seen specimens taken in Virginia and District of Columbia.

Trissacantha rugosa, sp. nov.

♂. Length, 2.5ᵐᵐ. Black, shining, pubescent; head above, collar and scutellum rugoso-punctate, the mesonotum smoother, the face near the eyes, striated. Antennæ 12-jointed, brown-black, longer than the body. Mesonotum smooth, shining, with two furrows and a few scattered punctures; middle spines of postscutellum twice as long and much stouter than the lateral spines, curved and rather blunt at apex; metathorax with prominent angles. Abdomen black, shining, striated. Wings hyaline, pubescent, the marginal vein six times as long as the short stigmal. Legs rufous, the coxæ black.

HABITAT.—Arlington, Va.

Type in Coll. Ashmead.

Described from a single specimen, which is readily distinguished from *T. americana* by the rugosity of the vertex, collar, and scutellum.

XENOMERUS Walker.

Ent. Mag., III, p. 355 (1836).

(Type *X. ergenna* Walk., ♂ ; *Teleas medon* Walk., ♀.)

(Pl. VIII, Fig. 3. ♂.)

Head large, transverse, broader than the thorax, the occiput slightly emarginated, with a delicate margin; ocelli three in a triangle, the lateral away from the margin of the eye; eyes rounded, pubescent.

Antennæ 12-jointed in both sexes, inserted on the clypeus, in ♀ terminating in a 5- or 6- jointed club, in ♂ pedicellate-nodose, with whorls of long hairs.

Maxillary palpi 3-jointed.

Mandibles bidentate.

Front wings as in *Teleas*, with a long marginal vein and a short stigmal.

Abdomen broadly oval, the first and second segments short, the first about twice as wide as long, both striated, the third segment large, the following very short.

Legs long, slender, pubescent; the tibial spurs weak, the tarsi long, slender; the basal joint of the hind tarsi more than twice as long as the second; claws simple.

The genus is readily distinguished in the male sex by the verticillate antennæ, and in the female from *Teleas*, *Prosacantha*, and allied genera by having distinct mesonotal furrows; from *Trissacantha*, which also has mesonotal furrows, it is distinguished by having a single postscutellar spine.

Xenomerus pallidipes, sp. nov.

♀. Length, 1.4ᵐᵐ. Polished black, shining, sparsely pubescent; head transverse, with the eyes slightly wider than the thorax, the latter slightly pubescent.

Antennæ 12-jointed, clavate, the club 6-jointed, the scape pale brown, the flagellum black; the pedicel is shorter than the first funicle joint; the first and second funicle joints longer than thick, the third and fourth small, transverse; club joints, except the last, transverse.

Thorax with two impressed lines subobsolete posteriorly; scutellum convex, separated from the mesonotum by a transverse line; metascutellum armed with a small acute tooth. Wings hyaline, ciliated. Legs pale brownish-yellow, the coxæ dusky basally. Abdomen longer than the head and thorax together, narrowed basally into a petiole and inserted far up on the metathorax behind the postscutellum, black and polished.

HABITAT.—Arlington, Va.

Type in Coll. Ashmead.

Described from a single specimen.

The insect described by me, in the beginning of my studies on this family, under the name *N. rubicola*, is a ♂ Pteromalid belonging to a genus unknown to me.

PROSACANTHA Nees.

Mon., II, p. 294 (1834); Förster, Hym. Stud., II, p. 103 (1856).

(*P. longicornis* Nees.)

Head transverse, the occiput slightly emarginated and delicately margined; ocelli 3, in a triangle, rather close together, the lateral very distant from the margin of the eye; eyes oval, pubescent.

Antennæ inserted on a clypeal prominence, 12-jointed, in ♀ terminating in a long, fusiform, 6-jointed club; the funiclar joints thicker than the pedicel, the first two joints longer than wide; in ♂ long, filiform, the joints long, cylindrical, the third angulated or emarginated from near the middle towards the base.

Maxillary palpi 3-jointed.

Mandibles curved, bidentate at tips, the teeth in the left mandible equal, in the right with the outer tooth the longer.

Thorax ovoid, the prothorax scarcely visible from above, narrowed into a slight neck anteriorly; mesonotum broader than long, without furrows, seldom occasionally with indications of furrows posteriorly. Scutellum large, semicircular, subconvex; metathorax armed with a large sharp spine or thorn; metathorax very short, the posterior angles usually acute or spined.

Front wings pubescent, ciliated, with a very long marginal vein that ends at two-thirds the length of the wing and is 6 or 7 times longer than the short stigmal vein, the postmarginal never developed.

Abdomen long-oval, depressed, longer than the head and thorax together, inserted high up on the metathorax, the first segment petioliform, much longer than wide, fluted or striated, the third segment always large and occupying more than half the rest of the abdomen, either smooth or striated, the following segments very short.

Legs long, slender, pubescent; the femora not especially thick; the tibial spurs weak; the tarsi long, slender, cylindrical, claws simple.

A genus most frequently confused with *Teleas*, but easily separated by the much slenderer hind legs and the weaker tibial spurs.

TABLE OF SPECIES.

FEMALES.

Antennæ not annulated with white.. 2
Antennæ annulated with white.
  First, second, and third funiclar joints white.......P. ANNULICORNIS, sp. nov.
2. Antennæ not wholly black, the scape rufous or yellow.......................... 4
Antennæ with the scape pale only at base...................................... 3
Antennæ wholly black.
  Legs wholly black.
    Thorax rugose, opaque.
      Face longitudinally striated........................P. NIGRIPES, sp. nov.
    Thorax finely punctate, shining.
      Face striate toward base of antennæ ....P. MELANOPUS, sp. nov.
    Thorax smooth, shining at the most microscopically punctate.
      Face smooth, shining........................P. PENNSYLVANICA, sp. nov.
  Legs not wholly black.
    Legs, except knees and tarsi, black or dark fuscous.
      Thorax rugoso-punctate; lower part of face striate.
        Scutellum rugose, opaque; wings abbreviated....P. UTAHENSIS, sp. nov.
        Scutellum smooth, polished......................P. CALIFORNICA, sp. nov.
      Thorax closely striately punctate; scutellum faintly punctate, opaque.
        Face smooth, shining...........................P. LEVIFRONS, sp. nov.
      Thorax finely rugose, subopaque; scutellum shining, faintly punctate.
        Face shining but sparsely finely punctate..............P. NANA, sp. nov.
3. Coxæ black; legs, unless otherwise mentioned, pale rufous.
  Thorax closely punctulate.
    Face highly polished; tibiæ and tarsi, fuscous ....P. CARABORUM, Riley.
  Thorax and scutellum longitudinally rugulose.
    Frons polished but with some scattered punctures.
                             P. PUNCTIVENTRIS, sp. nov.
  Thorax and scutellum finely rugoso-punctate, opaque.
    Frons shining, feebly shagreened, with some scattered punctures.
                             P. SCHWARZII, sp. nov.
  Thorax polished, sparsely punctate; scutellum polished.
    Frons polished; posterior tibiæ fuscous above..........P. PUSILLA, sp. nov.
  Thorax finely punctulate; head and scutellum smooth, shining; legs pale
    brownish-yellow..........................P. GRACILICORNIS, sp. nov.
Coxæ and legs pale rufous or brownish-yellow.
  Thorax and scutellum rugoso-punctate.
    Frons smooth, sparsely punctate...................P. MARYLANDICA, sp. nov.
    Frons coarsely longitudinally striate..................P. FUSCIPENNIS, Ashm.
  Thorax and scutellum shining, faintly punctate.
    Frons smooth, with some small, faint punctures....P. COLUMBIANA, sp. nov.
  Thorax finely punctulate with indications of furrows posteriorly; scutellum
    smooth, shining.
    Frons highly polished, impunctate...................P. BILINEATA, sp. nov.
  Thorax and scutellum finely rugose, opaque; pleura rufous.
    Frons shining, sparsely, faintly punctate..............P. PLEURALIS, sp. nov.
4. All coxæ, except sometimes the last pair, pale................................. 5
All coxæ black.
  Thorax and scutellum coarsely longitudinally rugulose.
    First funiclar joint one-third shorter than the second; legs rufous; three
    basal segments of abdomen striate.............P. COXALIS, sp. nov.

Thorax finely, closely punctate; the scutellum smooth, polished.
First and second funiclar joints long equal, about 5 times as long as thick;
legs pale brownish-yellow..............P. GRACILICORNIS, sp. nov.
5. Legs rufous.
Thorax and scutellum rugose, opaque.
Inner orbits and lower parts of face and the cheeks striate; wings hyaline.
P. LINELLII, sp. nov.
Thorax and scutellum sparsely punctate, shining.
Inner orbits and lower part of face striate; wings fuscous.
P. ERYTHROPUS, sp. nov.
Legs yellow.
Thorax and scutellum rugulose, opaque.
Head polished, inner orbits and lower part of face striate; wings hyaline;
coxæ yellow.............................P. FLAVICOXA, sp. nov.
Thorax and scutellum coarsely rugulose, opaque.
Head polished, inner orbits not striate, wings subfuscous, hind coxæ black.
P. AMERICANA, Ashm.
Thorax punctate, shining, with indications of furrows posteriorly; the scutel-
lum smooth.
Head polished, the lower part of face and cheeks striate; wings hyaline;
petiole of abdomen yellow............P. FLAVOPETIOLATA, sp. nov.

MALES.

Antennæ not wholly black... 2
Antennæ wholly black.
Legs wholly black; face longitudinally striated.
Antennæ shorter than body, first flagellar joint less than four times as long
as thick................................... P. NIGRIPES, sp. nov.
Legs brownish-yellow.
Antennæ much longer than the body, first flagellar joint 6 times as long as
thick.................................P. STRIATIFRONS, sp. nov.
2. First flagellar joint not as long as the scape................................... 3
First flagellar joint longer than the scape.
Thorax finely rugose, smoother on shoulders; scutellum smooth, with punctures
at base.
Legs brownish yellow, middle tarsi and posterior tibæ and tarsi, fuscous.
P. CARABORUM, Riley
3. Thorax closely punctate or rugose.. 4
Thorax smooth, shining, faintly punctate; legs brownish-yellow.
First flagellar joint 4 times as long as thick, a little shorter than the following:
Frons polished.
Third abdominal segment not aciculate or punctate; petiole black.
P. COLUMBIANA, sp. nov.
Third abdominal segment smooth, impunctate; petiole yellow.
P. FLAVOPETIOLATA.
Frons striated.
Third abdominal segment finely aciculated and punctate; petiole dull
rufous...............................P. MACROCERA, Ashm.
4. First flagellar joint shorter than the second.
Thorax and scutellum closely punctate, shining.
Lower part of the face striate; mandibles large, yellow.
P. XANTHOGNATHA, sp. nov.
Thorax rugose, opaque or subopaque, pubescent.
Frons longitudinally striate; mandibles and legs rufous.
P. MARYLANDICA, sp. nov.
First flagellar joint longer than the second, abdomen longitudinally striated.
P. STRIATIVENTRIS, sp. nov.

Prosacantha annulicornis, sp. nov.

♀. Length, 2 to 2.2ᵐᵐ. Head and abdomen black, shining; the thorax dull rufous, punctate, the metathorax and metapleura stained with black. Head transverse, the vertex subacute, face smooth, with a few scattered punctures, the orbits longitudinally aciculated. Antennæ 12-jointed, long; the scape long, rufous, extending one-third its length beyond the ocelli, its apex and the pedicel fuscous: the first three funicular joints are long, subequal in length, yellowish-white; the fourth funicular joint and the club, black; the fourth funicular joint is a little longer than thick and fully as thick as the preceding joints. Scutellum roughly punctate; the post-scutellum armed with an unusually long acute spine. Metathorax with the angles acute. Wings fuscous, paler basally. Legs pale rufous, the coxæ and trochanters yellowish, the tips of the femora and tibiæ a little fuscous. Abdomen as long as the head and thorax together, polished, the petiole and the second segment longitudinally striated.

HABITAT.—Washington, D. C.

Types in National Museum, American Entomological Society, and Coll. Ashmead.

Described from many specimens, taken in winter, by sifting.

Prosacantha nigripes, sp. nov.

♂ ♀. Length, 2 to 2.4ᵐᵐ. Black; thorax and scutellum finely rugose, subopaque; head transverse, vertex punctate, the occiput transversely aciculated, the face longitudinally striate, the cheeks with striæ converging toward the mouth. Mandibles rufous. Antennæ 12-jointed, black, scape not extending beyond the middle ocellus; pedicel small; the first three funicular joints nearly equal in length, about thrice as long as thick.

Posterior angles of metathorax with a short tooth. Legs entirely black. Wings fuscous. Abdomen slightly longer than the head and thorax together, polished, the petiole and second segment longitudinally striated, the disk of the third segment faintly longitudinally aciculated.

In the ♂ the antennæ reach to the middle of the abdomen. The flagellar joints very slightly subequal, the first the longest and thickest, four times as long as thick.

HABITAT.—The Dalles and Portland, Oregon.

Types in Coll. Ashmead.

Described from specimens received from Mr. H. F. Wickham.

Prosacantha striatifrons, sp. nov.

♂. Length, 2ᵐᵐ. Black; head broadly transverse, shining, vertex smooth, frons and face longitudinally striated; mandibles long, falcate, pale rufous; thorax and scutellum rugulose, opaque; legs brownish-yellow. Antennæ 12-jointed, filiform, black; first funicle joint 6 times as long as thick; second slightly shorter; third still shorter; fourth and following longer than third, but not quite as long as the second. Post-scutellar spine long, acute. Angles of metathorax acute. Wings sub-hyaline, the nervures pale brown. Abdomen broadly oval, black, shining, the first and second segments and the third, on disk, longitudinally striated.

HABITAT.—Jacksonville, Fla.

Type in Coll. Ashmead.

Described from a single specimen.

Prosacantha melanopus, sp. nov.

♀ . Length, 1.4ᵐᵐ. Black, shining; face perfectly smooth; thorax and scutellum faintly microscopically punctate, but not sufficient to destroy the smoothness or luster of the surface. Mandibles dull, rufous. Antennæ 12-jointed, short, black; scape long, pedicel two-thirds the length of the first funiclar joint; the first and second funiclar joints nearly equal, the second the smaller; third and fourth, minute, transverse; club 6-jointed, as long as the pedicel and all the funiclar joints together; the joints transverse, about twice as wide as long. Wings fuscous. Legs black; tip of trochanters and base of tarsi, piceous or brownish. Abdomen as long as the head and thorax together, smooth, shining; the petiole and first segment striated.

HABITAT.—Ottawa, Canada.

Type in Coll. Ashmead.

Described from a single specimen received from Mr. W. Hague Harrington.

Prosacantha pennsylvanica, sp. nov.

♀ . Length, 1ᵐᵐ. Black, shining, sparsely pubescent, the punctuation of the thorax very fine and faint. Antennæ 12-jointed, black, not extending beyond the base of abdomen; first funiclar joint scarcely as long as the pedicel, the second shorter, the third and fourth minute, transverse. Wings hyaline, strongly iridescent, with a faint fuscous tinge. Legs brown-black, trochanters, knees, tips of tibiæ and base of tarsi yellowish. Abdomen broadly oval, not as long as the head and thorax together, smooth, polished, with a fine pubescence; the petiole longer than thick, striated; the second segment with some striæ at base.

HABITAT.—Pennsylvania.

Type, ♀ in Berlin Museum.

Described from a single specimen labeled "Penna. Zimmermann."

Prosacantha utahensis, sp. nov.

♀ . Length, 1.5ᵐᵐ. Black; thorax and scutellum rugoso-punctate, opaque; head almost smooth, shining, the face with delicate striæ toward the orbits and on cheeks below the eyes. Mandibles rufous. Antennae 12-jointed, black; the pedicel half the length of the first funicular joint: pale at tip; second funicular joint one-third shorter than the first, third and fourth minute, transverse; club as long as the funicle, the joints transverse. Wings abbreviated, not extending beyond the base of second abdominal segment. Legs black, articulations of trochanters, knees, and the tarsi, rufous. Abdomen a little longer than the head and thorax together, shining; the first and second segments coarsely striated, the following segments shagreened, or with a scaly punctuation; venter except the first segment, feebly microscopically punctate.

HABITAT.—Park City, Utah.

Type ♀ in Coll. Ashmead.

Described from a single specimen collected by Mr. E. A. Schwarz.

Prosacantha californica, sp. nov.

♀. Length, 1.8ᵐᵐ. Black; thorax finely rugoso-punctate, opaque; the scutellum smooth, shining; head smooth, shining, the orbits and lower part of the cheeks, striated. Mandibles piceous. Antennae 12-jointed, black, the pedicel about half the length of the first funicular joint, the second funicular joint very slightly shorter than the first, the third and fourth, minute, rounded; club as long as the funicle and the pedicel together, the joints transverse. Metapleura polished. Hind angles of metathorax toothed. Wings subfuscous, fully developed, the margins ciliated. Legs black, the articulations and the tarsi fuscous. Abdomen oblong-oval, not longer than the head and thorax together, polished, impunctured, the first segment and the second at base alone, striated.

HABITAT.—Santa Cruz Mountains, Cal.

Type in Coll. Ashmead.

Described from 1 specimen.

Prosacantha laevifrous, sp. nov.

♀. Length, 1ᵐᵐ. Black; thorax closely, striately punctate, opaque; the scutellum feebly, minutely punctate, subopaque; head shining, the face polished, the cheeks with striae below the eye, the occiput closely punctate. Mandibles rufous. Antennae 12-jointed, short, black, the first and second funicular joints equal, only one-third longer than the pedicel, the third and fourth, minute, transverse, club longer than the funicle, the joints transverse. Postscutellar spine very minute. Legs fuscous, the trochanters, knees and tarsi, pale brown. Wings subfuscous, pubescent. Abdomen broadly oval, shining, the first and second segments striated, the following microscopically punctate.

HABITAT.—Washington, D. C., and Arlington, Va.

Types in Coll. Ashmead.

Described from 3 specimens.

Prosacantha nana, sp. nov.

♀ . Length, 1ᵐᵐ. Black; thorax finely rugose, subopaque, the scutellum smooth, shining, faintly punctate; head smooth, shining, the face polished, very sparsely, finely punctate above. Mandibles black. Antennæ 12-jointed, black, the pedicel two-thirds the length of the first funiclar joint, the second funiclar joint slightly shorter than the first, the third and fourth, minute, transverse, club as long as the funicle, excluding the pedicel, the joints transverse. Postscutellar spine very minute. Wings subfuscous. Legs black, the tarsi fuscous. Abdomen longer than the head and thorax together, polished, the first segment and the second basally, striated.

HABITAT.—Utah Lake, Utah.

Type in Coll. Ashmead.

Described from a single specimen, received from Mr. E. A. Schwarz.

Prosacantha caraborum, sp. nov. Riley.

(Pl. VIII, Fig. 4, ♀.)

"♀ . Length, 1.8ᵐᵐ. Black, shining, the thorax finely rugulose, the scutellum almost smooth. Head transverse, thrice as wide as long, the occiput longitudinally striated, the face smooth with a delicate carina extending from the front ocellus to the base of the antennæ; lower part of orbits and cheeks with striæ converging toward the mouth; mandibles pale rufous, deeply bifid at tips; antennæ 12-jointed, about as long as the body, slender, dark brown or black, the scape a little pale at extreme base; pedicel about as long as the third funicular joint; the first and second funicular joints elongate, equal, about twice as long as the pedicel, the fourth subequal with the third; club 6-jointed, slender, long, fusiform. Metanotum with silvery pubescence; tegulæ black; wings hyaline, the venation pale brownish-yellow. Legs reddish-yellow, the coxæ basally and the middle and posterior tibiæ and tarsi, fuscous. Abdomen as long as the head and thorax together, shining, with a sparse, fine pubescence toward the apex; the first three segments longitudinally striated.

"♂ . Smaller (1.6ᵐᵐ) and remarkable for the very long, filiform, hairy antennæ which are fully twice the length of the whole insect; they are black, the scape rufous, the pedicel round; the flagellar joints all exceedingly long and slender, longer than the scape, the third being angularly produced toward the base."—[From Riley's MS.]

HABITAT.—Washington, D. C., and Arlington, Va.

Types in National Museum and Coll. Ashmead.

Reared by Dr. Riley in June, 1884, from the eggs of a Carabid beetle (Chlænius impunctifrons), and also taken by me in Virginia by sweeping.

Prosacantha punctiventris, sp. nov.

♀. Length, 2.5ᵐᵐ. Black; thorax and scutellum longitudinally rugulose, subopaque; head very broad, shining; face smooth, with scattered punctures; lower part of face and cheeks striated. Antennae 12-jointed, black, the scape basally rufous; first flagellar joint twice the length of the pedicel the second one-third shorter than the first but a little stouter, the third and fourth transverse-quadrate; club shorter than the funicle. Mesopleura rugose. Postscutellar spine triangular, rugose. Lower part of metapleura striated. Angles of metathorax produced into triangular teeth. Wings subfuscous, the marginal vein less than five times as long as the stigmal. Legs pale rufous, the coxae black. Abdomen broadly oval, shining, the first segment coarsely striated, the disks of the second and third finely striated, the venter punctate.

HABITAT.—Fortress Monroe, Va., and District of Columbia.

Types in Coll. Ashmead.

Described from 5 specimens collected by Mr. E. A. Schwarz.

Prosacantha Schwarzii, sp. nov.

♀. Length, 1.5ᵐᵐ. Black; thorax and scutellum finely rugoso-punctate, opaque; head shining, the frons feebly shagreened, the face and cheeks strongly striated. Mandibles rufous. Antennae 12-jointed, black, the scape paler at base; pedicel pale at apex, a little shorter than the first funicular joint; the first and second funicular joints about of an equal length, the second the stouter; the third and fourth minute, transverse; club as long as the pedicel and funicle together. Mesopleura rugose. Post-scutellar spine triangular. Angles of metathorax acute. Wings subhyaline, the marginal vein 4 times the length of the stigmal. Legs brownish-yellow, the coxae black. Abdomen broadly oval, the first segment and the second, for two-thirds its length, coarsely striated; rest of the abdomen smooth, polished.

HABITAT.—Washington, D. C.

Type ♀ in Coll. Ashmead.

Described from a single specimen taken by E. A. Schwarz, and in honor of whom this species is dedicated.

Prosacantha pusilla, sp. nov.

♀. Length, 1ᵐᵐ. Black, polished, shining, the thorax feebly, sparsely punctate; frons polished, impunctate, lower part of face and cheeks striated. Mandibles and legs brownish-yellow, the coxae black. Antennae black, the scape pale at extreme base; pedicel and second funicular joint of an equal length, the first a little longer; third and fourth minute, transverse. Post-scutellar spine acute. Wings subhyaline, the marginal vein six times the length of the stigmal. Abdomen oblong-oval, highly polished, impunctured, the first segment and the second for two-thirds its length, coarsely striated, the third with a row of very short striæ at extreme base.

HABITAT.—Jacksonville, Fla.

Type in Coll. Ashmead.

Described from a single specimen.

Prosacantha gracilicornis, sp. nov.

♀. Length, 1.8ᵐᵐ. Black; head and scutellum smooth, polished; thorax finely, closely punctate, opaque; legs pale brownish-yellow, the coxæ black basally. Antennæ 12-jointed, brown-black, the scape pale at base, long and slender, about as long as the body; the scape extends slightly beyond the ocelli, slender, a little thickened at the middle; pedicel cyathiform; first and second funiclar joints long, equal, about five times as long as thick, the third half the length of second; the fourth still shorter, a little longer than thick; club 6-jointed, very slender, and about as long as the the first and second funiclar joints together, the joints only slightly wider than long. Wings hyaline, pubescent, the venation brown. Abdomen a little longer than the head and thorax together, the petiole longer than thick, finely pubescent and coarsely striated, the second segment coarsely striate at base only, the third with some fine longitudinal aciculations basally.

HABITAT.—Carolina.

Type in Berlin Museum.

Described from a single specimen, labeled simply "Carolina, Zimmermann."

Prosacantha marylandica, sp. nov.

♂ ♀. Length, 2ᵐᵐ. Black; thorax and scutellum rugoso-punctate; head smooth with some sparse punctures and some broad but not deep striæ on lower part of the face. Mandibles pale rufous. Antennæ black, the scape basally and the pedicel at apex yellowish; the first funiclar joint is twice the length of the pedicel and one-third longer than the second; third and fourth joints small, transverse. Legs, including coxæ, brownish-yellow. Post-scutellar spine large, triangular, nearly horizontal and projecting over the base of the scutellum. Wings subhyaline, the marginal vein at least six times as long as the stigmal. Abdomen oblong-oval, longitudinally striated, the fourth and following segments and the venter, punctate.

In the male the post-scutellar spine is not so decumbent, and does not project over the base of the abdomen; the third abdominal segment is smoother and shows only a few longitudinal striæ on the disk; while the antennæ are long, filiform, the first flagellar joint the longest, about five times as long as thick, the following a little shorter.

HABITAT.—Oakland, Md.

Types in Coll. Ashmead.

Described from specimens received from Mr. E. A. Schwarz.

Prosacantha fuscipennis Ashm.

Ent. Amer., III, p. 117; Cress. Syn. Hym. p. 313.

♂. Length, 2.2ᵐᵐ. Robust, black; thorax and scutellum coarsely rugoso-punctate, opaque; head large, broad, shining, vertex polished, the frons and face coarsely, longitudinally striate. Antennæ 12-jointed, long, filiform, brown-black, the scape towards basal half rufous; first flagellar joint the longest, the third, the shortest and emarginated at base; the joints after the third longer, very slightly increasing in length to the last. Legs rufous, the hind coxæ sometimes black. Post-scutellar spine large, triangular, acute. Wings fuscous or sub-fuscous, the marginal vein nearly six times as long as the stigma. Abdomen oblong-oval, polished, the first and second segments coarsely longitudinally striated, the third feebly longitudinally striated, the following smooth.

HABITAT.—Jacksonville, Fla.

Types in Coll. Ashmead.

Prosacantha columbiana, sp. nov.

♀. Length, 2.2ᵐᵐ. Black, shining, sparsely, faintly, minutely punctate; the head polished, lower part of orbits, on either side of the insertion of the antennæ and cheeks, striated. Mandibles reddish-yellow. Antennæ brown-black, the basal half of scape and tubercles pale rufous; first and second joints of funicle of an equal length, twice the length of the pedicel, the third and fourth transverse. Legs, including coxæ, brownish-yellow. Post-scutellar spine acute, triangular; metapleura smooth, polished. Wings subfuscous, hyaline toward base. Abdomen oblong-oval, polished, the first and second segments coarsely striated, the third more or less striated basally, and the rest of the segments smooth, impunctured.

HABITAT.—District of Columbia.

Types in Coll. Ashmead.

Prosacantha bilineata, sp. nov.

♀. Length, 1.5ᵐᵐ. Black, shining, the thorax, excluding the scutellum, minutely punctulate, with indications of furrows posteriorly; head highly polished, impunctate, with some striæ toward the mouth. Mandibles piceous. Antennæ 12-jointed, black, the extreme base of scape reddish; first funicular joint not quite twice as long as the pedicel, the second, a little shorter, third and fourth, small, transverse. Legs, including coxæ, pale rufous. Post-scutellar spine very acute. Metathorax with small acute teeth at the posterior angles, the metapleura crenulate. Wings hyaline, the venation yellowish. Abdomen oblong-oval, smooth, shining, the first segment and the second at base, striated.

HABITAT.—Washington, D. C.

Type in Coll. Ashmead.

Described from 4 specimens.

Prosacantha pleuralis, sp. nov.

♀. Length, 1.5ᵐᵐ. Black, opaque, finely rugose, the face smooth, shining, the pleura rufous, the mesonotum with slight indications of furrows posteriorly. Mandibles large, falcate. Antennæ 12-jointed, brown-black, the basal half of scape pale; first and second funicular joints nearly equal in length, the second, slightly the longer, third and fourth joints not minute, fully as thick at apex as the preceding joints, subquadrate. Legs, including coxæ, brownish-yellow. Post-scutellar spine rather large, acute, oblique. Wings subfuscous, hyaline at base, the venation brown. Abdomen oblong-oval, polished, shining, the first and second segments of an equal length, striated.

HABITAT.—District of Columbia.

Types in Coll. Ashmead.

Described from 2 specimens.

Prosacantha Linellii, sp. nov.

♀. Length, 2.2ᵐᵐ. Black; the thorax and scutellum finely rugose, opaque; the head broad, shining, inner orbits and lower part of face coarsely striated, the middle of face smooth, shining, with a few minute punctures, the cheeks with punctate striæ. Antennæ 12-jointed, brown-black, the scape towards base, brown; first flagellar joint about twice as long as the pedicel, the latter pale at apex; the second, one-third shorter than the first; third and fourth, transverse. Post-scutellar spine large, triangular. Wings hyaline, the venation pale brownish. Legs, including coxæ, reddish-yellow. Abdomen oblong-oval, shining, the first and second segments striated, the third, with some faint striæ toward base.

HABITAT.—Long Island, New York.

Type in Coll. Ashmead.

Described from a single specimen obtained from Mr. Martin Linell.

Prosacantha erythropus, sp. nov.

♀. Length, 2.2ᵐᵐ. Robust, black, shining, punctate; head very wide, inner orbits, cheeks, and lower part of face, striated; apical half of scutellum smooth, impunctate; scape, mandibles, and legs, reddish-yellow; flagellum brown-black. Antennæ 12-jointed, the flagellum stout; the second funicular joint one-half longer than thick, and a little shorter than the first, third and fourth transverse-quadrate. Post-scutellar spine large acute. Angles of metathorax spined. Wings fuscous. Abdomen oblong-oval, shining, the first and second segments and the basal two-thirds of the third longitudinally striated, the rest smooth.

HABITAT.—District of Columbia.

Types in Coll. Ashmead.

Described from two specimens.

Prosacantha flavicoxa, sp. nov.

♀. Length, 2.1ᵐᵐ. Black; head wide, sparsely punctate, shining, the inner orbits and lower part of face striated; thorax and scutellum rugulose, opaque, pubescent; scape and legs, including all coxae, bright yellow; flagellum black. Antennae 12-jointed; first funicle joint about twice as long as the pedicel, 2½ times as long as thick; second joint shorter; third and fourth minute, transverse; club 6-jointed, about as long as the funicle, the joints transverse, the second a little the longest joint. Postscutellar spine triangular, acute, horizontal. Angles of metathorax acutely spined. Wings clear hyaline, the nervures yellowish. Abdomen oblong-oval, polished, black, the first and second segments striated.

HABITAT.—Virginia.

Type in Coll. Ashmead.

Described from a single specimen.

Prosacantha americana Ashm.

Ent. Am., III, p. 116, ♂ ; Cress. Syn. Hym., p. 313.

♂. Length, 1.2ᵐᵐ. Black; head transverse but thick, polished, the face smooth, with some faint striae only below the eyes, the inner orbits not striated; thorax and scutellum coarsely rugulose, opaque; antennae long, filiform, the scape pale only at base; legs yellow, the coxae black, shining.

Post-scutellar spine small. Angles of metathorax subacute, small. Wings subhyaline, the nervures brown. Abdomen broadly oval, the first and second segments striated, the following smooth, polished, black.

HABITAT.—Jacksonville, Fla.

Type in Coll. Ashmead.

Prosacantha flavopetiolata, sp. nov.

♂. Length, 1.5ᵐᵐ. Black, shining, very sparsely punctate; head transverse, not thick, the face highly polished, impunctate; mandibles and legs brownish-yellow; thorax with faint indications of furrows posteriorly. Antennae long, filiform, 12-jointed, black, the scape yellowish beneath; the flagellar joints long, cylindrical, the second a little longer than the first, the third thickened toward tip. Post-scutellar spine small, triangular. Angles of metathorax not prominent or acute. Wings hyaline, the nervures pale brown, the marginal vein very long and slender, the stigmal very short but ending in a little knob. Abdomen broadly oval, black, shining, the petiole yellow, the first and second segments striated, rest of the segments polished, impunctured.

♀. Length, 1.2ᵐᵐ. Differs from male in having the thorax and scutellum closely punctulate, opaque, the scutellar spine larger and broad at base, while the antennæ are clavate, black, the scape yellow, the first and second funiclar joints equal, twice as long as thick, the third and fourth, small, transverse.

HABITAT.—District of Columbia and Arlington, Va.

Types in Coll. Ashmead.

Prosacantha macrocera Ashm.

Ent. Am. III. p 117, ♂ ; Cress. Syn. Hym., p. 313.

♂. Length, 1.8ᵐᵐ· Black, shining, faintly punctate; head wide, the frons and face striated; mandibles large, rufous; legs rufous. Antennæ 12-jointed, very long, cylindrical, brown-black, the scape rufous; first and third funiclar joints about equal in length, a little shorter than the second, the third angulated outwardly a little before the middle; fourth joint and the following, longer. Scutellum polished, sparsely punctate at base. Post-scutellar spine large, acute, erect, very slightly curved. Angles of metathorax acute. Wings subfuscous, fringed. Abdomen oblong-oval, shining, but minutely sparsely punctate, the first and second segments and the third, on disk, longitudinally striated or aciculated; the first segment brownish or dull rufous, rest of the segments black.

HABITAT.—Jacksonville, Fla.

Type in Coll. Ashmead.

Prosacantha xanthognatha, sp. nov.

♂. Length, 1.5ᵐᵐ. Black, shining; head very wide, polished, the frons smooth. The inner orbits and lower part of face striated; mandibles large, yellow, the teeth in the left mandible about equal, in the right with the outer the longer; thorax and scutellum closely punctulate or minutely rugulose, shining, the parapsidal furrows slightly indicated posteriorly; legs brownish-yellow. Antennæ 12-jointed, brown-black, the scape yellowish: first flagellar joint a little shorter than the second; fourth and following longer than the second. Post-scutellar spine triangular, acute, oblique. Angles of metathorax subacute. Wings subfuscous, not fringed, the nervures brown. Abdomen oval, black, shining, the first and second segments coarsely striated, the third and following punctate.

HABITAT.—Jacksonville, Fla.

Type in Coll. Ashmead.

Closely allied to *P. macrocera* but slightly smaller, the thorax more closely punctate, the antennæ shorter, the third flagellar joint not angulated, while the wings are not fringed.

Prosacantha striativentris, sp. nov.

♂. Length, 1.5ᵐᵐ. Black; the head and abdomen shining: the thorax opaque, minutely, closely punctulate; abdomen wholly longitudinally striated; legs brownish-yellow, the coxæ dusky basally. Head transverse, the face convex, polished, the orbits and lower part of face striated. Mandibles yellowish, the teeth black. Antennæ 12-jointed, filiform, brown-black, the scape pale toward the base; first funicular joint longer than the second, the third a little shorter than the second; the following a little longer than the third. Post-scutellar spine, small, acute. Angles of metathorax small but acute. Wings subhyaline, the nervures pale brown.

HABITAT.—District of Columbia.

Type in Coll. Ashmead.

Described from a single specimen. The striated abdomen and the relative length of the flagellar joints readily distinguish the species.

TELEAS Latreille.

Gen. Crust. et Ins., IV, p. 32 (1809).

(Type *T. clavicornis* Latr.)

Head transverse, the occiput margined; ocelli 3, rather close together in a triangle; eyes ovate.

Antennæ 12-jointed in both sexes, inserted on a clypeal prominence; in ♂ filiform, in ♀ ending in a 6-jointed club.

Maxillary palpi 4-jointed; labial palpi 3-jointed.

Mandibles falcate, bidentate at apex, the outer tooth long, acute.

Thorax short, ovoid, the prothorax not visible from above; mesonotum short, broader than long, without furrows; scutellum semicircular; post-scutellum spined; metathorax very short, the posterior angles rarely acutely toothed.

Front wings with a linear marginal vein, rarely five times as long as the short stigmal, postmarginal not developed.

Abdomen long-oval, depressed, inserted far above the hind coxæ, the first segment longer than wide, the third the longest and widest.

Legs slightly pilose, the posterior femora swollen, their tibiæ and first tarsal joint dilated, the tibial spurs distinct but short.

The swollen hind femora, their dilated tibiæ and tarsi, stronger tibial spurs, and the larger mandibles can be depended upon to distinguish the genus.

The species are rare and only a few have been described that really belong here.

TABLE OF SPECIES.

Winged.
  All coxæ and legs pale brownish-yellow 2
  All coxæ black, legs pale rufous or brownish-yellow.
    Thorax and scutellum coarsely rugose.
      Vertex transversely striated; three basal abdominal segments coarsely striated, the fourth basally with a fine, transverse, wavy sculpture; rim of scutellum yellow............T. LINEATICEPS, sp. nov.

Vertex not striated, but slightly sculptured; three basal abdominal segments striated, the fourth and following, finely punctate at base; rim of scutellum black...................T. COXALIS, sp. nov.

Vertex smooth; first abdominal segment alone striate; the second and third, polished, but with a microscopic wavy sculpture.

T. MANDIBULARIS Ashm.

2. Vertex of head rugose but not striated; three basal segments striated.

T. PALLIDIPES, sp. nov.

Teleas lineaticeps, sp. nov.

♂. Length, 2.80^{mm}. Black, pubescent; head transverse, the vertex and occiput transversely striated; the face, except a space at the middle, longitudinally striated. Antennæ 12-jointed, dark-brown, the scape pale at base; pedicel small, annular; first funiclar joint four times as long as thick (antennæ broken beyond). Thorax and scutellum coarsely rugose or scabrous. Posterior rim of scutellum yellow. Post-scutellar spine large, triangular, oblique. Metapleura smooth, polished. Legs pale rufous or brownish-yellow, the coxæ black, the posterior femora swollen, dusky, their tibiæ dilated at tip with two distinct spurs; tarsi dilated, spinous. Wings fusco-hyaline, pubescent. Abdomen black, shining, as long as the head and thorax together, narrowed at base; the first three segments wholly longitudinally striated; the fourth, with a fine, transverse wavy sculpture; the first, at extreme base, yellowish.

HABITAT.—District of Columbia and Virginia.

Types in Coll. Ashmead.

The larger size of this species, longer first funiclar joint of antennæ, coarser sculpture, yellow rim of scutellum, and stouter femora and tarsi, will at once separate it from the European *T. clavicornis* Latr., specimens of which are in my collection from France.

Teleas coxalis, sp. nov.

♀. Length, 1.80 to 2^{mm}. Black, shining, pubescent: head transverse, polished, the vertex very slightly sculptured but not striated; lower part of face and the cheeks coarsely striate, the latter rounded; thorax coarsely, lineatedly rugose. Mandibles very large, rufous, the teeth stout, the outer the longer. Antennæ 12-jointed, black, reaching to the middle of the abdomen, the scape pale at base; first funiclar joint long, twice as long as the scape; second, one-third shorter than first; third and fourth transverse. Post-scutellar spine large, triangular, oblique. Metapleura striated. Angles of metathorax produced into a tooth. Legs pale rufous, the coxæ black. Abdomen broadly oval, polished, the three basal segments striated, the third laterally smooth, the fourth and following segments very finely punctulate at base.

HABITAT.—Arlington, Va.

Type in Coll. Ashmead.

Described from two specimens.

Its smaller size, non-striated vertex, the lineated rugosities of the thorax, the length of the funiclar joints, and the sculpture of the thorax sufficiently distinguish the species.

Teleas mandibularis Ashm.

Prosacantha mandibularis Ashm., Ent. Am., III, p. 117.

♀. Length, 1.80ᵐᵐ. Robust, black; thorax rugose and covered with a fuscous pubescence; cheeks flattened; lower part of head below the eyes striated, rest of the head smooth, polished. Mandibles long, curved, brownish yellow, the teeth black, the outer tooth very long, acute. Antennæ 12-jointed, very short, black, not reaching beyond the tegulæ, the scape pale at base; first funiclar joint very little longer than the pedicel, the second slightly shorter; third and fourth minute, transverse; club scarcely longer than the funicle. Posterior rim of scutellum, black. Post-scutellar spine small. Wings fusco-hyaline, pubescent. Legs brownish-yellow, the coxæ black. Abdomen a little longer than the head and thorax together, smooth, shining, with a faint wavy-lined sculpture, the first segment alone striated.

HABITAT.—Jacksonville, Fla.

Type in Coll. Ashmead.

The short antennæ, relative length of the funiclar joints, and the sculpture of abdomen can be depended upon to distinguish the species.

Teleas pallidipes, sp. nov.

(Pl. VIII, Fig. 5, ♂.)

♂. Length, 2.20ᵐᵐ. Black, pubescent; head transverse, polished, the vertex almost smooth, the lower part of the face striate; thorax coarsely rugulose, the pronotum and pleura yellowish or rufous. Mandibles large, yellow. Clypeus yellow. Antennæ 12-jointed, very long, filiform, black, the scape and pedicel yellow; first funiclar joint two thirds the length of scape; the second and third one-third shorter than the first, the latter emarginated at base; the joints after the fourth longer than the first. Post-scutellar spine large, triangular, horizontal. Wings subhyaline, pubescent. Legs wholly yellow. Abdomen oblong-oval, narrowed at base, black, shining, the three basal segments striated.

HABITAT.—New Jersey.

Type in Coll. Ashmead.

Described from a single specimen. It is difficult to tell whether this species is the opposite sex of either of the two females described above, but as the coxæ are pale, and the sculpture is different, I surmise it to be a distinct species.

HOPLOGRYON Ashm., gen. nov.

(Type *H. minutissimus* Ashm.)

Head transverse; cheeks above flattened, broader at base, face subconvex; vertex subacute; ocelli 3, triangularly arranged, the lateral farther away from the margin of the eye than to the front ocellus; eyes ovate, villose.

Antennæ inserted on a tubercle just above the clypeus, 12-jointed in both sexes, in the ♀ terminating in a 6-jointed club, funiclar joints 3 and 4 minute; in the ♂ long, filiform, the flagellar joints all cylindrical, the third not angulated.

Maxillary palpi 3-jointed.

Mandibles bifid, the teeth equal.

Thorax subovoid, the mesonotum without furrows, rarely with traces of furrows posteriorly; scutellum short, semicircular, with a punctate frenum; post-scutellum produced into a small spine; metathorax short, the angles obtuse or acutely produced.

Wings occasionally wanting; when present with the venation as in *Telcas*.

Abdomen broadly oval or oblong-oval, the first segment transverse, rarely as long as wide, striated, the third segment very large.

Legs as in *Gryon*.

Differs from *Prosacantha* Nees by the short transverse first abdominal segment, and by having the teeth in both mandibles equal; and from *Gryon* Hal. by having the postscutellum spined or tubercular.

TABLE OF SPECIES.

FEMALES.

3. All coxæ black, or black basally.
  Head and thorax minutely closely punctulate, the face striolated.
    Legs fuscous; trochanters, knees, tips of tibiæ, and tarsi, yellow.
      Scape not pale beneath; third abdominal segment and the venter smooth,
        shining, impunctured ....H. OBSCURIPES, sp. nov.
    Legs rufous.
      Scape pale beneath; third abdominal segment distinctly punctate, the
        venter faintly punctate ....H. RUFIPES, sp. nov.

MALES.

Wingless.
  Thorax minutely rugose, punctate; vertex finely punctate, the face striolated.
  Antennæ very long, filiform, black; legs, except coxæ, rufous.
                                              H. SOLITARIUS, sp. nov.

Hoplogryon longipennis, sp. nov.

(Pl. VIII, Fig. 6, ♀.)

♀. Length, 1^{mm}. Black, opaque, minutely, closely punctate, with a sericeous pubescence; cheeks flat and narrowed back of the upper part of the eye; the face shining, the lower portion and around the mouth, striated. Mandibles pale. Antennæ 12-jointed, black, the scape pale beneath toward base; first and second funiclar joints almost equal, longer than the pedicel and very little longer than thick; third and fourth joints minute, transverse. Post-scutellar spine acute. Metathorax with acute angles. Legs-fuscous or brown, the trochanters, knees, tips of tibiæ, and tarsi pale; coxæ black. Wings subhyaline, ciliated; when folded, longer than the abdomen. Abdomen broadly oval, narrowed at base; first segment a little wider than long; it, as well as the second, striated; third segment faintly microscopically punctate, but shining.

HABITAT.—Ottawa, Canada.

Types in Coll. Ashmead.

Described from 2 specimens received from Mr. W. H. Harrington.

Hoplogryon minutissimus, Ashm.

Prosacantha minutissima Ashm., Ent. Am., III, p. 117, ♀.

♀. Length, 0.75^{mm}. Black, subopaque, closely, microscopically punctulate; face smooth, shining, with some striæ only between the eye and the mandibles. Mandibles brownish. Antennæ 12-jointed, black; the first and second funiclar joints subequal, a little longer than thick; third and fourth, minute, transverse. Post-scutellar spine small, but acute. Angles of metathorax not acute. Legs brown or fuscous, the trochanters, knees, tips of tibiæ, and the tarsi pale; coxæ black. Wings subhyaline, fringed; when folded scarcely extending beyond the tip of the abdomen. Abdomen broadly oval, polished, impunctured, the first and second segments striated.

HABITAT.—Jacksonville, Fla.

Type in Coll. Ashmead.

Allied to *IL. longipennis*, but smaller, more shining, the punctation finer, the wings shorter, the angles of the metathorax not acutely toothed, while the abdomen, except the first two segments, is impunc.. tured, the first segment being much wider than long.

Hoplogryon tibialis, sp. nov.

♀. Length, 1.20ᵐᵐ. Black, subopaque, very closely, minutely punctate; face smooth, polished, with some faint striæ on either side the antennæ. Mandibles pale rufous, the teeth black. Antennæ 12-jointed, black; first and second joints about equal, fully twice as long as thick; third and fourth minute, transverse. Post-scutellar spine acute. Angles of metathorax acute. Legs yellow, the coxæ black, while all the tibiæ and tarsi are fuscous. Wings smoky, with short cilia. Abdomen oval, much narrowed at base, highly polished, black; the first segment, and the second at base and at the middle, striated; the first segment is as long as wide.

HABITAT.—Virginia.

Type in Coll. Ashmead.

Described from a single specimen received from Mr. O. Heidemann. This species is quite distinct from the others by the length of the first and second funiclar joints, smoky wings, and the color of the legs.

Hoplogryon claripennis, sp. nov.

♀. Length, 1.20ᵐᵐ. Black, shining; head transverse, polished, with some faint striæ below the eyes; thorax faintly punctulate. the scutellum polished, impunctured; mandibles piceous. Antennæ 12-jointed, black; the scape at base and the pedicel at tip, pale; first and second funiclar joints nearly equal, the first more than twice as long as thick; third and fourth small, subquadrate. Post-scutellar spine, oblique, acute. Metathoracic angles produced into acute spines. Legs, including coxæ, brownish-yellow. Wings hyaline, scarcely tinged. Abdomen oblong oval, narrowed at base and highly polished, the first and second segments coarsely striated, the first nearly as long as wide.

HABITAT.—Virginia.

Type in Coll. Ashmead.

This pretty species was given to me by my friend Mr. O. Heidemann. Its shining surface, less distinctly punctured thorax, highly polished scutellum, and the length of the funiclar joints readily distinguish it.

Hoplogryon pteridis, sp. nov.

♀. Length, 0.80ᵐᵐ. Black, subopaque, closely microscopically punctulate; face smooth, polished; mandibles yellowish. Antennæ 12-jointed, black; first and second funiclar joints equal, scarcely longer than thick and very little longer than the pedicel; third and fourth minute, transverse. Post-scutellar spine distinct, acute. Angles of metathorax subacute. Legs, including the coxæ, yellow, the posterior pair slightly dusky at base. Wings subhyaline, when folded, not or scarcely longer than the abdomen and somewhat narrowed. Abdomen rotund, oval, polished. widened from the base of the first segment; first and second segment at base striated. the first, twice as wide as long.

HABITAT.—Fortress Monroe, Va., and District of Columbia.

Types in Coll. Ashmead.

This species was taken by Mr. E. A. Schwarz, at Fortress Monroe. The color of the legs and the shape of the abdomen and wings, readily separate the species from all the others.

Hoplogryon brachypterus Ashm.

Prosacantha brachyptera Ashm., Can. Ent., xx, p. 50.

♀. Length, 0.75 to 0.80ᵐᵐ. Black, subopaque, closely microscopically punctulate; the head polished. shining, the face having a delicate central line below the front ocellus; mandibles pale; legs brownish-yellow, reddish, brownish or fuscous, with the distal ends of femora and tibiæ and the tarsi, yellow or pale. Antennæ 12-jointed, black or brown-black; first and second funiclar joints subequal, a little longer than thick; third and fourth, very minute. Post-scutellar spine minute. Angles of metathorax not toothed. Wings subhyaline, much narrowed and shortened, when folded reaching scarcely beyond the middle of the abdomen, pubescent, but not distinctly fringed. Abdomen broadly oval, highly polished, black, the first segment wider than long, yellowish and striated.

HABITAT.—Ottawa, Canada.

Types in Coll. Ashmead.

For several specimens of this highly interesting species. I am indebted to Mr. W. H. Harrington. The abbreviated, narrowed, non-ciliated wings and the yellow petiole distinguish the species.

Hoplogryon obscuripes, sp. nov.

♀. Length, 0.80ᵐᵐ. Black, opaque. closely. minutely punctulate, covered with a fine sericeous pubescence; the face shining, striolated. Mandibles pale brown. Antennæ 12-jointed, brown-black: first and second funiclar joints subequal, the first about one and a half times as long as thick; third and fourth minute. transverse. Post-scutellar spine minute but acute. Metathoracic angles obtuse. Wings entirely wanting. Legs fuscous, the coxæ black or dusky basally, the trochanters, knees, tips of tibiæ, and tarsi, yellowish. Abdomen oval, polished, the first segment transverse, twice as wide as long, striated, the second striated only at the basal suture, the venter impunctate.

HABITAT.—Ottawa, Canada.

Types in Coll. Ashmead.

Described from two specimens received from Mr. W. H. Harrington.

Hoplogryon rufipes, sp. nov.

♀. Length, 0.80ᵐᵐ. Black, subopaque, microscopically punctulate, the face striolate. Mandibles pale. Antennae 12-jointed, black, the scape pale beneath; first and second funicular joints equal, 1½ times as long as thick; third and fourth minute, transverse. Post-scutellar spine minute, triangular. Angles of metathorax acute. Wings wanting. Legs rufous, the tarsi yellowish. Abdomen broadly oval, shining, the first and second segments transverse, longitudinally striated; the third and the venter microscopically punctate.

HABITAT.—District of Columbia.

Type in Coll. Ashmead.

Described from a single specimen collected by Mr. E. A. Schwarz. The color of the legs and the fine punctation of the third dorsal segment and the venter distinguish the species.

Hoplogryon solitarius, sp. nov.

♂. Length, 1.5ᵐᵐ. Black, subopaque, closely, minutely punctulate; head wide, 3½ times as wide as thick antero-posteriorly, the cheeks rounded, margined behind, the face smooth, the orbits and around the mouth striated. Mandibles rufous. Antennae 12-jointed, filiform, much longer than the body, black, the scape pale basally; first flagellar joint about half the length of the scape; second, very slightly longer than the first; third to the penultimate, nearly of an equal length, but slightly shorter than the first; last joint slightly longer. Scutellum smoother than the thorax, shining. Post-scutellar spine acute. Angles of metathorax acutely toothed. Wings wanting. Legs, except coxae, uniformly rufous. Abdomen oblong-oval, polished, very little longer than the head and thorax united, the first segment transverse, much broader than long, the second a little longer than the first, both coarsely striated.

HABITAT.—Ottawa, Canada.

Type in Coll. Ashmead.

Described from a single specimen received from Mr. W. H. Harrington.

On account of the very large size of this species, in comparison with the described wingless females, it is scarcely probable that it can be the opposite sex of any of them.

GRYON Haliday.

Ent. Mag., I. p. 271 (1833); Först. Hym. Stud., II. p. 101 (1856).

(Type *G. misellus* Hal.)

Head large, transverse, the occiput scarcely emarginated; ocelli 3, triangularly arranged, close together on vertex; eyes ovate, villose.

Antennæ 12-jointed in both sexes; in ♀ clavate, the club large, 6-jointed, the last two funiclar joints minute; in ♂ filiform.

Maxillary palpi 3-jointed.

Mandibles bifid, the teeth equal.

Thorax short, subovoid, the prothorax not at all visible from above; the mesonotum more than twice as wide as long, without furrows; scutellum short, transverse; postscutellum not spined, simple; meta-thorax very short, the angles subacute or acute.

Wings most frequently wanting, when present with the venation as in *Teleas*, the marginal vein rarely more than 4 times as long as the stigmal.

Abdomen broadly oval, the first and second segments short, trans-verse, of nearly an equal length, usually striated, the third very large, but broader than long.

Legs as in *Prosacantha*, the tibial spurs weak, the basal joint of hind tarsi a little more than thrice as long as the second.

Closely allied to *Prosacantha* and *Hoplogryon*, but separated from both by the simple, *not spined*, postscutellum.

Dr. Förster's description of this genus in the Hym. Stud. is scarcely correct and very misleading. I have recognized the genus by Walker's figure of the male in Ent. Mag., Vol. III, Pl. XIII, Fig. 5, which plainly shows its relation to *Prosacantha*, but is distinct in having the post-scutellum simple, not spined, and the basal abdominal segment much broader than long, not petioliform.

TABLE OF SPECIES.

FEMALES.

Wingless species 2
Winged.
  Head and thorax finely punctulate; metathoracic angles obtuse.
    Mesonotum with traces of furrows posteriorly.
      Coxæ and legs yellow ....G. FUMIPENNIS, sp. nov.
2. Head and thorax finely, closely punctulate, subopaque, the lower part of face with striæ converging toward the mouth.
  All coxæ black; first and second abdominal segments striated.
    Metathoracic angles obtuse or subacute.
      Legs pale rufous or brownish-yellow; abdomen shining, but microscopic-ally punctate ....G. BOREALIS Ashm.
      Legs fuscous, the trochanters, knees, and tarsi yellow; abdomen polished, impunctate...............................G. CANADENSIS Ashm.
    Metathoracic angles acute.
      Legs brownish-yellow; abdomen, above and beneath, punctate, subopaque.
        G. COLUMBIANUS, sp. nov.
  All coxæ pale; metathoracic angles obtuse.
    Legs pale yellow; abdomen smooth shining...........G. FLAVIPES, sp. nov.

Gryon fumipennis, sp. nov.

♀. Length, 1.40ᵐᵐ. Black, shining, minutely punctulate; face smooth, polished, with faint striæ at the mouth; mandibles and legs yellow. Antennæ 12-jointed, black, the scape yellowish toward the base; first two funiclar joints equal, about twice as long as the pedicel; third and fourth, minute, transverse. Mesonotum with traces of the parapsidal furrows posteriorly. Scutellum smoother than the mesonotum, but still slightly punctate. Metathorax laterally above and the metapleura polished, the angles obtuse. Wings smoky-hyaline, paler at base. Abdomen oval, smooth, polished, the first and second segments striated.

HABITAT.—District of Columbia.

Type in Coll. Ashmead.

Described from one specimen received from Mr. E. A. Schwarz.

Gryon borealis Ashm.

(Pl. VIII, Fig. 7, ♀.)

Acolus borealis Ashm., Can. Ent., xx, p. 50.

♀. Length, 1ᵐᵐ. Apterous; black, subopaque, closely microscopically punctate and with a fine sericeous pubescence; lower part of face striated. Mandibles pale. Metathorax with the angles subacute. Antennæ 12-jointed, black or brown-black; first funiclar joint longer and thicker than the pedicel; second one-third shorter than the first; third and fourth minute, transverse. Legs pale rufous or brownish-yellow, the coxæ black. Abdomen oval, microscopically punctate, but shining, the first and second segments striated.

HABITAT.—Ottawa, Canada.

Types in Coll. Ashmead.

Described from four specimens received from Mr. W. H. Harrington.

Gryon canadensis Ashm.

Acolus canadensis Ashm., Can. Ent., xx, p. 50.

♀. Length, 0.60ᵐᵐ. Apterous; black, subopaque, closely, minutely punctulate and pubescent; face polished, with striæ at base. Mandibles pale rufous. Antennæ 12-jointed, black, the scape pale at base; first and second funiclar joints very little longer than thick, the second slightly the smaller; third and fourth minute, transverse; club very large, the joints close set, transverse. Scutellum small, sublunate, shining. Metathorax short, narrowed posteriorly, the angles not at all prominent. Legs fuscous, the trochanters, knees, and tarsi, honey-yellow. Abdomen rotund-oval, smooth, shining, impunctate, the first and second segment very short, transverse, striolate, the second with a fringe of glittering hairs at apex.

HABITAT.—Ottawa, Canada.

Types in Coll. Ashmead.

Described from four specimens received from Mr. W. H. Harrington. Allied to *G. borealis*, but smaller, the legs differently colored, the metathoracic angles not at all prominent, while the abdomen is polished, impunctate.

Gryon columbianus, sp. nov.

♀. Length, 1.40ᵐᵐ. Subrobust, black, opaque, closely, minutely punctulate; head transverse, wider than thorax, the anterior orbits, broadly, and the cheeks, striated; mandibles pale rufous; scape and legs, except coxae, brownish-yellow or pale rufous; flagellum brown-black. Antennae 12-jointed, not quite as long as the body; first and second funicular joints about equal, twice as long as the pedicel; third and fourth minute, annular; club a little longer than the pedicel and funicle united, rather stout, the joints transverse-quadrate. Angles of metathorax acute; post-scutellar ridge emarginated at the middle. Abdomen broadly oval, closely punctate, subopaque, sericeous, the first and second segments with coarse striae, the apical margin of the second smooth, polished; first and second ventral segments striated, the third and following punctate.

HABITAT.—District of Columbia.

Types in Coll. Ashmead.

Described from several specimens.

Gryon flavipes, sp. nov.

♀. Length, 0.60ᵐᵐ. Black, shining, finely punctulate; face smooth, polished; legs pale yellow; metathorax with obtuse angles. Antennae 12-jointed, brown-black, the scape, beneath, toward base, pale; first funicular joint a little longer than the pedicel; second shorter than the first, not longer than thick; third and fourth, minute, transverse. Abdomen rotund-oval, smooth, shining, the first segment and the second at the suture, striated; venter piceous.

HABITAT —Ottawa, Canada.

Type in Coll. Ashmead.

Described from a single specimen received from Mr. W. H. Harrington. Differs from the other species in the smoothness of the face, paler legs, smoothness and sculpture of the abdomen, and the color of the venter.

Tribe IV.—SCELIONINI.

A tribe allied to the *Teleasini* and always with the abdomen distinctly carinated along the sides; but, except in a few genera, the abdomen is much more elongated, being pointed or fusiform, rarely oval, and extends beyond the tips of the wings when folded, the third segment the longest, or the second and third are about equal in length. The venation, however, is quite distinct; the postmarginal nervure, except in a few cases, is fully developed and usually longer than the marginal, while the stigmal is never very short. When the postmarginal nervure is absent the submarginal terminates in a stigma (*Baeoneura* and *Scelio*).

The parasitism of many of the genera is known, all being parasites in the eggs of various insects, but principally in those belonging to the Orthoptera and Hemiptera. *Baryconus* and *Cacus* attack the eggs of the White Tree-crickets, *Oecanthus niveus*, etc.; *Macroteleia*, those of *Orchelimum*; *Scelio*, various Locusts or Grasshoppers, *Dissosteira*, *Caloptenus* and *Acridium*; while *Hadronotus*, a genus easily mistaken for *Telenomus*, is parasitic on different Heteroptera-Hemiptera, belonging to the families Coreidæ, Pyrrhocoridæ and Reduviidæ.

It is believed the genera can be readily separated by the aid of the following table:

TABLE OF GENERA.

FEMALES.

Postmarginal vein wanting or never greatly developed, always shorter than the stigmal vein, the submarginal vein often never reaching the costa and terminating in a large stigma, the abdomen long, fusiform....4
Postmarginal vein always greatly lengthened, the submarginal never terminating in a stigma.
  Basal vein wanting.................3
  Basal vein present.
    First abdominal segment without a horn at base.................2
    First abdominal segment with a horn at base.
      Marginal vein short; abdomen long, pointed, fusiform, the first segment narrow, petioliform, the second and third segments nearly equal.
                          CALOTELEIA Westw.
      Marginal vein long; abdomen long, linear or subfusiform, the first segment quadrate or subquadrate.................BARYCONUS Först.
2. Abdomen long, pointed fusiform or linear, segments 2, 3, and 4 nearly equal.
  Mesonotum with 2 furrows.
    Metascutellum without a spine.
      Metanotum with no inclosed space at base.
        Marginal vein about twice the length of the stigmal.
          Mandibles 3-dentate.................MACROTELEIA Westw.
          Mandibles 2-dentate.................CALLISCELIO, gen. nov.
        Marginal vein very short.
      Metanotum with a large, semicircular inclosed space at base.
        Marginal vein punctiform.................CHROMOTELEIA, gen. nov.
  Abdomen oblong-oval or fusiform, but not especially lengthened.
    Metascutellum spined.
      Mesonotum with 2 furrows.
        Mandibles 2-dentate; abdominal segments 1 and 2 equal in length, the 3d longer.................OPISTHACANTHA, gen. nov.
      Mesonotum without furrows.
        Mandibles 2-dentate; abdominal segments 1 and 2 equal in length, the 3d longer.................(OPISTHACANTHA).
        Mandibles 3-dentate; segments 2 and 3 equal in length, the 1st shorter.
                          LAPITHA, gen. nov.
    Metascutellum not spined, simple.
      Marginal vein short, or not more than half the length of the stigmal, most frequently punctiform.
      Mesonotum without furrows.
        Head quadrate, mandibles 3-dentate.................CACUS Riley.

Mesonotum with two furrows.
  Antennæ with 5 or 6 jointed club.....................ANTERIS Förster.
  Antennæ filiform, without a club.....................APEGUS Förster.
3. Mesonotum with three distinct furrows.
  Metascutellum with 2 erect teeth......................HOPLOTELEIA, gen. nov.
  Mesonotum with 2 furrows.
    Abdomen very long, fusiform.
      Metathorax unarmed; mandibles 3-dentate.......MACROTELEIA Westw.
      Metathorax with 2 teeth (mandibles 3-dentate).............CACUS Riley
    Abdomen not very long, ovate or oblong-oval.
      Metathorax unarmed; mandibles 2-dentate............ANTERIS Förster.
  Mesonotum without furrows.
    Metascutellum spined...................................(OPISTHACANTHA).
    Metascutellum simple.
      Abdomen with a horn at base $\begin{cases} \text{Marginal vein short......(CALOTELEIA).} \\ \text{Marginal vein long........(BARYCONUS).} \end{cases}$
      Abdomen without a horn at base.
        Abdomen long, fusiform; mandibles 2-dentate.
          Abdominal segments normal; antennal club 6-jointed..CACUS Riley.
          Abdominal segments strongly constricted; antennal club oval, 5-jointed...............................CREMASTOBAEUS, gen. nov.
        Abdomen broadly oval, sessile, the second segment usually a little the largest....................................HADRONOTUS Förster.
4. Submarginal vein not reaching the costa, knobbed.
  Wings narrow, fringed; abdomen much depressed, long and pointed.
                                             BÆONEURA Först.
Submarginal vein reaching the costa often by a thickened stigma.
  Submarginal vein not terminating in a thickened stigma.
    Marginal vein very short, the postmarginal scarcely developed or shorter than the stigmal.
      Mesonotum with 2 furrows...............................IDRIS Förster.
  Submarginal vein terminating in a thickened stigma.
    Head without a frontal lamina or ledge; postmarginal vein never developed..5
    Head with a frontal lamina or ledge.
      Scutellum quadrate, the posterior angles acute; postscutellum with a large erect spine.............................ACANTHOSCELIO, gen. nov.
      Scutellum and postscutellum simple, not spined........SPARASION Jurine.
5. Mesonotum without furrows, or very rarely with 2 distinct furrows.
  Maxillary palpi short, 3-jointed.............................SCELIO Latreille.
  Mesonotum with 2 distinct furrows.
    Maxillary palpi long, 5-jointed......................SCELIOMORPHA, gen. nov.

MALES.

Postmarginal vein wanting or never greatly developed, always shorter than the stigmal vein, the submarginal vein often never attaining the costa and terminating in a large stigma, the abdomen long-fusiform... 5
Postmarginal vein always greatly lengthened, the submarginal vein reaching the costa and never terminating in a stigma.
  Basal nervure wanting.. 4
  Basal nervure present.
    Mesonotum without furrows................................... 3
    Mesonotum with 2 furrows.
      Metathorax with no inclosed space at base.
        Metascutellum not spined.
          Marginal vein punctiform or never longer than the stigmal.
            Mandibles 3-dentate............................CALOTELEIA Westw.
            Mandibles 2-dentate............................ANTERIS Förster.

Marginal vein longer than the stigmal.
Mandibles 3-dentate.
First flagellar joint scarcely longer than the third, the third excised
                                                    MACROTELEIA Westw.
First flagellar joint much longer than the third.
                                                    BARYCONUS Förster.
Metascutellum spined.................OPISTHACANTHA Ashm., gen. nov.
Metathorax with a large semicircular inclosed space at base.
Marginal vein punctiform; mandibles 3-dentate.
                                            CHROMOTELEIA, gen. nov.
Metascutellum not spined.
Marginal vein punctiform or never as long as the stigmal.
Mandibles 3-dentate.
First flagellar joint very long......................CALOTELEIA Westw.
First flagellar joint shorter than second.....................CACUS Riley.
Marginal vein long, always longer than the stigmal.
Mandibles 3-dentate...BARYCONUS Förster.
Metascutellum spined.
Marginal vein longer than the stigmal; mandibles 3-dentate.. LAPITHA, gen. nov.
Marginal vein shorter than the stigmal; mandibles 2-dentate.
                                        OPISTHACANTHA Ashm., gen. nov.
Mesonotum with three distinct furrows.
Metascutellum armed with 2 teeth; tip of abdomen ending in two short prongs.
                                        HOPLOTELEIA Ashm., gen. nov.
Mesonotum with two furrows.
Metathorax unarmed; mandibles 3-dentate...................(MACROTELEIA).
Metathorax with 2 small teeth; mandibles 2-dentate...........(CACUS Riley).
Mesonotum without furrows.
Metascutellum with a small acute spine.....OPISTHACANTHA Ashm., gen. nov.
Metascutellum simple.
Metanotum with 2 small teeth at apex.
Mandibles 2-dentate..CACUS.
Metanotum unarmed, simple.
Abdominal segments strongly constricted.
Antennae subclavate.......................CREMASTOBÆUS, gen. nov.
Abdominal segments normal.
Antennae filiform, submoniliform..................HADRONOTUS Förster.
Submarginal vein not reaching the costa, knobbed............BÆONEURA Först.
Submarginal vein reaching the costa.
Submarginal vein not terminating in a stigma.
Marginal vein very short, the postmarginal scarcely developed or shorter
        than the stigmal.
Mesonotum with 2 furrows.......................................IDRIS Förster.
Submarginal vein terminating in a thickened stigma.
Head without a frontal ledge or lamina; postmarginal vein not developed. 6
Head with a frontal lamina or ledge.
Scutellum quadrate, the angles acute; the postscutellum spined.
                                            ACANTHOSCELIO, gen. nov.
Scutellum and postscutellum normal...................SPARASION Jurine.
Mesonotum with 2 furrows.
Maxillary palpi long, 5-jointed; antennae long, setaceous, 12-jointed.
                                            SCELIOMORPHA, gen. nov.
Mesonotum without furrows, or rarely distinct.
Maxillary palpi short, 3-jointed; antennae short, fusiform, 10-jointed.
                                            SCELIO Latreille.

CALOTELEIA Westwood.

Trans. Lond. Ent. Soc., II, p. 55, Pl. 7. f. 10.

(Type *C. aurantia* Westw.)

Head transverse, subquadrate, the frons convex, the occiput roundedly emarginated; ocelli 3 in a triangle, the lateral close to the eye; eyes large, oval, bare or faintly pubescent.

Antennæ inserted just above the clypeus, 12-jointed in both sexes; in ♀ clavate, the club 5- or 6-jointed, the last two funiclar joints usually transverse or quadrate; in ♂ filiform, long; the flagellar joints long, cylindrical.

Mandibles 3-dentate.

Maxillary palpi 4-jointed; labial palpi 3-jointed.

Thorax ovate, the prothorax scarcely visible from above, except at the lateral corners; mesonotum usually entirely without furrows, although occasionally with 2 distinct furrows; scutellum semicircular; metathorax emarginate and carinated along the sides.

Front wings with the marginal vein short, punctiform or rarely half the length of the stigmal; the stigmal vein oblique, and usually ending in a little knob; postmarginal vein very long, basal vein distinct, rarely entirely absent.

Abdomen long, fusiform, pointed at tip and narrowed at base; the basal segment in the ♀ with a horn extended forward over the metathorax.

Legs rather long, the femora and tibiæ subclavate; the tibial spurs 1, 1, 1, the middle and hind pairs rather weak; tarsi long, the basal joint several times longer than the second; claws simple.

This genus was originally described by Prof. Westwood, from a single specimen found embedded in Gum animé, in the collection of Rev. F. W. Hope. The genus is not rare in South America and the West Indies, and five species have been discovered in our fauna.

It is closely allied to *Baryconus* Förster; but the marginal nervure is usually punctiform, or at least very short, and the petiole is longer and more slender.

TABLE OF SPECIES.

Body pale.. 2
Body black.
  Coxæ black, legs rufous or brownish-yellow.
    Three basal abdominal segments longitudinally striated, the following smooth,
      polished ....C. STRIATA, sp. nov.
    Two basal abdominal segments striated; first and second funiclar joints very
      long, slender; third and fourth stouter, the third longer than thick;
      the fourth quadrate......................C. HEIDEMANNII, sp. nov.
2. Brownish-yellow.
  Head pale, concolorous with the rest of the body.
    Abdomen banded with black or fuscous.............C. CINCTIVENTRIS Ashm.
    Abdomen not banded; apex of horn and tip of abdomen black or fuscous.
                                    C. RUBRICLAVA Ashm.
  Head black or fuscous ....C. MARLATTII, sp. nov.

Caloteleia striata, sp. nov.

♀ . Length, 2.6ᵐᵐ. Black, lustrous, rather closely punctate; head quadrate, with a polished, impunctate spot on the frons. Antennæ 12-jointed, brown, the base and tip of scape and the tip of pedicel, yellowish; pedicel rather long; first funiclar joint slightly longer and slenderer than the pedicel; second, about half the length of the first and a little stouter; third and fourth, transverse; club 6-jointed, all the joints transverse. Thorax without furrows. Legs rufous, the coxæ black. Wings hyaline, the venation brown; there is a distinct basal vein; the marginal is very short, about one-third the length of the stigmal, while the stigmal ends in a small knob, the postmarginal being very long. Abdomen long, pointed, fusiform, about thrice the length of the thorax, the basal segment with a distinct blunt horn at base, and it as well as the second and third segments are distinctly longitudinally striated; all the following segments are smooth, shining.

HABITAT.—Washington, D. C.

Type in National Museum.

Described from a single specimen.

Caloteleia Heidemanuii, sp. nov.

(Pl. ix. Fig. 1, ♀.)

♀ . Length, 2.4ᵐᵐ. Elongate, slender, black, shining, punctate; head subquadrate; mandibles and palpi yellow; legs rufous, the trochanters, knees, tips of tibiæ and tarsi yellow, the coxæ black, shining, impunctured. Antennæ 12-jointed, piceous, the tips of the scape, pedicel, and funiclar joints yellow, the club black; pedicel and first funiclar joint very long, nearly of an equal length, the latter the slenderer; second funiclar joint one-third shorter than the first; third and fourth stouter, the third longer than thick, the fourth quadrate; club a little thicker than last two funiclar joints, long, cylindrical, the joints subquadrate. Mesonotum without furrows. Mesopleura with a deep femoral groove, pubescent, sparsely punctate. Wings subhyaline, the marginal vein short, scarcely one-third the length of the rather long stigmal vein; basal nervure distinct. Abdomen long, pointed, fusiform, black, shining, the first and second segments striated, the following smooth, polished, impunctured.

HABITAT.—Virginia.

Type in Coll. Ashmead.

Described from a single specimen collected by O. Heidemann. Allied to *C. striata*, but quite distinct by the shape of the head, relative length of funiclar joints, and the sculpture of the abdomen.

Caloteleia cinctiventris Ashm.

Baeoneura cinctiventris Ashm., Ent. Am., III. p. 99, ♀ : Cress. Syn. Hym., p. 313.

♀ . Length, 1.80ᵐᵐ. Brownish-yellow, closely, minutely punctate; abdominal segments at base barred with black or fuscous; legs pale;

the femora and tibiae very slightly obfuscated. Head transverse, with a fuscous spot on vertex. Antennæ 12-jointed, the scape and funicle pale brown, the club black; first funiclar joint slightly shorter than the pedicel, or a little longer than thick; second, shorter; third and fourth minute, transverse; club fusiform, the joints short, transverse. Wings hyaline, with a band below the punctiform marginal vein. Abdomen long, fusiform, minutely punctate; the short horn on first segment black, shining; base of the following segments black or fuscous.

HABITAT.—Jacksonville, Fla.

Type in Coll. Ashmead.

This pretty little species is quite distinct from all the others, in size, in the banded anterior wings, and in the banded abdomen. It was originally described in the genus *Bæoneura* on account of the punctiform marginal vein.

Caloteleia rubriclava Ashm.

Aeolus rubriclavus Ashm., Ent. Am. III. p. 99; Cress. Syn. Hym., p. 313.

♀. Length, 3ᵐᵐ. Pale brownish-yellow; the short horn at base of abdomen black; eyes and ocelli brown. Antennæ 12-jointed, pale; the club large, subsolid, inarticulated, reddish-brown; first funiclar joint shorter than the pedicel; the three following very short. Mesonotum closely, minutely punctulate, with 2 delicate furrows. Wings hyaline, the marginal vein very short. Abdomen long, pointed-fusiform, the first three segments longitudinally striated, the third apically, and the following, smooth, shining.

HABITAT.—Jacksonville, Fla.

Type in Coll. Ashmead.

This species is quite distinct from the others in color, and in having a subsolid club.

Caloteleia Marlattii, sp. nov.

♂♀. Length, 2.8 to 3.2ᵐᵐ. Brownish-yellow, closely, minutely punctulate; the head black or fuscous, the lower part of cheeks and face often pale, especially in the female; the first and apical abdominal segments in the male black or fuscous.

Antennæ 12-jointed, brown-black; in ♂ filiform with the scape pale, the first funiclar joint a little longer and thicker than the pedicel, the second a little shorter than the third, and both shorter than the first, the joints beyond oval-moniliform, loosely joined; in ♀ terminating in a large 6-jointed club, the second funiclar joint about half the length of the first, the third and fourth subquadrate.

Mesonotum with 2 delicate furrows, punctate. Wings hyaline, pubescent, the marginal vein slightly thickened, short; the stigmal vein short and ending in a knob.

Abdomen long, fusiform, much narrowed at base, the first segment being long, petioliform, linear, the three basal segments longitudinally striated, the striations on the third fainter than on the other two, smooth laterally and at apex; rest of the abdomen smooth.

HABITAT.—Kansas, Illinois, and New Jersey.

Type in Coll. Ashmead and National Museum.

BARYCONUS Först r.

Hym. Stud., II, p. 101 (1856).

Head subquadrate or quadrate, roundedly emarginated posteriorly; ocelli 3 in a triangle, the lateral very close to the eye; eyes large, oval, pubescent.

Antennae inserted just above the clypeus, 12-jointed in both sexes, in ♀ terminating in a 6-jointed club, the last funiclar joint transverse; in ♂ long, filiform. the first flagellar joint longer than the third.

Maxillary palpi 4-jointed; labial palpi 3-jointed.

Mandibles tridentate.

Thorax ovoid, the prothorax not or scarcely visible from above; mesothorax with or without furrows; metathorax short, emarginate posteriorly, with delicate keels laterally.

Front wings with a long marginal and a very long postmarginal vein, the marginal being usually twice as long as the stigmal, oblique, clavate, sometimes a little curved; basal vein usually distinct, rarely subobsolete.

Abdomen long, linear, or fusiform, and not so much narrowed at base as in *Caloteleia*, the basal segments in ♀ with a horn at base extending forward over the metathorax, sometimes reduced to an elevated convex prominence.

Legs long, the femora and tibiae subclavate; the tibial spurs 1,1,1; all tarsi slender, much longer than their tibiae.

Förster's type of the genus appears never to have been described. The genus is closely allied to *Caloteleia*, but is readily separated by the longer marginal vein and the difference pointed out in the abdomen. It is parasitic on the eggs of white tree crickets, *Oecanthus*.

Baryconus œcanthi, sp. nov. Riley.

(Pl. IX, Fig. 2 ♀.)

"♂ ♀. Length, 2.5 to 3ᵐᵐ. Black, closely punctate, subopaque and sparsely covered with a sericeous down. Head quadrate; eyes pubescent; antennae 12-jointed, black; in ♂ filiform-submoniliform, the first and second funicular joints elongate, about equal, thrice as long as thick; third slightly shorter; fourth half the length of third; joints beyond to the last, moniliform; the last fusiform, twice the length of the penultimate; in ♀ terminating in a long 6-jointed club.

"Thorax punctate, without furrows; legs black, the base of the tibiæ and tarsi brownish; wings hyaline, pube..rent, the venation brown, the marginal vein longer than the stigmal, the latter ending in a small knob. Abdomen twice as long as the head and thorax together, linear-fusiform, lineatedly rugulose, the apex of the horn in the female polished.

"HABITAT.—Lincoln, Nebr.

"Type in National Museum.

"This species was reared by Prof. L. Bruner from the eggs of *Oecanthus niveus*, and probably is the insect that Mr. Howard Ayers treats under the genus *Teleas* in his biological study published in Memoirs Bost. Soc. N. H., vol. III, p. 225, 1884."—[*From Riley's MS.*]

MACROTELEIA Westwood.

Proc. Zoöl. Soc., 1835, p. 70.

(Type *M. cleonymoides* Westw.)

Head transverse, subquadrate, broader than the thorax, the frons convex, the occiput slightly emarginate; ocelli 3 in a triangle, the lateral touching the eye; eyes oval, bare.

Antennæ inserted just above the clypeus, 12-jointed in both sexes, in ♀ clavate, the club large, 6-jointed; in ♂ long, filiform, the first flagellar joint scarcely longer than the third, the third excised.

Maxillary palpi short, 4-jointed; labial palpi 3-jointed.

Mandibles tridentate.

Thorax ovate, the prothorax slightly visible from above; mesothorax with or without furrows; scutellum semicircular; metathorax not very short with two carinæ above, diverging posteriorly, and with delicate lateral carinæ.

Front wings with a long marginal vein nearly twice the length of the stigmal, the postmarginal greatly lengthened, the stigmal vein oblique, usually with a little knob and sometimes with a radial branch from its tip; basal vein sometimes present, usually obsolete.

Abdomen sessile, greatly elongated, fusiform or linear, projecting beyond the tip of the wings when folded, the first four segments nearly equal.

Legs as in *Baryconus*, the tibial spurs 1, 1, 1, distinct, the basal joint of hind tarsi less than thrice as long as the second.

Distinguished by the long marginal nervure and the greatly elongate, fusiform abdomen. Species occur with and without parapsidal furrows, and with and without a basal nervure, and these characters can be used to separate the genus into sections. If the species become numerous they might be entitled to generic value. The genus is parasitic on the eggs of the locustid genus *Orchelimum*.

TABLE OF SPECIES.

FEMALES.

Abdomen more than five times as long as the head and thorax united, the
three basal segments coarsely lineatedly sculptured, the rest
more finely striated and punctate; scape and legs yellow,

M. MACROGASTER, sp. nov.

Abdomen scarcely thrice as long as the head and thorax united.

Coxæ black; abdomen with a lineated sculpture basally, the apical segments almost smooth but still faintly punctate..M. FLORIDANA Ashm.

Coxæ pale; abdomen with a thimble-like punctation....M. VIRGINIENSIS, sp. nov.

Macroteleia maciogaster, sp. nov.

(Pl. IX, Fig. 6, ♂.)

♂ ♀. Length, ♀. 8ᵐᵐ; ♂, 4.60ᵐᵐ. Brown-black, pubescent, with
round punctures; six basal joints of ♀ antennæ and legs, except coxæ,
brownish-yellow; the pedicel and the first funiclar joint very long, the
latter the longer; second funiclar joint one-half the length of the first;
third and fourth, short, subequal; club rather slender, the joints, except the last, a little wider than long; the ♂ antennæ are filiform; the
pedicel greatly elongate, as long as the first and second funiclar joints
together, the joints beyond a little more than twice as long as thick.
Mesonotum with 2 furrows. Scutellum short, transverse, with a row
of punctures behind. Metanotum very short, the abdomen being inserted just back of the post-scutellum on a line before the insertion of
hind coxæ. The abdomen in both sexes is very long, in the ♀ meas-
uring 5ᵐᵐ in length and more than twice the length of the wings when
folded, with coarse, rounded punctures and little raised lineations;
after the third segment the punctures are smaller and the surface is
smoother. Wings fusco-hyaline, the marginal vein about 1½ times as
long as the stigmal, the latter ending in a small knob, the postmarginal very long.

HABITAT.—Texas.

Types in Coll. Ashmead.

Described from 1 ♂, 1 ♀ specimen.

Macroteleia floridana Ashm.

(Pl. IX, Fig. 6, ♀.)

Bæoneura floridana Ashm., Ent. Am., III, p. 99; Cress. Syn. Hym., p. 313.

♂ ♀. Length, 3.5 to 4ᵐᵐ. Slender, greatly elongated, black, rugoso-
punctate, and sparsely pubescent. Antennæ dark-brown, the scape
pale; first funiclar joint in ♀ longer than the pedicel, the second slightly
smaller, the third still shorter, the fourth wider; club 6-jointed, large.
In ♂ the pedicel and first funiclar joint about equal, the second short,
the third a little dilated, slightly emarginated at base, the following
submoniliform, slightly longer than thick. Mesonotum with 2 distinct furrows, the metanotum with 2 delicate keels on disk, sometimes
with a tuft of pubescence between. Wings hyaline, very slightly

tinged, pubescent; marginal vein about twice as long as the stigmal, the postmarginal very long. Legs brownish-yellow; in ♀ with the coxæ black. Abdomen, in ♀, long fusiform, in ♂ long, linear, scarcely widened at the middle, extending far beyond the wings when folded, and about thrice as long as the thorax; first and second segments striated and punctate, the following segments punctate and faintly rugose.

HABITAT.—Florida; Virginia.

Types in Coll. Ashmead.

Macroteleia virginiensis, sp. nov.

♀. Length, 4ᵐᵐ. Black, shining, with thimble-like punctation; head transverse-quadrate; eyes very large, rounded, pubescent. Antennæ 12-jointed, clavate, brown-black, the scape brownish-yellow; first funiclar joint long, one-third longer than the pedicel, the second two-thirds the length of the first, the third and fourth subequal, but still longer than wide; club 6-jointed, gradually thickened from the fourth funiclar joint, long, gradually tapering off at both ends.

Mesonotum with two distinct furrows. Scutellum smoother, sparsely punctate. Metathorax with a median keel. Wings hyaline, the marginal vein twice the length of the stigmal, the latter rather short, ending in a knob; basal nervure wanting. Legs, including coxæ, reddish-yellow. Abdomen pointed, fusiform, about twice as long as the head and thorax united, with close, thimble-like punctures, some of which are elongated and with raised lines, the first segment striolated, the suture between the first and second crenate; the first and second segments are about equal in length, a little shorter than the third and fourth, which are also equal.

HABITAT.—Arlington, Va.

Type in Coll. Ashmead.

Since this description was drawn up specimens of this species were reared by Dr. Riley, in the Department of Agriculture, from the eggs of *Orchelimum glaberrimum*.

CALLISCELIO Ashm., gen. nov.

(Type, *C. laticinctus*.)

Head transverse, the frons convex, the face not impressed, occiput and cheeks delicately margined; ocelli 3, large, in a triangle, the lateral touching the eye; eyes very large, oval, bare.

Antennæ inserted at the clypeus, 12-jointed in both sexes, in ♀ with a 6-jointed club, all the funiclar joints elongate; in ♂ filiform, the first flagellar joint long, joints 4 to 10 submoniliform.

Mandibles bidentate.

Maxillary palpi 4-jointed; labial palpi 3-jointed.

Thorax ovate, rounded before, the prothorax slightly visible from above, the mesonotum with two distinct furrows, scutellum semicircular, convex; metathorax short, carinated laterally.

Front wings with a distinct basal nervure, the marginal vein long, fully twice as long as the stigmal, the latter oblique, knobbed at tip, basal vein distinct.

Abdomen pointed fusiform, much narrowed at base, the tip projecting a little beyond the wings when folded, first segment linear, petioliform, not quite as long as the second, which is the longest segment; third segment two-thirds the length of the second; the fourth a little more than half as long as the third; fifth subequal with fourth; sixth conical, about as long as the third.

Legs long, slender, the tibial spurs 1, 1. 1, weak, the basal joint of hind tarsi nearly four times as long as the second.

Allied to *Macroteleia*, but separated by the bidentate mandibles, more transverse head, and narrower petiole.

Calliscelio laticinctus, sp. nov.

(Pl. ix, Fig. 7, ♀.)

♀. Length, 2.5ᵐᵐ. Head black; face, clypeus, mandibles, and palpi, pale; thorax rufous or brown, the metathorax black; legs yellowish, posterior coxæ and femora obfuscated above; abdomen fusiform, much longer than the head and thorax together, piceous-brown, the basal one-third of second segment, and basal one-half of third, yellow; petiole, the apical two-thirds of second, and the three apical segments black; the petiole is nearly three times as long as thick, of a uniform width throughout, longitudinally striated; the second segment is the longest, about one-half longer than the first, broadened at apex to three times its width at base, its basal half longitudinally aciculated; the third segment is two-thirds the length of the second; the fourth, two-thirds the length of the third; the fifth, a little more than half the length of the fourth; the sixth, conical, about as long as the third. Head transverse, finely punctate. Antennæ 12-jointed, brownish-yellow, the club 6-jointed, black; the first and second funiclar joints are long, cylindrical, subequal, the third two-thirds the length of the second, stouter, the fourth about one-half the length of the third and thicker. Thorax with small, sparse punctures. Wings with the basal half hyaline, the apical half, except the margin, fuscous; venation brown; the basal nervure distinct, the marginal nervure 3 times as long as the oblique stigmal, the latter terminating in a rounded knob, the postmarginal longer than the marginal.

HABITAT.—St. Vincent, West Indies.

Types in British Museum and National Museum.

Described from 6 ♀ specimens, collected by Herbert H. Smith.

CHROMOTELEIA Ashm., gen. nov.

(Type, *C. semjeyana*.)

Head large, transverse, rounded before, the occiput delicately margined, scarcely emarginate, the vertex not very broad; ocelli 3, triangularly arranged, the laterals touching the eye; eyes very large, ovate, bare.

Antennae inserted on the clypeus, 12-jointed in both sexes; in ♀ clavate, the club large, 6-jointed, the funiclar joints elongate, the first the longest; in ♂ long, filiform.

Maxillary palpi 4-jointed; labial palpi 3-jointed.

Mandibles tridentate.

Thorax ovate, produced into a little neck anteriorly, the prothorax visible from above as an arcuate line; mesothorax with 2 distinct grooved lines; scutellum semicircular, with a ridge behind; metanotum with a semicircular inclosed space at base, the sides carinated.

Front wings with the submarginal vein distant from the costa, curving, and joining it at about the middle of the wing; marginal vein punctiform; the postmarginal very long; stigmal rather long, oblique, ending in a little knob; basal vein distinct, with a basal cell.

Abdomen very long fusiform, extending beyond the tip of the wings when folded, the second, third, and fourth segments longer than the rest, the fourth a little the shortest.

Legs as in *Macroteleia*, tibial spurs 1, 1, 1, basal joint of hind tarsi about 4 times as long as the second.

A beautiful and distinct genus, remarkable for the large, semicircular inclosure on the metanotum, which, in connection with the venation, renders the genus easy of recognition.

Chromoteleia semicyanea, sp. nov.

(Pl. IX, Fig. 3, ♀.)

♂ ♀. Length, 4.5 to 5ᵐᵐ. Head and thorax cyaneous, punctate; abdomen sessile, very long, pointed fusiform, ochraceous, punctate, the first and second segments striated; first segment a little more than half the length of the second; second and third long, equal, the three following segments shorter, subequal, the last two very minute. Antennae black, the scape yellow; in ♀ ending in a 6-jointed club, the first funiclar joint the longest, about one-half longer than the second and not quite twice the length of the pedicel, the third funiclar joint subequal with the second, the fourth a little longer than thick and stouter than the third; in ♂ subfiliform, the first funiclar joint twice the length of the pedicel, the joints after the third, except the last, about equal in length, less than twice as long as thick, the last longer, ovate. Wings fuscous, the nervures brown, the marginal vein punctiform, the basal nervure distinct, the stigmal a little curved, ending in a small knob.

HABITAT.—St. Vincent, West Indies.

Types in British Museum and National Museum.

Described from 1 ♂ and 1 ♀ specimen collected by Herbert H. Smith.

OPISTHACANTHA Ashm., gen. nov.

Raia Ashm., MS. (Olim.).

(Type, *O. mellipes* Ashm.)

Head transverse, the occiput and cheeks delicately margined; ocelli 3, triangularly arranged, the lateral close to the eye; eyes large, oval, pubescent.

Antennæ inserted at the clypeus, 12-jointed in both sexes, in the ♀ short, clavate, the club 6-jointed, the last funiclar joint very minute; in ♂ filiform, submoniliform.

Mandibles bidentate.

Thorax short, ovoid, the prothorax not visible from above; mesonotum with 2 delicate but distinct furrows; scutellum semicircular; post-scutellum produced into a short, acute spine; metathorax very short, angles subacute.

Front wings ciliated, the submarginal vein reaching the costa a little before half the length of the wing; marginal vein short, not half the length of the stigmal; postmarginal greatly lengthened; the stigmal vein oblique, knobbed; basal vein and basal cell present.

Abdomen oblong-oval, the second and third segments the longest, the third a little longer than the second.

Legs slender, the tibial spurs 1, 1, 1, weak, the basal joint of hind tarsi about 4 times as long as the second.

In general appearance this genus closely resembles a Telenomid, but the venation, carinated abdomen, and the spined postscutellum readily distinguish it.

Opisthacantha mellipes, sp. nov.

(Pl. ix, Fig. 4, ♀.)

♂ ♀. Length, 1.40^{mm}. Black, subopaque, the punctuation so fine as to be scarcely perceptible. Head transverse, broader than the thorax. Eyes very large, pubescent, the space between them narrow. Lateral ocelli close to the eye. Mandibles 3-dentate, brown. Antennæ 12-jointed, brown-black, the scape yellow; in ♀ clavate, pedicel small, the first funiclar joint slightly longer than the pedicel, about twice as long as thick; the second and third moniliform, the fourth very minute; the club large, the joints scarcely separable, transverse; in ♂ filiform, the pedicel moniliform, the first and third funiclar joints about equal, long-oval, the second small, the joints from the fourth to the twelfth oval-moniliform, the last ovate, longer than the preceding. Mesonotum with two nearly parallel furrows. Scutellum subconvex, with a transverse line at base. Post-scutellum armed with an acute spine or thorn. Legs pale honey-yellow. Wings hyaline, the venation pale brown; the basal nervure delicate, but distinct; the marginal, about as long as the stigmal; the stigmal rather short, oblique, and ending in a knob; postmarginal vein very long. Abdomen long-oval, longer than the head and thorax together; black, with a brownish tinge at the middle, the first three segments longitudinally striated, the following segments smooth, polished.

HABITAT.—Washington, D. C.

Types in Coll. Ashmead.

Besides the above species, I have seen others from South America and the West Indies.

LAPITHA Ashm., gen. nov.

(Type *L. spinosa* Ashm.)

Head large, transverse, the occiput and cheeks delicately margined; ocelli large, in a triangle; eyes long-oval, bare.

Antennae inserted close to the clypeus, 12-jointed in both sexes; in ♀ clavate, the club 6-jointed; in ♂ filiform, the flagellar joints all long, cylindrical.

Maxillary palpi 4-jointed; labial palpi 3-jointed.

Mandibles 3-dentate.

Thorax ovoid, the prothorax not visible from above; mesonotum without distinct furrows, indicated only slightly anteriorly; scutellum large, semicircular, margined behind; metascutellum produced into an acute spine; metathorax short, emarginated, the sides carinated.

Front wings with the submarginal vein reaching the costa beyond the middle of the wing, the marginal vein long or as long as the stigmal; the latter clavate, slightly obliquely curved, the postmarginal longer than the marginal.

Abdomen fusiform, the first three segments nearly equal, the fourth a little more than half the length of the third, the fifth still shorter, the sixth and seventh short, equal.

Legs long, the femora and tibiae subclavate, the tibial spurs 1, 1, 1, distinct, the tarsi longer than their tibiae, the basal joint very long, about five times as long as the second joint.

Allied to *Opisthacantha*, but of a larger, more elongate form, the head differently shaped, the venation totally different, the basal nervure being more oblique, the stigmal nervure shorter, the postscutellar spine longer, more acute, while the abdomen is longer, fusiform, and more narrowed at base.

Lapitha spinosa, sp. nov.

(Pl. IX, Fig. 8, ♂.)

♂. Length, 3.5ᵐᵐ. Head and thorax finely, closely punctulate, brownish-yellow; metathorax with oblique carinae extending from base of postscutellum; postscutellum produced into an acute spine. Legs yellow. Abdomen fusiform (extending slightly beyond the tip of the wings when folded), black, closely punctate, sometimes the basal half of third segment yellow; the first and second segments are striate, the latter with punctures in the striae; first segment a little longer than wide, very slightly wider at apex than at base; second and third seg-

ments the longest, about equal in length; fourth, the length of the
first; fifth, two-thirds the length of the fourth; sixth, half the length
of fifth; seventh very short, smooth basally; eighth subtriangular,
margined. Antennæ filiform, dark brown, the scape and pedicel yellow;
second, third, and the last joint of flagellum about equal in length;
first and fourth about equal, shorter than the second; joints beyond
very slightly shorter. Wings hyaline, with a large smoky blotch below
the postmarginal vein; nervures fuscous; basal nervure distinct; marginal nervure as long as the shaft of the stigma, the latter oblique,
clavate at tip.

HABITAT.—St. Vincent, West Indies.

Types in National Museum and British Museum.

Described from 4 ♂ specimens collected by Herbert H. Smith.

CACUS, gen. nov. Riley.

(Type *C. æcanthi* Riley.)

"Head subquadrate, emarginate behind, the occiput feebly margined;
ocelli 3, triangularly arranged, the lateral close to the eye; eyes ovate,
bare.

"Antennæ inserted at the clypens, 12-jointed in both sexes; in ♀ clavate, the club 6-jointed, the last funicular joint usually transverse, the
others longer than wide; in ♂ filiform, the third funicular joint a little
dilated and emarginate toward base.

"Mandibles short, either tridentate or bifid at tip.

"Maxillary palpi 3-jointed; labial palpi 2-jointed.

"Thorax ovate, narrowed before, the prothorax slightly visible from
above; mesonotum most frequently without furrows, rarely with distinct furrows; metathorax not especially short, with two lateral keels,
a median prominence or carina, and with usually two small erect teeth
at apex, above the insertion of the abdomen.

"Front wings with the submarginal vein joining the costa a little beyond the middle of the wing, the marginal vein variable, scarcely half
the length of the stigma, or much longer; the latter oblique, terminating in a knob; the postmarginal vein very long, the basal vein subobsolete.

"Abdomen much as in *Baryconus*, but the female without the horn-
like projection at base.

"Legs as in *Baryconus*, the tibial spurs 1, 1, 1, the middle and hind
pairs weak, tarsi long and slender, the basal joint more than four times
the length of the second."—[*From Riley's MS.*]

It is probable that under this genus Dr. Riley has included two distinct
genera, one distinguished by having 3 dentate mandibles, the other in
having the mandibles bifid or 2-dentate, but otherwise they are so
similar that for the present I believe it best to follow him in consider-

ing the species congeneric. The type of the genus, *Cacus œcanthi* Riley, has 3-dentate mandibles. Those with 2-dentate mandibles are from South America and the West Indies, and are not included here.

The only species whose parasitism is known was bred by Prof. F. M. Webster from the eggs of *Oecanthus niveus* De Geer.

Cacus œcanthi, sp. nov., Riley.

(Pl. XVI, Fig. 6, ♀.)

"♂ ♀. Length, 2 to 2.20ᵐᵐ. Black, subopaque, closely punctate, and covered with a fine, sericeous down. Head quadrate, the cheeks large, swollen, with striæ converging toward the mouth; mandibles and palpi, pale; antennæ with the base and apex of scape and two last funicular joints pale; pedicel rather long, thicker than the first funicular joint; the latter longer and slenderer than the pedicel; second funicular joint one-third shorter than the first; third and fourth transverse. Thorax without mesonotal furrows; metathorax with a central carina and margined at apex; legs pale rufous, the coxæ slightly dusky basally; wings hyaline, the marginal vein about half the length of stigmal, the latter knobbed at tip, the postmarginal vein long. Abdomen long, sublinear, narrowed at base, about one-third longer than the head and thorax together, and extending beyond the tip of the wings when folded; first segment petioliform, striated, the second and third longitudinally shagreened, nearly equal in length, the following microscopically sculptured.

"The ♂ agrees with the female very closely in shape, but the three basal abdominal segments are striated and the antennæ are filiform, brownish-yellow, the first and second funicle joints being about equal, thrice as long as the pedicel, the third one-third shorter than the second, the following to the last very slightly shorter and almost equal in length, the last pointed, fusiform, and as long as the third flagellar joint.

"HABITAT.—Lafayette, Ind.

"Types in National Museum.

"Bred by Prof. F. M. Webster, May 31, 1881, from the eggs of *Œcanthus niveus*."—[*From Riley's MS.*]

Anteris Förster.

Hym. Stud., II, p. 101 (1856).

(Type *A. rufitarsis* Först.)

Head transverse, the face convex or subconvex, not, or but slightly, impressed above the antennæ, the occiput and cheeks delicately margined; ocelli 3, triangularly arranged, the lateral about their width from the border of the eye; eyes oval, usually pubescent.

Antennæ inserted at the clypeus, 12-jointed in both sexes, in ♀ clavate, the club 5 or 6 jointed, subsolid, in ♂ filiform, submoniliform, the second flagellar joint shorter than the first or third.

Mandibles bifid at tips.

Thorax ovoid, the prothorax rounded before, not, or scarcely, visible from above, except the lateral angles; mesonotum broader than long, with two distinct furrows; scutellum semicircular, subconvex, with a punctate frenum behind; metathorax short, carinated at apex.

Front wings with the submarginal vein reaching the costa a little beyond the middle of the wing, the marginal vein usually a little thickened, although linear and scarcely as long as the short stigmal, the latter knobbed at tip, the postmarginal vein very long; basal nervure wanting or subobsolete.

Abdomen oval or oblong oval, the first three segments nearly equal, the third, usually a little the longest.

Legs with all femora and the anterior tibiae clavate, tibial spurs 1, 1, 1, the basal joint of hind tarsi thrice, or a little more than thrice, as long as the second.

The distinct parapsidal furrows, transverse head, short or punctiform marginal vein, and absence of the horn at base of abdomen distinguish the genus.

Two species in our fauna can be thus tabulated:

Thorax and abdomen, except sometimes the tip, brownish-yellow.............. 2
Mostly black, the collar, pleura, and base of abdomen pale.
  Scutellum smooth, highly polished............................A. VIRGINIENSIS.
2. Head black or fuscous.
  Scape, mandibles, and legs yellowish ....A. NIGRICEPS.

Anteris virginiensis, sp. nov.

♂. Length, 1.20ᵐᵐ. Black, shining; head transverse; mesonotum very faintly punctate, in marked contrast to the highly polished scutellum; collar, pleura, and base of abdomen rufous or yellowish; mandibles rufous. Antennae 12-jointed, filiform; scape as long as the first three flagellar joints united; pedicel small, round; first funicular joint the longest, about twice as long as the second; the following joints about equal, longer than the second, but not quite as long as the first. Mesonotum with two distinct furrows. Scutellum semicircular, smooth, polished. Metathorax very short, polished, its posterior margin carinated. Legs, including coxae, honey-yellow. Wings hyaline, iridescent, the marginal vein thick, as long as the stigmal, the latter oblique, ending in a knob, the postmarginal about three times as long as the marginal, the basal vein subobsolete. Abdomen long-oval, as long as the head and thorax together, smooth, polished, the first segment and the second, at base, striated; the first segment is yellowish, the third the largest, about one-third longer than the second, or a little more than twice as long as the fourth.

HABITAT.—Virginia.

Type in National Museum.

Described from a single specimen taken by Mr. Theo. Pergande, October 10, 1880.

Anteris nigriceps, sp. nov.

(Pl. x. Fig. 2, ♀.)

♂ ♀. Length, 1.8 to 2.1ᵐᵐ. Head and thorax closely, minutely punctate; three basal abdominal segments striated. Head and abdomen black, the petiole and second segment at base, yellow, sometimes fuscous or black; mandibles pale, the teeth black; thorax and legs brownish-yellow; in ♀ the scutellum and metathorax fuscous or black, rarely so in ♂; mesonotum with 2 furrows. Antennæ 12-jointed, black, the scape yellow, sometimes the pedicel in female yellow; in the ♀ the first funiclar joint is long, longer than the pedicel, the second scarcely half as long as the first, the third very slightly shorter than second, the fourth transverse, club long, fusiform; in ♂ filiform, the first funiclar joint about twice as long as the pedicel, the third shorter, emarginated at base, the second shorter than the third, the joints after the third oblong oval, about twice as long as thick. Wings hyaline or subhyaline; the marginal vein punctiform or about twice as long as thick, the stigmal vein oblique, ending in a distinct rounded knob.

HABITAT.—Jacksonville, Fla., and Arlington, Va.

Types in Coll. Ashmead.

Described from several specimens.

APEGUS Förster.

Hym. Stud., ii, p. 101 (1856).

(Type *A. leptocerus* Först.)

Head transverse quadrate or subquadrate, the occiput concave, margined; ocelli 3, triangularly arranged, the lateral a little away from the margin of the eye; eyes large, oval, bare.

Antennæ inserted just above the clypeus, 12-jointed, filiform in both sexes.

Maxillary palpi 4-jointed; labial palpi 3-jointed.

Mandibles 3-dentate.

Thorax long ovoid, the prothorax visible from above, especially laterally; mesonotum with 2 distinct furrows, abbreviated anteriorly; scutellum short, with a row of punctures posteriorly; metathorax short, with the dorsum bicarinated.

Front wings with the marginal vein usually longer than the stigmal, the postmarginal greatly lengthened, the stigmal oblique, with a slight knob at tip; basal vein wanting.

Abdomen sessile, fusiform, always much longer than the head and thorax together, all the segments lengthened, the second and third the longest, the first and second with lateral dorsal carinæ.

Legs rather long, the femora clavate. tibiæ long, subclavate, tibial spurs 1, 1, 1, distinct, the tarsi lengthened, slender, the basal joint more than thrice as long as the second.

This genus is closely allied to *Macroteleia*, but is distinguished by filiform antennæ in both sexes.

Apegus elongatus Ashm.

Anteris elongatus Ashm., Ent. Am., III. p. 118; Cress. Syn. Hym., p. 314.

♂ . Length, 4.5ᵐᵐ. Elongate, linear, black; the head, thorax, and abdomen with round punctures, sparsely covered with a whitish pubescence; cheeks bulging, rugoso-punctate. Mandibles 3-dentate, pale brown. Mesonotum with 2 furrows. Scutellum short, with a row of large punctures posteriorly. Metathorax coarsely furrowed, with two carinæ on the disk and lateral keels. Antennæ 12-jointed, filiform, pale brownish-yellow, the tips fuscous; first funiclar joint a little longer than the pedicel; second shorter; third slightly emarginated at base, the following, except the last, moniliform, scarcely longer than thick, the last longer than the penultimate. Legs brownish-yellow, the posterior coxæ a little dusky at base. Wings fusco-hyaline, the venation brown; the submarginal vein is very long, the marginal about twice as long as the stigmal, the postmarginal very long. Abdomen fully twice as long as the head and thorax united, punctate, segments 1, 2. and 3 dorsally towards sides, carinated, the second and fourth segments about equal, the third a little longer.

HABITAT.—Florida.

Type in Coll. Ashmead.

HOPLOTELEIA Ashm., gen. nov.

(Type *H. floridana* Ashm.)

Head transverse-quadrate, the cheeks margined, the face with a deep impression above the antennæ; ocelli 3, in a triangle, the lateral close to the margin of the eye; eyes large, oval.

Antennæ inserted close to the mouth, 12-jointed in both sexes; in ♀ clavate, in ♂ filiform.

Maxillary palpi short, 4-jointed; labial palpi 3-jointed.

Mandibles 3-dentate.

Thorax ovoid, the prothorax slightly visible from above; mesonotum with 3 distinct furrows; scutellum large, semicircular; postscutellum with 2 erect teeth; metathorax short, the hind angles slightly prominent.

Front wings, when folded, not extending to the tip of the abdomen, the submarginal vein joining the costa a little beyond half the length of the wing, the marginal vein very short, almost punctiform, the postmarginal very long. the stigmal long, oblique, terminating in a small knob; basal vein obsolete.

Abdomen sessile, fusiform, the third segment slightly the longest; in ♀ pointed at apex, in ♂ truncate, with 2 prongs.

Legs moderately stout, pilose, the tibial spurs 1, 1, 1, well developed, the tarsi not, or but slightly, longer than the tibiae, the basal joint of hind tarsi about 3½ times as long as the second, with stiff bristles beneath.

The three impressed lines on mesonotum, the spined postscutellum, and the venation separate the genus at once from all others, although otherwise it resembles *Baryconus* and *Macroteleia*.

Besides the species described here from North America, I have another species in my collection from South America. Nothing is known of the habits of the genus.

Hoploteleia floridana Ashm.

(Pl. x. Fig. 1, ♂.)

Baryconus floridanus Ashm., Ent. Am., III, p. 118; Cress. Syn. Hym., p. 314.

♂ ♀. Length, 3.5 to 4^{mm}. Brown-black, coarsely rugoso-punctate, the middle of mesonotum smoother, closely punctate. Antennae 12-jointed: in ♂ filiform, brown, the scape yellowish: pedicel triangular: flagellar joints to fifth subequal, the first the longest, the third excised at base: joints from fifth to last a little longer than the fourth: the last 1½ times as long as the penultimate; in ♀ clavate, the club joints broader than long. Mesonotum with 3 distinct furrows. Pleura impressed, lineated. Scutellum rugose, with coarse, irregular punctures. Postscutellum with two short, erect teeth. Wings subhyaline, tinged with fuscous: in the ♀ the marginal vein is very short, about one-third the length of the stigmal, or nearly punctiform: in the ♂ longer, fully half the length of the stigmal; postmarginal vein very long: stigmal, oblique, not curved, ending in a small knob. Legs brownish-yellow, in ♀ with the coxae black. Abdomen fusiform, sessile, lineatedly rugose, the basal segment deeply fluted, in ♀ pointed at apex, in ♂ truncate and ending in two spines.

HABITAT.—Jacksonville, Fla., and Arlington, Va.

Types in Coll. Ashmead.

The types of this species and genus were collected by myself in Florida, but I have a ♂ taken in Virginia, and Mr. Herbert Smith has taken another species in South America, so that the genus is widely distributed.

CREMASTOBÆUS Ashm., gen. nov.

(Type *C. bicolor* Ashm.)

Head transverse, the face convex, the vertex not very broad, rounded off towards occiput, the latter a little emarginated: ocelli 3, triangularly arranged, the lateral touching the eye: eyes very large, round, hairy.

Antennae inserted close to the clypeus, 12-jointed in both sexes, in

♀ short, clavate, the club large oval, compact, 5-jointed; in ♂ subclavate, submoniliform.

Mandibles short, bidentate.

Maxillary palpi 4-jointed; labial palpi 3-jointed.

Thorax ovate, produced into a little neck anteriorly; the mesonotum subconvex, without furrows; the scutellum semicircular; the metathorax short, the angles not prominent.

Front wings with the submarginal vein joining the costa at about half the length of the wing, the marginal vein short, linear, the postmarginal very long, the stigmal short, oblique, clavate.

Abdomen fusiform, sessile, a little longer than the head and thorax together, with the segments strongly constricted at the sutures, the first the longest, the following subequal.

Legs with the femora and tibiæ subclavate, the tibial spurs well developed 1, 1, 1, the middle and posterior tarsi not longer than their tibiæ, the basal joint of hind tarsi about thrice as long as the second.

This genus is readily distinguished by the constricted abdominal segments. I have species from South America and have seen others from the West Indies, but no doubt the genus will be found to occur also in our fauna.

Cremastobæus bicolor, sp. nov.

(Pl. X, Fig. 3, ♀.)

♂ ♀. Length, 1 to 1.1ᵐᵐ. Head and thorax black, faintly microscopically punctate, the punctation not destroying the luster of the surface; eyes oval, pubescent; abdomen brownish-yellow, fusiform, the segments strongly constricted at the sutures, the sutures crenate; legs yellow. Antennæ brownish-yellow, the club oval-rotund, 5-jointed, black; the first funiclar joint is the stoutest and longest joint, the following to the club gradually subequal, the last two rounded, a little transverse. Wings hyaline, the marginal vein a little longer than the stigmal, the latter short, oblique, ending in a little knob; no trace of basal or anal nervures.

The ♂ is entirely black, with the scape, pedicel of antennæ, and the legs, yellow; the thorax more distinctly punctate; flagellum sub-filiform, brown, pubescent, very slightly thickened toward tip, the joints, after the first, submoniliform, scarcely longer than thick.   •

HABITAT.—St. Vincent.

Types in National Museum and British Museum.

Described from one ♀, three ♂ specimens collected by Herbert H. Smith.

HADRONOTUS Förster.

Hym. Stud., II, p. 101 (1856).

(Type H. laticeps Först.)

Head transverse, usually very wide, the frons convex, the face with an impression above the insertion of the antennæ, cheeks margined;

ocelli 3, triangularly arranged, the lateral contiguous to the eye; eyes large, oval, usually bare, sometimes pubescent.

Antennae inserted close to the mouth, 12-jointed in both sexes, in ♀ clavate, the club 6-jointed, in ♂ filiform, pubescent.

Maxillary palpi short, 4-jointed; labial palpi 3-jointed.

Mandibles bidentate.

Thorax oval, truncate anteriorly, the angles of the prothorax alone visible from above; mesonotum much wider than long, without furrows; scutellum rather large, semicircular, convex; metathorax very short, abrupt.

Front wings with the submarginal vein attaining the costa at about half the length of the wing, or a little before; marginal vein variable, punctiform, or rarely more than half the length of the stigmal, the latter rather long, oblique.

Abdomen broadly oval, sessile, carinated along the sides, the second segment the largest, the first and third about equal.

Legs with the femora and tibiae subclavate, the tibial spurs 1, 1, 1, distinct, the tarsi long, slender, the basal joint of hind tarsi about thrice as long as the second.

This genus closely resembles *Telenomus* and could only be confused with it, for, like that genus, it is found parasitic in the eggs of various Hemiptera. It is, however, readily distinguished by the distinctly carinated abdomen, the 12-jointed antennae in the females, the coarser sculpture, and the more sessile abdomen.

It occurs in all parts of the world and is particularly well represented in the South American and West Indian faunas, where no doubt it does good service in diminishing the number of Hemiptera to be found there.

The several species found in the United States may be readily separated by the aid of the following table:

TABLE OF SPECIES.

Species very coarsely rugoso-punctate 2
Species smooth, at the most faintly punctate or shagreened, sericeous.
  Coxae black or dusky, legs yellow, or pale brownish-yellow.
    Head and thorax shagreened, the scutellum finely punctate.
      Abdomen polished, the first and second segments faintly longitudinally
      aciculated, the second, without a row of punctures at base.
                            H. LARGI, sp. nov.
    Head distinctly punctate, the thorax faintly punctate.
      Abdomen closely, finely punctate, the basal and apical edges of all the segments smooth, polished, the first segment striate at base, the second with a row of striate punctures at base......H. LEPTOCORIS.E How.
    Head and thorax closely, finely punctulate, opaque.
      Abdomen smoother, shining toward apex, the first and second segment striated ....H. MYRMECOPHILUS, sp. nov.
2. Coxae and legs brownish-yellow.
    Abdomen with coarse, longitudinal cribrate rugosities on all the segments, smooth and polished along the sutures...........H. RUGOSUS How.

Abdomen cribrate rugose, the first segment with deep coarse striæ, the following smoother, the segments not smooth along the sutures.

H. FLORIDANUS Ashm.

Abdomen finely, evenly rugose, the first segment striate at base.

H. ANASÆ Ashm.

Coxæ black, the legs brownish-yellow.

Head and thorax rather coarsely rugose, the thorax and abdomen very finely rugoso-punctate, the mesonotum posteriorly slightly lineated; first abdominal segment and the second, at base only, striated.

H. RUGICEPS, sp. nov.

Hadronotus largi, sp. nov.

♂ ♀. Length, 0.80 to 1ᵐᵐ. Black, opaque; head transverse, finely shagreened; mandibles, palpi, antennal scape, apex of pedicel, and legs, honey-yellow; coxæ slightly dusky. Antennæ 12-jointed, clavate, brown-black, except as before mentioned; pedicel longer than the first funiclar joint; second and third funiclar joints not, or scarcely, longer than wide; the joints beyond all transverse. Thorax with fine wavy longitudinal rugæ; scutellum almost smooth, the punctation being microscopic. Wings hyaline, the venation yellowish, the marginal vein short, scarcely half the length of the oblique stigmal vein. Abdomen broadly oval, sessile, smooth, shining, the first and second segments faintly longitudinally aciculated.

In the ♂ the antennæ are filiform, honey-yellow, the pedicel longer than the first funiclar joint, the second funiclar small, the third enlarged, stout, the joints beyond transverse, twice as wide as long, subserrate, the last oval.

HABITAT.—Los Angeles, Cal.

Types in National Museum.

Described from several specimens bred by Mr. D. W. Coquillett, from the eggs of *Largus succinctus.*

Hadronotus leptocorisæ How.

Hubbard's Orange Insects, App., p. 215; Ashm. Ent., Am. III, p. 118; Cress. Syn. Hym., p. 314.

♀. Length, 1.4ᵐᵐ; expanse, 2.5ᵐᵐ. Head and thorax evenly covered with small round punctures, except in the facial impression, which is transversely striate; antennæ subclavate; lateral ocelli nearly touching the margins of the eyes. Mesonotum a trifle smoother than the head, and furnished with a very fine white pubescence. First segment of the abdomen dorsally longitudinally striate; remaining segments closely covered with fine round punctures; ventral surface sparsely punctate. Color brown; scape brown; all coxæ black; all trochanters, femora, tibiæ, and tarsi light brown; mandibles and wing veins light brown.

♂. Length, 1.3ᵐᵐ; expanse, 2.8ᵐᵐ. Antennæ filiform. In other characters resembling the ♀. (*Howard.*)

HABITAT.—Crescent City, Fla.

Types in National Museum.

Bred by Mr. H. G. Hubbard from the eggs of *Zelus bilobus* Say, wrongly determined as the eggs of *Leptocorisa tipuloides.*

Hadronotus myrmecophilus, sp. nov.

♂. Length. 1ᵐᵐ. Head and thorax finely closely punctulate, opaque; the apex of the abdomen smoother, shining, the two basal segments punctate and striate, the following almost smooth. Head very wide, the cheeks flat, delicately margined. Antennæ 12-jointed, filiform, the scape yellowish, the flagellum dark-brown, thicker than the scape and pedicel; first flagellar joint the longest, longer than the pedicel, second, half the length of the first, a little contracted, third, very slightly longer, a little dilated outwardly toward one side; following joints scarcely as long as thick. Legs brownish-yellow, the coxæ black. Wings subfuscous, the marginal vein punctiform, not longer than thick, the stigmal oblique, the postmarginal very long.

HABITAT.—Arlington, Va.

Type in Coll. Ashmead.

This species was taken from an ant's nest. It differs decidedly from the other species in the sculpture of the head and thorax and in the smooth, apical abdominal segments.

Hadronotus rugosus How.

Insect Life, Vol. I, p. 242.

♀. Length, 1.8ᵐᵐ. Antennæ arise immediately above the mouth; scape reaches nearly to anterior ocellus; pedicel subcylindric, as long as first funicle joint; funicle joints increasing regularly in width from joint 1 to basal joint of club; joint 1 of funicle twice as long as 2, the remaining joints subequal in length; joint 2 of club equal to joint 1; joint 3 longer than 2 and pointed. Head and face closely, deeply, and regularly punctate; facial impression shallow, with transverse punctures and with a distinct, central longitudinal carina. Mesonotum strongly punctate, the punctures of the scutum assuming a longitudinal direction. Dorsal surface of abdomen strongly longitudinally rugose, each joint smooth at extreme base and apex, the rugosities strongest upon joint 1, growing slightly fainter on succeeding joints; joint 2 longest, joints 1 and 3 shorter; venter with well-marked circular punctures. Entire surface of body with sparse whitish pilosity. General color, black; mouth parts, antennæ, and legs honey-yellow, except that the front coxæ, antennal club and pedicel, and first two funicle joints above are brownish.—(Howard.)

HABITAT.—Rockledge, Fla.

Types in National Museum.

Mr. Howard described this species from 3 ♀ specimens dissected from the supposed eggs of *Dysdercus suturellus*, sent to Dr. Riley by Mr. H. S. Williams, April 1, 1889. The eggs appear to me to be those of *Euthoctha galeator* Fabr., with which I am quite familiar, from my long residence in Florida.

Hadronotus floridanus Ashm.

Ent. Am., III, p. 118; Cress. Syn. Hym., p. 314.

♂ ♀. Length, 1.5ᵐᵐ. Head and thorax very coarsely rugose, the abdomen cribrate rugose, the first segment with coarse striæ; scape and legs brownish-yellow, the coxæ sometimes brownish-black.

Antennæ 12-jointed, in ♀ clavate, the pedicel not longer than the first funiclar joint, the second funiclar joint about half the length of the first, the third and following, transverse; in ♂ filiform, pale brownish-yellow, the first funiclar joint the longest, longer than the pedicel, the second one-third shorter than the first, the third slightly dilated toward one side at apex, the following a little longer than wide, the last pointed, almost as long as the first funiclar. Wings hyaline, the venation yellowish, the marginal vein very short. Legs, including coxæ, brownish-yellow.

HABITAT.—Jacksonville and Cocoanut Grove, Fla.

Types in Coll. Ashmead and National Museum.

Described from many specimens reared from the eggs of *Metapodius femoratus* Fabr.

This species is most closely allied to *H. rugosus* Howard, but quite distinct in the sculpture of the abdomen.

Hadronotus anasæ Ashm.

Telenomus anasæ Ashm., Bull. No. 14 U. S. Dept. Agric., p. 23; Cress. Syn. Hym., p. 314.

♂ ♀ . Length, 1.2^{mm}. Black, coarsely, irregularly rugoso-punctate, with a sparse whitish pubescence, the abdomen more evenly and less coarsely sculptured, somewhat lineated, the extreme apical edges of the segments smooth, polished, the first segment striate, the second a little longer than the first and the longest segment, the third a little shorter than the first. Head large and broad, about $3\frac{1}{2}$ times as wide as thick antero-posteriorly, and wholly rugose. Antennæ 12-jointed, in ♀ clavate, brown, the scape, pedicel, and sometimes the funiclar joints, yellow; the pedicel is one-half longer than the first funiclar joint, the second funiclar shorter than the first, the third and fourth transverse, club acuminate towards apex; in ♂ subfiliform or subclavate, the pedicel shorter than the first funiclar joint, the second and third subequal, shorter than the first; the following joints, except the last, which is ovate, are a little wider than long. Mandibles large, pale or yellowish. Wings hyaline, the marginal vein punctiform. Legs brownish-yellow, the coxæ sometimes dusky, more rarely black.

HABITAT.—Jacksonville, Fla.

Types in National Museum and Coll. Ashmead.

Described from many specimens reared by myself from the eggs of *Anasa tristis* De Geer. Dr. R. S. Turner reared, May 20, 1880, a large series from the same insect eggs at Fort George Island, Florida, and Miss Mary Murtfeldt reared August 2, 1882, a variety with black coxæ at Kirkwood, Mo.

Hadronotus rugiceps, sp. nov.

♀ . Length, 1.5^{mm}. Head and scutellum rather coarsely but shallowly rugoso-punctate; the thorax and abdomen finely rugoso-punctate,

pubescent, the mesonotum a little lineated posteriorly. Head large,
wider than the thorax, the cheeks margined. Antennæ 12-jointed,
dark-brown, the scape pale at base; first funiclar joint slender, cylindric, scarcely longer than the pedicel, the latter pale at tip; second
and third funiclar joints short, quadrate; the fourth wider; club 6-
jointed. Wings subfuscous, the marginal vein nearly as long as the
stigmal. Legs brownish-yellow, the coxæ black. Abdomen oval, the
first segment and the second, at base, striate, the following segments
all finely closely punctulate.

HABITAT.—Washington, D. C.

Type in Coll. Ashmead.

Described from a single specimen. The sculpture of the head and
scutellum and the length of the marginal vein readily distinguish the
species.

BÆONEURA Förster.

Hym. Stud., II, p. 100 (1856).

Head quadrate, anteriorly with a carina between the antennæ, frons
depressed; ocelli widely separated, the lateral close to the hind margin
of eye. Eyes large, oval, occupying the whole side of the head.

Antennæ 12-jointed, inserted at the clypeus, in ♀ clavate, in ♂
filiform.

Mandibles short and broad, bidentate.

Thorax ovoid, as in *Phanurus*, polished, without furrows, the scutellum
short, the metanotum divided by a central carina into two areas.

Wings very narrow and strongly fringed, with only a stigmal vein
that ends in a knob before attaining the costa.

Abdomen elongate fusiform, strongly depressed, subsessile, the first
segment narrowed, but wider than long, the second and third segments
subequal and the longest and broadest segments.

Legs normal.

A curious little genus, remarkable for the narrow and strongly
fringed front wings and the peculiar venation, strongly recalling some
of the forms in the Family *Mymaridæ*, with which it might easily be
confused. It also resembles *Phanurus*, in the Tribe Telenomini.

Förster says of it: "Not much larger than *Bæus*, although much
more elongate and narrower, stands the genus *Bæoneura*. In this
genus, however, we find the club distinctly jointed. It is readily distinguished from the following described genera by the small development of the submarginal vein, which passes from the base into the field
of the wing, but does not unite with the costa. In the formation of the
wings it forms, therefore, a fine transition to the Platygastroidea."

Förster's type seems not to have been described, and the following
species is, therefore, the first species to be described. Kirchner, in his
Cat. Hym. Eup., p. 193, mentions the fact, however, that there are two
species found in Europe, but gives no names.

Bæoneura bicolor sp. nov.

(Pl. ix. Fig. 6.)

♀. Length 1ᵐᵐ. Elongate, highly polished, black; abdomen, legs and antennæ, except the club, brownish or honey-yellow. Wings hyaline, narrow, strongly fringed. Club of antennæ brown, 5-jointed. Abdomen pointed fusiform, much longer than the head and thorax united, flat, or strongly depressed.

Habitat.—Arlington, Va.

Type in Coll. Ashmead.

A single specimen, taken by beating.

IDRIS Förster.

Hym. Stud., ii, p. 102 (1856).

(Type *I. flavicornis* Först.)

Head transverse or subquadrate, the cheeks and occiput delicately margined; ocelli 3, in a triangle, the lateral not touching the border of the eye; eyes large, oval, pubescent.

Antennæ inserted close to the mouth, 12-jointed in both sexes; in ♀ clavate, the club 6-jointed; in ♂ filiform.

Maxillary palpi 4-jointed; labial palpi 3-jointed.

Mandibles deeply bifid at apex.

Thorax ovoid, rounded before, the prothorax slightly visible from above; mesonotum a little broader than long with two distinct furrows, sometimes obsolete posteriorly; scutellum semicircular, with a punctate frenum; metathorax short, submarginated.

Front wings when folded extending to, or a little beyond, the tip of the abdomen, the submarginal vein long, reaching the costa beyond the middle of the wing, the marginal vein short, the postmarginal vein scarcely developed or never longer than the shaft of the stigmal vein, the latter rather short, oblique, and knobbed at tip; basal nervure wanting.

Abdomen oblong-oval, subpetiolated, the first and second segments about equal, the third the longest segment.

Legs very similar to *Hadronotus*.

Idris læviceps sp. nov.

♂. Length 1.50ᵐᵐ. Black, polished, the thorax faintly microscopically punctate, sparsely pubescent. Mandibles rufous, the teeth and apex black. Antennæ and legs reddish yellow. Mesonotum with two furrows. Wings subhyaline, fringed, the basal vein distinct, the marginal, twice as long as thick, the postmarginal but slightly developed, the stigmal short, terminating in a little knob. Abdomen black, polished, pubescent toward the apex, the first and second segments striated.

HABITAT.—Arlington, Va.

Type in Coll. Ashmead.

Described from a single specimen.

ACANTHOSCELIO Ashm., gen. nov.

(Type *A. americanus*.)

Head large, subquadrate, with a frontal ledge as in *Sparasion*, the occiput rounded, margined; ocelli 3, in a triangle, the lateral not quite touching the margin of the eye; eyes very large, rounded.

Antennæ as in *Scelio*.

Maxillary palpi 3-jointed.

Thorax ovoid, prothorax visible from above, only at the sides; truncate anteriorly; mesonotum convex, without furrows; scutellum large, quadrate, the posterior margin slightly arcuately emarginate, the angles acute; postscutellum produced into a large erect spine; metanotum short, abrupt, the angles prominent.

Front wings with the submarginal vein remote from the costa and curving and joining a punctiform marginal vein at about the middle of the wing; stigmal vein rather long, oblique, with a long radial branch, that forms a long, lanceolate marginal cell.

Abdomen sessile, oblong-oval, depressed, strongly carinated along the sides and composed of 6 segments, the first and second nearly equal, the third the longest.

Legs as in *Scelio*, the tibial spurs well developed.

The affinities of this remarkable genus are with *Scelio* and *Sparasion*; it agrees with the former in all the essential characters, except in having a frontal ledge, the structure of the meso- and meta-scutellum, and in venation; with the latter it agrees only in its cephalic characters. It is at once distinguished from both by the shape of the scutellum and the strong erect postscutellar spine, the shape of the scutellum being, indeed, unique in the family.

The genus is founded upon a male specimen, collected by Mr. Herbert Smith, at Chapada, in South America, and a female specimen in the Berlin Museum, labeled "Bogota."

Acanthoscelio americanus sp. nov.

(Pl. x, Fig. 6, ♂.)

♂. Length 3ᵐᵐ Opaque, coarsely rugose; antennæ, post-scutellar spine, legs, except the black coxæ, and the lateral margins of the abdomen, brownish-yellow; wings fuscous, yellowish-hyaline at base.

HABITAT.—Chapada, Brazil, South America.

Type in Coll. Ashmead.

Acanthoscelio flavipes, sp. nov.

♀. Length, 3.4ᵐᵐ. Black, coarsely, deeply punctate; antennæ, except the 6-jointed club, the palpi, the legs, including coxæ, metathorax, first abdominal segment and the lateral margins of the second and third, reddish-yellow. Head subglobose, as wide as the thorax, with a frontal ridge, coarsely rugoso-punctate. Eyes very large, round. Mandibles black, curved, with a tooth within. Antennæ 12-jointed, clavate, very short; first funiclar joint not longer than the pedicel; second not longer than wide; third and fourth transverse; club large, 6-jointed, black, the joints transverse. Thorax ovoid, the mesonotum without furrows. Scutellum quadrate, elevated posteriorly, the angles acutely produced. Metascutellum produced into a large, erect spine. Metathorax ridged at the sides with the angles toothed. Wings fuliginous, the basal one-third yellow, the venation similar to *Scelio*. Abdomen elongate oval, rugulose; first segment transverse, a little shorter than the second; third very slightly longer than the second; fourth very slightly shorter; fifth about one-half the length of the fourth; the following very short.

HABITAT.—Bogota.

Type in the Berlin Museum.

Described from a single specimen.

SPARASION Latreille.

Hist. Nat., III, p. 316 (1802); Förster Hym. Stud., II, p. 101 (1856).

(Type, *S. frontale* Latr.)

Head transverse, or subquadrate, with a frontal ledge or carina; ocelli 3, in a triangle but widely separated; eyes oval.

Antennæ in both sexes 12-jointed, inserted just above the clypeus, in ♀ clavate, the funiclar joints submoniliform; in ♂ long, subsetaceous, the flagellar joints all longer than thick.

Maxillary palpi elongate, 5-jointed; labial palpi 3-jointed.

Mandibles elongate, bidentate, the teeth subequal.

Thorax ovoid, narrowed into a little neck anteriorly: prothorax distinctly visible from above, dilated at sides; mesonotum usually with 2 furrows; scutellum large, semicircular; metathorax with the angles produced into short spines.

Front wings with the submarginal remote from the costa and terminating in a small stigma from which issues a short stigmal and a short postmarginal.

Abdomen long, ovate, depressed, sessile, the sides carinated, in ♀ with 6, in ♂ with 7 segments, the segments nearly equal.

Legs of moderate length, pilose, the tibial spurs 1, 2, 2, the claws simple.

The frontal ledge and the 5-jointed maxillary palpi readily distinguish the genus from *Scelio*, while the differences in the scutellum and post-scutellum separate it from *Acanthoscelio*.

TABLE OF SPECIES.

Wings fuscous or fuliginous 2
Wings hyaline.
  Scapulæ with a grooved line their whole length.
    Legs, except coxæ, honey-yellow: flagellar joints after the first twice as long
      as thick... *S. famelicum* Say.

2 Scapulae without a grooved line or only slightly indicated at the middle.
 Legs black, anterior and middle tibiae and all tarsi, honey-yellow, the posterior
  tibiae dusky at the middle.
 Frontal ledge broad, on the same plane with the vertex; postscutellum toothed;
  pedicel a little shorter than first flagellar joint; flagellar joints after
  the first only a little longer than thick; scapulae impunctate.
              S. pilosum Ashm., sp. nov.
 Frontal ledge not so broad or on the same plane with the vertex, more oblique;
  postscutellum not or but slightly toothed.
 Pedicel much shorter than the first flagellar joint, the flagellar joints after the
  first wider than long (♀); the flagellar joints after the first more
  than twice as long as thick (♂) .... *S. nigrum* Ashm., sp. nov.
 Pedicel longer than the first flagellar joint, the joints beyond about as long
  as thick, ♀ ; in ♂ the pedicel is scarcely half the length of first
  flagellar joint .... *S. pacificum* Ashm., sp. nov.

Sparasion famelicum Say.

Bos. Jour., I, p. 276; Lec. Ed. Say, II, p. 223; Cress. Syn. Hym., p. 248.

♀ . Length, 4.5ᵐᵐ. Elongate, black, subopaque, punctate; head with
a frontal ridge, rugose; thorax shining, with small, rounded punctures,
the parapsidal furrows distinct, the scapulae with a longitudinal grooved
line; scutellum margined posteriorly; metathorax narrowed toward
apex, broader at base than long, with two parallel median carinae, postscutellum not toothed, the posterior angles subprominent; abdomen
as long as the head and thorax together, oblong-oval, longitudinally
striated; wings hyaline, the venation brown black, the stigmal vein ending in a knob with a radial ray from its tip; postmarginal as long as
the stigmal; legs, except the black coxae, honey-yellow; antennae filiform, 12-jointed, fuscous, extending to the tip of the metathorax.

HABITAT.—Indiana and Fort George, Fla.

Described from a single specimen taken by Dr. R. S. Turner at Fort
George, Fla., and which I think without doubt is this long-lost species.

Sparasion pilosum sp. nov.

(Pl. x, Fig. 7, ♀ .)

♂ ♀ . Length, 4 to 5ᵐᵐ. Black, shining, very pilose, head rugose
from large, coarse punctures; the ledge very broad, on the same plane
with the vertex, with a broad sulcus along the anterior margin. Antennae 12-jointed, black, the pedicel and first flagellar joint more or
less piceous, the latter in ♀ one-third longer than the pedicel, in ♂ about
equal, the joints beyond a little longer than thick. Thorax sparsely
punctate, the parapsidal furrows distinct, the metathorax rugose, emarginate behind with a ∧-shaped carina, the angles slightly prominent.
Wings fuscous, a little paler toward base, the venation black, the stigma
quadrate, black. Legs black, the tibiae and tarsi, honey-yellow, the hind
tibiae dusky at the middle. Abdomen smooth, shining, longitudinally
striate, more coarsely striate in the male.

HABITAT.—Nevada and Washington.

Types in Coll. Ashmead.

Described from two specimens.

Sparasion nigrum, sp. nov.

♂ ♀. Length, 3 to 3.5^{mm}. Very close to *S. pilosum*, but quite distinct in the very short frontal ledge, and in its antennal characters. The ledge in the ♂ merely a slight transverse carina, but in the ♀ it is broader and slightly oblique, the head rugose but still not so coarsely rugose as in *S. pilosum*. In the ♂ the antennæ are long, filiform, the pedicel less than half the length of the first flagellar joint, the joints beyond more than twice as long as wide; in ♀ the pedicel is shorter than the first flagellar joint, the flagellar joints after the first distinctly wider than long. Abdomen in ♂ opaque, rugose, the rugosities longitudinally directed; in ♀ shining, punctate and striate.

HABITAT.—Pen Mar, Pa., and Oakland, Md.

Types in Coll. Ashmead.

Sparasion pacificum, sp. nov.

♂. Length, 3 to 3.5^{mm}. Black, shining, with long, sparse hairs; legs brownish-yellow, pilose, the coxæ and femora black; head rugose, anteriorly with a broad semicircular ledge, margined anteriorly and with a curved carina at base, conforming to the curve of the ledge. Antennæ 12-jointed, filiform, black, the pedicel and two basal joints of flagellum, brown or piceous; first flagellar joint in ♂ twice the length of pedicel, narrowed toward base, the joints beyond very little longer than wide, the joint ovate: in ♀ the pedicel is much longer than the first flagellar joint, the joints beyond about as long as thick. Thorax with sparse punctures, the collar dilated at sides to tegulæ, rugose anteriorly ; mesonotum with 2 furrows, ending just before reaching the scutellum ; scutellum subconvex, margined behind and with some punctures in the frenum; metathorax rugose, with prominent rounded angles. Wings fuscous, stigma and nervures brown-black, the stigmal vein slightly curved, ending in a knob, with a fuscous ray directed toward the apex of wing, forming a long open marginal cell. Abdomen oblong-oval, smooth, polished, pilose, the second and third sutures with punctures at bottom, the segments nearly equal, the first truncate and carinated at base.

HABITAT.—California.

Types in National Museum and Coll. Ashmead.

Three specimens in the National Museum were collected in August, by Koebele, in Marion County, Cal.; another in the Santa Cruz Mountains, while my collection contains but a single specimen from California, obtained by purchase.

SCELIOMORPHA Ashm., gen. nov.

(Type *S. longicornis* Ashm.)

Head transverse quadrate, the occiput and cheeks margined: ocelli 3, large, in a triangle, the lateral not touching the eye; eyes large, oval or rounded, hairy.

Antennæ in ♂ subsetaceous, 12-jointed, the flagellar joints all long cylindrical.

Maxillary palpi long, 5-jointed; labial palpi 3-jointed.

Mandibles arcuate, 3-dentate, the outer tooth large, the inner two teeth small, equal.

Thorax ovoid, the prothorax abruptly truncate before, dilated at sides; mesonotum a little wider than long, with two deep distinct furrows; scutellum large, rounded behind; metathorax rounded behind, the angles prominent.

Front wings as in *Sparasion*.

Abdomen fusiform, depressed, sessile, the sutures of the segments deep, the segments themselves nearly equal in length.

Legs much as in *Sparasion*, the tibial spurs being 1, 2, 2, but the hind femora are much more swollen.

This genus agrees more closely with *Sparasion* than any other genus, and like that it is very hairy; but the absence of the frontal ledge, the more distinctly pilose eyes, longer antennæ, and the swollen posterior femora separate it at once.

Sceliomorpha longicornis, sp. nov.

(Pl. x, Fig. 8, ♂.)

♂. Length, 3.5ᵐᵐ. Black, closely, rather coarsely, punctate, pilose; legs, except coxæ, reddish-yellow; wings fuliginous; antennæ 12-jointed, long, filiform, tapering to a point, extending to the middle of the abdomen; the pedicel annular, piceous; flagellar joints after the first about 3½ times as long as thick, the second slightly emarginate at base; abdomen striate and punctate, as in *Sparasion*.

HABITAT.—Santarem. (Herbert Smith.)

Type in Coll. Ashmead.

(?) Sceliomorpha bisulca, sp. nov.

♀. Length, 3.4ᵐᵐ. Brown-black, finely rugose, sparsely pubescent; scutellum and abdomen longitudinally lineately rugose; antennæ reddish-yellow, the club brown; legs, except black coxæ, brownish-yellow or reddish-yellow. Head transverse, reticulately rugose; the occiput roundly emarginate, not margined; the face subconvex; the cheeks short, margined. Eyes ovate. Mandibles small, triangular, rugose. Antennæ 12-jointed, very short, the scape reaching only a little beyond the middle of the face; pedicel longer than the first and second funicular joints united; all funicular joints transverse; club wider, the joints transverse. Pronotum produced into a collar anteriorly; mesonotum with 2 distinct furrows; scutellum semicircular margined behind; metathorax closely punctate, with a median furrow, the angles straight, but not prominent. Wings subfuscous, the marginal vein ending in an oblique stigma. Abdomen fusiform, depressed, sessile; the first and second segments about equal, half the length of the third; third and fourth equal, one-third shorter than the second; fifth and sixth very short; venter shining, faintly lineated.

HABITAT.—Texas and Florida.

Types in National Museum and Coll. Ashmead.

The specimen in the National Museum is in the old Belfrage collection, while that in my own collection was captured, by myself, holding on to the elytron of a short-winged locust, evidently with the intention of finding out where the eggs were to be deposited. The joints of the palpi could not be counted, and I do not feel certain that the insect belongs in this genus.

SCELIO Latreille.

Hist. Nat., XIII, p. 226, (1805); Förster Hym. Stud., II, p. 102.

Caloptenobia Riley, First Rep. U. S. Ent. Comm., p. 306.

(Type *S. rugulosus* Latr).

Head transverse, or subquadrate, the vertex broad, the occiput somewhat excavated and rounded; ocelli 3, in a triangle, the lateral being close to the eye; eyes oval or ovate.

Antennæ inserted close to the clypeus, in ♀ short, fusiform, subcompressed, 12-jointed, the 6 terminal joints forming a large club, the funiclar joints moniliform; in ♂ 10-jointed, short, subclavate, the joints short, submoniliform.

Maxillary palpi very short, inconspicuous, 3-jointed; labial palpi 3-jointed.

Mandibles long, arcuate, bidentate, the teeth subequal.

Thorax ovoid, the prothorax visible from above only at the dilated sides; mesonotum subconvex, broad, usually without furrows, rarely with distinct furrows; scutellum semicircular; metathorax short with the hind angles acute.

Front wings with the submarginal vein remote from the costal edge and terminating in a stigma, with or without a stigmal vein; sometimes there issues from the tip of the stigmal vein another delicate nervure that extends forward and forms a long, narrow, open marginal cell.

Abdomen sessile, long-ovate, or fusiform, depressed, the sides acutely margined, in ♀ 6, in ♂ 7 jointed.

The segments after the first, which is short and campanulate, nearly of an equal length, the third usually the longest.

Legs of moderate length; all femora and anterior tibiæ clavate, middle and posterior tibiæ subclavate; the tibial spurs 1, 1, 1, distinct; basal joint of hind tarsi not more than thrice as long as the second; claws simple.

The peculiar venation, the short, 3-jointed maxillary palpi, the short, fusiform, subcompressed antennæ, and the male having but 10-jointed antennæ readily distinguish the genus.

The parasitism of several of the species is known, and all are parasites in locust eggs, belonging to the family Acridiidæ.

Some of the species bear a remarkably close resemblance to the Eurytominæ in the family Chalcididæ; and, this remarkable mimicry caused Dr. Riley to erect a new genus, *Caloptenobia*, in that family for the reception of a species of *Scelio*, reared by Mr. Samuel H. Scudder from the eggs of a locust found in Massachusetts.

TABLE OF SPECIES.

Species with parapsidal furrows more or less distinct......................... 3
Species without parapsidal furrows, coarsely reticulately rugose.
  Coxæ and legs pale brownish-yellow.. 2
  Coxæ black or brown-black; rest of legs brownish-yellow.
    Wings fuscous with a stigmal vein and a radius.
      Scape and pedicel brownish-yellow, the flagellum brown-black; tegulæ rufous.
            S. FUSCIPENNIS Ashm.
    Wings hyaline; no stigmal vein and radius.
     ♀ with the scape beneath pale; pedicel ¼ longer than the first flagellar joint; second, third, and fourth flagellar joints very short transverse, the third a little the longest, 2¼ times wider than long.
     ♂ with the scape dark brown, flagellar joints, except first and last, about equal in length and width, the third scarcely wider than the fourth.
            S. HYALINIPENNIS Ashm.
2. Stigma with a stigmal vein.
  Scape and pedicel yellowish, the flagellum brown-black, the first flagellar joint shorter than the pedicel.
    Venter distinctly, coarsely striate, but not punctured ♀; venter with oblong punctures, the surface aciculated, antennæ pale brownish, the third flagellar joint much wider than the others ♂ ...S. OVIVORUS Riley.
Stigma without a stigmal vein.
  ♀ unknown.   ♂ with the third flagellar joint wider than either the second or fourth; venter distinctly striated.
    Antennæ brown-black; scape brownish-yellow; wings hyaline.
            S. PALLIDIPES, sp. nov.
    Antennæ wholly brownish-yellow; wings subfuscous..S. PALLIDICORNIS, sp. nov.
3. Coxæ dark brown or black; rest of legs brownish-yellow.
  Wings hyaline; no stigmal vein.
    Scape and pedicel brownish-yellow.......................S. ÆDIPODÆ, sp. nov.
    Antennæ wholly brown-black...............................S. OPACUS Prov.
Coxæ and legs pale brownish-yellow.
  Wings hyaline or but slightly tinged.
    Stigma without a stigmal vein; scape and pedicel, and sometimes the first two or three flagellar joints pale.
      Abdomen mostly rufous............................S. RUFIVENTRIS, sp. nov.
      Abdomen black, the venter piceous.....................S. CALOPTENI Riley.
    Stigma with a short stigmal vein; scape alone pale, although the pedicel and the first flagellar joint are sometimes pale beneath.
     ♀ postscutellum striate; abdomen delicately longitudinally striated, the extreme apex of the segments smooth polished; angles of metathorax on a line with the apex of the metonotum........S. LUGGERI Riley.
     ♀ postscutellum punctate; abdomen longitudinally lineated, the venter, except first and second segments, smooth; angles of metathorax rather prominent, covered with a white pubescence..S. FLORIDANUS, sp. nov.

Scelio fuscipennis Ashm.

Ent. Am., III, p. 119; Cress. Syn. Hym., p. 314.

♂ ♀. Length, 3.5 to 5.3ᵐᵐ. Black, coarsely rugose, the parapsidal furrows not at all indicated; scape, pedicel and legs, except the coxæ, brownish-yellow or reddish-yellow; wings fuscous, paler at base, the stigmal vein distinct with a long branch forming a marginal cell.

HABITAT.—Jacksonville and Fort George Island, Fla.

Types in Coll. Ashmead.

Common. I think this species will prove to be a parasite in the eggs of *Dictyopterus micropterus.*

Scelio hyalinipennis Ashm.

(Pl. x, Fig. 9, ♀.)

Ent. Am., III, p 119; Cress. Syn. Hym., p. 314.

♂ ♀. Length, 4 to 4.5ᵐᵐ. Black, coarsely rugoso-punctate, the mesonotum without trace of furrows; face striate above insertion of antennæ and below the eyes. Mandibles, scape, and pedicel beneath in ♀, and legs, rufous, or brownish-yellow; coxæ in ♂ black, frequently pale in ♀, or only dusky or blackish basally. Antennæ 12-jointed, brown-black, except as mentioned above, the flagellum fusiform, depressed, the pedicel about one-third longer than the first joint, the second, third, and fourth joints very transverse, 3 times as wide as long; in ♂ 10-jointed, flagellar joints 2 to 7, about of an equal length, scarcely twice as wide as long, the third but little wider than the fourth. Wings clear hyaline, with only a trace of the submarginal vein; stigma present, the submarginal nervure before reaching it obsolete or subobsolete, hyaline; no trace of a stigmal nervure. Abdomen imbricato-rugose, venter in ♀ often rufous.

HABITAT.—Jacksonville, Fla.

Types in Coll. Ashmead and in National Museum.

Scelio ovivorus Riley.

Caloptenobia orivora Riley, First Rep. U. S. Ent. Comm., p. 306.
Scelio famelicus Riley, nec Say, Second Rep. U. S. Ent. Comm., p. 270.
Scelio orivora Ashm., Ent. Am., III, p. 119.

♀. Length, 4.20ᵐᵐ. Black, very coarsely rugoso-punctate, the mesonotum without furrows; there is a large polished space on face above the antennæ, and on either side coarse striæ converge toward the mandibles; mandibles pale rufous. Antennæ 12-jointed, brown-black, the scape and pedicel pale rufous or brownish-yellow; pedicel very slightly longer than first funicular joint. The thorax is very coarsely rugose, the large punctures confluent; scutellum coarsely rugoso-punctate; metathorax with a finer sculpture, the angles dilated, prominent acute. Tegulæ yellowish. Wings subfuscous, the stigma of the submarginal vein large with a stigmal vein and a ray from its tip. Legs, including all coxæ, pale yellowish. Abdomen fusiform, the dorsum and venter coarsely longitudinally striated; the first segment is short, transverse, very little longer than half the length of the second; the third, the longest, the fourth, a little shorter than the third; the fifth, about as long as the second; the sixth, shorter.

♂. Length, 3.6ᵐᵐ. Head and thorax coarsely rugose, the latter without furrows; antennae pale brown, 10-jointed, the third flagellar joint longer and wider than the others; legs brownish-yellow; wings hyaline, the submarginal vein terminating in a rounded stigma with a short stigmal vein; abdomen distinctly striated, the apical edges of the segments smooth, impunctate.

HABITAT.—Massachusetts.

Types in National Museum.

Described from 3 ♀ specimens, now in poor condition, reared by Samuel H. Scudder from the eggs of *Dissosteira carolina*.

Scelio pallidipes, sp. nov.

♂. Length, 3.2ᵐᵐ. Head and thorax coarsely rugose, the latter without parapsidal furrows. Antennae brown-black, scape brownish, the third flagellar joint much wider than the others, twice as wide as long. Angles of metathorax rounded, not prominent. Wings clear-hyaline, the submarginal vein and the stigma hyaline, scarcely apparent; no stigmal vein. Legs wholly brownish-yellow. Abdomen striated, the first segment quadrate, separated from the second by a strong constriction; venter distinctly, but not coarsely, striated.

HABITAT.—Jacksonville, Fla.

Types in Coll. Ashmead.

The wholly brown-black flagellum and the difference in the metathoracic angles separate the species from *S. pallidicornis*, the only species with which it would be apt to be confused.

Scelio pallidicornis, sp. nov.

♂. Length, 4.1ᵐᵐ. Black, coarsely rugose; antennae pale or brownish-yellow, the pedicel shorter than the first flagellar joint, the latter obconical and the longest joint; the third joint the widest, about 1½ times as wide as long, or only slightly wider than the following joints. Angles of metathorax somewhat prominent, covered with a whitish pubescence. Wings subfuscous, hyaline at base, the stigma without a stigmal vein. Legs wholly brownish-yellow. Abdomen above coarsely striated, and with dilated punctures; beneath smoother, shining, but longitudinally striated, the striae faint or indistinct beneath the lateral carina.

HABITAT.—Jacksonville, Fla.

Type in Coll. Ashmead.

The large size, pale antennae, and color of wings at once separate the species from the others described here.

Scelio œdipodæ, sp. nov.

♂ ♀ . Length, 3.5 to 4ᵐᵐ. Black, rugose; scape, pedicel, and legs, except the coxæ, brownish-yellow; thorax with parapsidal furrows distinct, in female wanting anteriorly; tegulæ piceous; wings hyaline, the stigma reaching the costa; the stigmal vein wanting, or only slightly developed; metathorax truncate behind, the angles straight, not prominent; abdomen longer than the head and thorax united, striated; first segment well separated from the second by a strong constriction, twice as wide as long.

The antennæ in the male are brownish, the face vertically striated, the femora brownish, the tibiæ and tarsi yellowish.

HABITAT.—Arlington, Va.

Types in Coll. Ashmead.

Described from 1 ♂ and 2 ♀ specimens, reared from the eggs of a species of *Oedipoda*. The black coxæ and the distinct parapsidal furrows at once separate the species from *Scelio orirorus* Riley, and the grooved mesonotum from *S. hyalinipennis* Ashm., the rugosities being finer than in either of these species.

Scelio opacus Prov.

Acerota opaca Prov., Add. et Corr., p. 184; Cress. Syn., p. 219.

♀ . Length, 3.25ᵐᵐ. Black, opaque, with the feet red, more or less dusky. Head and thorax rugoso-punctate. the abdomen with the disks of the segments longitudinally aciculated. Wings hyaline. Abdomen sessile, the first segment being much narrower than the others. (*Prov.*)

HABITAT.—Cap Rouge.

Type in Coll. Provancher.

Unknown to me.

[Since this was written a male *Scelio*, labeled "*Acerota opaca*, Type, Provancher", has been deposited in the National Museum by D. W. Coquillett, of Los Angeles, Cal., and I have in consequence merely removed Provancher's brief description from the genus *Acerota*, in the subfamily *Platygasterinæ*, to its proper place in *Scelio*, having had no time to draw up a full description.]

Scelio rufiventris, sp. nov.

♀ . Length, 3.5ᵐᵐ. Brown-black, rugose, the abdomen mostly rufous, fuscous above, except along the sides; scape, pedicel, first funiclar joint, and the legs pale rufous. Thorax with distinct parapsidal furrows. Tegulæ black. Wings hyaline, the stigma without a stigmal nervure. Funiclar joints after the first very transverse, three times as wide as long, the pedicel being as long as the first, second, and third funiclar joints united. Abdomen pointed-fusiform, longer than the head and thorax together, shagreened, not distinctly striated, the venter smooth, wholly rufous; dorsum with a rufous streak along the margins.

HABITAT.—Fort George, Fla.

Type in Coll. Ashmead.

Described from a single specimen taken by Dr. R. S. Turner, in August, 1882.

The color and sculpture of abdomen sufficiently distinguish the species.

Scelio calopteni sp. nov., Riley.

"♀. Length, 3 to 3.4ᵐᵐ. Black, rugose; the mesonotum with faint but distinct furrows. Head with a smooth, shining space on face above the antennae, with striae on each side converging toward mouth; mandibles pale rufous; antennae 12-jointed, brown-black, the scape and pedicel yellow; pedicel longer than the first funicular joint, the joints after the first very transverse; club slightly compressed from above, the joints closely conjoined. Thorax with the tegulae pale rufous; wings hyaline, the submarginal vein ending in a rounded stigma, but without a stigmal vein; legs, including the coxae, pale brownish-yellow. Metanotum with 2 complete longitudinal subparallel median carinae. Abdomen fusiform, longer than the head and thorax together, closely, longitudinally aciculate; the venter piceous, finely aciculate at sides.

"Described from many ♀ specimens, bred June 27 and July 3, 1883, from the eggs of *Caloptenus atlantis*, collected by self and A. Koebele in New Hampshire."—[From Riley's MS.]

HABITAT.—Boscawen, N. H.

Types in National Museum.

This species is closely allied to *S. Luggeri*, and may be but a variety; but as the stigma is without a stigmal vein and the color of the antennae is slightly different, I believe it to be a distinct species.

Scelio Luggeri sp. nov., Riley.

"♀. Length, 3 to 3.2ᵐᵐ. Black, rugose, the mesonotum with two furrows; scape and legs, including coxae, brownish-yellow. Head with the antennae brown-black, the pedicel distinctly longer than the first flagellar joint, joints 2 to 4 very short, transverse. Metanotum with 4 complete longitudinal median carinae; wings subfuscous, paler at base, the stigma with a short stigmal vein. Abdomen fusiform, about one-third longer than the head and thorax together; longitudinally aciculate, the apex of all the segments with a smooth, polished space; venter black, aciculate over entire surface.

"Described from three ♀ specimens, reared in July, 1889, from the eggs of a *Caloptenus* sp. by Prof. O. Lugger."—[From Riley's MS.]

HABITAT.—Otter Tail County, Minn.

Types in National Museum.

Scelio floridanus, sp. nov.

♀. Length, 3 to 3½ᵐᵐ. Densely black, subopaque, with rather coarse reticulated punctures, the thorax with distinct furrows, the postscutellum highly ridged. Antennæ brown-black, the scape, legs, and mandibles, yellow; funicular joints transverse, the club large fusiform, wider than long; angles of metathorax prominent. Wings subhyaline, the venation yellowish, the submarginal vein ending in a slight stigma and an oblique stigmal vein, the latter with an indistinct radius. Abdomen fusiform, lineatedly rugose; first segment transverse-quadrate of an equal length with the fifth, the second, longer, the third, the longest segment, the fourth shorter than third, the sixth, subequal with the fifth, the seventh much shorter; the venter polished, the segments striated towards apex.

HABITAT.—Haw Creek and Jacksonville, Fla.

Types in National Museum and Coll. Ashmead.

Subfamily VI.—PLATYGASTERINÆ.

Head transverse, rarely quadrate. Ocelli 3, triangularly arranged. Mandibles bifid at tips. Maxillary palpi 2-jointed; labial palpi 1-jointed. Antennæ elbowed, clavate, most frequently 10-jointed in both sexes, rarely 8-or 9-jointed, inserted at the base of the clypeus. Pronotum never very large, scarcely visible from above, mesonotum most frequently transverse, with or without furrows; scutellum variously shaped, often with an awl-shaped tip or spined, flat, semicircular or pillow-shaped; metathorax short, with a median sulcus. Front wings most frequently entirely veinless, or with a submarginal vein terminating in a stigma before attaining the costa, the basal nervure rarely present; hind wings lanceolate, veinless. Abdomen petiolate or subpetiolate, ovate, oblong-oval or conic-ovate, depressed, very rarely greatly elongate, usually composed of 6 visible segments and always carinated at the sides, the second segment the longest. Legs long, the femora and tibiæ clavate, the tibial spurs 1, 1, 1, the tarsi, except in a single genus, *Iphetrachelus*, 5-jointed, the claws simple.

A very large and extensive group, at one time classified with the *Scelioninæ*, but readily distinguished by the 10-jointed clavate antennæ, the 2-jointed maxillary palpi, 1-jointed labial palpi, and the bifid mandibles; the wings, except in a few genera, being entirely veinless and wholly different from the Scelioninæ.

The group is divided into numerous genera, the species of which confine their attacks almost exclusively to the Dipterous families Cecidomyiidæ and Tipulidæ, the only records conflicting being two recorded by Ratzeburg. *Platygaster contorticornis* Ratzb. is said to have been bred from *Tortrix strobilana* and *P. mucronatus* Ratzb. from *Tortrix resinana*.

From our present extensive knowledge of the rearings of the Platygasterids it is, however, quite evident that these records are erroneous and these Tortricids must have been accompanied by Dipterous larvæ overlooked by Ratzeburg.

My *Platygaster aphidis* is recorded from an Aphis on *Chenopodium album*, but we know now that some Aphides are parasitized by Cecidomyiids and this apparent discrepancy is thus explained. The species is unquestionably a secondary parasite on some Cecidomyiid infesting or living with the Aphid.

The genus *Amitus* Haldeman (=Zaerita Förster) is, however, apparently a primary parasite on the *Aleyrodidæ* a family of Homopterous insects allied to the *Coccidæ*, unless the *Aleyrodidæ* have Dipterous parasites not yet discovered.

The Platygasterinæ may be divided into two tribes as follows:

Anterior wing with a distinct clavate submarginal vein....Tribe I.—INOSTEMMINI.
Anterior wing entirely veinless, rarely with indications of a submarginal vein, if present very short, faint, and never clavate........Tribe II.—PLATYGASTERINI.

TRIBE I.—INOSTEMMINI.

In this tribe but seven genera are known, all represented in our fauna, and distinguished by the aid of the following table·

TABLE OF GENERA.

FEMALES.

Tarsi 5-jointed 2
Tarsi 4-jointed, antennæ 8-jointed........................IPHETRACHELUS Hal.
2. Antennæ 10-jointed.. 3
Antennæ 9-jointed..ALLOTROPA Förster.
3. Wings without basal and median veins.................................... 4
Wings with basal and median veins.
  Mesonotum with faint or distinct furrows.
    Antennal club 3-jointed........................METACLISIS Förster.
    Antennal club 4-jointed ....MONOCRITA Förster.
4. Lateral ocelli nearer the apical ocellus than to the inner margin of eye.
  Mesonotum without or with delicate furrows; antennal club 4-jointed.
                                              ISOSTASIUS Förster.
  Lateral ocelli nearer the inner margin of eye than to the apical ocellus.
    First abdominal segment with a horn; mesonotum with faint furrows.
                                              INOSTEMMA Hal.
    First abdominal segment without a horn; mesonotum with distinct furrows; antennal club 4-jointed, the funiclar joints slender, cylindric.
                                              ACEROTA Förster.

MALES.

Tarsi 5-jointed.. 2
Tarsi 4-jointed; mesonotal furrows distinct; antennæ 10-jointed, with whorled hairs.. IPHETRACHELUS Hal.
2. Antennæ 10-jointed... 3
Antennæ 9-jointed, with whorled hairs......................ALLOTROPA Förster.

3. Wings without basal and median veins....................................... 4
  Wings with basal and median veins.
    Mesonotum with 2 faint furrows.
      Antennæ subclavate moniliform, first funiclar joint very minute.
        second somewhat larger, the following to 10th larger, gradu-
        ally thickened, the last larger, conical ....METACLISIS Förster.
      Antennæ submoniliform, first funiclar joint very minute, second
        larger, thickened, curved, the third small, triangular, the fol-
        lowing except the last transverse moniliform, the last conical.
                                                   MONOCRITA Förster.
4. Lateral ocelli nearer the apical ocellus than to the margin of eye.
  Mesonotum without furrows..................................ISOSTASIUS, Förster.
  Lateral ocelli nearer the margin of eye than to the apical ocellus.
    Mesonotum with faint furrows; antennæ moniliform, pubescent, the first
      two funiclar joints nearly equal, the second somewhat curved,
      the third small, triangular, the four following moniliform,
      the last conical...INOSTEMMA Hal.
    Mesonotum with 2 distinct furrows; antennæ filiform, pubescent, the
      second funiclar joint long, cylindrical, longer than the first,
      the third shorter than the first, the following oval, the last
      about thrice as long as thick ....ACEROTA Förster.

IPHETRACHELUS Haliday.

Ent. Mag., III, p. 273 (1836); Förster Hym. Stud., II, p. 106 (1856).

(Type, *I. lar* Hal.).

Head transverse, the frons subconvex, smooth; ocelli 3 in a triangle, the lateral close to the margin of the eye; eyes rounded.

Antennæ inserted just above the clypeus, in ♀ 10-jointed, very long, the terminal joints forming a club; scape long, subclavate, funiclar joints small, moniliform; in ♂ 9-jointed, the flagellar joints nodose-pedicellate, with whorls of sparse long hairs.

Maxillary palpi short, 2-jointed: labial palpi 1-jointed.

Mandibles bifid at tips.

Thorax oval, the prothorax slightly visible from above; mesonotum with or without furrows: scutellum gibbous, nearly round; metathorax short, rounded behind.

Front wings ciliated, the submarginal vein ending in a knob a little before the middle of the wing.

Abdomen long ovate, subsessile, subconvex, and narrower than the thorax, the second segment very large, occupying most of the surface, the following segments all short.

Legs slender, the femora subclavate, the tarsi 4-jointed, the basal joint of hind tarsi about as long as all the others together.

The 4-jointed tarsi, and the 9-jointed, nodose-pedicellate antennæ of the male readily distinguish the genus.

Iphetrachelus americanus Ashm

(Pl. XI, Fig. 1, ♂.)

Proc. Ent. Soc. Wash., II, p. 58, 1890.

♂. Length, 0.6ᵐᵐ. Black, shining, delicately microscopically punctate and faintly sericeous.

Antennæ and legs uniformly yellow. Antennæ nodose-pedicellate, with whorls of long hairs; pedicel rounded; second funiclar joint triangular. Thorax somewhat flattened on the disk, without distinct furrows, although there are two shallow longitudinal furrows on the shoulders. Scutellum semicircular, subconvex, separated from the mesonotum by a delicate transverse grooved line. Tegulæ yellow. Wings hyaline, with a long fringe at apex, the submarginal vein pale yellow, knobbed at tip. Abdomen long, oval, smooth, polished and black, except the first segment, which is yellow.

HABITAT.—Arlington, Va.

Type in Coll. Ashmead.

Described from a single specimen, taken by sweeping.

ALLOTROPA Förster.

Hym. Stud., II, p. 106 (1856).

(Type A. mecrida Walk.)

Head transverse; ocelli 3, the lateral rather close to the margin of the eye.

Antennæ inserted just above the clypeus, 9-jointed in both sexes, the flagellar joints in ♂ dentate, verticillate; scape stout, fusiform, pedicel small, globose, in ♀ clavate, ending in a 4-jointed club.

Maxillary palpi 2-jointed; labial palpi 1-jointed.

Thorax ovoid, the mesonotum with 2 faint furrows; scutellum short, semicircular; metathorax short, the posterior angles subacute.

Front wings pubescent, the submarginal vein short, ending in a knob.

Abdomen long-ovate, convex, narrower than the thorax, the first segment short, broad, the second large, the following very short.

Legs moderate, the femora clavate, tibiæ and tarsi slender.

This genus is distinguished at once from all the other genera in having 9-jointed antennæ in both sexes.

Allotropa americana Ashm.

(Pl. XI, Fig. 2, ♀.)

Can. Ent., XIX, p. 125; Cress. Syn. Hym., p. 248.

♀. Length 1.4ᵐᵐ. Black, closely minutely punctulate; head transverse, opaque, the lateral ocelli twice their width from the margin of the eye; mandibles rufous; antennæ 9-jointed, ending in a 4-jointed club, the joints serrate toward one side; scape rufous; pedicel and flagellum dark brown; first and second funiclar joint about equal, the third triangular or subtriangular. Thorax microscopically punctate, but shining in the middle, the mesonotum with two distinct furrows; scutellum finely punctulate, opaque, with an elevated margin behind. Tegulæ black. Wings hyaline, the submarginal vein knobbed. Legs reddish-yellow, the coxæ dusky. Abdomen oblong-oval, shining, the petiole and second segment at base striated along the sides, minutely punctulate, above smooth, shining, the following segments subopaque, very finely punctulate.

HABITAT.—Jacksonville, Fla.

Type in Coll. Ashmead.

METACLISIS Förster.

Hym. Stud., II. p. 106 (1856).

(Type *M. areolata* Hal.)

Head transverse, broader than the thorax; ocelli 3, in a triangle, the lateral ocelli about twice their width from the margin of the eye; eyes rounded.

Antennæ inserted just above the clypeus, 10-jointed in both sexes; in the ♀ not half the length of the body, ending in a 3-jointed club; funiclar joints very small, transverse; club joints large, broad, the last conical, somewhat larger than the penultimate; in ♂ moniliform, incrassated toward apex and longer than half the length of body; first flagellar joint minute; fourth, moderate, fifth, and following to the ninth, larger and broader, tenth, conical, much longer than the penultimate.

Thorax short, ovate, finely scaly; pronotum short, scarcely visible from above; mesonotum with indistinct furrows; scutellum small, flattened, semicircular; metathorax moderate.

Front wings with the submarginal vein terminating in a stigma, the basal nervure distinct, and with a basal cell.

Abdomen ovate or oval, twice the length of the thorax, the apex pointed, stylus-like in ♀; first segment very short; second very large; third, fourth and fifth short, about equal; fifth about as long as segments 3 4 and 5 united; sixth shorter; in ♂ the abdomen is scarcely longer than the thorax.

Legs clavate.

This genus and *Monocrita* Förster are the only genera in the family having a distinct basal nervure, which alone is sufficient to distinguish them from all others. The female in *Metaclisis* is separated from *Monocrita* by having a 3-jointed antennal club, the male by the smaller second funiclar joint.

But two species have been described.

Metaclisis belonocnemæ Ashm.

(Pl. XI, Fig. 3, ♀.)

Can. Ent., XIX, p. 125 ♀.—Cress. Syn. Hym., p. 248.

♀. Length, 1.5ᵐᵐ. Black, opaque, closely, finely punctate. Legs and antennæ, except the club, pale brownish yellow; club brown-black. Antennæ 10-jointed, the scape thick, clavate; pedicel longer and thicker than the first funiclar joint; first and second funiclar joints cylindric, equal or nearly so, the first very slightly the longer; third and fourth very minute transverse; club 4-jointed, the joints, except the last, transverse, the last conical, longer than the preceding. Parapsidal furrows distinct. Scutellum convex with a carina around the posterior margin. Tegulæ dark brown. Wings, hyaline, the submarginal vein yellowish.

Abdomen oval, as long as the thorax, smooth, shining, except the first segment, which is a little roughened from some striæ.

HABITAT.—Jacksonville, Fla.

Type in Coll. Ashmead.

This species was reared by myself from a Cynipid gall, *Belonocnema treatæ* Mayr, and it is undoubtedly parasitic on a Cecidomyiid inhabiting the gall.

Metaclisis erythropus, Ashm.

Can. Ent., XX, p. 51.

♂. Length, 2.5ᵐᵐ. Black; head opaque, coarsely rugose on the vertex and cheeks; face transversely aciculated, with a central depressed line extending forward from the middle ocellus. Antennæ 10-jointed, black; pedicel as long as the second funiclar joint; first funiclar joint small, second swollen; club 6-jointed, the joints longer than wide, subpedicellate, hairy, the last joint conical, longer than the preceding. Thorax finely reticulated, or scaly; parapsidal grooves distinct, converging and meeting at the base of the scutellum. Metathorax covered with a sericeous pubescence. Legs rufous, the coxæ black. Abdomen black, polished. Wings subhyaline (probably tinged from the cyanide bottle), faintly pubescent, the tegulæ rufo-piceous.

HABITAT.—Ottawa, Canada.

Type in Coll. Ashmead.

Described from a single specimen received from Mr. W. Hague Harrington.

MONOCRITA Förster.

Hym. Stud., II, p. 106 (1856)

(Type *M. atinas* Walk.)

Head transverse, wider than the thorax; ocelli 3, in a triangle, the lateral as far from the front ocellus as to the eye margin.

Antennæ inserted just above the clypeus, 10-jointed in both sexes, in ♀ terminating in a 4-jointed club, the scape slender, pedicel cyathiform, first funiclar joint small; in ♂ with a 6-jointed filiform club, the joints slightly pedicellate, oval, the last long conical; the first funiclar joint about twice as long as thick, the second stouter, obliquely truncate at tip and curved.

Thorax short, ovoid, convex; prothorax very short; mesonotum with distinct but delicate furrows; scutellum convex; metathorax short.

Front wings pubescent, ciliated, the submarginal vein ending in a knob, the basal nervure distinct, and with a distinct basal cell.

Abdomen ovate, convex, slightly shorter and narrower than the thorax; first segment small, the second very large, the third and following very short.

Legs clavate, the tibial spurs weak, the basal joint of hind tarsi twice the length of the second.

The two species in our fauna may be thus distinguished:

Legs black, the head and thorax microscopically shagreened.
   First and second abdominal segments not striated, the latter hairy at base.
M. NIGRIPES Ashm.
Legs brownish-yellow, the head and thorax smooth, impunctured.
   First abdominal segment and the second at base striated, not hairy.
M. CARINATA, sp. nov.

Monocrita nigripes Ashm.

Bull. No. 1, Col. Biol. Assoc., p. 8, 1890.

♂. Length, 1.80ᵐᵐ. Black, subopaque, with a fine, granulated, or shagreened sculpture; the face above the insertion of the antennæ with some transverse aciculations. Antennæ 10-jointed, black, the flagellum about twice the length of the scape, pedicel twice as long as thick, narrowed at base, the second joint minute, the third, slightly swollen outwardly, the following joints very slightly longer than wide, truncate at tips and rounded off at base, or cup-shaped, connected by a very short pedicel. Parapsidal grooves distinct, converging and almost meeting posteriorly. Scutellum high convex, with a deep, transverse depression across the base. Mesopleura smooth, polished, with a curved impression across the disk; metapleura densely covered with a griseous pubescence; the metathorax and petiole more sparsely pubescent. Legs black, the knees slightly, anterior legs beneath, and all the tarsi, more or less piceous, or reddish. Abdomen highly polished, oblong-oval, as long as the head and thorax together, the petiole not longer than wide, the second segment occupying fully three-fourths of its length, smooth, the following segments exceedingly short and about equal in length. Wings hyaline, pubescent, the submarginal and basal veins distinct, the former knobbed at tip.

HABITAT.—Greeley, Colo.

Type in Coll. Ashmead.

Nothing is known of the habits of this species, which was obtained from H. F. Wickham.

Monocrita carinata, sp. nov.

(Pl. xi, Fig. 4, ♀.)

♀. Length, 1.20^{mm}. Black, shining, subpubescent; head transverse, the face with a central longitudinal carina extending from the front ocellus to between the base of antennæ. Antennæ 10-jointed, the scape yellow, the flagellum brown and twice the length of the scape; pedicel long-oval; joint 1 of funicle small, the second enlarged; club 6-jointed, the joints, except the last transverse-quadrate, the last cone-shaped, slightly more than twice as long as the penultimate. Thorax with the parapsidal furrows delicately impressed posteriorly, obsolete anteriorly. Scutellum convex, smooth, with a carina surrounding the hind margin. Legs brownish-yellow, the femora and tibiæ very slightly embrowned, the coxæ black. Tegulæ piceous-black. Wings subhyaline, pubescent, the submarginal vein brown, ending in a large rounded knob and with a distinct basal nervure. Abdomen oval, smooth, shining, the first segment and the second at base, striated.

HABITAT.—District of Columbia.

Type in Coll. Ashmead.

Described from a single specimen, captured by Mr. E. A. Schwarz, May 5, 1890.

ISOSTASIUS Förster.

(Type *Platygaster punctiger* Nees).

Hym. Stud., ii, p. 106 (1856).

Head transverse, vertex subconvex, occiput not margined; ocelli 3, in a triangle, rather close together, the lateral distant from the margin of the eye; eyes oval. Antennæ inserted just above the clypeus, 10-jointed in both sexes, in the ♀ the 4 terminal joints forming a large club, the last joint of which is the largest; pedicel cyathiform, much larger than the first funiclar joint; the four funiclar joints are all small, moniliform; in ♂ with a 5-jointed club.

Thorax ovoid, the prothorax visible from above, rounded before; mesonotum smooth or punctate, with or without furrows; the scutellum rather high, cushion-shaped or convex, separated at base from the mesonotum by a deep transverse furrow; metathorax very short.

Front wings with the submarginal vein terminating in a small knob before reaching half the length of the wing.

Abdomen in ♀ conical-ovate, 6-jointed; in ♂ oval, 7-jointed, the basal segment short, broader than long, the second very long and occupying fully two-thirds of the whole surface, the following very short.

Legs clavate, the tibial spurs 1, 1, 1, the tarsi 5-jointed, the basal joint of posterior tarsi 2½ times as long as the second, or less than thrice as long, 2 to 4 subequal, the last a little longer than the second.

Isostasius musculus Ashm.

(Pl. xi, Fig. 5, ♀.)

Can. Ent., xix, p. 126; Cress. Syn. Hym., p. 249.

♀ . Length, 1.20ᵐᵐ. Black, shining; head transverse, closely, finely, microscopically punctate and subopaque, the lateral ocelli nearer to the front ocellus than to the margin of the eye. Antennæ 10-jointed, piceous, or dark brown, the scape clavate, the pedicel longer and thicker than the first two funiclar joints, the first and second funiclar joints cylindric, nearly equal, the third and fourth small, transverse; club enlarged, 4-jointed, the joints quadrate. Thorax with scarcely a trace of furrows. Scutellum transverse, convex. Mesopleura aciculated. Legs brown, the coxæ black, the trochanters, base and apex of femora, and tibiæ and all tarsi honey-yellow. Tegulæ black. Wings hyaline, the submarginal vein terminating in a small, black knob. Abdomen pointed ovate, polished, black, the basal segment rugose.

HABITAT.—Jacksonville, Fla.

Type in Coll. Ashmead.

Isostasius fuscipennis, sp. nov.

♀ . Length, 1ᵐᵐ. Black, shining; head anteriorly faintly shagreened, posteriorly transversely aciculated, the lateral ocelli about twice their width from the margin of the eye. Antennæ brown-black. Thorax polished, with faint traces of furrows only posteriorly, the base of the middle lobe thus formed projecting slightly upon the scutellum; mesopleura polished, impunctured, metapleura subopaque, shagreened; scutellum convex, shining, margined posteriorly with a fine whitish pubescence. Tegulæ black. Wings fuscous. Legs piceous, the anterior femora, tibiæ, and tarsi, yellowish; middle and hind tarsi, pale rufous.

Abdomen as long as the head and thorax, polished, the first segment striated, the second, at base sulcate.

HABITAT.—Washington, D. C.

Type in Coll. Ashmead.

Described from a single specimen, collected by Mr. E. A. Schwarz.

Isostasius arietinus Prov.

Add. et Corr., p. 183.

Bæoneura arietina, Prov., Add. et Corr., p. 403.

♀ . Long. 11 pce. Noir, avec la bouche, le scape des antennes et les pattes, d'un beau jaune-miel. Tête aussi large que le thorax, à antennes fortes, le scape fort, arqué, jaune ainsi que l'article qui le suit, le reste formant une forte massue brune recourbée en corne de bélier. Mesonotum avec 3 sillons bien distincts. Ailes hyalines, avec la sous-marginale brune se terminant dans un stigma grand et épaissi qui donne naissance à un radius se dirigeant vers le sommet de l'aile. Pattes jaunes, hanches noires. Abdomen sessile, poli, brillant, droit, tarière non sortante. La tête et le thorax sont très finement ponctués et les 2 ocelles basilaires sont plus rapprochés de l'apical que du coin interne des yeux. (*Provancher.*)

HABITAT.—Cap-Rouge, Canada.

Unknown to me.

INOSTEMMA Haliday.

Ent. Mag., 1, p. 270 (1833); Förster, Hym. Stud., II, p. 107 (1856).

(Type *I. Boscii* Jurine.)

Head transverse, the frons convex, the occiput often impressed, rarely distinctly margined; ocelli 3 in a triangle, the lateral being nearer to the margin of the eye than to the front ocellus; eyes rounded.

Antennæ inserted just above the clypeus, 10-jointed, the scape clavate, in the ♀ terminating in a 4-jointed club, in ♂ with a 5-jointed club, the last funiclar joint very minute.

Thorax ovate, the mesonotum with 2 delicate impressed lines, often obsolete anteriorly; the scutellum semicircular, subconvex, separated from the mesonotum by a straight impressed line; the metathorax short, sloping, unarmed.

Front wings with the submarginal vein abruptly terminating in a stigma before attaining the middle of the wing.

Abdomen in ♀ pointed-ovate, subcompressed below, 6-jointed, the first segment always with a horn extending forward over the thorax, the second very long, the following very short, about equal in length; in ♂ shorter, oblong oval, with the tip rounded, 7-jointed.

Legs clavate, the tibial spurs 1, 1, 1, the basal joint of posterior tarsi about three times as long as the second, 2 to 4 subequal, the last as long as the second.

The horn-like structure at the base of the abdomen renders the females in this genus easy of recognition, while in the males the position of the ocelli and the antennæ must be carefully examined or they will be apt to be confused with those in closely allied genera.

TABLE OF SPECIES.

FEMALES.

Head finely microscopically punctate, the thorax smooth, shining.............. 2
Head and thorax punctulate or shagreened, subopaque.
  Vertex impressed at the middle, frons with a median furrow.
    Antennæ wholly black.
      All coxæ and femora black (femora sometimes piceous), tibiæ and tarsi yellowish or brownish-yellow.
        Scapulæ with a distinct median impressed line; metapleura with silvery pubescence...I. CRESSONI Ashm.
        Scapulæ without an impressed line; tibiæ fuscous; metapleura sericeous
                                             I. HORNI Ashm.
  Vertex not impressed.
    Scape yellow.
      All coxæ black, club of femora fuscous, rest of the legs brownish-yellow.
        Scapulæ without an impressed line, abdominal horn extending only to the middle of thorax .... I. PACKARDI Ashm.
2. Vertex not impressed in the middle, frons without a median furrow.
  Antennæ brown-black.
    Coxæ and legs dull rufous or piceous, trochanters, knees and tarsi pale.
      Metapleura finely striated, subsericeous ....I. RILEYI Ashm.

Vertex impressed in the middle, frons with a medial furrow.
  Coxæ black, legs dull rufous, base of tibiæ and tarsi yellow.
  Metapleura very rough.............................. I. CALIFORNICA, sp. nov.
  Metapleura smoother, but delicately punctate, with a deep sulcus.
                                                       I. LINTNERI, sp. nov.

Inostemma Cressoni Ashm.

(Pl. XI, Fig. 6, ♀.)

Can. Ent., XIX, p. 127; Cress. Syn. Hym., p. 249.

♀. Length, 2 to 2.2ᵐᵐ. Black, subopaque, minutely punctulate or shagreened; head transverse, the vertex at the middle and posteriorly impressed, the face with a median depression above the insertion of the antennæ, the lateral ocelli touching the border of the eye. Antennæ black or brown-black, the extreme apex of the pedicel yellowish, first two funiclar joints slender, cylindrical, the third and fourth small; club 4-jointed, the second and third a little longer than wide, the last conical. Thorax with delicate but complete parapsides, the middle lobe a little concave at the middle for the reception of the abdominal horn, the scapulæ with a longitudinal grooved line; mesopleura faintly aciculated, deeply foveated, the fovea extending from beneath the tegulæ to between the middle and hind coxæ, metapleura with a whitish or silvery pubescence. Legs black or piceous, the tibiæ and tarsi pale brown or yellowish. Wings hyaline, the tegulæ rufo-piceous. Abdomen polished, shining, longer than the head and thorax together, pointed and curving upward at tip, the horn long and extending forward over the thorax to the vertex of head, second ventral segment with 2 aciculated sulci at base.

HABITAT.—Jacksonville, Fla.
Type in Coll. Ashmead.

Inostemma Horni Ashm.

Can. Ent., XIX, p. 126; Cress. Syn. Hym., p. 249.

♂ ♀. Length, 1 to 1.4ᵐᵐ. Black, subopaque, closely, finely punctulate, the thorax above lustrous; head transverse, the vertex impressed deeply at the middle, the face with a median impressed line. Antennæ black, the pedicel and first two funiclar joints a little pale at tip, subequal, the third and fourth, minute; club 4-jointed, the second and third joints a little wider than long. Thorax with complete parapsidal furrows, the scapulæ most frequently without a trace of an impressed line; mesopleura polished, foveated but not aciculated; metapleura pubescent. Legs black or piceous, the tarsi yellowish, the tibiæ paler basally; wings hyaline, the tegulæ black or piceous. Abdomen polished black, as long as the head and thorax together, the horn extending over the thorax to the vertex of head, finely, longitudinally aciculated and much thicker at tip than at base; first segment and second at base, striated.

In the ♂ the first and second funiclar joints are about equal, oval, the third small, the club 5-jointed, pubescent, the joints, except the last, not longer than wide, slightly pedicellate; lateral ocelli about their width from the eye margin; while the second abdominal segment at base has two oblong, nearly confluent, finely punctate foveæ, the first segment with three grooved lines above.

HABITAT.—Jacksonville, Fla.

Types in Coll. Ashmead.

Described from many specimens.

Inostemma Packardi Ashm.

Can. Ent., XIX, p. 127; Cress. Syn. Hym., p. 249.

♀. Length, 1.4ᵐᵐ. Black, subopaque, closely, finely punctulate or shagreened. Head transverse, the vertex not impressed, the lateral ocelli not touching the margin of the eye. Antennæ 10-jointed, brown-black, the scape yellowish, fuscous at the middle, the pedicel yellowish at tip; first two funiclar joints about equal, shorter than the scape, third and fourth minute, club 4-jointed, stout. Thorax shining, the parapsidal furrows very indistinct anteriorly, the scapulæ without a distinct impressed line. Wings hyaline. Legs brownish-yellow, the coxæ black, the thickened part of femora fuscous. Abdomen not longer than the head and thorax together, the horn not extending much beyond the middle of the thorax, obliquely truncate at tip, basal segment and the second at extreme base finely striated.

HABITAT.—Jacksonville, Fla.

Type in Coll. Ashmead.

Described from a single specimen. The short abdominal horn and the lateral ocelli not touching the margin of the eye readily separate the species.

Inostemma Rileyi Ashm.

Can. Ent., XIX, p. 127; Cress. Syn. Hym., p. 249.

♂ ♀. Length, 0.5 to 0.8ᵐᵐ. Black, polished; the head finely microscopically punctate, but shining, the vertex not impressed at the middle; the frons convex without a median groove, the lateral ocelli not touching the margin of the eye. Antennæ 10-jointed, dark brown, the apex of pedicel yellowish; second funiclar joint subequal with the first, very little longer than thick, third and fourth minute; club 4-jointed, stout, the joints, except the last, broader than long; in ♂ with a 5-jointed club, the joints a little longer than wide, slightly pedicellate, hairy, the last conical; funiclar joints 1 and 2 very little longer than thick, the second obliquely truncate at tip, the third, small, subtriangular. Thorax with the parapsidal furrows indistinct or wanting, the scapulæ without a grooved line. Wings hyaline, the tegulæ piceous black. Legs brownish-piceous, the trochanters, base of tibiæ, and tarsi yellowish. Abdomen not longer than the head and thorax together, the horn not extending over the vertex of the head; first segment and the extreme base of second striated.

HABITAT.—Jacksonville, Fla.

Type in Coll. Ashmead and National Museum.

Described from many specimens.

Inostemma californica, sp. nov.

♂ ♀. Length, 1 to 1.6ᵐᵐ. Black, subopaque, closely finely punctulate, the mesonotum less distinctly punctate, shining. Head transverse, the vertex at the middle posteriorly, impressed. Antennæ 10-jointed, black, the scape one-third shorter than the flagellum; pedicel longer than the first funiclar joint, first funiclar joint almost as long as the second and third together, second, third, and fourth joints all small, the last the smallest, club 4-jointed, the first the narrowest, second and third quadrate, the last conical, longer than the preceding. In ♂ the club is 5-jointed, the flagellum nearly twice as long as the scape, the pedicel as long as the first two funiclar joints, the third being minute. Thorax with 2 faint furrows. Mesopleura with a deep, glabrous impression on the disk, extending to the middle coxæ. Metapleura finely rugose, with raised lines. Tegulæ rufo-piceous. Wings hyaline, not fringed, the submarginal vein with the knobbed tip, black. Legs piceous-black, the trochanters, anterior tibiæ beneath and at tips, base of middle and posterior tibiæ, and all tarsi pale brownish-yellow. Abdomen slightly longer than the head and thorax together, pointed fusiform, smooth and shining; the basal segment finely rugose, in ♀ with a large horn extending forward over the thorax to the vertex of head, finely longitudinally aciculated, and becoming distinctly striated at base; the second segment is long, slightly longer than the five following segments united, with a fovea and some faint striæ at base. In the ♂ the basal segment has no horn and is longitudinally striated.

HABITAT.—Los Angeles, Cal.

Types in National Museum.

Bred by A. Koebele, from a Cecidomyiid gall on *Telypodium integrifolium*.

Inostemma Lintneri, sp. nov.

♀. Length, 1ᵐᵐ. Black, polished, the head subopaque, closely microscopically punctate, the vertex slightly impressed in the middle, the face with a median furrow, the lateral ocelli a little away from the margin of the eye. Antennæ 10-jointed, brown-black, the apex of pedicel yellowish; first funiclar joint slender, more than twice longer than thick; second two-thirds the length of the first; third and fourth minute; club 4-jointed, the joints, except the last, a little wider than long. Thorax smooth, shining, with distinct furrows, the scapulæ with a faint trace of an impressed line; mesopleura polished foveated; metapleura finely delicately punctate with a deep sulcus. Wings hyaline.

Abdomen polished, the first segment and the second at extreme base, striated, the horn extending forward to vertex of head.   Legs brownish-piceous, the coxae black, the trochanters, base of tibiae and tarsi yellowish, the anterior tibiae being mostly yellow except above at the middle.

HABITAT.—Washington, D. C.

Type in Coll. Ashmead.

ACEROTA Förster.

Hym. Stud. II, p. 107 (1856).

(Type not described.)

Head transverse, the vertex subconvex, the occiput margined; ocelli 3, in a curved line, the lateral distant from the margin of the eye, but slightly nearer to it than to the front ocellus; eyes rounded.

Antennae inserted just above the clypeus, 10-jointed, in ♀ terminating in a 4-jointed club, the pedicel longer than the first funiclar joint; in ♂ subfiliform, the pedicel not longer than the first funiclar joint, the club 5- or 6-jointed, the terminal joint more than twice as long as the penultimate and scarcely thicker than the first funiclar joint.

Thorax ovate, the mesonotal furrows distinct, deep, and entire; scutellum convex or subconvex, separated from the mesonotum by a transverse line, and with a frenum or carina posteriorly; metathorax rather short.

Front wings with the submarginal vein terminating in a stigma at about one-third the length of the wing.

Abdomen long-oval, the first segment not or scarcely longer than wide, striated, the second larger, occupying about half of the remaining surface, the following segments short.

Legs clavate, the tibial spurs 1, 1, 1, the basal joint of posterior tarsi, about 2½ times as long as the second, 2 to 4 subequal, the last not quite as long as the second.

Closely allied to *Inostemma*, but the female without a horn at the base of the abdomen, the mesonotal furrows more sharply defined, the second funiclar joint in the male longer and cylindrical.

TABLE OF SPECIES.

Head and thorax closely punctulate.
  Hind coxae alone black...2
  All coxae black.
    Mesonotum with a distinct longitudinal carina at the base of the middle lobe, with irregular raised lines on either side of it.
      Antennae brown-black........................A. CECIDOMYIÆ, sp. nov.
    Mesonotum with no raised lines at base of the middle lobe.
      Legs rufous, the femora and sometimes the tibiae fuscous or black.
        Face with fine longitudinal striae toward base of antennae, the middle of frons nearly smooth, shining; antennae dull rufous.
                                                A. FLORIDANA Ashm.

Face with no fine striæ toward base of antennæ, finely, closely punctate; antennæ brown-black, the scape black (♂).

A. MELANOSTROPHA Ashm.

2. Legs yellow.

Scape, pedicel and first funiclar joint yellow, rest of the antennæ fuscous or brown .... A. CARYÆ Ashm.

Acerota cecidomyiæ, sp. nov.

♂ . Length, 2ᵐᵐ. Black, closely punctulate, subopaque; thorax with distinct parapsidal furrows, the middle lobe posteriorly with a delicate longitudinal carina and some irregular raised lines on either side; scutellum closely punctate, acutely rimmed posteriorly. Antennæ 10-jointed, black, the flagellum twice as long as the scape; pedicel not longer than the first funiclar joint, second funiclar joint slightly longer than the first, third, smaller and not so thick; club 5-jointed, the joints, except the last, almost equal, the first being slightly the smallest, the last conical, slightly more than twice as long as the penultimate. Metapleura covered with a sparse whitish pubescence. Legs brown-black, the coxæ black. Tegulæ black. Wings hyaline, not fringed and scarcely pubescent, the submarginal vein brown, its knob piceous. Abdomen long-oval, very slightly longer than the thorax, polished, first segment as long as wide, striated, with a transverse depression at the middle.

HABITAT.—Lancaster, Los Angeles County, Cal.

Types in National Museum.

Bred by A. Koebele, August, 1887, from a Cecidomyiid gall on *Ephedra californica.*

Acerota floridana Ashm.

Can. Ent., XIX, p. 128; Cress. Syn. Hym., p. 249.

♀ . Length, 2ᵐᵐ. Subrobust, black, subopaque; the head and thorax closely microscopically punctulate; abdomen highly polished, the first segment wider than long, striated. Head transverse, the vertex broad, the ocelli in a slightly curved line, the lateral ocelli about their width from the eye margin. Antennæ dark brown; pedicel longer than the first and second funiclar joints together; second funiclar joint not wider than long, a little shorter than the first; third transverse and wider than the second; club joints stouter, wider than long. Thorax with distinct parapsidal furrows, subopaque, microscopically punctate; scutellum convex with a raised margin posteriorly; angles of metathorax prominent, densely covered with a silvery white pubescence, as well as the metapleura. Wings hyaline. Legs dull rufous, the coxæ black, the femora fuscous.

HABITAT.—Jacksonville, Fla.

Types in Coll. Ashmead.

Acerota melanostropha Ashm.

Monocrita melanostropha Ashm., Can. Ent., XIX, p. 126; Cress. Syn. Hym., p. 249.

♂ ♀. Length, 1.5 to 2ᵐᵐ. Black, subopaque, closely punctulate; abdomen, except the petiole which is striated, highly polished, impunctate; antennae brown-black; legs rufous; the posterior femora obfuscated; sometimes all the coxae and femora, and sometimes the tibiae, fuscous. Wings hyaline, the submarginal vein ending in a fuscous or blackish knob. Antennae 10-jointed; in the ♂ the pedicel slender and much shorter than the first funiclar joint; first and second funiclar joints stout, the second the shorter; third slenderer, about twice as long as thick; all the joints are covered with a fine, whitish pubescence. In the ♀ the pedicel is longer than the first and second funiclar joints, the club joints transverse. Head transverse, closely punctate, obliquely narrowed behind the eyes; the face shortened, its width between the eyes longer than from the vertex to the mandibles. Thorax trilobed, the middle lobe with two short indistinct lines anteriorly. Scutellum closely punctate, bounded by a carina behind. Metathorax carinated at sides.

HABITAT —Jacksonville, Fla.

Types in Coll. Ashmead.

Described from several specimens. The species was wrongly described under the genus *Monocrita.* The much longer funiclar joints, sculpture, and color of the legs separate it from *A. cecidomyiae,* while its subopaque, closely punctulate surface, the non-striated face, and the impressed lines on the middle lobe of the mesonotum separate it from *A. floridana* and *A. caryae.*

Acerota caryæ Ashm.

(Pl. XI, Fig. 7. ♂.)

Can. Ent., XIX, p. 128; Cress. Syn. Hym., p. 249.

♂ ♀. Length, 1.5 to 2.1ᵐᵐ. Much like *A. melanostropha;* the thorax smoother and more shining, very faintly microscopically punctate; the head much as in *melanostropha,* but the face always with a deep median furrow; antennae, except the 4-jointed club, and the legs, except the posterior coxae, wholly brownish-yellow; the second abdominal segment finely striated at base above; the pedicel and first and second funiclar joints are long and slender, cylindrical, while the club joints, except the last, are transverse.

HABITAT.—Jacksonville, Fla.

Types in Coll. Ashmead.

Described from many specimens. The species varies in size, but is constant in the color of the legs and antennae.

Tribe II.—PLATYGASTERINI.

To this tribe belong all species with veinless wings. The genera are more numerous and much more difficult to separate than in the Inostemmini, but it is believed the table below will be found all that is necessary to distinguish them. If, however, the student should be at fault the full generic description may be consulted.

TABLE OF GENERA.

FEMALES.

Scutellum not lengthened, semicircular, either flat or convex and unarmed.... 5
Scutellum lengthened, never semicircular, or when shortened it is compressed
    at sides and furnished with an awl-shaped thorn or tubercle at tip. 2
2. Scutellum with a strong awl-shaped thorn at tip................................. 3
Scutellum with a short thorn or tubercle at tip................................ 4
Scutellum lengthened, triangular, often produced into a long, acute spine.
  Thorax strongly compressed from the sides.
    Head large, rounded, or quadrate....................PIESTOPLEURA Först.
  Thorax not strongly compressed from the sides.
    Mesonotal furrows deep, parallel posteriorly............XESTONOTUS Först.
    Mesonotal furrows feebly impressed or wanting..........AMBLYASPIS Först.
3. Lateral ocelli nearer the margin of eye than to the apical ocellus; antennal club
    4-jointed...LEPTACIS Först.
Lateral ocelli not nearer the margin of eye than to the apical ocellus; antennal
    club 3-jointed...ISORHOMBUS Först.
4. Abdomen very much lengthened; antennal club 5-jointed. Lateral ocelli as near
    to the front ocellus as to the margin of the eye ..POLYMECUS Först.
Abdomen not especially lengthened; antennal club 4-jointed.
  Second ventral segment strongly compressed, sack-like; lateral ocelli their
    width from the eye margin....................SACTOGASTER Först.
  Second ventral segment normal; lateral ocelli close to the eye margin.
                            SYNOPEAS Först.
5. Scutellum convex 6
Scutellum cupuliform as in the Cynipid genus *Eucoila*.
                        CŒLOPELTA Ashm., gen. nov.
Scutellum not cupuliform, flattened.
  Mesonotal furrows wanting; antennae 10-jointed, the club 4-jointed.
                        ANOPEDIAS Först.
  Mesonotal furrows usually distinct; antennae 8-jointed, club not jointed.
                        AMITUS Hald.
6. Scutellum with a tuft of hair at tip...........................TRICHACIS Först.
Scutellum without a tuft of hair at tip.
  Abdomen very much lengthened. (Polymecus.)
  Abdomen not much lengthened.
    Margin of abdomen very broadly deflexed ....HYPOCAMPSIS Först.
    Margin of abdomen normal.
    Thorax short; the scutellum pillow-shaped, separated from the mesonotum
      by a deep furrow; mesonotal furrows rarely distinct or complete.
    No keel between the antennae.......................POLYGNOTUS Först.
    A sharp distinct keel between the antennae.
              ERITRISSOMERUS Ashm., gen. nov.
    Thorax more elongate; scutellum not separated from the mesonotum by a
      deep furrow. Mesonotal furrows distinct, complete.
    Lateral ocelli nearer the margin of the eye than to the apical ocellus;
      head transverse......................... PLATYGASTER Latreille
    Lateral ocelli nearer to the apical ocellus than to the margin of the eye;
      head cubital ....ISOCYBUS Först.

MALES.

Scutellum not lengthened, semicircular, either flat, convex or cupuliform, and
    unarmed.. 5
Scutellum lengthened, never semicircular, or when shortened it is compressed
    at sides and furnished with an awl-shaped thorn or tubercle at tip. 2
2. Scutellum with a strong awl-shaped thorn at tip............................... 3
Scutellum with a short thorn or tubercle at tip................................. 4
Scutellum lengthened, triangular, often produced into a long, acute spine.
  Thorax strongly compressed from the sides.
    Head large, rounded or quadrate...................... PIESTOPLEURA Först.
  Thorax not strongly compressed at sides.
    Mesonotal furrows deep, parallel posteriorly............ XESTONOTUS Först.
    Mesonotal furrows very feebly impressed or wanting..... AMBLYASPIS Först.
3. Lateral ocelli nearer the margin of the eye than to the apical ocellus.
                                       LAPTACIS Först.
Lateral ocelli not nearer the margin of eye than to the apical ocellus.
                                      ISORHOMBUS Först.
4. Abdomen much lengthened..................................... POLYMECUS Först.
Abdomen not especially lengthened.
  Ocelli their width from the margin of the eye.......... SACTOGASTER Först.
  Ocelli close to the eye margin..............................SYNOPEAS Först.
5. Scutellum convex... 6
Scutellum cupuliform as in Cynipid genus Eucoila.. CŒLOPELTA Ashm. gen. nov.
Scutellum quite flat.
  Mesonotal furrows wanting or distinct; antennæ 10-jointed.
    Scutellum separated from mesonotum by a delicate transverse grooved line,
    not foveate at base; antennæ subclavate, not verticillate.
                                ANOPEDIAS Först.
  Mesonotal furrows usually distinct; antennæ verticillate.
                                AMITUS Haldeman.
6. Scutellum with a tuft of hair at tip...................... TRICHACIS Först.
Scutellum without a tuft of hair at tip.
  Margin of abdomen very broadly deflexed ....HYPOCAMPSIS Först.
  Margin of abdomen normal.
    Thorax short; the scutellum pillow-shaped, separated from the mesonotum
    by a deep furrow; mesonotal furrows rarely distinct or complete.
    No keel between the antennæ...................... POLYGNOTUS Först.
    A sharp distinct keel between the antennæ, third joint strongly dilated.
                    ERITRISSOMERUS Ashm. gen. nov.
    Thorax more elongate; scutellum not separated from the mesonotum by a
    deep furrow; mesonotal furrows distinct, complete, rarely incom-
    plete.
    Lateral ocelli nearer the margin of the eye than to the apical ocellus.
                    PLATYGASTER Latr.
    Lateral ocelli nearer to the apical ocellus than to the margin of the eye.
                    ISOCYBUS Först.

PIESTOPLEURA Förster.

Hym. Stud. ii, p. 144 (1856); *Catillus* Först. loc. cit., p. 107.

(Type *P. Catillus* Walk.)

(Pl. xi, Fig. 8, ♀.)

Head large, rounded, twice the breadth of the thorax, the occiput margined; ocelli 3, in a triangle, the lateral close to the margin of the eye.

Antennæ inserted just above the clypeus, 10-jointed in ♀, with a 4-jointed club, the pedicel much larger than the funiclar joints, the club joints transverse; in ♂ the first and third funiclar joints are much thickened, about as thick as the pedicel, the second, as long as the two preceding together, spindle-shaped, not especially thickened, club 4-jointed, the joints long, cylindric, slightly pedicellate, pilose.

Thorax strongly compressed, 2½ times as long as wide and highly convex, higher than wide, mesonotal furrows but slightly impressed, obsolete anteriorly, scutellum with a spine or small thorn.

Front wings pubescent, ciliated, entirely veinless.

Abdomen long-ovate, 6-jointed and more pointed at apex in the ♀, 7-jointed and rounded at apex in ♂, the second segment very long, occupying fully more than half the whole surface.

Legs clavate.

In the shape of the head and the strongly compressed thorax this genus is quite distinct from those that follow.

Piestopleura maculipes Ashm.

Can. Ent., xix, p. 128 ♀ (*Catillus*); Cress. Syn. Hym., p. 249.

♀. Length, 0.8mm. Polished black; antennæ and legs, rufous; club of antennæ 4-jointed, black or brown-black; pedicel long; funiclar joints small. Head much wider than the thorax. Thorax with indications of furrows posteriorly. Scutellum ending in a small spine, subpubescent. Metathorax very short, pubescent, with a prominent median keel. Abdomen about as long as the head and thorax together, oblong, the petiole striated, subpubescent. Wings hyaline.

HABITAT.—Jacksonville, Fla.

Type in Coll. Ashmead.

XESTONOTUS Förster.

Hym. Stud., ii, p. 107 (1856).

(Type *X. refulgens* Först.)

Head transverse, the occiput margined, the face subconvex; ocelli 3, triangularly arranged, the lateral nearer to the margin of the eye than to the front ocellus; eyes oval.

Antennæ inserted just above the clypeus, 10-jointed in ♀, with all the joints lengthened and thickened toward the apex, without a distinct club; in ♂ the first flagellar joint very small, the second very much thickened, the third the length and thickness of the pedicel, the five following joints cylindrical and slightly pedicellate.

Thorax ovate, the mesonotum with 2 distinct furrows, the scutellum triangularly lengthened, acute.

Front wings veinless.

Abdomen long ovate, the second segment very large, the following short, the first short, narrowed; in ♀ segments 3 and 6 united are only two-thirds the length of the second.

Legs clavate.

(?) Xestonotus andriciphilus Ashm.

(Pl. XI, Fig. 9, ♀.)

Can. Ent., XIX, p. 128, ♀ ; Cress. Syn. Hym., p. 249.

♀. Length, 1.8ᵐᵐ. Black; face finely punctate; antennæ and legs brownish-yellow. Mesonotum with 2 sharply defined parallel furrows. Scutellum not greatly prolonged, but subcompressed at sides. Wings hyaline.

HABITAT.—Jacksonville, Fla.

Type in Coll. Ashmead.

Originally described from one specimen reared from the Cynipid oak gall, *Andricus blastophagus* Ashm.

AMBLYASPIS Förster.

Hym. Stud., II, p. 107 (1856).

(Type *A. aliena* Först.).

Head transverse, the vertex subacute; the occiput delicately margined; ocelli 3, in a triangle, the lateral very close to the margin of the eye.

Antennæ inserted just above the clypeus, 10-jointed, the scape very long, subclavate and curved, in ♀ with a 4-jointed club, the first two joints loosely joined, the last two usually closely joined; in ♂ with a 5-jointed club, the joints oval and loosely joined or pedicellate.

Thorax ovate, slightly compressed at sides, the mesonotum convex, entirely without furrows or these only faintly traceable posteriorly; scutellum high, elongate, triangular, and usually produced into a long acute spine, extending high over the metathorax and scarcely separated from the mesonotum; metathorax usually densely pubescent or woolly.

Abdomen subovate or oblong-oval, not or scarcely longer than the thorax, the first segment petioliform, fluted and pubescent, the second segment large, occupying fully half of the remaining surface, without foveolæ at base, the following segments short, equal in length.

Legs long, the femora and tibiæ strongly clavate, the tibial spurs 1,

1, 1, distinct, the tarsi 5-jointed, longer than the tibiæ, the basal joint of hind tarsi three times as long as the second.

As here defined, this genus is divisible into two sections by the shape of the scutellum; in one section the scutellum is triangular, not produced into a long acute spine; in the other the scutellum is produced into a long acute spine projecting far over the metanotum, and it seems to comprise species that by some authors are included in the genus *Leptacis*. But *Leptacis*, as understood by me, is quite different, the scutellum being produced into a short curved spine or tubercle, and, as defined in this work, agrees more nearly with the genus *Ceratacis* Thomson.

TABLE OF SPECIES.

FEMALES.

Scutellum not or faintly pubescent, the apex produced into an acute spine......2
Scutellum covered with a short dense pubescence, apex not produced into an acute spine, triangular.
 Legs brown, trochanters, base of tibiæ, and tarsi honey-yellow: antennæ dark brown, wings with a short fringe.
                 A. OCCIDENTALIS, sp. nov.
 Legs brownish yellow; antennæ dark brown, scape pale beneath at base; wings with a long fringe..........................A. CALIFORNICUS, sp. nov.
 Legs and antennæ pale yellow, the flagellum brown; wings with a short fringe; petiole pale...A. PETIOLATUS, sp. nov.
2. Spine of scutellum long, extending far over the metathorax, yellow; wings with long ciliæ.
 Legs reddish-yellow, tips of posterior femora and tibiæ brown; tibial spurs short; petiole black, pubescent......................A. AMERICANUS Ashm.
 Legs uniformly reddish-yellow or yellow; legs not especially lengthened, the hind tarsi shorter than their tibiæ.
  Head highly polished; middle and posterior tibial spurs being long; the petiole yellow, pubescent......................A. MINUTUS, sp. nov.
  Head rugulose, opaque; middle and posterior tibial spurs very short, scarcely developed; petiole black, pubescent..........A. RUGICEPS, sp. nov.
  Head highly polished; femora and tibiæ brown or fuscous; the legs lengthened, the hind tarsi longer than their tibiæ; petiole black, pubescent.
                 A. LONGIPES Ashm.

MALES.

Scutellum scarcely pubescent, produced into an acute spine 2
Scutellum covered with a short, dense pubescence, the apex not acutely produced.
 Legs pale, brownish-yellow; the flagellum dark brown.
                A. CALIFORNICUS, sp. nov.
2. Antennæ and legs honey-yellow or reddish-yellow.
 Club joints not much longer than wide...................A. MINUTUS, sp. nov.
 Club joints five or six times longer than wide..............A. LONGIPES Ashm.
 Tips of posterior femora and tibiæ fuscous; club joints about five times as long as wide...A. AMERICANUS.

Amblyaspis occidentalis, sp. nov.

♀. Length, 0.8ᵐᵐ. Polished black, impunctured; antennæ brown-black, the scape pale at base; legs brownish, trochanters, base of tibiæ

and tarsi, honey-yellow. Antennæ 10-jointed; pedicel a little shorter than the first two funiclar joints together; first and second funiclar joints equal, third much shorter and more slender, fourth transverse; club 4-jointed, the last two joints closely united, the first two about equal, a little wider than long, rounded off at base. Scutellum triangular, pubescent, very slightly impressed on each side at base, but medially subconvex and not separated from the mesonotum. Metathorax and metapleura pubescent. Wings hyaline. Abdomen as long as the head and thorax together black, polished, the petiole and base of second segment pubescent.

HABITAT.—Riley County, Kans.

Type in Coll. Ashmead.

Described from a single specimen received from Mr. C. L. Marlatt.

Amblyaspis californicus, sp. nov.

♂ ♀. Length, 1.60ᵐᵐ. Black, smooth, shining; scutellum, metapleura, hind coxæ, and petiole, rather densely pubescent. Antennæ 10-jointed, dark brown, the scape brownish-yellow, more or less dusky above toward apex; pedicel longer than the first funiclar joint, the latter twice as long as thick, the second slightly shorter, the third still shorter, the fourth minute; club 4-jointed, stouter than the funicle, the first and second joints quadrate, slightly pedicellate, the two following closely joined, forming a cone. Thorax narrowed towards the head, the collar forming a slight neck, smooth, convex, without furrows. Legs, except the black coxæ, brownish-yellow, or honey-yellow, pubescent, the trochanters long. Wings long, hyaline, fringed.

Abdomen as long as the thorax, smooth, polished, the petiole pubescent. In the ♂ the pedicel is as long as the second funiclar joint, the first funiclar joint only half as long as the second; the club 6-jointed cylindrical; the joints twice as long as thick, all slightly pedicellate.

HABITAT.—Marin County, Cal.

Types in National Museum.

Described from 1 ♂, 1 ♀ specimen received from A. Koebele.

Amblyaspis petiolatus, sp. nov.

♂ ♀. Length, 0.8ᵐᵐ. Polished black, impunctured; antennæ, except club, legs, and petiole bright yellow; flagellum in ♂ brown.

In the ♀ the pedicel is almost as long as the first and second funiclar joints together; second funiclar joint about two-thirds the length of the first, the first being twice as long as thick; third and fourth small; club 4-jointed, the first two quadrate, the last two closely joined, conical; in ♂ the flagellum is pale brown, the club joints not wider than long, slightly pedicellate. Lateral ocelli twice their width from the eye margin. Thorax convex (in ♂ piceous at sides) without furrows. Scutellum triangular, not acutely spined at tip, more or less pubescent, with a fovea on each side at base. Metathorax pubescent. Wings

hyaline with a short fringe. Abdomen ovate or elliptic-oval, the petiole longer than thick, pale, pubescent.

HABITAT.—Jacksonville, Fla.

Type in Coll. Ashmead.

Amblyaspis americanus Ashm.

Can. Ent., XIX, p. 129; Cress. Syn. Hym., p. 249.

♀. Length, 1ᵐᵐ. Polished black, impunctured; antennæ and legs honey-yellow or brownish-yellow; the tips of posterior femora and tibiæ dusky; club brown, the joints, except the last, not longer than wide. Lateral ocelli only their width from the eye margin. Scutellum produced into a long, acute, yellowish spine. Metathorax and petiole covered with a white pubescence. Wings hyaline, longly fringed. Abdomen oval, polished black, the petiole a little longer than wide, densely pubescent.

HABITAT.—Jacksonville, Fla.

Type in Coll. Ashmead.

Amblyaspis minutus, sp. nov.

♀. Length, 0.6 to 0.8ᵐᵐ. Polished black, impunctured; lateral ocelli close to the border of the eye; antennæ (except club) and legs, yellow or reddish-yellow, with the tips of posterior femora and tibiæ sometimes dusky. Scutellum acutely spined, the spine yellow. Funicle very slender, the first, third, and fourth joints minute, the second very long and slender; club stout, the joints, except the last, not, or scarcely, longer than wide. Wings hyaline, strongly fringed. Abdomen oval, the petiole yellowish, pubescent.

HABITAT.—Washington, D. C.; St. Louis, Mo.

Types in Coll. Ashmead and National Museum.

The specimens in the National Museum were reared by Dr. Riley, at St. Louis, Mo., July 10, 1870, from *Cecidomyia* sp. in squash.

Amblyaspis rugiceps, sp. nov.

♂ ♀. Length, 0.8ᵐᵐ. Polished black, impunctured; frons and face transversely rugulose; antennæ and legs yellow or reddish-yellow, the thickened parts of middle and posterior femora and tibiæ sometimes dusky or brown; club usually obfuscated or brown. Lateral ocelli about twice their width from the eye margin. Club joints a little longer than wide. Scutellum acutely spined, yellowish at tip, foveated on each side at base, subpubescent. Metathorax covered with a white pubescence. Wings hyaline, strongly ciliated. Abdomen oval, the petiole not longer than thick, striated and pubescent.

The face in the ♂ is very rugose, the club joints about 2½ times as long as thick, the second funicular joint a little swollen, the first very minute and closely joined to the second, while the legs are paler and more uniformly yellow.

HABITAT.—District of Columbia.

Type in Coll. Ashmead.

Amblyaspis longipes Ashm.

(Pl. xi. Fig. 10, ♂.)

Can. Ent., xix, p. 128; Cress. Syn. Hym., p. 249.

♂. Length, 2ᵐᵐ. Polished black, impunctured; scape and legs pale brownish-yellow or yellow; flagellum brownish-black, pilose; funicle very long and slender, as long as the long scape, the first and last joints short, the second greatly elongated; club 5-jointed, the joints all long, 5 or 6 times as long as thick, subclavate. Head transverse, the vertex bounded behind by a delicate transverse carina; lateral ocelli close to the eye margin; face flat, smooth; mandibles pale. Thorax long, convex, without furrows; scutellum very long, produced into a long, acute, yellow spine, its tip extending over the base of the abdomen; two large pubescent foveæ on either side at base; metathorax pubescent. Tegulæ black. Wings hyaline. Legs very long, the hind pair especially long, honey-yellow, tibial spurs distinct, the tarsi very long, slender. Abdomen oval, the petiole about twice as long as thick, striated, pubescent above and beneath, body of al. domen pubescent at base beneath.

HABITAT.—Jacksonville, Fla.

Type in Coll. Ashmead.

LEPTACIS Förster.

Hym. Stud. ii. p. 107.   (1856.)

(Type *L. tipulæ* Kirby.)

Cerataeis Thoms. Öfvers, 1858, p. 69.

Head transverse, the frons subconvex, the occiput straight, margined; ocelli 3, subtriangularly arranged, the lateral nearer to the eye than to the front ocellus; eyes oval.

Antennæ inserted just above the clypeus, 10-jointed in both sexes; in ♀ terminating in a 4-jointed club, the pedicel much longer than thick, funiclar joints 1, 3, and 4 very small, short, the second lengthened; in ♂ ending in a 5-jointed club, the joints of which are usually elongate; joint 1 of funicle small, the second elongate, somewhat swollen, the third smaller.

Thorax ovoid, highly convex, the prothorax visible as an arcuate line, the mesonotum with or without furrows, the scutellum subtriangular, convex at the middle, depressed and with two large transverse or oblique foveæ at base, the apex armed with a more or less curved thorn, rarely reduced to a tubercle, the metathorax short, the metapleura usually covered with a dense silvery, or hoary, pubescence.

Front wings long, pubescent, and veinless.

Abdomen in ♀ pointed-ovate, in ♂ oval or long ovate, the first segment wider than long, the second very long, occupying most of the surface, the following all short.

Legs clavate, the basal joint of hind tarsi three or more times longer than the second.

This genus and *Amblyaspis* are often confused together. If in my definition of *Amblyaspis* I have included both *Leptacis* and *Amblyaspis*, then this genus must be known as *Ceratacis* Thomson.

TABLE OF SPECIES.

FEMALES.

Mesonotal furrows wanting or only indicated posteriorly......................... 2
Mesonotal furrows complete, distinct.
    Head rugose, the thorax minutely punctate, a distinct line on shoulders.
        Antennæ and front legs brownish-yellow, club brown black.
                L. RUGICEPS, sp. nov.
    Legs brownish-piceous, the trochanters, base of tibiæ and tarsi honey-yellow.
    Head closely, finely punctate, the thorax minutely punctate, no line on shoulders.
        Antennæ brownish-yellow, club brown.............L. PUNCTATUS, sp. nov.
        Antennæ wholly brown-black......................L. FLORIDANUS, sp. nov.
2. Frons smooth or microscopically punctate..................................... 3
Frons transversely striated; thorax smooth.
    Antennæ and legs brownish-yellow or pale rufous, club brown.
        L. STRIATIFRONS, sp. nov.
3. Frons microscopically punctate... 4
Frons smooth; coxæ black.
    Abdomen twice the length of thorax, subcompressed.
        L. LONGIVENTRIS, sp. nov.
    Abdomen not longer than the thorax, not compressed.
        L. BREVIVENTRIS, sp. nov.
4. Coxæ black.
    Legs black or brownish-piceous, trochanters, knees, tips of tibiæ, and tarsi
        yellow.......................................L. CYNIPIPHILUS Ashm.
Coxæ brownish-yellow; legs and antennæ, except club, yellow.
        L. FLAVICORNIS, sp. nov.

MALES.

Mesonotal furrows wanting or only indicated posteriorly..................... 2
Mesonotal furrows distinct, complete.
    Head rugose, thorax minutely punctate...................L. RUGICEPS, sp. nov.
    Head finely punctate, thorax minutely punctate.
        Antennæ and the middle legs brownish-yellow, the hind legs brownish-
            piceous........................ ....L. PUNCTATUS, sp. nov.
    Head and thorax smooth, shining, the parapsidal furrows delicate.
        Antennæ black; legs, except trochanters and tarsi, fuscous.
        L. FLORIDANUS, sp. nov.
2. Frons smooth; coxæ black.
    Legs brownish-piceous.
        Scape and funicle brownish-yellow, the second funiclar joint swollen.
        L. BREVIVENTRIS, sp. nov.
    Legs brownish-yellow or yellow.
        Flagellum brown-black, the club joints $2\frac{1}{2}$ or 3 times as long as thick, much
            narrowed basally.........................L. CYNIPIPHILUS Ashm.
        Flagellum yellow, the club joints twice as long as thick, cylindric.
        L. FLAVICORNIS, sp. nov.
Frons distinctly punctate.
    Scape, pedicel, and legs yellow.......................L. PUNCTICEPS, sp. nov.
Frons microscopically punctate; coxæ black.
    Antennæ and legs, except anterior tibiæ and tarsi, which are honey-yellow,
        brownish-piceous or fuscous...............L. PUBESCENS, sp. nov.

Leptacis rugiceps, sp. nov.
(Pl. xii, Fig. 1, ♀.)

♂ ♀. Length, 1 to 1.2ᵐᵐ. Black; antennæ, except club, front legs, and base of middle tibiæ and all tarsi, yellow; rest of the legs fuscous or black. Head transverse, opaque, the frons and face rugose, the lateral ocelli close to the eye. Antennæ 10-jointed, the club 4-jointed, black, the joints, except the last, which is conical and about twice as long as the preceding, are scarcely longer than thick. Thorax subopaque, minutely punctate, with 2 distinct furrows, and a distinct grooved line on the shoulders. Scutellum pubescent, with 2 large foveæ at base, a slight median carina, and terminating in a long awl-shaped spine or thorn. Metathorax very short, with a median carina and covered with a silvery pubescence. Wings hyaline. Abdomen oval, shorter than thorax, polished, the petiole wider than long, rugose, densely pubescent.

In the ♂ the scape and funicle are brownish yellow, the 4-jointed club brown-black; the pedicel is very long and slender, almost as long as all the funiclar joints united, last two funiclar joints short, not longer than thick; club joints, except the last, quadrate.

HABITAT.—Jacksonville, Fla., and Arlington, Va.

Types in Coll. Ashmead.

Described from several specimens.

Leptacis punctatus, sp. nov.

♂ ♀. Length, 0.8 to 1ᵐᵐ. Black, subopaque, closely, minutely punctate, the thorax smoother; parapsidal furrows delicate but complete; no grooved line on the shoulders; antennæ, except club, and legs, yellow; middle and posterior legs more or less fuscous, or brown; club 4-jointed, brown-black, the joints, except the last, wider than long, the last conic, 1½ times as long as the penultimate; funicle slender. Scutellum with two large foveæ at base, pubescent, ending in a slightly curved, awl-shaped spine. Metathorax and petiole densely pubescent. Abdomen oval, polished, shorter than the thorax.

HABITAT.—Florida, District of Columbia, and Virginia.

Types in Coll. Ashmead.

Common. Allied to L. rugiceps, but smaller, more evenly punctate, and the head not so rugose.

Leptacis floridanus, sp. nov.

♂ ♀. Length, 1.1ᵐᵐ. Black, subopaque, faintly, microscopically punctulate, the thorax almost smooth, with two distinct furrows; antennæ wholly brown-black, the club joints twice as long as thick, loosely joined. Scutellum foveated at base, the awl-shaped spine rather short. Metathorax with a silvery pubescence. Wings hyaline. Legs brown or fuscous, the trochanters, base of tibiæ, and tarsi yellowish. Abdomen ovate, not quite as long as the thorax, the petiole very short, transverse, densely covered with a silvery white pubescence.

HABITAT.—Jacksonville, Fla.

Types in Coll. Ashmead.

Leptacis striatifrons. sp. nov.

♀. Length, 1.2ᵐᵐ. Black, shining; frons transversely striated; thorax smooth, with indications of furrows posteriorly; antennæ, except the club, and legs wholly brownish-yellow; club 4-jointed, brown, the joints (except the conical last joint) not, or scarcely, longer than thick. Scutellum bifoveated at base and ending in a long awl-shaped thorn Metathorax covered with a dense white pubescence. Wings hyaline. Abdomen subovate, shorter than the thorax, black, polished, impunctured, the petiole very short, transverse, pubescent.

HABITAT.—Jacksonville, Fla.

Type in Coll. Ashmead.

Leptacis longiventris, sp. nov.

♂ ♀. Length, 1ᵐᵐ. Polished black, impunctured; tarsi rufo-piceous. Head as wide as the thorax from tegulæ to tegulæ. Mandibles black. Antennæ 10-jointed, in ♀ with a 4-jointed club, the joints, except the last, a little wider than long, the last conical, twice as long as the preceding; funicle slender, the second joint long and slender, the first and last two, short. Thorax without distinct furrows, a slight elevation or prominence just in front of the scutellum, and with a sparce pubescence on either side. Mesopleura smooth, polished. Scutellum with a tubercle at tip, two oblique depressions at base, a median carina, and sparsely covered with a silvery pile. Metathorax at base, metapleura, and base of abdomen densely covered with a silvery pubescence. Tegulæ black. Wings hyaline, iridescent, but slightly pubescent and without a distinct fringe. Abdomen pointed-ovate, twice the length of the thorax, subcompressed, and in shape not unlike the ♀ in *Eurytoma*, but with the lateral carina quite distinct; the first segment is densely pubescent, the second occupies most of the surface, the third, fourth, fifth, and sixth very short, equal, the seventh conical, as long or a little longer than the four preceding together.

In the ♂ the antennæ are bristly, the pedicel oval, the first funiclar joint minute, rounded, the second, longer than the pedicel, swollen and a little curved, the club 6-jointed, the first joint oval, about half the length of the second; joints 2-5 long oval, the last cone-shaped, one-third longer than the preceding and thinner. Tip of anterior tibiæ and tarsi dark honey-yellow. Abdomen oblong-oval, depressed.

HABITAT.—District of Columbia and Virginia.

Types in Coll. Ashmead and National Museum.

Many specimens. The peculiar shape of the abdomen and the carinated scutellum readily distinguish the species.

Leptacis breviventris, sp. nov.

♂ ♀. Length, 0.65 to 0.80ᵐᵐ. Black, shining, impunctured; the face, very slightly shagreened just above the insertion of the antennæ; lateral ocelli close to the border of the eye; antennæ and legs brown, the trochanters, base of tibiæ, and tarsi yellowish; anterior and middle legs sometimes yellowish. Antennæ 10-jointed, in ♀ with a 4-jointed club, the joints, except the last, not longer than wide, usually wider than long, the last ovate; funicle slender, joints 1, 3. and 4 small, joint 2 elongate; in ♂ with a 4-jointed club, the joints slightly pedicellate and covered with sparse white hairs: second funiclar joint swollen, longer than the pedicel; first funiclar joint very small, closely joined to the second; joints 3 and 4 small. Thorax convex. polished, without furrows; scutellum foveated at base, pubescent at sides, and terminating in a tubercle or very short spine, which is only twice as long as thick; metapleura covered with a dense white pubescence. Abdomen broadly oval, highly polished, black, a little shorter than the thorax, the petiole very short and transverse, pubescent.

HABITAT.—District of Columbia, Virginia, and Maryland.

Types in Coll. Ashmead.

Described from several specimens.

Leptacis cynipiphilus

Ashm., Can. Ent., XIX. p. 129, ♀ ♂ ; Cress., Syn. Hym., p. 249.

♂ ♀. Length, 1.5 ᵐᵐ. Polished black; antennæ, except club and the legs, brownish-yellow; coxæ black; in ♀ with the legs fuscous, the trochanters, base of tibiæ. and tarsi yellow; club brown-black; mandibles pale.

Head a little wider than the thorax. Antennæ 10-jointed; the funicle slender, the second joint longer than the first, the third and fourth, small, but a little thicker than the second; club 4-jointed, the joints, except the last, as wide as long; in ♂ with the flagellum black, pilose, the club joints from two and one-half to three times as long as thick; the first funiclar joint as long as thick; the second more than twice as long, and stouter.

Thorax smooth, without furrows. Scutellum with 2 broad foveæ at base, pubescent and delicately margined at sides, ending in a small tubercle or a short awl-shaped spine. Metapleura and petiole densely pubescent. Tegulæ black. Wings hyaline with a short, sparse pubescence. Abdomen oval, a little shorter than the thorax, the petiole very short, transverse, covered with a dense white, glittering pubescence, the pubescence extending on to the base of the second abdominal segment.

HABITAT.—Jacksonville, Fla.

Types in Coll. Ashmead.

Described from several specimens.

Leptacis puncticeps, sp. nov.

♂. Length, 0.8ᵐᵐ. Black, polished; head finely punctate; scape, pedicel, and legs yellow, the posterior femora and tibiæ fuscous or brown. Head transverse, the frons convex. Antennæ 10-jointed, the scape clavate; flagellum brown-black, pilose, the first joint minute, the second about as long as the pedicel but stouter, the third small, the following joints oval, loosely joined. Thorax without furrows, polished; the scutellum terminating in an awl-shaped spine; the metapleura covered with a silvery pubescence. Wings hyaline, the tegulæ piceous. Abdomen oval, not quite as long as the thorax; the petiole short, covered with a glittering white pubescence.

HABITAT.—Jacksonville, Fla.

Types in Coll. Ashmead.

Leptacis flavicornis, sp. nov.

♂ ♀. Length, 0.8 to 0.9ᵐᵐ. Polished black; antennæ, except the club in the female, and legs, bright yellow. Thorax smooth, shining, impunctured, without furrows; scutellum foveated at base and ending in a minute tubercle; metapleura and the very short petiole with a silvery white pubescence. The antennæ in the ♀ end in a 4-jointed club, the joints of which, except the last, are not longer than wide; the funicle is slender, the second joint a little longer than the first, the third and fourth being small; in ♂ the antennæ are entirely yellow; the club 5-jointed, the joints, except the last, being only a little longer than wide; the last being conical, twice as long as the preceding; the first funiclar joint is small, closely joined to the second, the latter longer than the pedicel and slightly thickened, the third small, contracted. Wings hyaline. Abdomen oval, shorter than the thorax, pubescent at base.

HABITAT.—Washington, D. C., and Jacksonville, Fla.

Types in Coll. Ashmead.

Leptacis pubescens, sp. nov.

♂. Length, 1.5ᵐᵐ. Black, shining, covered with a fine pubescence; scape and legs, except the anterior tibiæ and tarsi, which are honey-yellow, dark brown or piceous; flagellum black, the club joints pedicellate, pilose, the first funiclar joint a little shorter than the pedicel, the second long, fully twice as long as the pedicel, the third small, the first 4 joints of club twice as long as thick, the last three times as long as thick. Thorax without furrows; the scutellum ends in a short tubercle and is foveate at base, the metathorax densely pubescent. Wings hyaline, the tegulæ black. Abdomen oblong-oval, the petiole wider than long, elevated above.

HABITAT.—District of Columbia.

Type in Coll. Ashmead.

Described from a single specimen taken by Mr. E. A. Schwarz.

ISORHOMBUS Förster.

Hym. Stud., II, p. 107 (1856).

Head transverse, the vertex flattened, the occiput with a sharp margin; ocelli 3, triangularly arranged, the lateral far away from the margin of the eye or nearer to the front ocellus than to the eye margin.

Antennæ inserted just above the clypeus, 10-jointed in both sexes, in ♀ terminating in a 3-jointed club. in ♂ with a 5 jointed club, the joints of which are longer than thick. the first funiclar joint very minute.

Thorax ovate, the prothorax distinct, the mesonotum much longer than wide, with furrows, the scutellum subconical or subpyramidal, ending in a little spine at tip, and more or less pubescent, the metathorax short, with a median carina.

Wings when folded extending to, or a little beyond, the tip of the abdomen, pubescent and veinless.

Abdomen long, in ♀ pointed-ovate, in ♂ oblong-oval, the first segment longer than wide, the second very large, occupying fully two-thirds of the whole surface.

Legs clavate, the basal joint of hind tarsi twice as long as the second.

Distinguished by the 3-jointed club in the female, the shape of the scutellum, the long mesonotum, and the position of the ocelli.

Isorhombus hyalinipennis Ashm.

(Pl. XII, 2, ♂.)

Can. Ent., XIX, p. 129; Cress. Syn. Hym., p. 249.

♀. Length, 1.5ᵐᵐ. Black, shining, impunctured, the occiput alone finely transversely aciculated; antennæ, except the 3-jointed club and legs, brownish-yellow or yellow; the coxæ and club black. Head transverse, the lateral ocelli about their width from the border of the eye, the mandibles and palpi pale or yellowish.

Antennæ 10-jointed; scape subclavate, as long as the pedicel and funicle united; pedicel not quite as long as the first and second funiclar joints united; second funiclar joint very little longer than the first; third and fourth small, transverse; fifth much wider, transverse; club 3-jointed, black, the joints loosely joined, the first two subquadrate, very slightly serrate toward one side at apex, last joint oblong. Thorax convex, without furrows or only slightly indicated posteriorly, the little lobe thus formed projecting slightly upon the scutellum. Scutellum subconvex, foveated at base and with only a slight tubercle at tip, the tubercle being subobsolete. Metathorax and petiole subpubescent. Wings hyaline. Abdomen ovate, depressed, polished, impunctured, the petiole scarcely as long as wide.

♂. Length, 1ᵐᵐ. Differs only in the antennæ, the club being 5-jointed, pale, the first and third funiclar joints small, the second being stouter and about thrice as long as thick.

HABITAT.—Jacksonville, Fla.

Types in Coll. Ashmead.

Isorhombus arizonensis. sp. nov.

♀ . Length, 2ᵐᵐ. Black, highly polished; parapsidal furrows obsolete anteriorly. Head very wide, not very thick antero-posteriorly, the lateral ocelli far away from the margin of the eye. Mandibles piceous. Antennae 10-jointed, rufo-piceous, the flagellum twice the length of the scape; pedicel a little longer than thick; joint 1 of funicle very minute, closely joined to the second, second about twice the length of third; club 5-jointed, cylindric, the joints, except the last, nearly equal, longer than thick, the last cone-shaped, a little longer than the penultimate. Scutellum conically elevated posteriorly, depressed at base, sparsely pubescent and with a small acute spine at tip. Metathorax, hind coxae and petiole, pubescent. Legs brownish-yellow, the coxae black. Tegulae black. Wings hyaline, pubescent, extending, when folded, scarcely beyond the tip of the abdomen. Abdomen long-oval, polished, about as long as the head and thorax together, the basal joint very little longer than wide, densely pubescent, the margin very wide.

HABITAT.—Fort Huachuca, Ariz.

Type in National Museum.

Described from a single specimen, reared May 8, 1883, from a Cecidomyiid gall on an unknown plant, sent to the Department by Mr. H. K. Morrison.

POLYMECUS Förster.

Hym. Stud., II. p. 144 (1856).

Ectadius Förster, loc. cit., p. 108.

Epimeces Westwood (*pars*).

(Type *P. craterus* Walk.)

Head transverse, the occiput delicately margined; ocelli 3, in a curved line, the lateral ocelli only their width from the eye margin.

Antennae 10-jointed in both sexes, the scape long, subclavate; in the ♀ the club is 4 or 5 jointed, when 4-jointed the last funicle joint is shorter than the preceding, when 5-jointed the last two funicular joints about equal or the last the longer; in ♂ with a 6-jointed, bearded club, the second funicular joint being somewhat swollen and curved or twisted, the first being small.

Thorax ovate, the prothorax distinct from above, the mesonotum much longer than wide (nearly twice as long as wide), usually with 2 distinct furrows; scutellum convex, often, but not always, with a tubercle or spine at tip, the basal part of the middle lobe of mesonotum projecting slightly upon it at base; metathorax short, with 2 carinae down the center, the metapleura densely pubescent.

Front wings pubescent, veinless, rarely with the submarginal nervure visible, when present not knobbed at tip; in ♀ the wings, when folded, do not extend to the apex of the long abdomen.

Abdomen in the ♀ greatly lengthened, usually twice or more than twice the length of the head and thorax together, sinuated or slightly contracted beyond the middle, the apical portion being narrowed and pointed; the second segment, as usual, is the longest, but the following are also lengthened; in the ♂ oblong-oval as long as the head and thorax together, the second segment very long, the following all short, the terminal segment curving downward and with a distinct margin.

Legs clavate, the basal joint of hind tarsi not quite twice the length of the second.

The very long, pointed abdomen distinguishes the genus.

TABLE OF SPECIES.

FEMALES.

Head finely punctate, the thorax smooth, impunctured 3
Head and thorax finely microscopically punctate or shagreened.
  All coxæ black...2
  Hind coxæ alone black; legs rufous.
    Second funiclar joint as long as the pedicel.............P. CANADENSIS Ashm.
    Second funiclar joint shorter than the pedicel..........P. AMERICANUS Ashm.
2. Femora black or brown black, tibiæ and tarsi rufous.
    Fifth abdominal segment much longer than the last.......P. NIGRIFEMUR Ashm.
    Fifth abdominal segment shorter than the last......P. VANCOUVERENSIS, sp. n.
  Legs piceous, anterior tibiæ, base of middle and posterior tibiæ and all tarsi honey-yellow; antennæ black.
    First funiclar joint contracted, not longer than thick, collar not striate at sides ....P. LUPINICOLA, sp. nov.
    First funiclar joint longer than thick; collar striate at sides.P. PICIPES, sp. nov.
  Legs brownish-yellow; second funiclar joint about half the length of the pedicel.................. ....P. PALLIPES Ashm.
  Legs honey-yellow; scape and the pedicel at tip, yellow.P. MELLISCAPUS, sp. nov.
3. Abdomen 2½ times as long as the head and thorax together; very strongly depressed, when viewed from the side cultriform. Coxæ piceous; legs bright yellow.............................P. COMPRESSIVENTRIS, sp. nov.
  Abdomen 1½ times as long as the head and thorax together, not depressed.
    Coxæ and legs, yellow...P. AURIPES, sp. nov.
    Coxæ black; tip of anterior femora and tibiæ, base of middle and posterior tibiæ and all tarsi, honey-yellow..................P. ALNICOLA, sp. nov.

Polymecus canadensis Ashm.

Ectadius canadensis Ashm., Can. Ent., xx, p. 51.

♀. Length, 3^{mm}. Polished black; the head posteriorly very finely shagreened and delicately transversely striated; lateral ocelli about twice their width from the eye; face polished, impunctured. The front ocellus with a transverse furrow at base before; mandibles black. Antennæ 10-jointed, brown-black, the scape tinged with rufous; pedicel as long as the second funiclar joint but not so thick; first funiclar joint small; club 6-jointed, the first two joints a little more slender than the following, the first, a little longer than the second; joints 3, 4, and 5 about equal in length, 1½ times longer than thick, the upper, outer angle of each joint very acute, the last joint conical, longer than the preceding. Mesonotum with two distinct furrows. Metathorax, including the pleura and the first abdominal segment, both above and beneath, densely pubescent. Tegulæ rufo-piceous. Wings clear hyaline. Legs rufous, the posterior coxæ basally piceous. Abdomen greatly elongated, more than twice the length of the head and thorax united, narrowed into a long tail from the apex of the second segment. the three terminal segments subequal in length.

HABITAT.—Ottawa, Canada.

Type in Coll. Ashmead.

Described from a single specimen received from Mr. W. Hague Harrington.

Polymecus americanus Ashm.

Epimecus americanus Ashm., Can. Ent., XIX, p. 129.
Ectadius americanus Cr. Syn. Hym., p. 249.

♀. Length, 1 to 1.8ᵐᵐ. Polished black; the head behind finely, opaquely sculptured; face highly polished; mandibles rufous. Antennæ 10-jointed, rufous or piceous, sometimes brown-black; pedicel a little longer than the second funicular joint, but more slender; first funicular joint minute; club subclavate, the first and second joints unequal in length, the first slightly the longer and slenderer; joints 3, 4, and 5 not longer than wide at tip, the outer posterior angle acute, last joint conical, a little longer than the preceding. Mesonotum with two delicate but distinct furrows; tegulæ black. Wings hyaline, very slightly tinged. Legs dark rufous to piceous, the trochanters, knees, anterior tibiæ, and the tarsi. paler. Abdomen about one and a half times as long as the head and thorax united, or twice the length of the thorax, the third segment a little longer than the fourth. the others subequal in length, the last being about one-third the length of the fourth, pointed and obliquely truncate from above: the penultimate is twice as long as wide.

HABITAT.—Jacksonville and Fort George Island, Fla.

Types in Coll. Ashmead.

Polymecus pallipes Ashm.

(Pl. XII. Fig. 3. ♀.)

Ectadius pallipes Ashm., Bull. No. 1, Col. Biol. Assoc., p. 9, 1890.

♀. Length, 2.6ᵐᵐ. Black, subopaque, finely granulately sculptured; face smooth, polished, microscopically transversely aciculated just above the insertion of the antennæ. Mandibles rufous. Antennæ 10-jointed, the flagellum one and a half times as long as the scape; the scape, pedicel. and the three or four following joints pale brownish-yellow, the joints beyond black or brown-black; the pedicel is thrice as long as thick, the first joint of flagellum small, annular. the three following joints cylindrical, less than twice as long as thick, the four terminal joints much thicker and larger, the last being twice as long as thick, the others a little shorter.

The parapsidal grooves are distinct, converging posteriorly; the middle lobe thus formed projects slightly on to the scutellum. Scutellum sub-convex. Pleura smooth, polished, the mesopleura with a grooved fur-row extending obliquely to the base of middle coxae; metapleura densely pubescent; the petiole and the hind coxae at base are also slightly pubescent. Legs uniformly pale brownish-yellow, with all coxae black. Abdomen about twice as long as the head and thorax together, and projecting considerably beyond the tips of the wings when folded. Wings hyaline, pubescent.

The ♂ measures but 2^{mm}. in length; the face is sculptured as the rest of the body, and the antennae are wholly pale brown; otherwise similar to the ♀, except the following structural differences: The an-tennae are shorter, the club 6-jointed instead of 4-jointed, the pedicel only a little longer than wide; the first joint of flagellum is small but triangular, the second enlarged, swollen, nearly as broad as long; the following 6 joints, which constitute the club, are narrower and cylin-drical, the first the shortest, the terminal one the longest, being about twice as long as the preceding. The abdomen is not much longer than the head and thorax together, and the wings, when folded, project be-yond its tip. It might easily be mistaken for a genuine *Platygaster*.

HABITAT.—Greeley, Colo.

Types 5 ♀, 3 ♂ specimens in Coll. Ashmead.

Obtained through H. F. Wickham.

Polymecus nigrifemur Ashm.

Ectadius nigrifemur Ashm., Bull. No. 1, Col. Biol. Assoc., p. 10.

♀. Length, 3 to 3.2^{mm}. Black; sculptured as in previous species, except that the face is not so smooth. Antennae wholly black, the pedicel about twice as long as thick; otherwise the joints are similar to *pallipes*. All coxae and femora black, the tibiae and tarsi, reddish; sometimes the anterior femora are also red, but usually they are piceous or obscured above; the tibiae, too, are sometimes more or less dusky. Abdomen more than twice longer than the head and thorax combined, the petiole being more pubescent than usual, the sides being almost as densely pubescent as the metapleura. Wings hyaline.

The ♂ is but 2^{mm} in length, and agrees in color and sculpture with the ♀, except that the tibiae and tarsi are darker than in that sex; structurally it is like the ♂ of *pallipes*.

HABITAT.—Greely, Colo.

Types 4 ♀, 3 ♂. specimens in Coll. Ashmead.

Obtained from Mr. H. F. Wickham.

Polymecus vancouverensis, sp. nov.

♀. Length, 2.8ᵐᵐ. Black, subopaque, closely, microscopically shagreened. Antennæ black, the antennal tubercles yellow; the first and
second club joints quadrate, the third, fourth, and fifth not longer than
wide at tip, the upper outer angle acute, the last joint conical, twice
the length of the preceding. Legs piceous-black; the tibiæ fuscous,
tarsi pale brown, the trochanters and base of tibiæ yellow. Tegulæ
black. Wings clear hyaline. Abdomen about twice the length of the
head and thorax combined, the third segment half the length of the
fourth, the fifth and sixth of an equal length, one-third longer than the
fourth, the sixth conical.

HABITAT.—Vancouver Island.

Type in Coll. Ashmead.

Described from a single specimen received from Mr. W. Hague Harrington.

Polymecus lupinicola, sp. nov.

♂ ♀. Length, 1 to 1.6ᵐᵐ. Black, shining, impunctured. Head transverse, as broad as the widest part of the thorax, the vertex posteriorly
faintly aciculated. Mandibles black. Antennæ 10-jointed, black, the
flagellum rather slender, not quite twice as long as the scape; pedicel
more than twice as long as thick at tip; funicle joints slender, cylindric, the first nearly twice as long as thick, the second longer, the
third slightly shorter than the second; club 5-jointed, slightly and
gradually thickened toward tip, the first joint twice as long as thick,
the second, third, and fourth nearly equal in length, but not twice as
long as thick, the last cone-shaped, very little longer than the preceding. Mesonotal furrows distinct posteriorly, becoming obsolete anteriorly, the middle lobe projecting slightly on the scutellum, the lateral
lobes with a sparse tuft of pubescence at base. Scutellum convex, smooth, shining, with a small tubercle at tip. Metathorax
pubescent. Legs black, the tarsi fuscous or brown. Wings hyaline,
pubescent, but not fringed. Abdomen longer than the head and
thorax together, pointed at apex, contracted from the apex of the
second segment; the fourth segment is one-third longer than the third;
the fifth and sixth nearly twice the length of the fourth; segments
3 and 4 with a transverse row of punctures; fifth aciculated except
at base.

In the ♂ the legs are black, except the anterior tibiæ and all tarsi
which are pale brown; the abdomen is oblong-oval, not longer than
the thorax, the first segment and the second with the foveolæ at base
striate or aciculate, otherwise mostly polished; tegulæ rufo-piceous;
antennæ black; pedicel shorter than the second funiclar joint; first
funiclar small, subtriangular; second swollen, a little curved; club
6-jointed, the joints loosely joined, hairy, the first, the smallest, the

last the largest, intermediate joints a little longer than thick; tubercle at tip of scutellum subobsolete; mesonotal furrows complete.

HABITAT.—San Francisco, Cal.

Types in National Museum.

Described from ♂ and ♀ specimens, reared November, 1885, from a Cecidomyiid gall on *Lupinus athorea*, collected by Albert Koebele.

Polymecus picipes, sp. nov.

♀. Length, 1.8 to 2ᵐᵐ. Black, shining; head transverse, the vertex posteriorly distinctly shagreened; face smooth, polished, except a few transverse lines just above the insertion of the antennae; legs variable, from a rufo-piceous to almost black, the trochanters, tips of anterior tibiae, base of middle and posterior tibiae, and all tarsi pale brownish or honey-yellow. Antennae 10-jointed, brown-black, the first funiclar joint a little longer than thick, the second transverse, the third quadrate, club joints oblong. Mesonotal furrows complete, the middle lobe projecting slightly on the scutellum, the lateral lobes with no tufts of pubescence at base. Collar and mesopleura striated. Scutellum highly convex, subopaque, without a tubercle at tip. Metathorax bare or faintly pubescent. Tegulae black. Wings hyaline.

Abdomen more than twice as long as the thorax, the petiole coarsely striated, the base of second segment with two aciculated foveolae at base.

HABITAT.—District of Columbia.

Types in Coll. Ashmead.

Several specimens.

Polymecus melliscapus, sp. nov.

♀. Length. 25ᵐᵐ. Black, subopaque, very faintly, microscopically shagreened; the head behind and the face transversely aciculated. Scape, pedicel, and legs brownish-yellow; flagellum black; pedicel about as long as the first and second funiclar joints united; joints 3, 4, and 5 of club not longer than wide at apex, the last conical, scarcely twice as long as the penultimate. Metathorax and petiole, above and below, densely pubescent. Tegulae black. Wings hyaline, pubescent. Abdomen twice the length of the head and thorax united, the third segment one-third the length of the fourth, the fifth and sixth equal in length, a little longer than the fourth, the sixth being conical.

HABITAT.—Washington, D. C.

Type in Coll. Ashmead.

Described from a single specimen collected by Mr. E. A. Schwarz.

Polymecus compressiventris, sp. nov.

♀. Length, 2.6ᵐᵐ. Polished, black, impunctured; the antennae and legs golden-yellow, the flagellum very slightly obfuscated at tip. The pedicel is oval, not quite as long as the first and second funiclar joints united; first funiclar joint narrowed, but twice as long as thick; the second stouter, obconic; third and fourth about equal, obconic, but longer than the second; the three following joints bell-shaped, loosely joined, the last conical, longer than the preceding. Thorax with two deep furrows. Tegulæ piceous. Wings hyaline. Abdomen a little more than three times the length of the thorax, very long and acute, and strongly compressed from above and below; when viewed from the side, knife-shaped.

HABITAT.—Washington, D. C.

Type in Coll. Ashmead.

Described from a single specimen collected by Mr. E. A. Schwarz. The very strongly depressed or flat abdomen and the length of the antennal joints at once separate this species from all other described forms, and I am not certain but that they are of sufficient importance to establish a new genus.

Polymecus auripes, sp. nov.

♀. Length, 1ᵐᵐ. Black, shining; the head opaque, closely, microscopically punctate; antennae and legs bright golden yellow. Thorax without furrows, or these only faintly indicated posteriorly; scutellum pubescent, foveated across the base and terminating in an awl-shaped spine; metathorax and base of abdomen densely pubescent. Tegulæ prominent, black. Wings hyaline, pubescent. Abdomen a little longer than the head and thorax united, gradually contracted into a tail from the apex of the second segment; the third segment one-half the length of the fourth; the fifth twice as long as thick, and as long or a little longer than the third and fourth segments united; the sixth or last segment conical, a little longer than the fifth.

HABITAT.—Virginia.

Type in Coll. Ashmead.

The small size and the color of the legs sufficiently distinguish the species.

Polymecus alnicola, sp. nov.

♂ ♀. Length, 1.4 to 1.8ᵐᵐ. Black, shining, the head and dorsum of thorax finely, microscopically punctate. Head transverse. Mandibles piceous. Antennae 10-jointed, black, the tip of the pedicel honey-yellow; the flagellum is about twice as long as the scape; pedicel twice as long as thick; first funicular joint slender, slightly longer than thick; second twice as long as thick, stouter; third and fourth about equal, slightly thicker than the second; club joints, except the last, equal, a little longer than thick, the last cone-shaped, slightly longer than the preceding. Mesonotal furrows complete, the middle lobe not projecting on to the scutellum, the lateral lobes with no tufts of pubescence at base. Scutellum highly convex, subopaque, the tip with the tubercle subobsolete. Metathorax sparsely pubescent. Legs black, tips of anterior femora, their tibiae and base and tips of middle and posterior tibiae and all tarsi honey-yellow. Wings hyaline.

Abdomen a little longer than the head and thorax together, shaped as in *P. lupinicola.*

In the ♂ the trochanters, knees, tips of tibiæ and tarsi are honey-yellow; abdomen as in *lupinicola*; tegulæ rufo-piceous; antennæ black, the first funiclar joint minute, the second, swollen, curved; club 6-jointed, the joints, except the last, very little longer than thick: scutellum not tubercular at tip.

HABITAT.—District of Columbia.

Types in National Museum.

Described from ♂ and ♀ specimens, reared April 30 and May 1, 1884, from a Cecidomyiid gall, *Cecidomyia serrulata* O. S., found on alder.

SACTOGASTER Förster.

Hym. Stud. ii, p. 108, 1856.

Epimeces Westw. (*pars*).

(Type *E. ventralis* Westw.)

Head transverse, the vertex somewhat acute, the occiput delicately margined; ocelli 3, small, triangularly arranged, the lateral distant from the margin of the eye.

Antennæ 10-jointed in both sexes, in ♀ terminating in a 4-jointed club; in ♂ with a 5-jointed, hairy club, the joints twice as long as thick, the funicle cylindrical, the pedicel shorter than the first two funiclar joints united.

Thorax subovoid, the mesonotum smooth, a little longer than wide, without furrows, or the furrows only delicately indicated posteriorly; scutellum convex, ending in a thorn, bifoveated at base; metathorax very short with a median carina, the metapleura with a silvery or hoary pubescence.

Front wings veinless, pubescent.

Legs clavate, the basal joint of hind tarsi more than three times as long as the second.

The females in this genus are readily separated from all others, by the inflated second ventral segment of the abdomen, the males by the position of the lateral ocelli.

The two species in our fauna may be thus separated:

TABLE OF SPECIES.

Head closely, finely, microscopically punctate, shining.
  Thorax microscopically punctate, parapsidal furrows distinct posteriorly; scutellum, metapleura and base of abdomen with a silvery pubescence; tail not longer than the first and second segments together.
                                    S. ANOMALIVENTRIS Ashm.
Head polished impunctate.
  Thorax polished, without a trace of the furrows; scutellum striated: metapleura and base of abdomen bare; tail nearly twice as long as the first and second segments together......................S. HOWARDII Ashm.

Sactogaster anomaliventris.

Ashm. Can. Ent., xix. p. 130, ♀ ; Cress. Syn. Hym., p. 219

♂ ♀ . Length, 0.6 to 1ᵐᵐ. Black, shining; the head finely. microscopically punctate; antennae and legs black or brown-black; trochanters, base of tibiae and anterior tibiae at tip, and all tarsi paler or yellowish. The lateral ocelli are away from the margin of the eye; the basal three club-joints are transverse. the first the narrowest. last joint conical: the parapsidal furrows are distinct posteriorly; the scutellum. metapleura and base of abdomen are covered with a silvery pubescence; the scutellum ends in an awl-shaped spine: wings hyaline; the inflated second abdominal segment is as long as the tail.

The ♂ differs from the ♀ in having an oval abdomen, which is shorter than the thorax, and without the inflated second ventral segment; the antennal club is 4-jointed, the basal three joints being equal. quadrate. the funicle slender, while the thorax is subopaque from a faint, microscopic punctuation.

HABITAT.—Jacksonville, Fla.

Types in Coll. Ashmead.

Many specimens. Not rare in April in the flower of gall-berry (*Ilex glaber*), and possibly a parasite on some Cecidomyiid larvæ inhabiting these flowers.

Sactogaster Howardii Ashm.

(Pl. xii. Fig. 4. ♀.)

Can. Ent., xx, p. 52.

♀ . Length, 2ᵐᵐ. Polished black. impunctured; the face convex, highly polished: mesonotum without a trace of furrows; scutellum striated, the spine about twice as long as thick; metapleura at base a little wrinkled; antennæ and legs dark rufous, the middle and posterior femora and tibiæ at tips fuscous. Wings hyaline. Abdomen about twice as long as the head and thorax united, the tail being about twice as long as the inflated second segment; the apical lateral parts of the fourth and the fifth segments wholly opaque, sculptured, otherwise the segments are smooth and shining.

HABITAT.—Washington, D. C.

Type in Coll. Ashmead.

Dedicated to my friend, Mr. L. O. Howard, of the U. S. Department of Agriculture. In the highly polished surface, entire absence of parapsidal furrows, striated scutellum, the bare metapleura, and the long tail, it is quite distinct from all other described forms in this peculiar genus.

SYNOPEAS Förster.

Hym. Stud. ii. p. 108 (1856).

(Type *S. prospectus Först.*)

Head transverse, the occiput margined; ocelli three, triangularly arranged; the lateral far away from the margin of the eye.

Antennae 10-jointed in both sexes, the scape much lengthened, clavate; club in female 4-jointed, the funiclar joints slender; in male 5-jointed, the joints cylindrical.

Thorax ovoid, convex, the mesonotum with or without furrows; scutellum broad, subconvex, bifoveolated at base, the tip ending in a minute tubercle, rarely entirely wanting, and usually pubescent; metathorax very short, channeled at the middle, the metapleura usually densely pubescent. Wings, veinless, pubescent.

Abdomen, oblong-oval, the apex a little more pointed in the ♀, the first segment a little wider than long, pubescent, the second very large.

Legs clavate, the basal joint of hind tarsi twice or a little more than twice as long as the second.

TABLE OF SPECIES.

FEMALES.

Mesonotum without furrows, or these only indicated posteriorly 2
Mesonotum with delicate but complete furrows.
  Coxæ black.
    Legs brown black.
      Antennæ black, first and third funiclar joints small...S. NIGRIPES, sp. nov.
    Legs pale rufous, posterior femora dusky.
      Antennæ pale brown, the scape rufous ....S. RUFIPES, sp. nov.
  Coxæ pale.
    Legs pale brown.
      Antennæ dark brown, first and second funiclar joints slender, cylindric,
        nearly equal.....................................S. INERMIS Ashm.
2. Mesonotal furrows indicated posteriorly.
  Coxæ brownish-piceous.
    Legs yellowish brown or brownish yellow, the posterior femora and tibiæ at
      tip obfuscated, with sometimes all femora dusky.
      Antennæ pale brown, tip of scape and the club dusky, joints 1 and 3 of
        funicle shorter than the second.............S. CORNICOLA, sp. nov.
    Legs, including coxæ, uniformly brownish yellow, rarely with posterior
      femora dusky.
      Antennæ dark brown, the first three funiclar joints longer than thick, the
        first the shortest ....S. ANTENNARLÆ, sp. nov.
Mesonotal furrows entirely wanting.
  Coxæ black.
    Legs reddish yellow or honey yellow, the middle femora and the posterior
      femora and tibiæ dusky or black.
      Scape rufous, the flagellum brown .... S. RUFISCAPUS, sp. nov.

Synopeas nigripes, sp. nov.

♂. Length, 1^{mm}. Black, shining, with a microscopic sculpture; the face highly polished, with a median impressed line; lateral ocelli twice their width from the margin of the eye. Antennæ 10-jointed, black, the flagellum thickened toward the apex, covered with sparse white hairs; pedicel as long as the first and second funiclar joints together, the first funiclar joint small, rounded; club joints, except the last, longer than wide, the last ovate. Mesonotum with 2 delicate furrows; scutellum convex, with a subobsolete tubercle at tip; metapleura opaque, sparsely pubescent; metanotum subpubescent. Legs entirely black, the tarsi piceous. Tegulæ black. Wings clear hyaline. Abdomen ovate, petiolate, as long as the thorax, the petiole and the base of second segment striated, the third, fourth, fifth, and sixth segments with a transverse row of fine punctures.

HABITAT.—Washington, D. C.

Type in Coll. Ashmead.

Synopeas rufipes, sp. nov.

(Pl. XII, Fig. 5, ♀.)

♂. Length, 1.5ᵐᵐ. Subrobust, black, polished; the head posteriorly transversely striated, the face very faintly punctured; lateral ocelli one and a half times their width from the margin of the eye. Antennæ 10-jointed, pale brownish yellow, the 5-jointed club black; pedicel longer than the second funiclar joint, first and third funiclar joints small, the second the smaller, the club joints oval-moniliform. Mesonotum, with the parapsidal furrows distinct posteriorly, the base of the middle lobe projecting slightly upon the scutellum. Tegulæ black. Wings hyaline. Legs rufous, the coxæ black. Abdomen ovate, a little longer than the thorax, the petiole and the foveolæ at base of the second segment striated.

HABITAT.—Arlington, Va.

Type in Coll. Ashmead.

Synopeas inermis Ashm.

Bull. No. 1, Coll. Biol. Assoc., p. 10.

♀. Length, 1.4ᵐᵐ. Black, alutaceous shining. Antennæ 10-jointed, reddish, the pedicel twice as long as thick, the three following joints very slender, cylindrical, the first two about equal in length and twice as long as thick, the third hardly half as long as the second, the fourth short but stouter, not longer than thick; the four following joints, comprising the club, are much stouter: the first of these is the shortest, the second and third about twice as long as thick, the fourth, or terminal joint, being the longest and more than twice as long as thick; these joints are slightly pedicellate.

Thorax with grooves only faintly indicated. Mesopleura smooth and only slightly impressed on the disk. Metapleura and petiole hairy. Legs, including coxæ, honey yellow, the tarsi, except the terminal joint, pale or whitish. The abdomen is about one-third longer than the head and thorax together, prolonged into a point at apex, the second segment occupying fully more than one-half of its whole surface, the segments beyond nearly equal. Wings hyaline, pubescent.

HABITAT.—West Cliff, Colo.

Type in Coll. Ashmead.

Described from a specimen received from Mr. T. D. A. Cockerell.

Synopeas cornicola, sp. nov.

♂ ♀. Length, 1 to 1.4ᵐᵐ. Polished, black, impunctured. Head a little wider than the thorax, the vertex posteriorly not aciculated. Antennæ 10-jointed, pale brown, the scape and the club dusky or fuscous; the first and third funiclar joints longer than thick, but still shorter than the second; club 4-jointed, a little thickened toward apex. Thorax ovoid, smooth, polished, with a very sparse pubescence, thickest near the scutellum; parapsidal furrows distinct only posteriorly, anteriorly obliterated; scutellum a little elevated posteriorly, very pubescent, the tubercle covered with hairs; metapleura woolly. Tegulæ rufopiceous. Wings hyaline, pubescent. Legs pale brownish-yellow, the coxæ brown, the posterior femora and tibiæ at tips, obfuscated. Abdomen oblong-oval, as long as the head and thorax together, the petiole striated, sparsely pubescent, the second segment with 2 oblong foveolæ at base, one on each side.

Habitat.—Kirkwood, Mo.

Types in National Museum.

Described from specimens reared April 3, 1887, by Miss Mary Murtfeldt, from a Cecidomyiid gall on *Cornus paniculata*, at Kirkwood, Mo.

Synopeas antennariæ, sp. nov.

♂ ♀. Length, 0.8 to 1.1ᵐᵐ. Polished, black, impunctured; vertex of head posteriorly with a few faint aciculations. Antennæ 10-jointed, dark brown, the flagellum subclavate, the first funiclar joint small but still longer than thick and a little longer than the third, the club 5-jointed, the joints, except the last, scarcely twice as long as thick. Thorax with the mesonotal furrows indicated only posteriorly; scutellum subconvex, smooth, not pubescent, the spine or tubercle wanting; metapleura sparsely pubescent. Tegulæ brown or rufous. Wings hyaline, legs reddish-yellow, the coxæ brownish, the posterior femora sometimes dusky. Abdomen long-ovate, distinctly longer than the head and thorax together, the petiole grooved and with a sparse pubescence.

In the ♂ the abdomen is oblong-oval, about as long as the head and thorax united; the legs uniformly bright yellow; the antennæ pale brown, the first funiclar joint very small, rounded, the second a little swollen and curved, about as long as the pedicel, while the club joints are all longer than thick.

Habitat.—Milwaukee, Wis.

Types in National Museum.

Described from many specimens, in both sexes, reared May 31, 1888, by Mr. William M. Wheeler, from *Cecidomyia antennaria* Whlr.

Synopeas rufiscapus, sp. nov.

♂ ♀. Length, 1 to 1.1ᵐᵐ. Polished, black, impunctured; lateral ocelli about twice their width from the margin of the eye, but still a

little nearer to the margin of the eye than to the front ocellus. Scape
and legs reddish yellow, the coxæ black, the posterior femora and the
tips of the posterior tibiæ fuscous. The pedicel is much stouter than
the funicle; the first two funiclar joints about equal, cylindric, the
third and fourth small, rounded; club 4-jointed, the first three transverse, the last ovate. Mesonotum without furrows; scutellum foveated
at base; metathorax short, pubescent. Tegulæ black. Wings hyaline.
Abdomen sessile, short-oval, a little shorter than the thorax, the base
densely pubescent.

In the ♂ the legs, except the coxæ, are wholly reddish-yellow; the
antennæ long, the scape, pedicel, and first funiclar joint yellow, the following joints brown-black; the second funiclar joint is as long as the
pedicel and first funiclar joint united, stout, the following joints more
slender, cylindric, and about 2½ times as long as thick, the apical joint
a little thicker than the basal; scutellum with a minute tubercle at
tip; metathorax and base of abdomen covered with a dense whitish
pubescence; while the abdomen is oval, much shorter than the thorax.

HABITAT.—Jacksonville, Fla.

Types in Coll. Ashmead.

Synopeas melanocerus Ashm.

Can. Ent., XIX, p. 130, ♀ ♂ ; Cress. Syn. Hym., p. 249.

♂. Length, 2.1ᵐᵐ. Black, polished; face faintly alutaceous, with a
central furrow. Lateral ocelli as near to the front ocellus as to the
margin of the eye. Antennæ 10-jointed, long, filiform, black, the joints
after the fourth long, cylindrical, about 4 times as long as thick; pedicel as long as the first and second funiclar joints united; the first funiclar joint small, the second thickened and curved. Mesonotum with
2 distinct furrows. Metathorax very sparsely pubescent. Tegulæ
piceous. Wings hyaline. Legs piceous, the coxæ black, the anterior
legs reddish-yellow. Abdomen as long as the head and thorax united,
oblong-oval, the petiole and the foveolæ at base of the second segment
striated.

HABITAT.—Jacksonville, Fla.

Type in Coll. Ashmead.

CŒLOPELTA Ashm. gen. nov.

(Type C. mirabilis Ashm.)

Head transverse, the occiput scarcely margined, the lateral ocelli as
near to the front ocellus as to the margin of the eye.

Antennæ in ♂ 9-jointed, ending in a 4-jointed club; the scape very
slender, not extending beyond the ocelli; pedicel very small; first funiclar
joint a little shorter than the second, the second fully twice as long as
thick; the third subtriangular; club joints elliptic-oval, loosely joined.

Thorax ovoid, convex, the mesonotum very little longer than wide,

with faint indications of parapsidal furrows posteriorly; scutellum
cupuliform, similar to the Cynipid genus Eucoila; metathorax pubescent, with lateral carinæ.

Wings veinless, slightly fringed.

Abdomen ovate, petiolate, the petiole longer than thick, striated.

Legs clavate, the tibial spurs weak, scarcely developed, the tarsi 5-
jointed, a little longer than the tibiæ, the basal joint of hind tarsi fully
3 times as long as the second.

A remarkable genus readily distinguished by the cupuliform scutellum and affording additional evidence of the close affinities existing between the *Proctotrypidæ* and the parasitic *Cynipidæ*.

Cœlopelta mirabilis, sp. nov.

(Pl. XII. Fig. 6, ♂.)

♂. Length, 0.8^{mm}. Black, polished; antennæ brown, the scape
yellow; legs reddish yellow, the coxæ black; metathorax with a silvery
pubescence; wings hyaline, iridescent, the hind wings rounded at
apex, with long cilia; abdomen ovate, polished, the petiole subopaque,
striated and bare.

HABITAT.—St. Vincent, W. I.

Type in National Museum.

Described from a single specimen collected by Herbert H. Smith.

ANOPEDIAS Förster.

Hym. Stud., II, p. 108 (1856).

Head transverse, the occiput not or indistinctly margined; ocelli 3,
triangularly arranged, the lateral usually wide from the eye margin.

Antennæ 10-jointed in both sexes, club in the ♀ 4-jointed, the joints
being but slightly thicker than the funicle, cylindric, much longer than
thick, the first funicular joint small, the second, third, and fourth nearly
equal; in ♂ ending in a 5- or 6-jointed club, the joints loosely joined, the
last the largest and thickest, ovate or cone-shaped, the first funicular joint
very minute.

Thorax ovoid, the mesonotum rather short, convex, without furrows,
or with the furrows exceedingly delicate and so fine as to be overlooked
without a strong lens; scutellum flat or subconvex, unarmed at apex
and without foveæ at base, separated from the mesonotum by a delicate,
transverse, impressed line; metathorax very short, with 2 dorsal carinæ
and lateral carinæ.

Wings veinless.

Abdomen in ♀ ovate or pointed, in ♂ oblong-oval, longer than the
thorax, or the thorax and head united, the first segment petioliform,
the second, very large, with 2 foveolæ at base.

Legs clavate, the tarsi longer than their tibiæ, the basal joint of hind
tarsi a little more than twice the length of the second.

Distinguished at once by the shape of the antennae, and the flat or subconvex scutellum, which is not foveated at base, being separated from the mesonotum by a delicate transverse line.

Förster indicated no type. Thomson's definition of the genus is quite different from mine.

Anopedias error Fitch.

(Pl. XII, Fig. 7, ♀.)

Platygaster error Fitch, Sixth N. Y. Rep., p. 76, Pl. 1, Fig. 4.

♂ ♀. Length, 1 to 1.2ᵐᵐ. Polished black, impunctured. Head transverse, as wide as the thorax across from wing to wing, the vertex posteriorly polished, without trace of aciculations. Antennae 10-jointed, black, the flagellum subclavate; pedicel pale at tip, as long as the first two funiclar joints together; first funiclar joint minute; the second, third, and fourth nearly equal; club 4-jointed, very slightly thicker than the funicle, all the joints distinctly longer than wide, the first the shortest, the last the longest. Thorax long-ovate, polished, impunctured, the parapsidal furrows very faint; scutellum flattened, or subconvex, separated from the mesonotum by a very delicate, transverse, impressed line; mesopleura smooth; metapleura strongly sericeous. Tegulae black. Wings hyaline, iridescent, pubescent and fringed. Legs brown-black, the trochanters, tip of anterior tibiae, and all the tarsi paler. Abdomen oblong-ovate, polished, very slightly longer than the thorax, the petiole coarsely grooved, the second segment with some longitudinal striae at base.

The male is smaller, the abdomen oval, shorter than the thorax, while the antennae have a 6-jointed, slightly pedicellate club. The first funiclar joint is very minute, rounded, closely connected with the second, the second a little curved and thickened, truncate at tip; the first club joint is the smallest, the others very gradually increase to the last, oblong-oval in shape, the last being larger and thicker, fusiform, and nearly twice as long as the preceding.

HABITAT.—New York, Washington, D. C., Arlington, Va., and Lafayette, Ind.

Specimens in National Museum.

Described from ♂ and ♀ specimens, reared June 14, 1884, by Mr. F. M. Webster, from *Diplosis tritici*.

I had previously identified a species of *Polygnotus* as *Platygaster error* Fitch, but a more careful examination of Fitch's figure and description satisfies me now that I was mistaken, and that the species described here is really his *P. error*.

The identification by Dr. Fitch of the fragments of an insect sent to him by Mr. Herrick and reared from an Hemipterous egg (*Nabis*), as his *Platygaster error*, was certainly erroneous, since it was undoubtedly nothing but a species of *Telenomus*, as Mr. L. O. Howard has already pointed out.

Anopedias pentatomus, sp. nov.

♀. Length, 0.8mm. Polished black, impunctured, except the meta-thorax and petiole; parapsidal furrows entirely wanting; scutellum subconvex, separated from the mesonotum by a transverse grooved line; metapleura with a dense sericeous down; petiole sparsely pubescent, striated; legs and antennae, except the 5-jointed club which is brown-black, honey-yellow, the femora and tibiae tinged with red.

The head is broadly transverse, much wider than the thorax, the face convex, highly polished, the lateral ocelli twice their width from the border of the eye. Antennae 10-jointed, the scape clavate, very slender at base, strongly curved; pedicel as long as the 3 funiclar joints together; first funiclar joint very minute, the second and third equal, not longer than thick; club 5-jointed, the joints a little longer than thick. Abdomen ovate, narrowed at base, the base of the second segment with two long, striated foveolae at base.

HABITAT.—Arlington, Va.

Types in Coll. Ashmead.

Anopedias incertus.

Ashm., Can. Ent., XIX, p. 138, ♀ ; Cress., Syn. Hym., p. 249.

♀. Length, 1.6mm. Polished black, feebly sericeous; parapsidal furrows faintly indicated posteriorly; scutellum ending in a small tubercle; metapleura covered with a dense silvery white pubescence; legs and antennae brownish-yellow or reddish, the 4-jointed club and the coxae black, the middle and posterior femora obfuscated.

The head is transverse, obliquely narrowed behind the eyes, alutaceous, the ocelli not quite touching the border of the eye. Antennae 10-jointed, the funicle long and slender, the first joint as long as the long and slender pedicel, the following joints shorter, subequal, club joints, except the last, quadrate, the last conical, nearly twice as long as the penultimate. Abdomen oblong-oval, polished, the broadest part a little wider than the thorax, the first segment short, densely pubescent above.

♂. Length, 1.4mm. Agrees with the ♀ except that the antennae and legs, including all coxae, are brownish-yellow, the club dusky, the abdomen oval, narrower than the thorax.

HABITAT.—Jacksonville, Fla.

Types in Coll. Ashmead.

AMITUS Haldeman.

Sill., Am. Jour., 2d Ser., IX, p. 109, 1850.
Zaerita Förster, Kleine Mon., p. 46.

(Type A. aleurodinus Hald.)

Head transverse, the frons subconvex, the occiput not margined; ocelli 3, subtriangularly arranged, the lateral distant from the margin of the eye; eyes oval.

Antennæ inserted just above the clypeus; in ♀ 8-jointed, the scape subclavate, lengthened, extending far above the vertex; pedicel long, thicker than the first funicular joint; funicular joints subequal, gradually becoming wider and shorter, the first the longest and slenderest; club thicker than the last joint of funicle and as long as the two preceding joints together, unjointed, but under a high power 3 indistinct joints can be detected; in ♂ 10-jointed, verticillate from the third joint; pedicel short and thick; first flagellar joint as long as the pedicel but thinner and obconic; second nearly twice the length of first, thicker at apex than at base; third to last cylindrical and subequal; last joint elongate, conical.

Thorax robust, subovoid, convex, the mesonotum broader than long with two delicate grooved lines; scutellum large, semicircular, convex, separated from the mesonotum by a transverse grooved line at base; metathorax exceedingly short, abrupt.

Wings broad, ciliated, veinless, although a trace of the submarginal vein can be detected when the wing is viewed through transmitted light.

Abdomen broadly ovate or subcordate, sessile, about as long as the thorax, the first segment very short, transverse; the second very large, occupying most of the surface.

Legs rather long. The femora clavate, tibiæ subclavate, tarsi slender, 5-jointed, longer than their tibiæ, the basal joint of hind tarsi 2½ times as long as the second.

Amitus aleurodinis Hald.

(Pl. xii, Fig. 8, ♀.)

Amitus aleurodinis Hald., Sil. Jour. Sci., 2d ser., ix, p. 110, 1870; Cr., Syn., Hym., p. 250,
Elaptus aleurodis Forbes, 14th Ill. Rep., 1884. p. 110, Pl. ii, Fig. 6, ♀
Alaptus aleurodis Cr., Syn. Hym., p. 250.

♂ ♀. Length 0.75 to 1ᵐᵐ. Polished black; legs variable in color, most frequently brownish-yellow or pale rufous, the coxæ and femora sometimes dusky or blackish; sometimes only the hind femora and coxæ black.

Antennæ in ♀ 8-jointed, as long as the body, terminating in an unjointed club, brownish-yellow, the club brown; the pedicel and first funicular joint long, the former a little the longer and thicker, the following joints to club subequal in length but gradually thickened; in ♂ 10 jointed, yellow, the flagellar joints with verticillate hairs. Thorax with 2 delicate furrows. Metapleura pubescent. Wings hyaline, ciliated.

Abdomen broad, about as long as the thorax: the first segment very short rugose: the second, except foveolæ at base, and the following, smooth, polished.

HABITAT.—Pennsylvania, District of Columbia, and Illinois.

Types in Coll. American Entomological Society.

The types I have seen; they are still preserved in the Coll. American Entomological Society, although in poor condition. The National

Museum contains a fresh specimen, reared March 29, 1882, by Dr. Riley from *Aleurodes* sp. occurring on *Acer dasycarpum* at Arlington, Va., while my collection contains a single specimen captured with the sweeping net.

Prof. Haldeman gives the following interesting facts respecting it: "Parasitic on the larva of *Aleurodes corni* Hald., of which it destroys a great many. I found it with that insect beneath the leaves of *Cornus sericea* on the margin of a water course. It leaps, walks, and flies with facility, and when touched simulates death. The antennæ are kept in a constant state of vibration. I have kept them a week or more, living in confinement. The ova (crushed from the ovaries) are fusiform, rounded at one extremity and produced at the other like the neck of a flask."

The insect described by Prof. Forbes as *Elaptus aleurodis*, reared from *Aleurodes aceris* Forbes, is evidently identical.

Geoffroy has described an *Aleurodes aceris* in Europe, and I would here suggest the name *A. Forbesii*, for Prof. Forbes's species.

TRICHACIS Förster.

Hym. Stud., II, p. 108 (1856).

(Type, T. pesis Walk.)

Head transverse; the frons subconvex, the occiput not or only delicately margined; ocelli 3, in a triangle, the lateral ocelli their width away from the margin of the eye.

Antennæ 10-jointed in both sexes, in ♀ ending in a 5-jointed club, the first and third funiclar joints minute, the second lengthened, as long as the pedicel and first funiclar joint together; first club joint the shortest, the last the largest, ovate; in ♂ with the club thinner, the joints more elongate, cylindrical, the first and second funiclar joints long, not small, or the first very minute, the second swollen.

Thorax ovate, the pronotum produced into a slight neck anteriorly; mesonotum convex, about twice as long as wide, with 2 deep furrows; scutellum somewhat elevated, with a tuft of hair at tip, but without a thorn or tubercle; metathorax very short, with two median keels, the pleura pubescent.

Wings ciliated, veinless, the submarginal vein traceable basally.

Abdomen in ♀ conic-ovate, 4-jointed, the first segment petioliform, striated, the second as long as the two following together, with 2 foveolæ at base, the third less than one-third the length of the fourth, with transverse rows of small punctures, the fourth or last conical, compressed from above and below, margined, the apical half very flat or subconcave, with striæ at base followed by a slight smooth prominence; in ♂ oblong-oval, 8-jointed, the last two segments very minute, segments 3, 4, and 5 nearly equal, usually with transverse rows of punctures.

Legs clavate, the basal joint of hind tarsi more than twice as long as the second.

Distinguished by the tuft of hair at tip of scutellum, antennal characteristics, and the long, distinctly furrowed mesonotum.

TABLE OF SPECIES.

Coxæ black, or at least the hind pair black.
  Legs and scape of antennæ rufous.
    Flagellum brown black; pedicel scarcely as long as the first funiclar joint; fourth
        ventral segment not punctate, the fifth not striate. T. RUFIPES, sp. nov.
  Posterior femora, and tips of tibiæ brown.
    Flagellum brown; pedicel much longer than the first funiclar joint, the second
        longer than the pedicel, the third small; fourth ventral segment
        punctate, the fifth striate.....................T. RUBICOLA, sp. nov.
  Legs brownish-piceous, anterior pair, base of middle and posterior tibiæ and tarsi
        honey-yellow.
    Antennæ brown, first funiclar joint very small, the second elongate; ventral
        segments with a row of punctures at base.. T. ARIZONENSIS, sp. nov.
  Legs pale brown, the trochanters, anterior tibiæ, and base of middle and posterior
        tibiæ honey-yellow.
    Antennæ brown, the scape and pedicel yellow, first and third funiclar joints
        small, the second elongate; apical ventral segment striated.
                                            T. BRUNNEIPES Ashm.
Coxæ pale.
  Legs piceous or brownish, base of middle and posterior tibiæ, anterior tibiæ, and
        tarsi pale.
    Antennæ dark brown, the first and second funiclar joints about equal, the third
        and fourth minute; ventral segments impunctured.
                                            T. VIRGINIENSIS, sp. nov.

Trichasis rufipes, sp. nov.

(Pl. XII. Fig. 9, ♀.)

♂ ♀. Length, 2 to 2.5ᵐᵐ. Black, shining; the thorax with a sparse,
fine pubescence. Head transverse, with some microscopic punctures
on the crown; lateral ocelli with a slight curved depressed line behind;
face highly polished. Antennæ 10-jointed, the scape rufous, the flagellum brown-black; pedicel of equal length with the second funiclar joint;
first and third funiclar joints smaller than the second; club 5-jointed,
slightly thickened toward apex, the first joint a little longer than thick,
narrowed toward base, the second, third, and fourth joints quadrate,
a little rounded basally, the last joint cone-shaped, one-third longer
than the preceding. Thorax elongate-ovate with distinct furrow; scutellum depressed across the base, pubescent and with a tuft of hairs at
tip; metapleura sericeous; tegulæ black. Wings subhyaline, pubescent, not fringed or the fringe exceedingly short. Legs yellowish-red,
the coxæ black. Abdomen longer than the head and thorax together,
pointed at apex, the petiole longer than thick, finely rugose, with a
middle carina, the second segment very long with two long foveolæ at
the base, the tip with sparse white hairs.

In the ♂ the second funiclar joint is elongate the club 6-jointed;
otherwise closely resembling the ♀.

HABITAT.—District of Columbia, Virginia, Florida, and Missouri.

Types in Coll. Ashmead and National Museum.

Described from several specimens. A single ♀ specimen is in the National Museum, reared by Dr. Riley at St. Louis, Mo., from acorns infested with *Balaninus nasicus* and *Blastobasis glandulella;* but the acorns must have also contained Cecidomyiid inquilines.

Trichasis rubicola, sp. nov.

♂ ♀. Length, 1.4 to 1.6^{mm}. Polished black, impunctured; head with a few transverse aciculations on the crown. Antennae 10-jointed, the scape yellow, the flagellum pale brown; pedicel oval; first and third funicular joints very minute; second much elongated and thick; club 5-jointed, the joints longer than thick. Thorax ovate, with distinct furrows, the base of the middle lobe elevated slightly upon the scutellum; scutellum with depressions at sides and base, and with a tuft of pubescence at tip; metapleura sericeous; tegulae black. Wings hyaline. Legs yellowish, the coxae black, the posterior femora and tibiae toward tips, dusky. Abdomen longer than the head and thorax together, pointed at tip, the petiole fluted, the second segment with 2 foveolae at base, the third with some punctures, the fourth, striated.

In the male, the fourth, fifth, and sixth abdominal segments have a transverse row of punctures; club of antennae 6-jointed, darker colored, the joints cylindrical, at least thrice as long as thick.

HABITAT.—Cadet, Mo., and District of Columbia.

Types in National Museum and Coll. Ashmead.

Described from many specimens. The National Museum contains specimens, reared by Dr. Riley, June 16, 1883, from a Cecidomyiid stem-gall on Blackberry at Cadet, Mo.; and others reared June 9, 1886, from a Cecidomyiid gall on *Vernonia noveboracensis*, collected at Washington.

Trichasis arizonensis, sp. nov.

♂ ♀. Length, 1.2 to 1.6^{mm}. Very closely allied to *T. rubicola*, but differs in the vertex of the head being distinctly aciculated, the antennae being wholly brown-black, the legs darker, brownish or fuscous, the anterior legs, base of middle and posterior tibiae and tarsi pale or honey-yellow; while the 5-jointed club in the ♀ is more slender than in *T. rubicola*, with the joints at least twice as long as thick. The sculpture in both species is similar.

HABITAT.—Mount Graham, Ariz.

Types in National Museum.

Described from specimens reared from a Cecidomyiid gall on wild sunflower, received from Mr. H. K. Morrison.

Trichasis brunneipes Ashm.

Can. Ent., XIX, p. 131, ♀ ; Cress. Syn. Hym., p. 250.

♀. Length, 2^{mm}. Elongate, polished, black, impunctured; head transverse, the lateral ocelli a little more than their width from the margin of the eye. Antennae 10-jointed, cylindrical, the scape and pedicel yellow, the flagellum brownish-black; the first funiclar joint is very small, the second elongate, twice the length of the pedicel, the third obconic, shorter than any of the following joints, which are fully twice as long as thick, cylindric and loosely joined, the last being conical and a little longer than the preceding joint. Thorax long, with 2 complete furrows; scutellum with a tuft of pubescence at apex covering a minute tubercle; metathorax pubescent. Legs brownish-yellow, the anterior pair yellowish, the middle and posterior coxæ and thickened parts of their femora and tibiæ brownish. Wings hyaline, pubescent. Abdomen long, conically pointed, polished, the petiole striated, the third segment with a transverse row of punctures, the fourth striated; the apical dorsal valve flattened, smooth, delicately margined at sides.

HABITAT.—Jacksonville, Fla.

Type in Coll. Ashmead.

Trichasis virginiensis, sp. nov.

♀. Length, 1.1ᵐᵐ. Polished, black, impunctured; head transverse, shining, the vertex separated from the occiput by a transverse ridge, the lateral ocelli being nearer to the front ocellus than to the margin of the eye. Antennæ 10-jointed, brownish-yellow, the club black; first and second funiclar joints equal, longer than thick, the third small, the fourth triangular; the club a little thickened towards apex, the first three joints transverse, rounded at base, truncate at tip, the first the smallest, the last joint conical, nearly twice as long as the penultimate; all the club joints slightly pedicellate. Thorax ovate, polished, with two distinct furrows which are very delicately impressed anteriorly; sides of collar visible in front of the tegulæ; scutellum, with finely rugose, pubescent foveæ at base, a convex, shining prominence medially and a tuft of pubescence at tip, inclosing a small tubercle; metapleura pubescent. Wings subhyaline, probably from the fuscous pubescence. Legs brownish-yellow, the coxæ and femora with a reddish tinge. Abdomen pointed ovate, polished, petiolated; the petiole roughened, the apical segment conical, as long as the fourth and fifth together; the second segment with 2 striated foveolæ at base.

HABITAT.—Arlington, Va.

Types in Coll. Ashmead.

Trichacis auripes Prov.

Add. et Corr., p. 403.

♀. Long. 15 pce. Noire, polie, brillante; le scape avec les pattes d'un beau jaune d'or. Le funicule des antennes brun foncé. Ailes sans nervures distinctes, grandes, plus longues que l'abdomen. Le thorax rétréci en avant, le mésothorax avec deux sillons très apparents; l'écusson convexe, terminé par une touffe de poils grisâtres, séparé du mésothorax par une double fossette large et profonde. Les hanches noires, les cuisses postérieures renflées en massue et plus ou moins obscures. Abdomen subsessile, poli, brillant, déprimé, arrondi à l'extrémité, le 2ᵉ segment le plus grand. (Provancher.)

HABITAT.—Cap. Rouge, Canada.

Not recognized.

HYPOCAMPSIS Förster.

Hym. Stud., ii, p. 108, (1856.)

(No type described.)

Head transverse, frons convex, the occiput not margined; ocelli 3, in a triangle, the lateral nearer to the front ocellus than to the margin of the eye.

Antennæ 10-jointed in both sexes, in ♀ subclavate, or with a 4- or 5-jointed club, all the joints cylindrical, the joints 2 to 6 longer than thick; in ♂ with the first funiclar joint short, cup-shaped, the second very strongly thickened, the last 6 joints cylindric, with short hairs, the last the longest.

Thorax subovoid, the mesonotum short, convex, with two impressed lines; scutellum subelevated, rounded behind, foveated at base; metathorax very short, the pleura pubescent.

Wings veinless, pubescent.

Abdomen nearly elliptical, the margins broadly deflexed, the first segment longer than thick, the last 4 segments together not quite as long as the second.

Legs clavate.

Distinguished by the broad, deflexed margins of the abdomen. Förster's type appears not to have been described, although Thomson has recognized the genus in Sweden and describes three species therein.

Hypocampsis pluto Ashm.

(Pl. xii, Fig. 10, ♀.)

Can. Ent. xix, p. 131, ♀; Cress. Syn. Hym., p. 250.

♀. Length, 1.5^{mm}. Wholly black, polished, the tarsi alone slightly piceous. Head transverse, wider than thorax, the lateral ocelli more than their width from the margin of the eye; the eyes large, long-oval. Antennæ 10-jointed, the last funiclar joint thicker than the preceding; club 4-jointed (5-jointed if we count the thickened last funiclar joint which might really be considered as belonging to it), the joints all longer than thick, cylindric, the last conical and longer than the penultimate. The thorax is roundedly narrowed anteriorly with 2 delicate furrows; scutellum rounded, highly convex; the metathorax short, strongly sericeous. Wings hyaline, pubescent, the tegulæ black. Abdomen subovate, pointed at tip, narrowed toward base, very little longer than the thorax, with broad lateral margins.

HABITAT.—Jacksonville, Fla.

Type in Coll. Ashmead.

ERITRISSOMERUS Ashm., gen. nov.

(Type E. cecidomyiæ.)

Head broadly transverse, with a sharp acute process between the antennæ, the lateral ocelli away from the eye margin, but still nearer to it than to the front ocellus.

Antennæ 10-jointed, in both sexes, ending in a 6-jointed club, the second funiclar joint in the male very broadly dilated, the first being very minute and closely joined to, and not separable from, the second.

Thorax ovoid, convex, the mesonotal furrows distinct, the scutellum rounded, the metathorax very short, bicarinated.

Wings veinless, almost bare of pubescence.

Abdomen oval, the petiole short.

Legs clavate, the hind tarsi longer than their tibiæ, the basal joint long.

A genus allied to *Platygaster*, but readily separated in the female by the acute carina between the antennæ, and in the male by the very broadly and roundedly dilated second flagellar joint, the first and second being conjoined.

Only a single species is known.

Eritrissomerus cecidomyiæ, sp. nov.

(Pl. xiii, Fig. 1, ♀.)

♀. Length, 2.1ᵐᵐ. Black, subopaque, finely sculptured; head transverse, the vertex and cheeks rugulose; frons with an impressed line in front of the front ocellus, the face transversely striated; lateral ocelli a little more than their width away from the margin of the eye; mandibles rufo-piceous. Antennæ 10-jointed, the scape, except at both ends, and the 6-jointed club, brown-black; antennal tubercle, distal ends of scape, the pedicel and funicle honey-yellow; the pedicel as long as the first and second funiclar joints united, the first funiclar joint slender, closely united with the second and longer than thick, the second obliquely truncate at tip; club 6-jointed, submoniliform, the joints closely joined, except the last, joints not longer than thick. Thorax with two distinct furrows, the scutellum convex, margined at sides, closely punctate, the mesopleura impressed at the middle, the convex piece beneath the tegulæ striated, the metapleura bounded above by a distinct carina, striate and pubescent. Wings clear hyaline, the tegulæ piceous. Legs blackish, the anterior pair, except the coxæ, base of middle and posterior tibiæ, honey-yellow, the middle and posterior tarsi brownish. Abdomen oval, scarcely longer than the thorax, with the basal one-third of the second segment and the petiole striated.

In the ♂ the antennæ are wholly brownish-yellow, with the second funiclar joint very broadly roundedly dilated, the first, second, and last joints of the club being distinctly longer than thick, while all the trochanters and the knees of middle and hind legs are honey-yellow.

HABITAT.—Jacksonville, Fla.

Types in Coll. Ashmead.

Described from several specimens, reared from a Cecidomyiid gall on Hickory.

POLYGNOTUS Förster.

Hym. Stud. ii, p. 108 (1856).
(Type *P. striolatus* Nees.)

Head broadly transverse, the vertex broad, the occiput margined, ocelli 3 in a triangle, the lateral close to the margin of the eye.

Antennæ 10-jointed, the scape subclavate, slightly curved at base, in both sexes terminating in a 5-jointed club, the funiclar joints vary but the first in both sexes is usually minute, the second and third subequal, cylindric, in ♀, while in the ♂ the second is enlarged, swollen, the third scarcely half as large, thinner and obliquely truncate at apex.

Thorax subovoid, convex, the mesonotum rarely 1½ times as long as wide, with or without furrows; scutellum transversely elevated, convex or cushion-shaped, separated from the mesonotum by a transverse depression or furrow; metathorax short, the metapleura bounded above by a carina.

Wings veinless.

Abdomen, in ♀, pointed-ovate, depressed, 6-jointed, the first joint petioliform, striated, the second longer than the following together, smooth, with 2 oblong foveolæ at base, in ♂ oblong oval, 7- or 8- jointed.

Legs clavate, the basal joint of hind tarsi about thrice as long as the second.

Distinguished from *Platygaster* by the shorter, rarely distinctly grooved mesonotum, broader head, different antennæ, and the higher, more convex, or cushioned-shaped scutellum, which has a deep transverse impression all across the base.

TABLE OF SPECIES.

FEMALES.

Mesonotal furrows entire 6
Mesonotal furrows indicated only posteriorly 3
Mesonotum without trace of furrows, or the faintest trace of them posteriorly.
  Vertex not transversely aciculated 2
  Vertex transversely aciculated.
    Coxæ black.
      Legs black or brown-black.
        Antennæ black; pedicel as long as the two following joints united and much thicker, the tip pale; first joint of funicle small, second and third larger, nearly equal, the second slightly the larger, fourth transverse; club joints, except the last, as broad as long.

                                           P. SALICICOLA

        Antennæ brown-black; pedicel much longer than the two following joints united and much stouter; joints one and two of funicle very small, nearly equal, the third slightly larger, fourth transverse; club joints, except the last, only slightly longer than thick.

                                           P. DIPLOSIDIS

    Coxæ dark rufous or piceous.
      Legs rufous, the tarsi yellowish, femora and tibiæ often dusky.
        Scape black, sometimes pale at base and apex, flagellum dark brown; pedicel as long as the two following joints united and stouter; joints one and three of funicle minute, the second slightly larger, fourth longer than wide; club joints except the last, very slightly wider than long .... P. BACCHARICOLA Ashm.

      Front legs, all trochanters and tarsi honey-yellow, sometimes the front femora and tibiæ dusky, middle and posterior femora and tibiæ dark brown or black; sometimes all the legs pale brown.

Antennæ brown, the scape variable, sometimes almost black; pedicel longer than the two following joints united, not especially stout; joint 1 of funicle very minute, second larger and a little thicker than the third, fourth broader; club joints, except the last, not especially widened, slightly longer than wide ....P. SOLIDAGINIS Ashm.

2. Legs dark rufous, the tarsi pale.

Antennæ brown; funiclar joints 1 to 3 very small, slightly increasing in size, fourth larger; club joints, except the last, scarcely longer than wide..P. PINICOLA.

3. Vertex not transversely striated or aciculated, smooth, or at the most very faintly aciculated 4

Vertex posteriorly strongly transversely striated or aciculated.

Vertex impressed at the middle, subangulated just over the eyes.

Legs and antennæ black or brown-black..................P. STRIATICEPS.

Vertex not impressed at the middle.

Coxæ black.

Legs and antennæ black, trochanters, tip of anterior tibiæ, and tarsi honey-yellow.

Dorsal abdominal segments 3, 4, and 5 impunctured, joint 1 of funicle very minute, pale, second slightly longer than the third.

Club joints a little longer than thick, cup-shaped, outwardly subserrate ....P. ATRIPLICIS.

Club joints wider than long, outwardly not serrated....P. ARTEMISIÆ.

Dorsal abdominal segments 3, 4, and 5 with a row of punctures.

P. VIRGINIENSIS.

Legs brown or piceous, trochanters, knees, tips of anterior tibiæ, and tarsi pale.

Head very wide.

Apical ventral segments not punctate.

Tegulæ black; joint 1 of funicle very minute, second longer than the third; club not especially widened, the joints scarcely broader than long..P. ALNICOLA.

Tegulæ rufo-piceous; joint 1 of funicle small, second and third nearly equal; club very wide, the joints wider than long, serrate toward one side ....P. TUMIDUS.

Apical ventral segments punctate.

Head four or more times as wide as thick.

Club joints longer than wide.......................P. LATICEPS.

Head not especially wide.

Tegulæ rufo-piceous; antennæ brown; pedicel not long, cylindric, twice as long as thick; joint 1 of funicle small, not longer than thick, second larger and a little longer than the third; club joints, except the last, scarcely longer than wide....P. HIEMALIS Forbes.

Tegulæ black; antennæ brown-black; pedicel long, cylindric, fully thrice as long as thick, second longer and thicker, third a little longer then the second; club slender, the joints distinctly longer than wide ....P. PROXIMUS.

Coxæ rufo-piceous or pale.

Legs and antennæ dark brown; trochanters, base of tibiæ, and tarsi pale.

Tegulæ rufo-piceous; joint 1 of funicle very minute, second and third larger, about equal in length, the second the thicker; club joints, except the last, as wide as long.

Four last dorsal segments with a row of punctures.......P. VITICOLA.

4. Vertex angularly produced just over the eyes.

Legs and antennæ black; tarsi piceous.......................P. UTAHENSIS.

Vertex not angularly produced.

Coxæ rufo-piceous or pale 5

Coxæ black.

Legs black or brown-black, the tarsi usually pale.

Metapleura bare, aciculated.

Club stout; the joints, except the last, as wide as long.

P. CYNIPICOLA Ashm.

Metapleura covered with a sericeous pubescence.

Tegulæ rufo-piceous.

First three funiclar joints slender, cylindric, of equal thickness, the second slightly longer than the third; club slender, the joints distinctly longer than thick .... P. EUROTLÆ.

First funiclar joint small, the second and third larger, equal; club slender, the joints, except the last, not longer than thick ..... P. RUBI.

Tegulæ black.

First funiclar joint very small, second and third longer, equal, the fourth stouter; club very slender, the joints, except the last, scarcely longer than thick; a tuft of hairs at base of parapsides.... P. ASYNAPTÆ.

First funiclar joint small, the second and third longer, the second the larger, the fourth stouter, a little longer than the second; club not so slender, the joints, except the last, slightly wider than long; no tuft of hairs at base of parapsides .... P. HUACHUCÆ.

5. Tegulæ rufo-piceous.

Legs rufo-piceous, the front tibiæ and all tarsi and knees honey-yellow.

Antennæ brown-black; joint 1 of funicle minute, second slightly longer and thicker than the third, fourth as long as second, but stouter; club joints slightly pedicellate, a little longer than thick; no tuft of hairs at base of parapsides..................... P. ACTINOMERIDIS.

Antennæ brown; joint 1 of funicle very minute, second and third longer, about equal, fourth stouter; club joints longer than thick; a tuft of hairs at base of parapsides .... P. VERNONLÆ.

Tegulæ black.

Legs rufo-piceous, trochanters, knees, tips of tibiæ, and tarsi pale or honey-yellow.

Antennæ dark brown; joint 1 of funicle minute, not rounded, second a little larger, third shorter than second; club joints as wide as long; a slight tuft of hairs at base of parapsides......... P. ASTERICOLA.

6. Vertex posteriorly microscopically shagreened or punctate.

Coxæ black.

Legs black, tarsi fuscous.

Antennæ black, slender; funicle joints, first and second small, about equal, scarcely longer than thick, third and fourth larger, equal; club very slightly thicker than the last funiclar joint, a little thicker toward apex; the joints, except the last, one and a half times as long as thick; wings hyaline.................... P. COLORADENSIS.

Antennæ very slender, long, subclavate, black; joint 1 of funicle twice as long as thick, second nearly twice as long as the first, third and fourth shorter but stouter; the club joints thrice as long as thick; wings hyaline...................................... P. FILICORNIS.

Vertex posteriorly smooth, shining.

Coxæ and legs black, the tip of anterior tibiæ and tarsi fuscous.

Antennæ black, the pedicel 2½ times as long as thick; first three funiclar joints about of an equal length, the third a little the stoutest; club 5-jointed, very little longer than thick, except the last, the three middle joints outwardly subserrate; wings slightly smoky.

P. CALIFORNICUS.

MALES.

Mesonotum with the furrows complete 5

Mesonotum with furrows indicated only posteriorly 3

Mesonotum without trace of furrows or the faintest trace posteriorly.

    Vertex not or very faintly transversely aciculated 2

    Vertex transversely aciculated or striated posteriorly.

        Coxæ black.

            Legs black or brown-black.

                Antennæ black; pedicel longer than the first two funiclar joints; first funiclar joint very minute close to the second; the second as wide as long; third narrow, a little longer than thick; the fourth transverse, smaller than the first club joint......................P. SALICICOLA.

        Coxæ rufo-piceous.

            Legs dark rufo-piceous, tarsi honey-yellow, femora and tibiæ dusky at the middle.

                Scape black, flagellum dark brown........................P. SOLIDAGINIS.

                Antennæ brown-black; pedicel about as long and thick as the second funiclar joint; club 6-joint oval-moniliform, slightly pedicellate, hairy.

                                      P. BACCHARICOLA.

                Antennæ dark brown; pedicel almost as long as three funiclar joints united; first funiclar joint minute, close to the second; the second twice as long as the first; the third half the length of fourth; first three club joints equal, oval, the last conic, narrower and longer than the preceding...P. DIPLOSIDIS.

            Legs rufous, the tarsi pale.

                Antennæ brown; pedicel about as long as the first two funiclar joints united; the first funiclar joint minute, closely united with the second; the second subtriangular or twisted, dilated towards one side; club 6-jointed, oval-moniliform, slightly pedicellate, hairy..P. PINICOLA.

2. Coxæ black.

    Legs dark brown, tarsi whitish.

        Pedicel a little longer than the second funiclar joint, the first very small; club 6-jointed, the joints, except the last, moniliform, the last cone-shaped, nearly twice as long as the preceding ....P. PROXIMUS.

    Legs black.

        Pedicel as long as the second funiclar joint; first funiclar joint small, rounded; the second shorter than the first club joint; club 6-jointed, the joints, except the last equal, a little longer than thick, the last long fusiform, thicker, and twice as long as the preceding..P. ASYNAPTÆ.

    Coxæ pale or brown; antennæ dark brown.

        Legs brownish-yellow, posterior femora dusky.

        Pedicel as long as the first and second funiclar joints united; the tip yellow; first joint of funicle very minute, the second a little curved; club 6-jointed, a little thickened toward apex: the joints, except the last moniliform; the last conic, not twice as long as the preceding..

                                P. ACTINOMERIDIS.

3. Coxæ brown or rufo-piceous... 4

    Coxæ black.

        Vertex posteriorly transversely aciculated.

            Legs and antennæ black, tarsi pale.

                Pedicel as large as the first and second funicle joints together, the first small, rounded, the second wider than long, a little dilated towards one side, club 6-jointed, the joints oval, gradually increasing in size, loosely joined, hairy, the last fusiform, twice as long as the preceding ....,............................. P. STRIATICEPS.

Pedicel smaller than the second funicular joint, the first funicular as long as
thick, close to the second, the second longer than wide, and a little
thicker than any of the others, club 6-jointed, the joints, except the
last, not longer than wide, the last cone-shaped, twice as long as
the preceding, all with glittering white hairs....... P. ATRIPLICIS.

Pedicel as long as the second funicular joint, the first rounded, the second
curved, twice as long as thick, club 6-jointed, with white hairs, the
joints, except the last, oval, loosely joined, the last cone-shaped,
twice as long as the preceding......................... P. VITICOLA.

Trochanters, anterior knees and tibiæ at tips, all tarsi and the base of middle
and posterior tibiæ, honey-yellow.

Pedicel longer than the second funicular joint, which is dilated apically, the
first funicular joint small, club 6-jointed, the joints, except the first
and last, a little longer than thick, the first as wide as long, the last
conical, twice as long as the preceding.........P. HIEMALIS Forbes.

Legs rufo-piceous, anterior tibiæ and the tarsi honey-yellow.

Pedicel as long as the first and second funicular joints together, the first
small, the second a little longer than wide at apex, the third shorter,
club 5-jointed, the joints, except the last, quadrate...P. CYNIPICOLA.

Pedicel not as long as the first and second funicular joints united, the first
small, rounded, the second nearly twice as long as thick, club
6-jointed, hairy, the joints, except the last, oval, the last fusiform,
twice as long as the preceding......................... P. ALNICOLA.

4. Legs brown, middle and posterior femora dusky.

Pedicel longer than the second funicular joint, the first small, rounded, the
second curved, dilated at tip, club 6-jointed, the joints, except the
last, moniliform, loosely joined, the last fusiform, twice as long as
the preceding.................................... P. ASTERICOLA.

Pedicel as long as the first and second funicular joints united, the first very
minute, the second a little thicker than long, not dilated at tip, club
6-jointed, the joints, except the last, moniliform, the last thicker,
long-oval, not twice as long as the preceding........P. VERNONIÆ.

Legs dark rufous.

Pedicel as long as the second funicular joint, the first small, rounded, the
second thickened, slightly curved, club 6-jointed, the joints, except
the last, longer than thick, the last conic, one-third longer than the
preceding....................................P. EUURÆ.

Legs pale brownish-yellow, the posterior femora and tibiæ slightly dusky.

Antennæ, except the club, pale brownish-yellow; pedicel as long as the first
three funicular joints united, first and second small, the third dilated,
club 6-jointed oval-moniliform, the last joint twice as long as the
preceding ....P. FLORIDANUS.

Vertex not punctate or shagreened, shining................................. 6
5. Vertex finely closely punctate or shagreened, opaque.

Coxæ black.

Vertex angularly produced over the eyes.

Legs black, the tarsi rufous...............................P. ANGULATUS.

Vertex not angularly produced over the eyes, normal.

Legs dull rufous, posterior femora piceous, tarsi yellowish, sides of pronotum striated; joints of club, except the last, transverse-moniliform,
the last fusiform, twice as long as the penultimate; second funicular
joint very slightly dilated ....P. STRIATICOLLIS.

Sides of pronotum perfectly smooth, polished; joints of club, except the
last, transverse-moniliform, the last fusiform; second funicular joint
much dilated...............................P. LEVICOLLIS.

6. Head and thorax polished.

    Antennæ and legs black, pedicel about as long as the first and second funiclar joints united, pale at tip, the first small, the second a little curved, club 6-jointed, the joints, except the last, a little longer than thick, the last cone-shaped, one-third larger than the penultimate.

                                                            P. COLORADENSIS.

Antennæ and legs rufo-piceous.

    Pedicel not as long as the first and second funiclar joints united, the second large, dilated, curved, the joints of club longer than wide, the last twice as long as the penultimate....................P. FILICORNIS.

    Pedicel as long as the first and second funiclar joints united, the second swollen, the joints of the club twice as long as thick, the last not twice as long as the penultimate.................P. CALIFORNICUS.

Polygnotus salicicola, sp. nov.

♂ ♀. Length. 0.80 to 1.40ᵐᵐ. Polished black; head thrice as wide as thick antero posteriorly, the vertex posteriorly strongly transversely acienlated, face smooth. Antennæ 10-jointed, black, the flagellum not quite twice as long as the scape; pedicel as long as the first and second funiclar joints together but stouter, the tip pale; joint 1 of funicle very small, yellowish, not as long as thick; second scarcely twice as long as thick, third shorter, fourth no longer than the third but a little stouter, rounded basally; club joints, except the last which is longer than the preceding, as wide as long, truncate at tips and rounded basally. Thorax ovoid, the collar anteriorly slightly microscopically shagreened, mesonotal furrows most frequently entirely wanting, seldom slightly indicated posteriorly, the base of the parapsides, just in front of the scutellum, delicately punctate and sparsely pubescent. Scutellum transversely highly, convex, polished. Metapleura nearly bare, delicately sculptured. Tegulæ black. Wings hyaline, iridescent, pubescent. Legs black, the tibiæ and tarsi piceous or brown-black. Abdomen very slightly larger than the thorax, polished, the petiole fluted at base, the second segment with two slight foveolæ at base which are faintly acienlated.

The ♂ is usually the smaller, the abdomen being scarcely as long as the thorax, and broadly rounded behind, while the differences in the antennæ readily distinguish it. The flagellum is covered with a rather dense, short, white pubescence; the pedicel oval, not longer than the second funiclar joint; first funiclar joint moniliform; second thickened at tip, and longer than any of the club joints except the last; club joints, except the last, not longer than thick, the last pointed, fusiform, twice as long as the penultimate.

HABITAT.—Los Angeles, Cal.

Types in National Museum.

Bred by A. Koebele from a Cecidomyiid gall on the midrib of willow.

Polygnotus diplosidis, sp. nov.

♀. Length, 1ᵐᵐ. Polished, black: head as in *P. salicicola*, but the vertex posteriorly is not so strongly aciculated. Mandibles yellow. Antennæ 10-jointed, brown-black; pedicel longer and stouter than the two following joints together; first three funiclar joints about equal in thickness, the first and second small, about equal in length, the third very slightly longer; club 5-jointed, joint 1 not quite as thick as the following, 2, 3, and 4 about equal, a little longer than thick, cup-shaped, the last cone-shaped, 1½ times as long as the preceding. Thorax ovoid, not much longer than wide, without a trace of the furrows, polished. Scutellum transversely, highly convex, smooth, shining. Metapleura sericeous. Tegulæ black. Wings hyaline, pubescent. Legs brown-black, the base of tibiæ and tarsi paler. Abdomen as long as the thorax, polished, the petiole striated, pubescent beneath, the second segment with two striated foveolæ at base.

HABITAT.—New Brunswick, N. J.

Type in National Museum.

The specimens were reared February 12, 1891, by Prof. John B. Smith, from a Cecidomyiid, *Diplosis* sp. found on pine.

Polygnotus baccharicola Ashm.

(Pl. XIII, Fig. 2, ♀.)

Can. Ent. XIX, p. 132; Cress. Syn. Hym., p. 250.

♂ ♀. Length, 1 to 1.20ᵐᵐ. Black, shining; head broadly transverse, the vertex posteriorly transversely aciculated; face smooth, with some aciculations just above the insertion of the antennæ. Mandibles piceous. Antennæ 10-jointed, black to brown-black, the scape sometimes pale at base and apex, with the pedicel and first and second funiclar joints sometimes pale; pedicel in ♂ as long as the first and second funiclar joints together; joint 1 of funicle very minute, second as thick as long; third smaller; club 5-jointed, the first joint longer than wide; the following joints to the last a little wider than long; in ♀ the club is 4-jointed, the joints stouter than in the ♂. Thorax short, ovoid, smooth, and shining, without a trace of the mesonotal furrows.

Scutellum transversely, highly convex. Metapleura sericeous. Tegulæ piceous. Wings hyaline. Legs, including coxæ, rufo-piceous; trochanters, base and tips of tibiæ, and the tarsi honey-yellow. Abdomen longer than the thorax, polished; the petiole striated, the second segment with two striated foveolæ at base.

HABITAT.—Florida.

Types in Coll. National Museum and Coll. Ashmead.

Many specimens reared by myself at Jacksonville, Fla., from *Cecidomyia baccharicola* Ashm. MS.

Polygnotus solidaginis Ashm.

Can. Ent. xix, p. 131; Cress. Syn. Hym., p. 250.

♂ ♀. Length, 1 to 1.60ᵐᵐ. Black, polished; head very wide, the vertex posteriorly strongly transversely striated; face smooth, polished. Mandibles rufous. Antennæ 10-jointed, brown, the scape often black; pedicel in ♂ longer than the first two funiclar joints; joint 1 of funicle very small, yellowish; second much larger than the third, and thicker; club joints, except the first and last, hardly longer than thick, the first wider than long, the last cone-shaped and one-half longer than the preceding; club in ♀ 4-jointed, the last funiclar joint obconic. Thorax ovoid, smooth, usually without a trace of furrows posteriorly, and sparsely pubescent, especially near the scutellum. Scutellum high, convex. Metapleura sericeous. Tegulæ piceous or black. Wings hyaline, pubescent. Legs, including coxæ, variable, from a pale rufous to rufo-piceous, sometimes only the tarsi are pale; sometimes trochanters, bases and tips of tibiæ, and the tarsi honey-yellow, and sometimes the middle and posterior femora are black. Abdomen longer than the thorax, the petiole striated; the striæ from the basal foveolæ on the second segment extend to the middle of the segment.

HABITAT.—Florida and Missouri.

Types in Coll. Ashmead and National Museum.

Many specimens. First bred at Jacksonville, Fla., by myself from *Cecidomyia nebulosa* Ashm. MS. Dr. Riley has also reared it at Bushberg, Mo., from a Cecidomyiid gall on *Solidago*, September 21, 1876, while Miss Murtfeldt bred it from the same gall at Kirkwood, Mo., September 13, 1885.

Polygnotus pinicola, sp. nov.

♂ ♀. Length, .80 to 1ᵐᵐ. Black, shining; head wider than the thorax, smooth, highly polished, the vertex posteriorly and the occiput not acienlated. Antennæ 10-jointed, brown; first three funiclar joints small, slightly increasing in length; club 5-jointed, the joints, except the last, scarcely longer than wide; in ♂ the first funiclar joint is very small, the second slightly curved, dilated, and truncate at apex; the club 6-jointed; the joints, except the last, moniliform, slightly pedicellate; the last fusiform, much longer than the preceding. Thorax short, ovoid, shining, without a trace of the mesonotal furrows. Scutellum high, transversely convex. Metapleura nearly bare. Tegulæ piceous. Wings hyaline. Legs, including coxæ, dark rufous, the tarsi and sometimes the tip of anterior tibiæ, honey-yellow or whitish. Abdomen not longer than the thorax, smooth, the petiole and the second segment at base striated.

HABITAT.—Washington, D. C.

Types in National Museum.

Described from several specimens, reared May 14, 1879, from a Cecidomyiid, *Cecidomyia pini-inopis* O. S., found on pine needles.

Polygnotus striaticeps, sp. nov.

♂ ♀. Length, 0.60 to 1.20ᵐᵐ. Black, shining; head transverse, a little more than twice as wide as thick antero-posteriorly, and angularly produced over the eye; vertex impressed at the middle with strong transverse striæ occupying the whole occiput and extending forward as far as the front ocellus; face smooth. Mandibles black. Antennæ 10-jointed, black; pedicel about as long as the first and second funiclar joints united; joint 1 of funicle very small, 2 and 3 equal, very slightly longer than thick, 4 as long as thick at tip, narrowed toward base; club 4-jointed, the joints, except the last, a little longer than thick, the last slightly longer than the preceding, cone-shaped; in ♂ the second funiclar joint is curved, dilated and truncate at tip; the club 6-jointed, slightly thickened towards the tip, the joints oval-moniliform, hairy, slightly pedicellate, the last joint nearly twice as long as the preceding. Thorax ovoid, highly polished, the mesonotal furrows distinct for nearly half the length of the mesonotum posteriorly. Scutellum high, transversely, convex. Metanotum with 2 carinæ on the disk; metapleura bare, aciculated in the ♀, sericeous in ♂. Tegulæ black. Wings hyaline. Legs black, the anterior and middle tarsi white. Abdomen in ♀ as long as the head and thorax together, shorter in the ♂, the petiole striated, the second segment striated at base, the striæ extending along the sides to the middle of the segment.

HABITAT.—San Diego and Los Angeles County, Cal.

Types in National Museum.

Described from several specimens, reared in July, 1886, by A. Koebele, from a Cecidomyiid gall on an evergreen shrub (Bigeloria or Artemisia sp.) taken at Newhall, Los Angeles County, Cal.; also one female, labeled as having been reared from Aspidiotus on Bigeloria, by the same observer at San Diego, Cal.

It is scarcely necessary for me to state that this latter statement is erroneous, as we know positively no Platygasterid is parasitic on Coccids.

Polygnotus atriplicis, sp. nov.

♂ ♀. Length, 1 to 1.40ᵐᵐ. Black, polished; vertex posteriorly transversely aciculated, the face smooth. Mandibles piceous. Antennæ 10-jointed, black; pedicel as long as the first and second funiclar joints together, a little pale at tip; first funicular joint small, pale; second and third cylindric, the third shorter than the second; club 5-jointed, the first joint a little narrower than the following, contracted toward the base, the other joints to the last longer than thick, cup-shaped, outwardly at tips subserrate, the last cone-shaped, longer than the preceding. Thorax ovoid, polished, the mesonotal furrows indicated posteriorly for more than half the length of the mesonotum. Scutellum transversely high, slightly pubescent posteriorly. Metapleura sericeous. Tegulæ piceous. Wings hyaline, iridescent, sparsely pubescent. Legs black, the anterior tibiæ toward tips and all tarsi honey-yellow. Abdomen not quite as long as the head and thorax together, the first segment and the second at base striated.

In the male the abdomen is oblong-oval, broadly rounded behind, the antennæ black, with the pedicel more or less pale and shorter than the second funiclar joint; first funiclar joint small, rounded; second much dilated, truncate at tip, narrowed at base; club 6-jointed, bristly, the joints, except the last, oval-moniliform, the last cone-shaped, nearly twice as long as the penultimate joint.

HABITAT.—Los Angeles and San Bernardino County, Cal.

Types in National Museum.

Described from several specimens, reared by A. Koebele, during April and May, from a Cecidomyiid gall on *Atriplex canescens*, found in California.

Polygnotus artemisiæ, sp. nov.

♂ ♀. Length, 1 to 1.40ᵐᵐ. Black, polished; vertex posteriorly transversely aciculated. Mandibles piceous. Antennæ 10-jointed, the scape black, the flagellum dark-brown; pedicel as long as the first and second funiclar joints together; first funiclar joint very small; second thickened, longer than thick; third narrower and shorter than the second; club 5-jointed, the first joint shorter than the following and not quite so thick, rounded at base; the following joints to the last very slightly wider than long, rounded at base, truncate at tip, the last cone-shaped, longer than the preceding. In the male the pedicel is not longer than the second funicle joint, which is slightly curved and dilated and truncate at tip; club 6-jointed, hairy, the joints, except the last, moniliform, slightly pedicellate, the last cone-shaped, almost twice as long as the preceding. Thorax ovoid, shining, the mesonotal furrows distinct posteriorly. Scutellum polished, tranversely highly convex. Metapleura sericeous. Tegulæ black. Wings hyaline, pubescent. Legs black to brown-black, the trochanters, tips of tibiæ and tarsi honey-yellow. Abdomen as long as the head and thorax together, polished, the petiole striated, the second segment with 2 striated foveolæ at base.

HABITAT.—Lancaster, Los Angeles County, Cal.

Types in National Museum.

Described from several specimens, reared by A. Koebele, December 6, 1887, from a Cecidomyiid gall found on *Artemisia californica*.

Polygnotus virginiensis, sp. nov.

♀. Length, 1.2ᵐᵐ. Polished black. Head about 3 times as wide as long antero-posteriorly, the occiput transversely striated, the face smooth, shining, impunctured; legs and antennæ black, articulations of legs and tarsi brown. Funicle slender, the first joint small, the second a little larger than the third; club joints about twice as long as thick or nearly so. Thorax with distinct furrows posteriorly, the middle lobe projecting slightly upon the base of the high, convex scutellum, foveated on either side. Wings hyaline, pubescent. Abdomen conically pointed, longer than the head and thorax together, the petiole striated, segments 3, 4, and 5 with a transverse row of punctures.

♂. Length, 0.8 to 1ᵐᵐ. Agrees with the ♀, except the antennæ are quite different, the club being 5-jointed, dilated or thickened toward the tip, the joints, except the last, being transverse, the funiclar joints 1 and 2 subequal; the abdomen broadly ovate, not longer than the head and thorax together, its broadest part being wider than the thorax, the petiole and the base of second segment beneath with a tuft of woolly pubescence.

HABITAT.—District of Columbia and Arlington, Va.

Types in Coll. Ashmead.

Described from many specimens.

Polygnotus alnicola, sp. nov.

♀. Length, 1.20ᵐᵐ. Black, shining; head 3½ times as wide as long antero-posteriorly, the vertex posteriorly transversely aciculated, the face smooth, polished. Antennæ 10-jointed, dark brown; pedicel longer than the first and second funiclar joints together, pale brown; first funiclar joint very minute; second a little longer than thick, third not longer than thick; club 5-jointed, the first joint a little longer than wide, slightly narrowed toward base, the second, third, and fourth quadrate, the last ovate, longer than the preceding. Thorax ovoid, polished, the mesonotal furrows indicated only posteriorly and very faintly. Scutellum high, transversely, convex. Metapleura sericeous. Tegulæ black. Wings hyaline, with pale pubescence. Legs brown, the trochanters, tips of tibiæ and tarsi pale. Abdomen a little longer than the thorax, the petiole and the second segment at base striated.

HABITAT.—District of Columbia.

Types in National Museum.

Described from 4 specimens, reared July 31, 1886, from a Cecidomyiid gall in the flower bud of alder.

Polygnotus tumidus. sp. nov.

♀. Length, 1ᵐᵐ. Black, polished; head nearly 4 times as wide as long antero-posteriorly, the vertex transversely striated. Mandibles brown. Antennæ 10-jointed, brown-black; pedicel not quite as long as the first and second funiclar joints united; first funiclar joint small; second and third nearly equal; club 5-jointed, very wide; the second, third, and fourth joints wider than long and at tips toward one side a little serrated, at base rounded, last joint more slender and longer than the preceding. Thorax ovoid, highly polished, the mesonotal furrows only faintly indicated posteriorly. Scutellum high, convex, polished.

Metapleura bare, faintly striated. Tegulæ rufous. Wings hyaline. Legs brownish-piceous, the trochanters, tips of anterior tibiæ, and all tarsi paler. Abdomen polished, not longer than the head and thorax together, the petiole striated, the second segment, at base on either side, with two striated foveolæ, the striæ extending to the middle of the segment.

HABITAT.—Washington, D. C.

Type in National Museum.

Bred February 25, 1881, from *Cecidomyia symmetrica* O. S., a gall common on the leaves of various oaks.

Polygnotus laticeps, sp. nov.

♂ ♀. Length, 1.2 to 1.9ᵐᵐ. Polished black, impunctured; head very wide, fully 4 times as wide as long antero-posteriorly; the occiput transversely aciculated, the face flat and highly polished, the lateral ocelli about twice their width from the margin of the eye. Antennæ and legs black or brown-black, trochanters, base of tibiæ, and tarsi paler brown. Thorax rounded before, rather short, with distinct parapsidal furrows posteriorly, the middle lobe projecting a little upon the base of the scutellum, the scutellum highly convex, polished, deeply foveated along the base; metathorax short, the pleura faintly striated or pubescent. Abdomen broadly ovate, the apical, ventral, and dorsal segments with transverse rows of punctures. Wings hyaline. The antennæ in the ♂ terminate in a 5-jointed club, the joints, except the last, being as broad or a little broader than long; in the ♀ the joints are longer than wide.

HABITAT.—Jacksonville, Fla.

Types in Coll. Ashmead.

Polygnotus hiemalis Forbes.

Platygaster hiemalis Forbes, Psyche, Vol. 5, p. 39 (1888).

♂ ♀. Length, 0.80 to 1.40ᵐᵐ. Black, polished; head about two and a half times as wide as long antero-posteriorly, the vertex posteriorly only faintly aciculated, the face smooth, polished. Antennæ 10-jointed, brown-black, the flagellum twice as long as the scape; pedicel as long as and much stouter than the first two funicular joints; first funicular joint small, not longer than thick, yellowish basally; second larger and a little longer than the third; club 5-jointed, the joints, except the last, a little longer than wide, the last cone-shaped, one-half longer than the preceding. In the male the second funicular joint is thickened, curved, and as long as the pedicel, the latter whitish or yellowish at tip; the first funicular joint small, contracted at base; club 6-jointed, villose, the joints oblong, slightly pedicellate, the first, the shortest, narrowed basally, the last ovate, not quite twice as long as the penultimate. Thorax ovoid, polished, the mesonotal furrows delicate but distinct posteriorly, in the ♂ almost obliterated, the middle lobe projecting slightly upon the scutellum. Scutellum very high, transverse, convex. Metapleura subsericeous. Tegulæ rufo-piceous. Wings hyaline, pubescent. Legs dark brown to piceous, trochanters, tips of anterior femora and tibiæ, base of middle and posterior tibiæ, and all tarsi brownish-yellow or honey-yellow, sometimes the posterior femora black. Abdomen in the ♀ about as long as the head and thorax together, in ♂ shorter; in both sexes, the petiole and the foveolæ at base of the second segment striated.

HABITAT.—Western States.

Types in the Illinois State Laboratory of Natural History and National Museum.

This species seems to have been first reared by Dr. Riley, from specimens of the Hessian fly (*Cecidomyia destructor* Say), August 16, 1876, received from Blair, Nebr. It has, however, since been bred from the same fly, by various persons in the Western States. Prof. Forbes reared it in 1888 at Champaign, Ill.; Prof. Cook, of Agricultural College, Mich., in 1890; and Prof. Webster at Laporte, Ind., in 1889.

I know of no specimens reared in the Eastern States. Can the species be moving Eastward?

Polygnotus proximus, sp. nov.

♂ ♀. Length, 0.80 to 1.20ᵐᵐ. Black, polished; head thrice as wide as long antero-posteriorly, the vertex posteriorly transversely aciculated, the face polished. Mandibles piceous. Antennæ 10-jointed, brown-black, much slenderer than in *P. hiemalis*, the flagellum slightly more than twice the length of the scape: pedicel long, cylindric, as long as the first and second funiclar joints together: first funiclar joint longer than thick, whitish at base; second much longer and thicker; third shorter than the second and constricted at apex: club 5-jointed, slender, the joints nearly twice as long as thick, the last fusiform, longer than the preceding. In the male the pedicel is oval, not quite as long as the first and second funiclar joints together; first funiclar joint small, moniliform; second larger, dilated and truncate at tip: club 6-jointed, hairy, the joints, except the last, moniliform, not longer than thick, slightly pedicellate, the last cone-shaped, nearly twice as long as the preceding. Thorax ovoid, polished, the mesonotal furrows indicated only posteriorly, the middle lobe projecting a little on the base of the scutellum, the parapsides at base with a tuft of pubescence. Scutellum high, transverse, convex, with a little pubescence on either side posteriorly. Metapleura sericeous. Tegulæ black. Wings hyaline, iridescent, pubescent. Legs brownish-piceous, the anterior knees, tips of anterior tibiæ and all tarsi, honey-yellow. Abdomen in ♀ pointed at tip, slightly longer than the head and the thorax together, in ♂ oblong-oval, not longer than the thorax, the petiole striated, the second segment with two deep elongated foveolæ at base, the space between them showing some longitudinal striæ, otherwise the whole surface is highly polished, black.

HABITAT.—District of Columbia.

Types in National Museum.

Described from several specimens bred by Dr. Riley from the Cypress Cecidomyiid (*Cecidomyia c.-ananassa* Riley); no date of rearing is given.

Polygnotus viticola, sp. nov.

♂ ♀. Length, 1.20 to 1.40ᵐᵐ. Black, polished; head a little more than thrice as wide as long antero posteriorly, the vertex posteriorly transversely aciculated, the face smooth. Mandibles rufous. Antennæ 10-jointed, brown or brown-black, the flagellum not quite twice as long as the scape; pedicel hardly as long as the first and second funiclar joints together and not as thick at tip as the second funiclar joint; first funiclar joint minute, the second and third about of an equal length, but the second the thicker; club 5-jointed, the joints briefly pedicellate, the first four joints as wide as long, cup-shaped, the last cone-shaped, longer than the penultimate.

In the ♂ the flagellum is covered with short, white, stiff hairs; the second funiclar joint is larger and stouter than the pedicel; the club 6-jointed, the joints, except the last, oval-moniliform, pedicellated, the last cone-shaped, one and one-half times as long as the penultimate Thorax ovoid, polished, the mesonotal furrows very faintly indicated posteriorly, more distinct in the female, the basal point of the middle lobe projecting slightly upon the scutellum, the lateral lobes at base slightly pubescent. Scutellum transverse, high, convex, very sparsely pubescent. Metapleura sericeous. Tegulæ rufo-piceous. Wings hyaline, pubescent. Legs, including coxæ, rufo-piceous or brownish, the tarsi paler, or yellowish. Abdomen ovate, scarcely longer than the head and thorax together, pointed at tip, the petiole striated, the second segment with two long, striated foveolæ at base, one on either side, the third and fourth, with a transverse row of faint punctures, the fifth and following segments closely punctate; the third, fourth, and fifth ventral segments also show rows of punctures. In the ♂ the abdomen is shorter, oblong-oval, the third, fourth, fifth, and sixth segments with a single row of punctures at base, while the corresponding ventral segments show rows of very faint punctures.

HABITAT.—Washington, D. C.

Types in National Museum.

Described from several specimens, reared March 31, 1882, from a Cecidomyiid gall on the petiole of a grapevine, collected on the grounds of the Department of Agriculture.

Polygnotus utahensis, sp. nov.

♂ ♀. Length, 1ᵐᵐ. Black, polished; head thrice as wide as thick antero-posteriorly, the vertex in ♀ angularly produced over the eye, but not in the ♂, posteriorly not aciculated, or only faintly aciculated in ♀, perfectly smooth in ♂, the face smooth. Antennæ 10-jointed, black or brown-black; pedicel a little longer than the second funiclar joint; first funiclar joint very small, rounded; second and third about equal; club 5-jointed, the joints, except the last, longer than thick. In the ♂ the second funiclar joint is distinctly longer than the pedicel; club 6-jointed, the joints about twice as long as thick, pedicellate.

Thorax ovoid, polished, with the mesonotal grooves slightly indicated posteriorly and with some pubescence just in front of the scutellum. Scutellum in ♂ subconvex, pubescent with a slight median carina; in ♀ high, transversely, convex, polished. Metapleura in ♀ bare, finely sculptured; in ♂ sericeous. Tegulæ rufo-piceous. Wings hyaline, pubescent. Legs black or brown-black, tarsi piceous, in ♂ the knees, tips of anterior tibiæ, and all tarsi, honey yellow. Abdomen not longer than the thorax, polished, the petiole and foveolæ at base of second segment, striated.

HABITAT.—Pariah, Utah.

Types in National Museum.

Described from two specimens, 1 ♂, 1 ♀, received from A. L. Siler, and bred during July, 1881, from a Cecidomyiid gall on *Artemisia 3-dentata*.

Polygnotus cynipicola, sp. nov.

♂ ♀. Length, 0.60 to 0.80ᵐᵐ. Black, polished; head 3½ times as wide as thick antero-posteriorly, the vertex posteriorly very faintly aciculated, the face smooth, polished. Antennæ 10-jointed, brown-black; pedicel as long as the second funiclar joint, piceous; first funiclar joint very minute, yellowish, third, longer than the second; club 5-jointed, broad, the joints, except the last, slightly wider than long; in ♂ the club joints are longer than wide, covered with white bristles, the last pointed, fusiform. Thorax ovoid, polished, the mesonotal furrows faintly indicated posteriorly, the basal tips of the middle lobe slightly elevated. Scutellum high, transversely convex, sparsely pubescent posteriorly. Metapleura bare, aciculated, in ♂ slightly sericeous. Tegulæ black. Wings hyaline, pubescent. Legs black or brown-black, the tip of anterior tibiæ and all tarsi, honey-yellow. Abdomen not longer than the thorax, polished, the petiole and the foveolæ at base of second segment striated, the third and following segments with transverse rows of punctures.

HABITAT.—Arlington, Va.

Types in National Museum.

Described from several specimens, bred July 3, 1883, from a Cynipid gall, *Neuroterus batatus* Fitch.

Polygnotus eurotiæ, sp. nov.

♀. Length, 1 to 1.20ᵐᵐ. Black, polished; head thrice as wide as thick antero-posteriorly, the vertex perfectly smooth. Antennæ 10-jointed, black; pedicel slightly longer and thicker than the first and second funiclar joints together, the three funiclar joints cylindrical, of a uniform thickness, the first, the shortest, the second, the longest, but only a little longer than the third; club 5-jointed, slender, the joints all distinctly longer than thick, about 1½ times as long as thick, the last long, conical. Thorax ovoid, polished, the mesonotal grooves distinct posteriorly for half the length of the mesonotum, the base of the lateral lobes with some short striæ. Scutellum high, transversely, convex, slightly pubescent posteriorly. Metapleura sericeous. Tegulæ rufo-piceous. Wings hyaline, pubescent. Legs brown-black, anterior knees, tips of tibiæ and all tarsi honey-yellow. Abdomen as long as the thorax, polished, the petiole and foveolæ at base of the second segment striated.

HABITAT.—San Bernadino County, Cal.

Types in National Museum.

Described from two specimens bred April 17, 1887, by A. Koebele, from a Cecidomyiid gall on *Eurotia canata*.

Polygnotus rubi, sp. nov.

♀. Length, 1ᵐᵐ. Black, polished; head thrice as wide as thick antero posteriorly, the vertex not aciculated. Antennæ 10-jointed, brown, the scape and pedicel yellowish; pedicel as long as the first and second funiclar joints together; first funiclar joint very minute; second and third about equal; club 5-jointed, slender, the joints, except the last, not longer than thick, cup-shaped. Thorax ovoid, polished, the mesonotal furrows only slightly indicated posteriorly, the basal tip of the middle lobe slightly elevated, the parapsides at base with a small tuft of pubescence. Scutellum high, transversely convex, posteriorly pubescent. Metapleura sericeous. Tegulæ rufo-piceous. Wings hyaline, pubescent. Legs black, or brown-black, the tips of anterior femora, their tibiæ, and all tarsi, honey-yellow. Abdomen polished, about as long as the head and thorax together, pointed at tip, the petiole and the foveolæ at base of second segment striated.

HABITAT.—Arlington, Va.

Types in National Museum.

Described from two specimens, bred March 30, 1886, from *Cecidomyia farinosa* O. S. found on blackberry.

Prof. John B. Smith has also reared the same insect from a gall on blackberry at New Brunswick, N. J.

Polygnotus asynaptæ, sp. nov.

♂ ♀. Length, 0.80 to 1.20ᵐᵐ. Black, polished; head in ♀ about 3½ times as wide as thick antero posteriorly, but in the ♂ scarcely thrice as wide as thick, the vertex posteriorly not or very faintly aciculated. Antennæ 10-jointed, black, the flagellum slender, twice as long as the scape; pedicel about as long and thick as the second club joint; first funiclar joint very minute, second and third, longer, equal; club 5-jointed, the first joint a little shorter and narrower than the second; second, third and fourth equal in length, very slightly longer than thick, the last cone-shaped, longer than the preceding.   In the ♂ the pedicel is not longer than the second funiclar joint; the first funiclar joint is annular, fully as wide as the pedicel, the second funiclar joint is a little shorter than the first club joint; club 6-jointed, cylindrical, the last joint enlarged, fusiform, all the other joints pedicellated, about equal in length, cylindric, about twice as long as thick. Thorax ovoid, highly polished, the mesonotal furrows distinct posteriorly for half the length of the mesonotum, the parapsides at base with slight tufts of glittering white hairs. Scutellum high, convex, sparsely pubescent posteriorly. Metapleura bare, not perfectly smooth. Metanotum very short, with a deep central groove. Tegulæ black. Wings hyaline, covered with a short pubescence.

Legs black or piceous-black, the tip of anterior tibiæ and the anterior and middle tarsi brownish. Abdomen polished, pointed in the ♀ and scarcely longer than the head and thorax united; in the ♂ oblong-oval; in both sexes the petiole is bare, fluted, the second segment with two long, striated foveolæ at base.

HABITAT.—Maywood, Ill.

Types in National Museum.

Described from several specimens, reared January 22 and February 1 and 3, 1890, by Prof. O. S. Westcott, from a willow Cecidomyiid (*Asynapta*, sp.).

Polygnotus huachucæ, sp. nov.

♂ ♀. Length, 0.80 to 1.40ᵐᵐ. Black, polished; head about thrice as wide as thick antero posteriorly, the vertex posteriorly showing faint traces of aciculations. Mandibles piceous. Antennæ 10-jointed, black, the flagellum one and a half times as long as the scape; pedicel a little longer than the first and second funiclar joints together; first funiclar joint very minute, pale; second and third longer, the second the larger; club 5-jointed, the first joint smaller than the three following, which are a little wider than long, rounded at base, truncate at tip, the last cone-shaped, a little longer than the preceding. In the ♂ the flagellum is not quite twice as long as the scape, the pedicel equal in length with the second funiclar joint, first funiclar joint annular, second not quite as long as thick, obliquely truncate at tip, club 6-jointed, covered with a short, grayish pubescence, the joints submoniliform, the last cone-shaped, not quite twice as long as the preceding. Thorax ovoid, polished, the mesonotal furrows indicated only posteriorly, the lateral lobes posteriorly bare, without tufts of hairs.

Scutellum smooth, convex. Metapleura subsericeous, aciculated. Tegulæ black. Wings hyaline, strongly iridescent, sparsely pubescent. Legs piceous-black, the anterior tibiæ and all tarsi whitish or honey yellow. Abdomen as usual in the sexes, in the ♀ pointed and as long as the head and thorax together, the petiole and the foveolæ at base of the second segment striated.

HABITAT.—Fort Huachuca, Ariz.

Types in National Museum.

Described from several specimens, reared June 5, 1882, from a Cecidomyiid stem gall on sunflower, and from a Cecidomyiid pod-like gall on an unknown plant, sent to the Department by H. K. Morrison.

Polygnotus actinomeridis, sp. nov.

♂ ♀. Length, 0.60 to 1.20^{mm}. Black, polished; head 2½ times as wide as thick antero-posteriorly, the vertex posteriorly very faintly aciculated, the face smooth, with transverse aciculations above the insertion of the antennae. Mandibles piceous. Antennae 10-jointed, dark-brown, the scape paler toward base; pedicel as long as the second and third funiclar joints together; first funiclar joint very small; second stouter and longer than the third; third a little longer than the first; club 5-jointed, the joints longer than thick. In the ♂, the second funiclar joint is dilated and truncate at tip, the club 6-jointed, the joints oval-moniliform, the first the smallest, the last enlarged, fusiform, stouter, and twice as long as the preceding. Thorax ovoid, polished, the mesonotal furrows distinct posteriorly; no tufts of pubescence at base of parapsides. Scutellum high, convex, polished, with a slight silky pubescence. Metapleura subsericeous, aciculated. Tegulæ rufo-piceous. Wings hyaline, pubescent. Legs dark-rufous, the anterior tibiae, and all tarsi honey-yellow. Abdomen about as long as the head and thorax together, polished, the petiole and the foveolæ at base of second segment striated, the ventral segments 3, 4, 5, and 6 with a transverse row of fine punctures.

HABITAT.—Washington, D. C.

Type in National Museum.

Described from 4 specimens reared April 23, 1884, from a Cecidomyiid gall on *Actinomeris squarrosa*.

Polygnotus vernoniæ, sp. nov.

♂ ♀. Length 0.60 to 0.80^{mm}. Black, polished; head nearly thrice as wide as thick antero-posteriorly; the vertex not aciculated, the face smooth. Mandibles brown. Antennae 10-jointed, brown or dark brown; pedicel longer than the first and second funiclar joints together; first funiclar joint very minute, pale, second and third about as long as thick, equal; club 5-jointed, the joints very slightly longer than thick, the middle joints wider than the basal and apical joints. In the ♂ the second funiclar joint is dilated, the club 6-jointed, the last enlarged, oblong, the other joints moniliform. Thorax ovoid, polished, the mesonotal furrows only slightly indicated posteriorly, the basal tip of the middle lobe slightly elevated, upon the base of the scutellum; the parapsides with a slight tuft of pubescence at base. Scutellum high, transverse, convex. Metapleura covered with a white, glittering pubescence. Legs brownish-piceous; front legs, tips of middle and posterior tibiæ, and all tarsi honey-yellow. Abdomen pointed in ♀, a little longer than the head and thorax together; shorter and oblong-oval in ♂, the petiole and foveolæ at base of second segment striated; the third and following segments, in the ♂, with a row of punctures at base.

HABITAT.—Arlington, Va.

Types in National Museum.

Described from several specimens, reared June 15 and 17, 1886, from a Trypetid gall on *Vernonia noveboracensis*.

Polygnotus floridanus, sp. nov.

♂. Length, 1.4ᵐᵐ. Black; antennæ, except the club, and the legs, except the posterior pair, yellow; club black; posterior coxæ, femora, and tibiæ dusky; wings hyaline, almost devoid of pubescence, the tegulæ black.

The head is wide between the eyes and highly polished; mandibles piceous; palpi pale. Antennæ 10-jointed, the pedicel long, as long as the 3 funiclar joints united, the third funiclar joint the widest, club joints black, moniliform, the last fusiform, twice as long as the preceding.

Mesonotal furrows indicated posteriorly; scutellum cushion-shaped, the metapleura covered with a white pubescence.

Abdomen pear-shaped, not as long as the thorax, smooth, polished, except the petiole, which is wider than long, striated, and some striæ at base of second segment.

HABITAT.—Jacksonville, Fla.

Type in Coll. Ashmead.

Distinguished by the color of the legs and the peculiarity of its antennæ, the pedicel being unusually long for a species in this genus.

Polygnotus euuræ. sp. nov.

♂. Length, 1.60ᵐᵐ. Black, polished; head two and a half times as wide as thick antero-posteriorly, the vertex posteriorly not or very faintly aciculated. Mandibles rufous. Antennæ 10-jointed, brown-black; pedicel one and a half times as long as thick; first funiclar joint small, rounded; second longer than thick, slightly curved; club 6-jointed, the joints, except the last, oval-moniliform, slightly pedicellate, the last cone-shaped, a little stouter and about twice as long as the preceding. Thorax ovoid, polished, the mesonotal furrows indicated posteriorly. Scutellum transverse, convex, sparsely covered with a fine silky pubescence. Metapleura sericeous. Tegulæ rufous. Wings hyaline, pubescent. Legs, including coxæ, dark rufous. Abdomen oblong-oval, about as long as the thorax, the petiole and the foveolæ at base of second segment striated.

HABITAT.—St. Louis, Mo.

Types in National Museum.

Described from 2 ♂ specimens, bred February 24, 1872, by Dr. C. V. Riley, from the Tenthredinid gall *Euura s.-nodus* Walsh.

The species is unquestionably parasitic on the inquilinous Cecidomyious flies known to inhabit this gall, and not on the Tenthredinid.

Polygnotus angulatus, sp. nov.

♂. Length, 1.2ᵐᵐ. Black, shining: head wide, shagreened, the vertex angularly produced over the eyes, transversely aciculated; lateral ocelli far away from the eye. Antennæ 10-jointed, black, the pedicel as long as the first and second funiclar joints united, first funiclar joint minute, the second dilated toward apex, the apex truncate: club 6-jointed, the first joint slightly the smallest, the others, except the last, only a little longer than thick.

Mesonotum with two distinct furrows; scutellum transversely elevated, convex; metathorax very short, the pleura pubescent. Wings hyaline, the tegulæ black. Legs black, trochanters, anterior tibiæ toward apex, and all tarsi pale brown, the tarsi not longer than their tibiæ, the basal joint as long as the three following joints united. Abdomen oblong-oval, not longer than the thorax, or very slightly longer, smooth, polished, the petiole transverse, striated, the second segment with striate foveolæ at base.

HABITAT.—Jacksonville, Fla.

Types in Coll. Ashmead.

Polygnotus striaticollis, sp. nov.

♂. Length, 1.4ᵐᵐ. Black, shining; the head transverse, shagreened; the lateral ocelli away from the eye margin; thorax with two distinct furrows; legs rufo-piceous, articulations, base of tibiæ and tarsi yellowish or pale brown; wings hyaline, tegulæ blackish. The antennæ are 10-jointed, brown; the pedicel much longer than the first and second funiclar joints, which are small, the second being wider at apex than long; club 6-jointed, pilose, the joints loosely joined and all, except the last, transverse-moniliform. Pronotum at sides distintly striated, the mesopleura polished with a femoral impression, the scutellum highly convex, the metathorax sparsely pubescent, the pleura bounded by a carina above. Abdomen ovate, as long as the thorax, the petiole and the second segment at base striated.

HABITAT.—Jacksonville, Fla.

Type in Coll. Ashmead.

Polygnotus lævicollis, sp. nov.

♂. Length, 1ᵐᵐ. Black, shining; the head transverse with the lateral ocelli twice their width from the eye margin; antennæ brown, the scape toward apex and the pedicel at tip yellowish; legs rufous; wings hyaline; mesonotal furrows distinct. The antennæ are 10-jointed, the scape clavate, as long as the flagellum; pedicel obconic, as long as the first and second joints united, the latter closely joined, swollen or dilated; club 6-jointed, all the joints, except the last, transverse, the last conical, a little more than twice the length of the penultimate. Pronotum perfectly smooth, highly polished, the metapleura with a fine silvery pubescence. Abdomen oblong oval, polished, the petiole short, striated.

HABITAT.—District of Columbia.

Type in Coll. Ashmead.

Taken by Mr. E. A. Schwarz.

Polygnotus astericola, sp. nov.

♂ ♀. Length, 0.80 to 1ᵐᵐ. Black, polished; head thrice as wide as thick antero-posteriorly, the vertex not or very faintly aciculated, the face smooth, polished. Mandibles rufo-piceous. Antennæ 10-jointed, dark brown; pedicel twice as long as thick, first funicular joint minute, pale, a little longer than thick; second larger, very little longer than the third; club 5-jointed, rather slender, the joints about as wide as long. In the ♂ the second funicle joint is as long as the pedicel, a little curved, dilated and truncate at tip, the club 6-jointed, the joints oval-moniliform, slightly pedicellate, the last cone-shaped, a little stouter and twice as long as the preceding. Thorax ovoid, polished, the mesonotal furrows indicated posteriorly, the basal tip of the middle lobe very slightly elevated, the parapsides with a tuft of pubescence at base. Metapleura sericeous. Tegulæ black. Wings hyaline, pubescent. Legs, including coxæ, rufo-piceous, the trochanters, knees, tips of tibiæ, (the anterior tibiæ almost entirely) and all tarsi, honey-yellow. Abdomen not longer than the head and thorax together, oblong-oval, broader than, and not so pointed as in *P. cernoniæ*, the basal segment and the foveolæ on the second striated; the ♂ abdomen shorter and without row of punctures on the ventral segments.

HABITAT.—Holderness, N. H.

Types in National Museum.

Described from several specimens, reared May 24, 1884, from a Cecidomyiid gall on Aster, collected by A. Koebele.

Polygnotus coloradensis, sp. nov.

♂ ♀. Length, 1ᵐᵐ. Black, polished; head thrice as wide as thick antero-posteriorly, the vertex very faintly, microscopically shagreened, the face smooth, highly polished, the lateral ocelli as far from the margin of the eye as to the front ocellus. Mandibles piceous-black. An tennæ 10-jointed, black, the pedicel stouter than any of the club joints and about 2½ times as long as thick; first and second funiclar joints small, the second slightly the longer; the third and first joints of club about equal, longer than the second, the club very gradually thickened toward the tip, the joints, except the last, 1½ times as long as thick, the last joint stouter and longer than the penultimate, ovate. In the ♂ the pedicel is as long as the first and second funiclar joints together, pale at tip, the first funiclar joint very minute, the club 6-jointed, cylindrical, the joints nearly of an equal length, about twice as long as thick, the last conic, one-half longer than the preceding. Thorax ovoid, polished, the mesonotal furrows complete, distinct; the middle lobe posteriorly extending slightly upon the base of the scutellum; no tufts at the base of the lateral lobes. Scutellum transversely, convex, shining, very slightly pubescent. Metapleura covered with a whitish pubescence. Legs black or brown-black, the tarsi paler; sometimes the tip of anterior tibiæ, knees, and all tarsi, honey-yellow. Abdomen as long as the head and thorax together, in the ♀ pointed at tip, in the ♂ rounded, the petiole and the rather deep foveolæ at base of the second segment striated.

HABITAT.—Fort Garland, Colo.

Types in National Museum.

Described from several specimens reared June 25, 1883, from a Cecidomyiid gall on sage bush, collected by L. Bruner.

Polygnotus filicornis, sp. nov.

♂ ♀. Length, 1 to 1.5ᵐᵐ. Polished, black, impunctured; head transverse, about 3 times as wide as long antero-posteriorly, the occiput faintly alutaceously sculptured, the lateral ocelli a little more than their width from the margin of the eye. Antennæ 10-jointed, very long and slender, subclavate, reaching beyond the middle of the abdomen; pedicel slender, nearly as long as the first and second funiclar joints together; funiclar joints slender, and merging so gradually into the club joints that the club can scarcely be separated, the last four joints about 2½ times as long as thick. Thorax polished, with two distinct furrows, mesopleura deeply impressed at the middle; scutellum highly convex; metathorax sparsely pubescent. Wings hyaline. Legs black, tips of anterior tibiæ and the tarsi pale brownish or fuscous. Abdomen pointed-ovate, smooth, shining, striated at base, about as long or a little longer than the head and thorax together.

HABITAT.—District of Columbia.

Types in Coll. Ashmead.

Polygnotus californicus, sp. nov.

♂ ♀. Length, 1.2 to 1.5ᵐᵐ. Very close to *P. coloradensis*, but with the following differences: The vertex posteriorly shows faint traces of aciculations; in ♀ the first three funiclar joints are very nearly of an equal length, the third, slightly the thickest; in the ♂ the second funiclar joint is stout and slightly curved, the club joints being not more than 1½ times as long as thick; the three middle joints outwardly subserrate, the last conically pointed, much longer than the preceding; the legs are black, the tip of anterior tibiæ and tarsi fuscous, while the wings are slightly smoky.

HABITAT.—San Francisco and Alameda, Cal.

Types in National Museum.

Described from many specimens, reared by A. Koebele, January 10 and 23, 1883, and July 16 and December 17 and 19, 1885, from a Cecidomyiid gall found on *Baccharis pilularis*.

PLATYGASTER Latreille.

Gen. Crus. et Ins., IV, p. 31 (1809); Förster Hym. Stud., II, p. 108 (1856).

(Type *P. ruficornis* Latr.)

Head transverse, rarely subquadrate, the vertex somewhat narrowed, convex, the lateral ocelli usually far away from the eye margin but still nearer to it than to the front ocellus.

Antennæ 10-jointed, the scape subclavate, the flagellum ending in a 6-jointed, filiform club, the second joint in the male a little swollen and curved.

Thorax ovate, usually more elongated than in *Polygnotus*, the collar usually distinct, the mesonotal furrows distinct, the scutellum convex, unarmed, the metathorax short, carinated.

Wings veinless, pubescent.

Abdomen oblong-oval or ovate, the petiole usually distinct, at least as long as wide, or a little longer, striated, the second segment very long, striate at base.

Legs long, clavate; the tibiæ slender, the tarsi very long, the hind tarsi being much longer than their tibiæ.

The genera *Platygaster* and *Isocybus* are exceedingly closely allied, but the head in *Isocybus* is more quadrate and the lateral ocelli are nearer to the front ocellus than to the margin of the eye.

I am not fully satisfied that they should be kept separate, although, so far, the position of the lateral ocelli seems sufficient to separate them.

The head in *Platygaster Herrickii* Packard and the position of the ocelli is, however, very similar to *Isocybus*.

TABLE OF SPECIES.

FEMALES.

Coxæ black.
  Legs yellow or golden yellow, antennæ yellow, the club fuscous.
    Vertex rugose, with raised lines in the middle; face closely punctulate, with some transverse lines on lower part.
      Pedicel very long; first funiclar joint a little longer than wide, the second very long, twice as long as the first; club 6-jointed, the joints, except the last, transverse..................................*P. CARYÆ* Ashm.

Legs piceous or rufo-piceous, anterior tibiæ and base of middle and posterior tibiæ
     pale or honey-yellow.
  Head including face, finely, closely punctate.
    Pedicel about as long as the first two funiclar joints together, the first funiclar
       joint subtriangular, small, the second much thicker and about as
       long as the longest club joint..................P. HERRICKII Pack.
  Head finely, closely punctate, the face with a large polished space in the middle,
    transversely striated on lower part.
    Pedicel not as long as the first two funiclar joints together, the first funiclar
       joint a little longer than thick, narrowed, the second thicker and
       slightly shorter than any of the following..P. OBSCURIPENNIS Ashm.

MALES.

Coxæ black.
  Legs yellow.
    Head finely rugoso-punctate.
      Face minutely punctate, with no grooved line; antennæ brownish-yellow,
         the second funiclar joint as long as the long pedicel; club joints not
         longer than thick...............................P. CARYÆ Ashm.
      Face highly polished, with a central grooved line; antennæ brownish-yellow,
         the pedicel as long as the first and second funiclar joints together,
         the first very small, second a little thickened and curved; club joints
         oval-moniliform, the last fusiform, nearly twice as long as the pre-
         ceding...................................P. FLORIDENSIS Ashm.
  Legs piceous or rufo-piceous; antennæ black or brown-black.
    Head closely punctured or shagreened.
      Face punctate, no transverse striæ above antennæ; club joints twice as long
         as thick, subpedicellate.......................P. HERRICKII Pack.
      Face highly polished, with transverse striæ above antennæ.
        Second funiclar joint not swollen; club joints broader than long; wings
          dusky...................................P. OBSCURIPENNIS Ashm.
        Second funiclar swollen; wings hyaline.
         Club joints about twice as long as thick............P. APHIDIS, sp. nov.
         Club joints thrice as long as thick...................P. GRACILIS Ashm.
      Face transversely aciculated.
        Mesonotum and all pleura longitudinally striated..P. ACICULATUS, sp. nov.

Platygaster caryæ, sp. nov.

♂ ♀. Length, 1.50 to 2ᵐᵐ. Black, shining; vertex finely rugose,
the face finely, closely punctate, with some transverse aciculations just
above the insertion of the antennæ. Mandibles pale brown. Antennæ
10-jointed, yellow; pedicel long, as long as the second funiclar joint,
first elongated, a little shorter than the second; club 6-jointed, fuscous,
the joints, except the last, wider than long, truncate at tip, rounded
at base. Thorax finely punctulate, with two distinct furrows, the middle lobe posteriorly with a slight central line. Scutellum transverse,
convex. Metapleura and hind coxæ beneath sericeous. Tegulæ piceous. Wings hyaline. Legs bright yellow or brownish-yellow, the
coxæ black, the posterior femora and sometimes the tibiæ dusky. Abdomen oblong-oval, not longer than the thorax, the first segment and
the second, at base, striated, sparsely pubescent.

In the ♂ the antennæ are always pale, the club very slightly darker; the second funiclar joint elongated, somewhat thickened and truncate at tip, the first only a little longer than thick; the club-joints close, not loosely joined, a little longer than thick.

HABITAT.—Jacksonville, Fla.; District of Columbia, and St. Louis, Mo.

Types in Coll. Ashmead and National Museum.

The species is a common parasite of Cecidomyiid galls on hickory trees.

Platygaster floridensis Ashm.

(Pl. xiii, Fig. 3, ♀.)

Can. Ent., xix, p. 132, ♀.; Cr. Syn. Hym., p. 250.

♂ ♀. Length, 1 to 1.5ᵐᵐ. Polished, black; head transverse, as wide as the thorax, the vertex posteriorly aciculated, the face smooth, highly polished; mandibles yellowish. Antennæ 10-jointed; in ♂ pale brownish-yellow, the pedicel as long as the second funiclar joint, the first very small, closely joined to the second, the second swollen, slightly curved, truncate at tip, club 6-jointed, the joints, except the last, submoniliform, the last cone-shaped, twice as long as the preceding. Thorax ovoid, smooth, the parapsidal furrows only slightly indicated; scutellum convex, slightly pubescent; metapleura sericeous; tegulæ piceous. Wings hyaline, pubescent. Legs pale brownish-yellow, the hind coxæ piceous or fuscous. Abdomen oval, not longer than the thorax, the petiole roughened, pubescent. In the ♀ the abdomen is a little longer, more pointed; the antennæ, except the club, yellow, club dark brown, 6-jointed, moniliform, the last joint fusiform; the pedicel is about as long as the first funiclar joint, but thicker; funiclar joints cylindrical, the first more than twice as long as thick, the second very much lengthened; legs yellowish, the hind coxæ and femora and tibiæ toward tips fuscous or blackish.

HABITAT.—Jacksonville, Fla.

Types in Coll. Ashmead.

Platygaster Herrickii Pack.

Third Rep. U. S. Ent. Comm., p. 220; Riley, Proc. U. S. Nat. Mus., viii, p. 420, Pl. xxiii., Fig. 6, ♂; Ashm., Can. Ent., xix, p. 132; Cress. Syn. Hym., p. 250.

Incurhynchus incurus Prov. Add. p. 176, ♂.

♂ ♀. Length 1.50 to 1.80ᵐᵐ. Black, shining, finely punctate, or microscopically shagreened; head transverse, punctate, the face more finely punctate; mandibles rufous. Antennæ 10-jointed, black; pedicel as long as the first two funiclar joints together; the first funiclar joint subtriangular, small, the second thicker and as long as the longest club joint; club 6-jointed, the joints, except the last, very nearly equal in length, the first slightly the narrowest, less than twice as long as thick, and truncate at tip, the last cone-shaped, narrower and slightly longer than the preceding. In ♂ the antennæ are slightly longer, cylindrical, with a short whitish pubescence; the pedicel is longer than the second funiclar joint; the first very minute, closely joined to the second; the second swollen, nearly as broad as long; the club 6-jointed, the first joint the shortest, the last the longest, cone-shaped, the intermediate joints twice as long as thick, subpedicellate.   Thorax ovate, a little more than twice as long as broad; parapsidal furrows deep, distinct; scutellum rounded, convex; the metathorax short, emarginate behind.   Wings hyaline, veinless, although sometimes the base of the submarginal vein is quite distinctly visible as a yellowish streak.   Legs piceous-black or rufo-piceous; sometimes the base of the tibiæ and the base of the tarsi yellowish; tarsi most frequently fuscous.

HABITAT.—Western and Northern States.

Types in National Museum.   A common parasite of the Hessian fly (*Cecidomyia destructor* Say).

From Mr. W. H. Harrington, I received a ♂ specimen of this species labeled "*Aneurhynchus aneurus* Prov. Type".

Platygaster obscuripennis, sp. nov.

♂ ♀.   Length 1.40 to 1.80^{mm}.   Very closely allied to *P. Herrickii*, but a little more slender, the face with a large polished space in the middle. Mandibles pale.   Antennæ slender, the flagellum dark brown; pedicel not quite as long as the first two funiclar joints together, but longer than the second joint alone, the tip yellowish; the first funiclar joint small, but longer than thick; second twice as long as thick; club 6-jointed, the first narrower than the others, the joints beyond to the last about equal, longer than thick, the last conic, longer than the preceding.   Sides of the parapsides smooth with a distinct grooved line parallel with the pronotal suture.   Wings dusky, or subhyaline.   Abdomen not longer than the thorax, the foveolæ at base of second segment and the first segment both above and beneath covered with a silky pubescence.   The ♂ differs decidedly from *P. Herrickii* in having the first five joints of the club wider than long, the last cone-shaped, about twice as long as the preceding, while the second funiclar joint is but slightly swollen and slightly curved.

HABITAT.—Ottawa, Canada.

Types in Coll. Ashmead.

Described from specimens received from Mr. W. H. Harrington.

Platygaster aphidis sp. nov.

♂.   Length, 1.6^{mm}.   Black, shining; the head posteriorly almost smooth, not distinctly punctate; face polished.   Antennæ 10-jointed, black; pedicel as long as the first and second funiclar joints together, the first small, subtriangular, closely united to the second, the second somewhat swollen and slightly twisted; club 6-jointed, the joints about twice as long as thick, subpedicellate. Thorax long-ovoid, smooth, shining, the furrows distinct; scutellum smooth, shining, convexly high; metapleura bare; tegulæ black. Wings hyaline, nearly bare or but slightly pubescent. Legs brown-black, the tarsi paler, all the coxæ distinctly black. Abdomen oblong-oval, nearly twice as wide as the thorax, the petiole rugose, subpubescent.

HABITAT.—Richfield Springs, N. Y.

Type in National Museum.

Described from 1 ♂ specimen, reared February 9, 1887, from an Aphis on *Chenopodium album*. The broad abdomen and the length of the club joints readily distinguish the species. The Aphis from which this species was bred was undoubtedly infested with Cecidomyious parasites, upon which the Platygasterid is a secondary parasite.

Platygaster gracilis Ashm.

Can. Ent., XIX, p. 132; Cress. Syn. Hym., p. 250.

♂. Length, 1.5ᵐᵐ. Black, shining; the head posteriorly microscopically punctulate; face smooth, highly polished, with a slight central impressed line from the middle ocellus. Antennæ 10-jointed, black; pedicel long, yellowish at extreme tip; first funiclar joint small, second somewhat swollen, slightly curved and shorter than the first club joint; club 6-jointed, cylindrical, the joints twice as long as thick. Thorax long-ovoid, smooth, shining, the furrows very distinct; metapleura sericeous; tegulæ black. Wings long, hyaline, pubescent. Legs rust-brown, the anterior pair more yellowish, the middle and posterior femora and tibiæ dusky toward tips; all coxæ black. Abdomen oblong-oval, narrowed toward base, as long as the thorax, the petiole more or less pubescent.

HABITAT.—Jacksonville, Fla.

Type in Coll. Ashmead.

Platygaster aciculatus, sp. nov.

♂. Length, 1.4ᵐᵐ. Black, shining; vertex, occiput, and cheeks finely shagreened; face transversely aciculated; sides of prothorax, mesonotum, mesopleura beneath the wings and the metapleura all distinctly longitudinally aciculated; the parapsides, scutellum, and the lower portion of mesopleura smooth, polished. Antennæ 10-jointed, rufo-piceous; the flagellum subclavate; pedicel longer and stouter than the first and second funiclar joints together, the latter scarcely longer than thick, the following joints to the last transverse, the last short, conic. Parapsidal furrows deep, distinct. Scutellum elevated, cushion-shaped. Legs rufo-piceous; tips of anterior femora and their tibiæ and the articulations of the middle legs yellowish. Wings clear hyaline, entirely devoid of pubescence. Abdomen oval, smooth, polished, the petiole not longer than thick, striated and pubescent, the second segment with some striæ at base.

HABITAT.—Pennsylvania.

Type in Berlin Museum.

Described from a single specimen labeled "Penn., Zimmermann."

ISOCYBUS Förster.

Hym. Stud., II, p. 108 (1856).

(Type, *I. grandis* Nees.)

Head quadrate or subquadrate, the vertex broad, the cheeks full, the occiput slightly emarginate, not or very delicately margined; ocelli 3, triangularly arranged, the lateral being placed far away from the margin of the eye; eyes ovate.

Antennæ inserted at the clypeus, 10-jointed, the scape subclavate, slightly bent; in the ♀ the flagellum is subfiliform, the six terminal joints thicker than the preceding, submoniliform, the first and second joints very slender; in the ♂ the second funiclar joint is usually slightly swollen and slightly curved.

Thorax long-oval, the prothorax distinctly visible from above, the mesonotum with two distinct furrows (rarely indistinct or subobsolete), the scutellum convex or slightly elevated posteriorly, rounded and unarmed at apex, distinctly separated from the mesonotum by a furrow and with two oblique foveæ in the furrow; metathorax as long as wide.

Front wings rather long, veinless, except sometimes the submarginal vein, which is however always pale and never knobbed at tip; usually it is only visible as a hyaline streak.

Abdomen oval, ovate, or oblong-oval, usually as long as the head and thorax united; the first segment is longer than wide, rugose or striated, the second, very large, occupying two-thirds of the surface, the following segments all short.

Legs rather long, the femora strongly clavate.

TABLE OF SPECIES.

FEMALES.

Mesonotum without furrows.. 2
Mesonotum with distinct furrows.
  Coxæ black.
    Legs and antennæ, except club, pale brownish or honey yellow.
      Pedicel as long as the first funiclar joint, the second funiclar joint shorter than the first...............................I. NIGRICLAVUS Ashm.
      Pedicel shorter than first funiclar joint.
        Pleural piece beneath the anterior wing smooth, not striated.
            I. PALLIPES Say.
        Pleural piece beneath the anterior wing striated.... I. CANADENSIS Prov.
2. Legs and antennæ, except the club, bright yellow.
    Abdomen nearly twice as long as the head and thorax united.
        I. LONGIVENTRIS Ashm.

MALES.

Flagellum brownish-yellow or pale 2
Flagellum brown.
    Pedicel and second funiclar joint equal, the first shorter.
    Pleural piece beneath the anterior wing smooth, not striated; terminal club
        joint only ½ longer than the penultimate......I. NIGRICLAVUS Ashm.
2. Pedicel a little shorter than the second funiclar joint.
    Pleural piece beneath the anterior wing smooth, not striated; terminal club joint
        twice as long as the penultimate; first funiclar joint twice as long as
        thick.. P. PALLIPES Say.
    Pleural piece beneath the anterior wing striated; terminal club joint not twice
        as long as the penultimate; first funiclar joint as wide as long.
                                                 P. CANADENSIS Prov.

Isocybus nigriclavus Ashm.

Bull. No. 1, Col. Biol. Assoc., p. 10.

♀. Length, 3.4 to 4^{mm}. Black, closely, finely punctate, the face rugulose. Head subquadrate, the occiput concave. Antennæ 10-jointed, brownish-yellow, the 6 terminal joints, constituting the club, black; the pedicel is more than twice as long as thick, the two following funiclar joints a little shorter, cylindric, the second obliquely truncate at tip, a little shorter than the first; the joints of the club a little stouter and a little longer than thick, the last being the longest and less than twice as long as thick; parapsidal grooves distinct, converging but not quite meeting posteriorly. Scutellum convex, sparsely covered with a fuscous pubescence, and separated from the mesonotum by a transverse furrow at base. The mesopleura alone smooth and shining, with a large, deep fovea on the disk. Metapleura and metathorax rather densely pubescent, the former divided by a longitudinal grooved line or impression. Legs brownish-yellow; all coxæ black. Abdomen a little longer than the head and thorax together, polished, the petiole roughened, pubescent, and about one-third longer than wide. Wings dusky hyaline, pubescent.

The ♂ differs in having the second funiclar joint distinctly longer than the first, the club being paler, with the joints about twice as long as thick.

HABITAT.—Greeley, Colo.

Types in Coll. Ashmead.

Described from several specimens.

Isocybus pallipes Say.

(Pl. XIII, Fig. 5, F.)

Platygaster pallipes Say, Lec. Ed. Say., Vol. 1, p. 383; Ashm. Can. Ent., XIX, p.
    132; Cr. Syn. Hym., p. 250.

♂ ♀. Length, 3.5^{mm}. Black, very finely, closely punctate; legs honey-yellow, the coxæ black, pubescent; thorax long, the mesonotum with 2 distinct furrows; mesopleura polished, not striated posteriorly; metathorax with a grooved central ridge, the pleura pubescent. Antennæ 10-jointed; in the ♂ honey-yellow, with a 6-jointed cylindrical club, the flagellum one and two-thirds the length of the scape, the pedicel a little more than half the length of the first funiclar joint, the second funiclar joint cyathiform, longer than the first; club joints submoniliform, longer than thick; in the ♀ black, the pedicel nearly as long as the first funiclar joint, the first, one-third longer than the second, the second cylindric, the flagellar joints a little longer than thick, the last conical, longer than the preceding. Tegulæ rufo-piceous. Wings hyaline. Abdomen polished, about as long as the head and thorax united, much depressed, widest toward the apex and obtusely rounded, narrowed toward the base; petiole longer than thick, finely rugose, with a V-shaped carinated space above; second segment elongate, with 2 finely shagreened foveæ at base.

HABITAT.—Indiana, Canada, Western States, and Texas.

Specimens in National Museum and Coll. Ashmead.

Isocybus canadensis Prov.

Platygaster canadensis Prov., Add., p. 181.
Monocrita canadensis Ashm., Can. Ent., XIX, p. 126; Cr. Syn. Hym., p. 250.

♀. Length, 3ᵐᵐ. Black, minutely, closely, rugosely punctate; antennæ and legs, except coxæ, brownish-yellow. The pedicel and second funiclar joint are about equal, two-thirds the length of the first funiclar joint, the second funiclar joint being obliquely truncate at tip; club 6-jointed, rather stout, the first two joints not longer than thick, the three following a little longer than thick, the last a little longer than the penultimate. Thorax trilobed, minutely, rugosely punctate, the middle lobe smoother anteriorly, the lobes posteriorly and the high convex scutellum covered with a sparse fuscous pubescence. Mesopleura deeply impressed or foveated at the middle, smooth, shining, except the piece beneath the anterior wing, which is distinctly striated posteriorly. Wings subfuscous. Abdomen oblong-oval, narrowed at base, the petiole fluted, opaque, pubescent; body of abdomen smooth, polished, the second segment at base with two foveolæ, pubescent both above and beneath.

In the ♂ the antennæ are usually wholly yellow, the second funiclar joint being longer and thicker than the pedicel, the first small, subtriangular, not longer than thick, the club joints about twice as long as thick, the last not quite twice as long as the penultimate.

HABITAT.—Ottawa, Canada.

Types, ♂ and ♀, in Coll. Ashmead.

Provancher described this species from the male sex alone, a type specimen of which was kindly sent me by Mr. Harrington.

The fuscous streaks in the wings, resembling nervures, misled me into describing it as a species of *Monocrita* before I had seen the type.

Isocybus longiventris Ashm.

Can. Ent., Vol. xix, p. 130; Cress. Syn. Hym., p. 219.

♀. Length, 1ᵐᵐ. Polished black, impunctured; head large, quadrate, convex before, the lateral ocelli far away from the eye margin; antennae, except the club and the legs bright yellow; club brown, 5-jointed. Thorax twice as long as the head is long antero-posteriorly, without furrows; scutellum convex, the metathorax pubescent. Abdomen long, acuminate, almost twice as long as the head and thorax together. Wings hyaline.

HABITAT.—Jacksonville, Fla.

Type in Coll. Ashmead.

Described from a single specimen. As I have already indicated in my original description, this insect is not a genuine *Isocybus*, and is placed here only provisionally or until more specimens are taken and the species can be properly studied. In the shape of the abdomen it recalls *Polymecus*, but the quadrate head and the position of the ocelli will exclude it from that genus.

Subfamily VII.—HELORINÆ.

Head transverse, the vertex broad. Ocelli 3, in a triangle. Eyes large, oblong-oval. Mandibles depressed, the apices acute, the exterior margin rounded, the interior bidentate. Maxillary palpi 6-jointed. Labial palpi 3-jointed. Antennae porrect, 15-jointed, inserted on the middle of the face. Pronotum distinctly visible from above, the sides impressed, striated, and anteriorly produced into a slight neck; mesonotum with two furrows; scutellum small, semicircular, subconvex, transversely impressed across the base; metathorax short, obtusely rounded posteriorly. Front wings rather broad at tips, with a large, oblong, stigmated marginal vein, two basal cells, a triangular closed marginal cell and closed discoidal cells; the basal vein does not attain the submarginal, but is abruptly bent backward, forming with the median vein a triangular cell, and with the submedian vein a subquadrate discoidal cell; hind wings also broad, with a distinct submarginal vein that curves obliquely backward to the hind margin before attaining one-third the length of the wing.

Abdomen petiolated, obconic; the petiole very long, swollen toward base; the second segment very large, the following short. Legs moderate, the posterior tarsi lengthened a little longer than their tibiae; tibial spurs 1, 2, 2, the claws pectinate.

This subfamily is at once distinguished by the venation, the basal nervure being abruptly bent, intersecting with the base of the recurrent nervure and forming a triangular cell; the large conical-shaped petiole, and the pectinate claws.

Only one genus is known, *Helorus* Latreille, although Abbé Provancher has described in his "Additions à la Faune Hymenopterologique de la Province de Québec," page 154, a genus, *Ropronia*, with one species, *pediculata*, which he at first placed in the family *Braconidæ*, but which subsequently in the same work, page 406, he removes to the *Helorinæ*. The genus is unknown to me in nature, but from his description and figure of the anterior wing, I believe it to be a Braconid, and it is in consequence not included here.

HELORUS Latreille.

Hist. Nat., XIII. p. 230 (1802).

Syn. Copelus Prov., Faune Ent. Can., II. p. 540 (1883).

(Type *H. anomalipes* Panz.)

This genus is sufficiently described in the characters given for distinguishing the group. It is parasitic in the cocoons of *Chrysopa*, a Neuropterus insect, and, so far, only a single species has been detected in our fauna.

Helorus paradoxus Prov.

(Pl. XIII, Fig. 5, ♀.)

Copelus paradoxus Prov., Nat. Can., XII, p. 207, ♀.
Helorus paradoxus Prov., Faun. Ent. Can., II, p. 540; Cress., Syn. Hym., p. 251.

♀. Length, 4.5^{mm}. Black polished, shining; the head transverse, punctate, the face with a sparse whitish pubescence. Mandibles pale rufous, the tips black. Palpi piceous. Thorax smooth, its dorsum with a few minute punctures; the pronotum strongly impressed and striated at sides; mesopleura rugulose anteriorly, smooth posteriorly with fine punctures toward the base of the middle coxæ, sparsely pubescent; metathorax rugose. Tegulæ pale rufous. Wings hyaline, with piceous-black nervures. Leg pale rufous, the coxæ black, the middle and posterior femora toward base rufo-piceous. Abdomen shining black, polished, the apical segments and the venter with fine punctures, the petiole rugose, with some raised longitudinal lines above.

HABITAT.—Cap Rouge, Canada; and Montana.

Specimens in National Museum, Coll. American Entomological Society, and Coll. Ashmead.

This species comes nearest to the European *H. anomalipes* Panz., but it is slenderer, the scutellum smoother, the petiole rougher, and the legs paler than in that species.

Subfamily VIII—PROCTOTRYPINÆ.

Head transverse or quadrate. Ocelli 3, in a triangle. Mandibles edentate, acute at apex. Maxillary palpi 5- or 4-jointed; labial palpi very short, 3-jointed. Antennæ porrect, 13-jointed in both sexes, with a ring-joint. Pronotum distinct, narrowed before; mesonotum elongate, seldom with furrows; scutellum convex, foveated at base; metathorax usually longer than wide, rounded posteriorly or subtruncate and produced beyond the insertion of coxæ. Front wings with a triangular or semicircular stigmated marginal vein situated much beyond the middle of the wing, a very short, sometimes almost obsolete, marginal cell, and a closed costal cell; and often with traces of discoidal nervures; hind wings broad, veinless. Abdomen petiolated or subpetiolated, conicovate, the second segment very large; in the male terminating in two prongs; in the female prolonged into a long, tubular case, inclosing the ovipositor. Legs rather long, the femora clavate, tibial spurs 1, 2, 2, well developed, the tarsi long, 5-jointed, claws simple.

This group, from which the family derives its name, was first described by Latreille in 1796. It is at once distinguished by the edentate mandibles and the shape of the abdomen; in the female the abdomen being prolonged into a long, tubular case, inclosing the ovipositor, while in the male it terminates in two prongs.

The *Proctotrypinæ* attack apparently only Dipterous and Coleopterous larvæ living in fungi.

The only instance of the rearing of a species in America is by Prof. Comstock, who reared *Proctotrypes obsoletus* Say from the Coleopteron *Stelidota strigosa*. In Europe, the habits of about a dozen species are known, all bred from Coleopterous and Dipterous larvæ living in fungi.

The genera are not numerous, and may be recognized by the aid of the following synoptical table:

TABLE OF GENERA.

FEMALES.

```
Abdomen terminating in a stylus.
   Wingless............................................................ 2
   Winged; maxillary palpi 4-jointed, long, the last joint linear.
      Parapsidal furrows distinct.............................Disogmus Förster.
      Parapsidal furrows wanting..............................Proctotrypes Latr.
2. Maxillary palpi 3-jointed, short, the last joint subclavate.
      Metathorax smooth......................................Codrus Jurine.
```

MALES.

```
Abdomen terminating in two prongs; winged.
   Maxillary palpi 4 jointed, long, the last joint linear.
      Parapsidal furrows distinct.............................Disogmus Förster.
      Parapsidal furrows wanting..............................Proctotrypes Latr.
   Maxillary palpi 3-jointed, short, the last joint subclavate.
      Metathorax smooth......................................Codrus Jurine.
```

DISOGMUS Förster.

Hym. Stud., II, p. 99 (1856).

(Type *D. areolator* Hal.)

(Pl. XIII, Fig. 6, ♀.)

Head transverse or subquadrate.

Antennæ 13-jointed, long, the joints cylindrical, pubescent, or pilose, some of the joints in the male sometimes dentate.

Mandibles conical.

Maxillary palpi 4-jointed, long, the third joint the longest, the second dilated.

Thorax long, the prothorax rounded before, the mesonotum with two deep furrows, the metathorax areolated.

Front wings with an oblong stigma, the marginal cell rather large.

Abdomen much as in *Proctotrypes*, the cauda usually shorter and more slender.

Legs as in *Proctotrypes*, except that the tibial spurs are smaller.

Distinguished from *Proctotrypes* by the distinct mesonotal furrows, the dilated second joint of the maxillary palpi, and the shape of the stigma.

The genus is unknown, as yet, out of the European fauna.

PROCTOTRYPES Latr.

Préc., p. 108 (1796); Förster, Hym. Stud., II, p. 99; = ? *Serphus* Schrank, Schrift. d. Berl. Naturf. Fr., I (1780).

(Type *P. gravidator* Linn.)

Head transverse or quadrate, the occiput margined; ocelli 3, prominent, in a triangle, rather close together; eyes ovate or long-oval.

Antennæ inserted on the front between the eyes, 13-jointed with a ring-joint; the scape is short, oval; the pedicel very minute, annular, more or less hidden within the scape and only visible as a ring-joint, hence the genus has been described as having but 12-jointed antennæ; the flagellar joints vary from long cylindrical joints to short, or moniliform joints, and sometimes in the males some of the joints are dentate.

Mandibles acute at tips, edentate.

Maxillary palpi long, 4-jointed, the last joint linear; labial palpi 3-jointed, the last joint fusiform.

Thorax elongate, the prothorax always visible, depressed above and produced into a neck anteriorly; mesonotum long, highly convex, without furrows, scutellum convex, foveolated at base; metathorax longer than high, sloping or obtusely rounded posteriorly and produced beyond the insertion of the coxæ; spiracles oval or linear.

Front wings with a triangular stigma at about two-thirds the length of the wing with a distinct but very short marginal cell, its length rarely more than half the length of the stigma; costal cell closed; all other cells and nervures entirely, or subobsoletely, obliterated; if present indicated only by fuscous streaks.

Abdomen petiolated, ovate, slightly compressed, the petiole short, the second segment very large occupying most of its surface, in the ♀ terminating in a long cauda; in ♂ ending in two prongs or spines.

Legs long, slender, the femora slightly swollen; tibial spurs 1, 2, 2; tarsi long, slender; claws long, curved, simple.

Our species in this genus are numerous, but may be readily determined by the aid of the following table:

TABLE OF SPECIES.

Head quadrate.. 6
Head transverse.
  Head and thorax black.. 2
  Head and thorax pale or rufous, the metathorax sometimes black, wings subfuscous, the venation distinct. .
    Metathorax coarsely rugose, with longitudinal raised lines, but without a distinct central longitudinal carina; cauda as long as the abdomen.
                                        P. CAUDATUS Say.
    Metathorax coarsely rugose, not longitudinally striated, but with a distinct central longitudinal carina; cauda half the length of abdomen.
                                          P. PALLIDUS Say.
2. Abdomen black.. 3
  Abdomen rufous or honey yellow, the apex or petiole alone black.
    Metathorax rugose without a distinct central longitudinal carina.
      Rugosities longitudinal ♂ ....P. CAUDATUS Say.
      Rugosities not longitudinal ♂ ....P. RUFIGASTER Prov.
    Metathorax rugose, with a distinct longitudinal carina.
      All coxæ black.
        Petiole not as long as thick.
          Four terminal abdominal segments black ♂ ......P. TERMINALIS, sp. nov.
          Three terminal abdominal segments black ♂ ....P. LINELLII, sp. nov.
      All coxæ pale.
        Petiole twice as long as thick.
          The sixth abdominal segment dusky...........P. MELLIVENTRIS Ashm.
        Petiole not as long as thick.
          Four or five terminal abdominal segments black........(P. TERMINALIS.)
3. The discoidal nervures more or less distinct by fuscous streaks.
  Coxæ black or black basally.
    Metathorax coarsely rugose with a central carina; antennæ with joints 1 to 5 dentate, ♂ ....P. CALIFORNICUS Holmgr.
    Metathorax coarsely rugose, but without a central carina; antennæ simple, ♂ ....P. OBLIQUUS, sp. nov.
  The discoidal nervures entirely wanting.
    Marginal cell always short... 4
    Marginal cell large, as long as the stigma.
      Legs, including coxæ, pale yellow, the inner posterior tibial spur nearly as long as the basal joint of the tarsi, ♂ ♀ ....P. FLAVIPES Prov.
4. Cauda short, not or scarcely one-third the length of abdomen................. 5
  Cauda as long as abdomen.
    Legs, including coxæ, yellow, clypeus and mandibles light rufous.
                                  P. CLYPEATUS, sp. nov.
5. Metanotum with 3 carinæ, inclosing 2 large smooth areas above; apex and sides rugose.
  Coxæ black or dusky above.
    Antennæ brown, not extending to apex of thorax, basal two or three yellow.
                             P. ABRUPTA Say.
  Coxæ pale.
    Middle carina of metathorax ending at the superior edge of the rounded posterior face.

Antennæ honey yellow, slightly dusky at apex and reaching the base of the abdomen; second flagellar joint about two-thirds the length of the first, about 2½ times as long as thick, the four following joints equal, as long as the second; cauda short.......P. OBSOLETUS Say.

Middle keel extending to the base of the petiole.

Second flagellar joint half the length of the first, 2½ times as long as thick, the following joints, except the last subequal; cauda short.

P. BELFRAGEI, sp. nov.

Metanotum on each side of the median keel rugose.

Middle keel ending at the superior edge of the posterior face.

Second flagellar joint thrice as long as thick, two-thirds the length of first, following joints except the last gradually shortening; cauda short.

P. TEXANUS sp. nov.

Middle keel extending to the base of the petiole.

Second flagellar joint 4 times as long as thick; abdomen ending in two spines, ♂..................................P. CAROLINENSIS, sp. nov.

6. Head with a frontal carina between the antennæ.

Metathorax not twice as long as high 7

Metathorax twice as long as high.

Head a little longer than wide.

Antennæ and legs rufous, the coxæ black basally.

Antennæ longer than the head and thorax, the first flagellar joint one-third longer than the second, the others at least thrice as long as thick, ♀ ....P. LONGICEPS, sp. nov.

7. Antennæ and legs, unless otherwise mentioned, rufous or reddish yellow, coxæ more or less dusky basally.

Head a little larger than wide; antennæ brown.

First flagellar joint not longer than the second, shorter than the scape, the others about twice as long as thick, the last twice as long as the penultimate, ♀ ....P. CANADENSIS, sp. nov.

Head not longer than wide.

First flagellar joint, not or scarcely longer than the second, the others three times as long as thick, ♀. (Coxæ sometimes wholly pale.)

P. SIMULANS, sp. nov.

First flagellar joint distinctly longer than the second.

Flagellar joints 2 to 6 at least thrice as long as thick, ♀.

P. MEDIUS, sp. nov.

Flagellar joints 3 to 10 only a little longer than thick, ♀.

P. QUADRICEPS, sp. nov.

All coxæ black, the femora piceous.

Antennæ slightly thickened towards apex, dusky, the first flagellar joint one-third longer than the second, joints 2 to 10 scarcely longer than thick, ♀ ....P. FEMORATUS, sp. nov.

Proctotrypes caudatus Say.

(Pl. XIII, Fig. 7, ♀.)

Lec. Ed. Say's Works, I, p. 21; Ashm., Ent. Am., III, p. 98; Cress., Syn. Hym., p. 248.

P. crenulatus Patton, Can. Ent., XI, p. 61 (1879); Ashm. loc. cit., p. 99; Cr. loc. cit. (supra), p. 248.

♂ ♀. Length, 7 to 10ᵐᵐ. Ovipositor as long as or a little longer than the abdomen. Reddish testaceous; metathorax black, upper part of mesopleura and the sutures often black; head and thorax smooth, shining; collar a little wrinkled or grooved at sides; metathorax elongated, rounded behind, longitudinally rugulose and reticulated with coarse punctures; a row of crenate punctures along the hind margin of the mesopleura. Antennæ 13-jointed, black, the basal joint reddish, robust, partially inclosing the small second joint. Wings subfuscous, the nervures fuscous, the marginal cell very short, about one-fourth the length of the stigma, the internal venation distinct as fuscous streaks. Abdomen not as long as the thorax, the petiole very short, grooved, black; the second segment at base with a median furrow, and three grooves on either side; terebra fully as long as the abdomen, a little curved, longitudinally striated along the sides.

In the ♂ the head and thorax are most frequently black, although sometimes more or less piceous or reddish, the coxæ black or dusky, the tarsi most frequently fuscous or at least tinged, the abdomen dusky at apex and ending in two short prongs; otherwise as in the ♀.

HABITAT.—United States generally.

Specimens in National Museum, Royal Berlin Museum, Coll. American Entomological Society, and Coll. Ashmead.

A common species, somewhat variable in size and color. I have seen Patton's type of *P. crenulatus* in Coll. American Entomological Society, and there is no doubt of its being a synonym of this species.

Dr. Clarence M. Weed has sent me quite a large series collected in Ohio, and I have seen specimens from various parts of the country.

Proctotrypes pallidus Say.

Lec. Ed. Say's Works, 1, p. 382; *idem.*, 11, p. 725; Ashm. Ent. Am., 111, p. 99; Cress. Syn. Hym., p. 248.

♂ ♀. Length, 6 to 8ᵐᵐ. Reddish-testaceous, closely resembling *P. caudatus*, but with the metathorax more finely rugose, with a median carina, the sculpture not longitudinal, the antennæ wholly testaceous or brown, the second abdominal segment with fine striæ at extreme base.

In the ♂ only the meso- and metapleura are blackish, the coxæ pale, the tarsi not dusky, apex of abdomen very slightly dusky, often concolorous with the rest of the abdomen, while the prongs are shorter than in *caudatus*.

HABITAT.—United States.

Specimens in Coll. American Entomological Society, National Museum, and Coll. Ashmead.

Proctotrypes rufigaster Prov.

Nat. Can., XII, p. 263; Faun. Ent. Can., p. 561, ♂ ♀; Ashm. Ent. Am., 111, p. 99; Cress. Syn. Hym., p. 248.

♂. Length, 5ᵐᵐ. Head and thorax black; legs reddish-yellow, the tarsi slightly dusky; abdomen reddish-yellow, the petiole, extreme apex of second and following segments black. Antennæ long, cylindrical, black, the basal joint rufous beneath. Thorax smooth; collar at sides, mesopleura beneath the tegulæ and anteriorly and posteriorly, striated; metathorax coarsely rugose, without a median carina. Wings subhyaline, as in *P. caudatus.* Petiole short, fluted, rugose above.

HABITAT.—Ottawa, Canada.

Types in Coll. Provancher, Ashmead, and Harrington.

Proctotrypes terminalis, sp. nov.

♂. Length, 5.5 to 6.5ᵐᵐ. Head, antennæ, and thorax entirely black; legs rufous, the coxæ black; second abdominal segment, except the apex, red. Antennæ extending to middle of abdomen, slender, cylindrical, the flagellar joints all long, pubescent. Prothorax at sides anteriorly, mesopleura beneath the tegulæ anteriorly and at posterior margin striated or wrinkled; metathorax rugose, with a slight median carina. Wings as in *P. caudatus* Say. Petiole very short, fluted. Hind tibial spur less than one-third the length of the basal tarsal joint.

HABITAT.—Washington, D. C., and Columbus, Ohio.

Types in National Museum and Coll. Ashmead.

My specimens were presented to me by Dr. C. M. Weed.

Proctotrypes Linellii, sp. nov.

♂. Length, 4ᵐᵐ. Closely allied to *P. terminalis,* agreeing with it in color and sculpture, except that only the 3 terminal abdominal joints are black, the metathorax more coarsely rugose, with the median carina quite distinct and extending from base to apex, while the hind tarsi are slightly dusky, the tibial spur being one-third the length of the basal tarsal joint. The wings are hyaline, the marginal cell nearly obsolete, not as long as the width of the radius, the latter prolonged into the disk of the wing with a small fuscous spot at its origin.

HABITAT.—Long Island.

Type in Coll. Ashmead.

Described from a single specimen collected by Mr. Martin Linell.

Proctotrypes melliventris Ashm.

Ent. Am., III. p. 99 ♂ ; Cress. Syn. Hym., p. 248.

♂. Length, 4.1ᵐᵐ. Head and thorax black, polished, the metathorax rugose, with a longitudinal median carina; legs and abdomen reddish-yellow; the petiole and the apex of abdomen black; petiole a little more than twice as long as thick, with longitudinal furrows; tip of abdomen ending in two prongs. Antennæ 13-jointed (with a ring-joint), black, reaching to the base of the second abdominal segment, cylindrical, pubescent, very slightly thinner at apex; the first and last flagellar joints the longest, unequal, the first a little more than 4 times as long as thick, the last 6 times as long as thick, the intermediate joints very slightly decreasing in length. Wings hyaline with a faint fuscous tinge, the venation brown-black, the radial cell very short, less than ⅓ the length of the stigma, with a small dusky streak below the base of the radius; disk of wings with traces of the cubitus and anal nervures. otherwise veinless. Hind tarsi as long as their tibiæ, the tibial spur ⅓ the length of the basal tarsal joint.

HABITAT.—Jacksonville, Fla.

Type in Coll. Ashmead.

Proctotrypes californicus Holmg.

Kongl. sv. Freg. Eug. Resa Ins., p. 431; Cr. Syn. Hym., p. 248.

♂. Length, 4 to 4.5^{mm}. Polished black; the mandibles, antennæ beneath and legs, except coxæ, rufous. Metathorax finely rugose, with a median carina. Joints 1 to 5 of flagellum dentate beneath. Wings subhyaline. the discoidal nervures distinctly visible as fuscous streaks, marginal cell one-third the length of the stigma. Abdomen black, polished, not longer than the thorax, the petiole very short, rugose, striated at sides and beneath, the second segment with some striæ at extreme base, terminal segment ending in two short prongs.

HABITAT.—California, Canada, and Virginia.

Specimens, agreeing in all particulars with Holmgren's description, are in my collection. The species is easily recognized by the dentated flagellar joints, being the only species in our fauna thus distinguished.

Proctotrypes obliquus, sp. nov.

♂. Length, 4.5^{mm}. Polished black. Head very broad, more than thrice as wide as thick antero-posteriorly. Eyes large. ovate. Mandibles piceous, black. Palpi very long, pale brownish. Antennæ 13-jointed, very long, filiform, pubescent. the flagellar joints nearly of an equal length, about four times as long as thick. Metathorax gradually sloping off posteriorly and produced into a point far beyond the insertion of the hind coxæ, coarsely reticulately rugose; at the base are two large foveæ which are connected by a sulcus or a grooved line with the spiracular foveæ. Tegulæ brown. Wings subhyaline, the stigma and radius brown-black; the marginal cell is a little longer than half the length of the stigma. Legs pale brownish-yellow, the hind coxæ black, the others black only at base; the tibial spurs are long, the inner spur of hind tibiæ being more than half the length of the basal tarsal joint, all tarsi longer than tibiæ. Abdomen subcompressed, black, shining, and composed of but 3 visible segments, the apex when viewed from the side being obliquely truncated, the usual two projecting spines wanting.

HABITAT.—Texas.

Type in National Museum.

Described from a single specimen in Belfrage collection.

Proctotrypes flavipes Prov.

Nat. Can., XII, p. 264; Faun. Ent. Can., II, p. 562 ♀.

Megaspilus lucens Prov., Faun. Ent. Can., II, p. 808 ♀.

Proctotrupes flaripes Prov., Add. et Corr., pp. 462 and 471.

♀. Length. 3.5 to 4^{mm}. Polished black: mandibles. tegulæ, and legs, yellow. Metathorax finely rugose, with two large smooth areas at base above. Wings hyaline, the nervures and stigma yellowish, the discoidal nervures entirely wanting; costal cell very wide; stigma semicircular, the marginal cell as long as the stigma. Inner spur of posterior tibiæ fully two-thirds the length of the basal tarsal joint. Abdomen black, polished, as long as the thorax, the ovipositor long, nearly half the length of the abdomen. Antennæ rather short, brownish-yellow, paler beneath and toward base, filiform, the first and second funiclar joints equal, the following shorter.

HABITAT.—Ottawa, Canada.

Specimens are in my collection, received from Mr. W. H. Harrington. The species is quite distinct from all others in our fauna in the size of the marginal cell and the long tibial spur of posterior legs. It seems to agree quite closely with the European *P. calcar* Haliday.

Proctotrypes clypeatus, sp. nov.

♀. Length, 3.5ᵐᵐ, to the tip of ovipositer 4.5ᵐᵐ. Polished black; clypeus and mandibles pale rufous; legs, including coxæ, yellowish; ovipositor longer than the abdomen, the basal two-thirds reddish yellow. Antennæ pale brownish-yellow, not longer than the thorax, the first flagellar joint a little longer than the second, joints 2 and 3 about equal, fourth slightly shorter, the following to the last about twice as long as thick, the last one-half longer than the penultimate, ovate. Metathorax rugose, with two smooth areas at base above. Wings hyaline, the discoidal nervures wanting; stigma large, brown-black; marginal cell less than one-half the length of the stigma. Tibial spurs of posterior legs about one-half the length of the basal tarsal joint. Abdomen polished black, about as long as the thorax, the petiole scarcely apparent, second segment at extreme base with a long foveola on either side.

HABITAT.—Ithaca, N. Y.

Type in Coll. Ashmead.

Described from a single specimen received from Mr. F. H. Chittenden, who informs me he reared it October 15, 1884, from a large yellow, rather woody, fast-growing tree fungus, from which Melandryid, Mycetophagid, Staphylinid and Scaphidiid beetles were obtained.

Proctotrypes abruptus Say.

Bost. Jour., I, p. 278 ♀; Lec. Ed. Say's Works, II, 725; Ashm. Ent. Am., III, p. 98; Cress. Syn. Hym., p. 248.

♀. Length, 2.5 to 3ᵐᵐ. Polished black; mandibles black or piceous; legs reddish yellow, the coxæ sometimes black or black toward base; ovipositor about ¼ the length of the abdomen or very slightly longer than the basal joint of hind tarsi. Antennæ brown, not longer than the thorax, slightly thickened toward base, the basal joint or sometimes the 3 or 4 basal joints, yellow; first flagellar joint about one-half longer than the second, or the second joint is two-thirds the length of the first, the following joints to the last very gradually shortening, the penultimate being but slightly longer than thick, the last being twice as long as the penultimate, fusiform. Metathorax rugose, with 3 carinæ, inclosing two large smooth areas at base. Abdomen as long as the thorax, black, shining, the ovipositor about ⅓ its length. Wings hyaline, without discoidal nervures, the stigma large, the marginal cell ⅛ its length, not petiolate at base.

HABITAT.—Jacksonville, Fla.; Virginia, and Indiana.

Specimens are in the National Museum, Coll. American Entomological Society, and Coll. Ashmead.

Proctotrypes obsoletus Say.

Bost. Jour., I, p.277, ♀ ; Lec. Ed. Say, II, p. 725; Ashm. Ent. Am., III, p.98; Cress. Syn, Hym., p. 248.

♀. Length, 4ᵐᵐ. Polished black; antennæ brownish-yellow; mandibles yellowish; legs reddish-yellow; wings hyaline. Antennæ 13-jointed, very slightly thickened towards tips, second flagellar joint two-thirds the length of the first, the four following equal, about as long as the second, 2½ times as long as thick, the terminal joint fusiform, almost as long as the two preceding joints together, or fully as long as the first. Metathorax rugose, the middle carina extending to the posterior face. Abdomen longer than the head and thorax united, emarginate at base, the petiole extremely short, the cauda very slightly longer than the basal joint of hind tarsi.

In the male the antennæ are filiform, pale brownish, the first flagellar joint as long as the scape, the joints very slightly increasing in length from the second, broken off from the sixth.

HABITAT.—District of Columbia and Indiana.

Specimens in National Museum and Coll. Ashmead.

A single specimen of what I believe to be this species was reared by Prof. Comstock, December 9, 1879, from *Stelidota strigosa*.

Proctotrypes Belfragei, sp. nov.

♀. Length, 5ᵐᵐ. Polished black, with a sparse fuscous down. Head transverse, thrice as wide as thick antero-posteriorly; palpi pale; mandibles rufo-piceous. Antennæ 13-jointed, pale brown, darker towards tips, extending to tegulæ, cylindrical, the scape oval, the pedicel annular, the first and the last flagellar joints equal, the joints after the first very gradually shortening, the three preceding the last not more than 1½ times longer than thick. Metathorax at sides coarsely rugose, the disk smooth, shining, with 3 carinæ, the middle of which extends to the apex, the lateral abbreviated. Tegulæ and legs reddish yellow, the posterior coxæ behind black; tibial spurs short. Wings hyaline, the venation, except costal nervure, piceous-black; the marginal cell is very short, one-third the length of the stigma; no traces of nervures in the discoidal region. Abdomen black, polished, the petiole very short, the base of second segment with some raised lines; cauda short, impunctured, curved, very slightly longer than the basal joint of the hind tarsi.

HABITAT.—Texas.

Type in Coll. National Museum.

Described from a single specimen in Belfrage collection.

Proctotrypes texanus, sp. nov.

♀. Length, 3ᵐᵐ. Polished black; mandibles, palpi, and legs, including coxæ, yellow; antennæ longer than the thorax, slender, filiform, brownish-yellow, pubescent; first funicle joint slightly longer than the second, the following joints to the last very gradually shortening, the penultimate being about 4 times as long as thick, the last about one-half longer than the penultimate. Metathorax finely rugose, with a median carina extending to apex. Inner spur of posterior tibiæ about one-half the length of the basal joint of tarsi. Wings hyaline, the stigma brown, the marginal cell about one-third the length of the stigma, petiolated at base. Abdomen a little shorter than the thorax, black, shining, the petiole distinct, fully as long as thick, fluted, second segment at extreme base striated; prongs very short.

HABITAT.—Texas.

Type in Coll. Ashmead.

Described from a single specimen.

Proctotrypes carolinensis, sp. nov.

♂. Length, 5ᵐᵐ. Polished black. Head twice as wide as thick antero-posteriorly with a frontal carina extending from the front ocellus to between the base of the antennæ. Mandibles rufous, the tips black. Palpi long, yellowish. Antennæ 13-jointed, filiform, brown-black, the scape yellow; the flagellar joints very gradually shortening to the last, the last equal with the second, the first joint is the longest and a little more than 4 times as long as thick. Metathorax rugose, rounded off posteriorly with a single central longitudinal carina extending quite to its apex. Tegulæ yellowish. Wings hyaline, the venation brown-black, the marginal cell short, less than one-half the length of the stigma. Legs yellow; the coxæ behind at base dusky; the tibiæ and tarsi are long, about of an equal length; the posterior tibial spurs not quite half the length of the basal tarsal joint. Abdomen black, polished, not longer than the thorax, composed of five segments, petiole distinctly grooved, second segment at base grooved, terminal segment ending in two prongs.

HABITAT.—North Carolina.

Type in National Museum.

Proctotrypes longiceps, sp. nov.

♀. Length, 7^{mm}. Polished black; antennæ yellowish, the apical joints a little dusky, the first flagellar joint the longest, the following to the last slightly shortening, the penultimate being scarcely 3 times as long as thick, the last one-half longer than the preceding; palpi yellowish; legs reddish-yellow, the coxæ slightly dusky basally. Metathorax twice as long as high, finely rugulose, smoother above, with a median carina. Wings hyaline, the discoidal nervures visible as fuscous streaks, the stigma fuscous, the marginal cell one-third the length of the stigma, petiolated. Abdomen black, shining, longer than the head and thorax together, distinctly petiolated, the petiole more than twice as long as thick, finely striated; ovipositor not longer than the basal joint of hind tarsi. The femora are swollen, the tibiæ subclavate, the inner spur of posterior tibiæ scarcely one-third the length of the basal tarsal joint.

HABITAT.—Ottawa, Canada.

Type in Coll. Ashmead.

Described from a single specimen received from Mr. W. Hague Harrington.

Proctotrypes canadensis, sp. nov.

♀. Length, 3^{mm}. Polished black; antennæ brown, the first flagellar joint not longer than the second, shorter than the scape, the following joints, except the last, about 2½ times as long as thick, the last fully twice as long as the penultimate; legs reddish-yellow, the coxæ dusky basally. Metathorax a little longer than high, rugose, smooth on dorsum toward base, with a median carina. Wings hyaline, without discoidal nervures, the stigma yellowish, the marginal cell less than half the length of the stigma, petiolated. Abdomen black, shining, not longer than the thorax, the petiole scarcely as long as thick, rugose, the second segment at base with numerous striæ; ovipositor very short, two-thirds the length of the basal joint of hind tarsi. The femora are not so much swollen, the tibiæ long and slender, the inner spur of hind tibiæ not more than one-fourth the length of the basal tarsal joint.

HABITAT.—Ottawa, Canada.

Type in Coll. Ashmead.

Described from a single specimen, received from Mr. W. Hague Harrington. Its smaller size, shorter metathorax and ovipositor, relative length of the antennal joints and the shorter tibial spurs, easily separate the species.

Proctotrypes simulans, sp. nov.

♀. Length, 4^{mm}. Polished black, antennæ rufous, the first flagellar joint scarcely longer than the second, but as long as the scape, the following joints after the third about 2½ times as long as thick, the last less than twice as long as the penultimate; legs, including coxæ, reddish-yellow. Metathorax 1½ times as long as high, rugose, smooth toward base above, with a median carina. Wings hyaline, the discoidal ner- vures traceable as fuscous streaks, stigma brown, the marginal cell less than half the length of the stigma, petiolated. Abdomen black, polished, about as long as the thorax, petiolated, the petiole not longer than thick, striated, extreme base of second segment striated, ovipositor not longer than the basal joint of hind tarsi.

HABITAT.—Arlington, Va.

Type in Coll. Ashmead.

Described from a single specimen captured by myself. Since this was written I have received another specimen, taken by Mr. E. A. Schwarz, at Fort Pendleton, W. Va., agreeing in every particular with the above description, except the coxæ are concolorous with the legs.

Proctotrypes medius, sp. nov.

♀. Length, 5.5ᵐᵐ. Polished, black; antennæ brownish-yellow, rather stout, the first flagellar joint the longest, a little longer than the scape and pedicel together, or one-third longer than the second, the joints after the fourth about 3 times as long as thick, the last longer, as long as the second; legs, including coxæ, reddish-yellow, the inner spur of posterior tibiæ fully one-third the length of the basal tarsal joint. Metathorax less than twice as long as high; coarsely rugose, with a median carina and two large, smooth areas toward base. Wings hya- line, the discoidal nervures very faintly traceable, the stigma brown, the marginal cell about one-fourth the length of the stigma. Abdomen black, polished, longer than the thorax, petiolated, the petiole longer than thick, striated; ovipositor as long as the basal joint of hind tarsi, stout.

HABITAT.—Ottawa, Canada.

Type in Coll. Ashmead.

Described from a single specimen received from Mr. W. Hague Har- rington. Approaches nearest to *P. longiceps*, only the head is a little wider than long, the metathorax shorter, more coarsely rugose, the petiole shorter, while the tibial spurs are longer.

Proctotrypes quadriceps, sp. nov.

♀. Length, 4ᵐᵐ. Polished black; legs reddish-yellow, the coxæ a little dusky basally; inner spur of hind tibiæ a little less than half the length of the basal tarsal joint; palpi whitish or pale yellowish; meta- thorax scarcely longer than high, rugose, with a median carina and two large, smooth areas on either side of it, inclosed at sides by delicate lateral carinæ. Antennæ robust, filiform, reddish-brown, scarcely longer than the head and thorax together, the 3 basal joints yellowish, the first flagellar joint scarcely as long as the last or very little longer than the second. Wings hyaline, the discoidal nervures subobsolete, stigma brown, the marginal cell petiolated, one-third the length of the stigma. Abdomen polished, black, not longer than the head and thorax together, the petiole not as long as wide, striated, second segment with sulci at base; ovipositor not longer than the basal joint of hind tarsi.

HABITAT.—New Jersey.

Type in Coll. Ashmead.

Described from a single specimen.

Proctotrypes femoratus, sp. nov.

♀. Length, 2.1ᵐᵐ. Polished black; legs yellowish, the coxæ black or piceous, the femora swollen, piceous or rufo-piceous; inner spur of hind tibiæ one-third the length of the basal tarsal joint; palpi yellowish; metathorax not longer than high, closely punctulate, the dorsum smooth and polished, with a median carina. Antennæ brown, not longer than the head and thorax together, slightly thickened toward the tips, the scape piceous, pedicel yellow; the first flagellar joint is two-thirds the length of the last joint, or very little longer than the second, the joints after the second scarcely longer than thick. Wings hyaline, the discoidal nervures entirely wanting, the stigma brown-black, the marginal cell petiolated one-third the length of the stigma. Abdomen polished, black, not longer than the head and thorax together; the petiole not as long as thick, rugose, second segment striated above at extreme base; ovipositor scarcely longer than the basal joint of hind tarsi.

HABITAT.—Wyoming.

Type in Coll. Ashmead.

Described from a single specimen obtained through Mr. H. F. Wick-

CODRUS Jurine.

Hym., p. 308 (1807); Thoms. Öfv., 1857, p. 421.

(Type *C. apterogynus*, Hal.)

(Pl. XIII, Fig. 8, ♀.)

Head subquadrate.

Antennæ 13-jointed, in ♀ very slightly thickened toward the tips; in ♂ setaceous, pubescent, the flagellar joints longer.

Mandibles acute at tips, edentate.

Maxillary palpi short, 3-jointed, the last joint subclavate; labial palpi very short, 2-jointed.

Thorax elongate, the prothorax narrowed and rounded before, mesonotum highly convex, without furrows, metathorax a little longer than high, slightly depressed above, smooth and shining.

Front wings with a subtriangular stigma and a very minute marginal cell, without traces of nervures in the discoidal region; the ♀ apterous.

Abdomen much as in *Proctotrypes*, the dorsum subdepressed, and the cauda very short.

Legs as in *Proctotrypes*.

The short 3-jointed maxillary palpi and the subdepressed smooth met-athorax distinguish this genus from *Proctotrypes* and *Disogmus*. No species is known from North America.

Subfamily IX.—BELYTINÆ.

Head transverse or subglobose. Ocelli 3, in a triangle, rarely want-ing. Eyes most frequently hairy. Mandibles usually short, acute at tips, with a tooth within, rarely falcate and crossing each other at tips. Maxillary palpi 4- or 5-jointed; labial palpi 3-jointed. Antennae por-rect, inserted on a frontal prominence, in males 14-jointed, in females 14- or 15-jointed, filiform setaceous, or subclavate, or more rarely clavate-moniliform, the scape long. Pronotum distinctly visible from above and narrowed into a short neck at the junction with the head; mesonotum usually as broad as long, with deep furrows, rarely entirely without furrows; scutellum convex, deeply foveated at base; meta-thorax short, usually carinated, rarely spined, posteriorly truncate or emarginate. Front wings with a closed costal cell, a single basal cell, and a radial or marginal cell, the latter either closed or open, with a branch of a vein interstitial with the second abscissa of the radius and extending backwards into the discoidal field of the wing; it is quite rare for the marginal cell to be entirely wanting; hind wings always with a basal cell. Apterous forms rare. Abdomen distinctly petiolated, oblong-oval, ovate, conic-ovate, or pyriform, and composed of from 3 to 8 segments, the second segment always large. Legs rather long and slender, the tibial spurs 1, 2, 2, the tarsi long, slender, 5-jointed, claws simple.

An extensive and but slightly studied group, closely related to the *Diapriinae* and formerly confused with them. A Belytid may, however, always be distinguished from a Diapriid by having a distinct basal cell in the hind wings, and by the 3-jointed labial palpi; also, except in a few cases, by the venation of the front wings, which have a distinct basal cell and usually a distinct marginal cell.

The exotic genus *Monomachus* Westwood, at present placed with the *Evaniidae*, should probably be placed in this group; but as the genus is known to me only from the description I can not tell positively with-out seeing specimens for study.

Nothing is known of the habits of the species composing this group, although Nees von Esenbeck and others believe they undergo their transformations within the larvae of Diptera that inhabit fungi. From their close structural resemblance to the *Proctotrypinae* this supposi-tion is probably correct.

Our entomologists should give more attention to the rearing of insects infesting fungi, not only for the purpose of throwing light upon the obscurity that enveils these insects, but upon those of other groups, and it is hoped the near future will bring forth some results from those so situated as to make observations upon fungi-feeding insects.

The following table includes all the genera at present known to me:

TABLE OF GENERA.

FEMALES.

Antennæ 14-jointed..7
Antennæ 15-jointed.
  Abdomen with 7 or 8 dorsal segments.....................................3
  Abdomen with 3 or 4 dorsal segments.
    Marginal vein as long as, or scarcely longer than, the marginal cell........2
    Marginal vein more than twice as long as the marginal cell.
                                       MACROHYNNIS Förster.
2. First funiclar joint nearly as long as all the rest together, the intermediate joints transverse-moniliform.............................DIPHORA Förster.
  First funiclar joint not unusually lengthened.
    Abdomen with 3, seldom with 4 dorsal segments, the second not greatly lengthened, the third long and strongly compressed laterally; marginal vein not shorter than the marginal cell; antennæ filiform, pubescent, the funiclar joints all long...................LEPTORHAPTUS Förster.
    Abdomen with 3 segments, the second very much lengthened, almost reaching to tip of abdomen, the third issuing from it like a short stylus; marginal vein usually distinctly shorter than the marginal cell; antennæ filiform, pubescent, the 5 or 6 terminal joints oval, the others long......................MIOTA Förster.
3. Abdomen with 7 dorsal segments; antennæ clavate-moniliform, the first funiclar joint slightly longer than the pedicel, all others to the last moniliform, the last enlarged, oval; first abscissa of radius straight, or at least not very oblique.....................ACROPIESTA Förster.
  Abdomen with 8 dorsal segments.
    Eyes bare...6
    Eyes hairy.
      Middle carina of metathorax not divided................................4
      Middle carina of metathorax divided, or wanting.............BELYTA Jurine.
4. Postscutellum with a strong thorn or spine....................OXYLABIS Först.
  Postscutellum without a thorn or spine.
    Third dorsal segment of abdomen not, or very little, longer than the fourth...5
    Third dorsal segment of abdomen much longer than the fourth.
      Mandibles short, small; marginal vein as long as marginal cell; antennæ filiform, pubescent, the last flagellar joint more than twice as long as thick.................................CINETUS Jurine.
      Mandibles long, falcate; marginal vein shorter than the marginal cell; last funiclar joint not more than twice as long as thick.
                                  XENOTOMA Förster.
5. Marginal cell closed.
  First abscissa of radius straight from the margin, shorter than the marginal vein; funiclar joints only slightly shortened toward the tips of antennæ....................................ZELOTYPA Förster.
  First abscissa of radius oblique, longer than the marginal vein; funiclar joints strongly shortened toward the tips of antennæ, much wider than long.............................PANTOCLIS Förster.
  Marginal cell open.
    Stigmal and postmarginal veins much shortened, the stigmal given off at almost a right angle.............................ZYGOTA Förster.

Stigmal and postmarginal veins much shortened, the stigmal given off at a very oblique angle; antennæ clavate, moniliform, the first funiclar joint only a little longer than thick and much smaller than the pedicel.
ACLISTA Förster.

Stigmal vein very short, with an uncus, marginal vein as long as the basal nervure, mandibles conical, not rostriform; scape at tip produced into a little spine; palpi 4-jointed..............SYNACRA Förster.

6. Mesonotum with 2 furrows; marginal cell long, open; antennæ clavate-moniliform, the first funiclar joint slightly longer than the pedicel.
PSILOMMA Först.

Mesonotum without furrows......................................ISMARUS Hal.

7. Ocelli wanting; wingless.............................ANOMMATIUM Förster.
Ocelli present.
Marginal cell distinct, closed; antennæ filiform or subclavate.
ANECTATA Förster.
Marginal cell scarcely discernible or wanting; antennæ subclavate-moniliform, pubescent, the first funiclar joint smaller than the pedicel.
PANTOLYTA Först.

MALES.

Petiole of abdomen not, or scarcely, longer than the metathorax............... 2
Petiole of abdomen almost twice as long as the metathorax.
. Marginal vein twice as long as the marginal cell.........MACROHYNNIS Först.
Marginal vein much larger than the stigmal and about as long as the marginal cell.
Second abdominal segment compressed laterally; petiole smooth above.
Antennæ filiform, the scape as long as the first funiclar joint, the latter strongly emarginated at base.................LEPTORHAPTUS Först.
Antennæ filiform, pubescent, the scape not as long as the first funiclar joint, the latter slightly emarginated at base.........MIOTA Först.
Second abdominal segment not compressed laterally, the abdomen becoming more flattened behind the segment; the petiole above more or less furrowed; scape longer than the first funiclar joint.
CINETUS Jurine.
Marginal vein as long as, or a little longer than, the stigmal, but much shorter than the marginal cell; mandibles falcate........XENOTOMA Först.

2. Middle carina of metathorax not divided..................................... 3
Middle carina of metathorax divided or absent.
Marginal vein scarcely longer than the stigmal, marginal cell long, postmarginal vein greatly lengthened; antennæ filiform, all the joints long, cylindric, the first funiclar joint emarginate at base......BELYTA Jurine

3. Postscutellum with a strong spine.........................OXYLABIS Förster.
Postscutellum without a spine.
Eyes not hairy.. 4
Eyes hairy.
Marginal cell closed.
Scape with the apical margin on one side produced into a tooth.
ACROPIESTA Först.
Scape not produced on one side into a tooth.
Marginal vein not or scarcely longer than the first abscissa of radius, the latter oblique.
Last ventral segment very straight and punctured.
Anterior tibiæ normal...............................ANECTATA Först.
Anterior tibiæ bent, with a median spined process....ZYGOTA Först.
Last ventral segment somewhat bent, not punctured..PANTOCLIS Först.

Marginal vein at least twice as long as the first abscissa of radius, the latter straight, in a right angle with the costa or only slightly oblique.

ZELOTYPA Först.

Marginal cell open, or wanting.

Marginal cell wanting.

Basal vein distinct; antennae filiform, pubescent, the first flagellar joint twice as long as the pedicel, slightly emarginate at base.

PANTOLYTA Först.

Marginal cell more or less distinctly present.

Marginal cell much lengthened; marginal vein hardly longer than the first abscissa of radius; antennae filiform pubescent, all the joints lengthened, the first flagellar joint emarginate at base; anterior tibiae strongly bent, outwardly produced towards one side into a tooth or spine ....ZYGOTA Först.

Marginal cell not much lengthened; first abscissa of radius very oblique; anterior tibiae simple; antennae thick, filiform, densely pubescent, the first flagellar joint not longer than the second, emarginate at base.

ACLISTA Först.

4. Mesonotum with two furrows; marginal cell long, open.

Antennae stout, filiform, the first flagellar joint longer than the second, the second slightly emarginate at base, the joints after second scarcely twice as long as thick.....................................PSILOMMA Först.

Mesonotum without furrows.

Antennae filiform, the first flagellar joint shorter than the second..ISMARUS Hal.

MACROHYNNIS Förster.

Hym. Stud., II, p. 131 (1856).

A genus unknown to me, and the type, if still in existence, has never been described. Dr. Förster in speaking of it says:

In the genus *Macrohynnis* we have before us, on account of its peculiar venation, a very striking form which can scarcely be confounded with any other.

The marginal nervure, for instance, is fully twice as long as the rather short marginal cell, and both combined present exactly the appearance of a plow-share of simple construction; while the backward directed branch of the radius, if continued would cross the basal nervure.

All joints in the female antennae are elongate, cylindrical, and so strongly lengthened as to be readily confounded with those of the male were it not for the excision of the first flagellar joint and the pointed abdomen, which betray its sex. The scape is short in both sexes; in the female at the most as long as the first joint of the flagellum, while in the male it is usually somewhat shorter. The lateral angles of the metanotum are slightly projecting. The abdominal petiole is longer than the metanotum, although not abnormally long, and slightly furrowed above. The second segment of the abdomen which viewed laterally appears to be slightly compressed, together with the remaining segments, have a pear-shaped appearance particularly pronounced in the case of the female, while in the male the tip of the abdomen is curved downward, giving the apex rather a more blunt appearance. The sutures between the segments are very fine and the segments themselves are strongly shortened, so that they are almost transversely linear.

DIPHORA Förster.

Hym. Stud., II, p. 130 (1856).

(Type *D. Westwoodii* Förster.)

This genus is likewise unknown to me. Dr. Förster, *op. cit.*, p. 141, mentions as the type, *Diphora Westwoodii*, but gives no description and it has not since, to my knowledge, been described. In his remarks he has the following to say respecting the genus:

If we consider, alone, the shape of the antennæ, we have undoubtedly in the genus *Diphora* the most remarkable form in the Belytoidæ. Not only does the scape attain a considerable length, but the first flagellar joint is also as long as the scape, while the other joints, with the exception of the last, are very short, being even broader than long, on account of which the flagellum has a moniliform appearance; we might therefore easily be led to the conclusion that the antennæ had a double flagellum. The mesonotal furrows are deep and distinct. The abdomen is composed of three segments, the petiole is short and stout, the second segment very large, while the third, which is separated from the second by a distinct suture, attains the length of the petiole, and from its apex projects a short point or nipple, as if from a tube. The marginal cell of the wings is completely closed; the marginal vein very short, even shorter than the stigmal branch, which forms a very acute angle. The post-marginal vein extends but a short distance beyond the apex of the marginal cell. The marginal cell is strongly elongated and narrow, the radius of which has a short blurred and but slightly curved stump, which, in its extension, does not cross the basal nervure.

LEPTORHAPTUS Förster.

Hym. Stud., ii, p. 131. (1856.)

(Type *L. abbreviatus* Först.)

Head transverse, the occiput slightly impressed at the middle, not or indistinctly margined; ocelli 3, prominent, close together in a triangle; eyes rounded, hairy.

Antennæ inserted on a frontal prominence, long, filiform, cylindrical in both sexes, seldom a little thickened toward tips; in ♀ 15-jointed, the scape very long, slender, reaching far above the ocelli, as long as the first two or three flagellar joints together, the pedicel oval, the first flagellar joint the longest, the following gradually shortening, the last being a little longer than the penultimate; in ♂ 14-jointed, the first flagellar joint nearly, or quite, as long as the scape, strongly excised at base.

Thorax ovate, the mesonotum with two profound furrows, the scutellum convex, broadly foveated across the base, the metathorax longitudinally carinated.

Front wings with the marginal vein reaching the costa at about the middle of the wing, a marginal vein as long as, or a little longer than, the triangular closed marginal cell, the latter with a backward directed vein from the stigmal, and a distinct basal cell.

Abdomen longer than the head and thorax together, composed apparently of but 3 segments, the petiole being unusually long, body of ♀ conic, ovate, of ♂ pear-shaped.

Legs long, slender, pilose, or pubescent, the tibial spurs more strongly developed than in *Miota*.

TABLE OF SPECIES.

FEMALES.

Body of abdomen black, the tip alone rufous.
  Antennæ as long as the body, very slender, brownish-yellow, fuscous toward tips,
    the second flagellar joint about 5 times as long as thick.
                                                            L. CONICUS, sp. nov.
Body of abdomen rufous.
  Antennæ extend only to the base of the abdomen, rufous, the 7 terminal joints
    moniliform... L. RUFUS, sp. nov.

MALES.

Entirely rufous..L. RUFUS, Ashm.
Entirely black..L. CONICUS, sp. nov.

Leptorhaptus conicus, sp. nov.
(Plate XIV, Fig. 1, ♀.)

♂ ♀. Length, 3 to 4ᵐᵐ. Black, shining; antennæ, mandibles, palpi, tegulæ and legs brownish-yellow; the antennæ toward tips fuscous; tip of abdomen rufous. Antennæ in ♀ 15-jointed, as long as the body, slender, filiform, the scape very long, the pedicel rounded, the first flagellar joint half the length of the scape, the joints beyond to the last very gradually subequal, the last a little longer than the penultimate, thrice as long as thick. In the ♂ the antennæ are 14-jointed, the scape much shorter, the first flagellar joint is about two-thirds the length of the scape excised at base. Thorax with two furrows, the scutellum with a deep fovea at base, the metathorax carinated and rugose at the sides. Abdomen long, conic-ovate, the petiole thrice as long as thick, fluted. Wings hyaline, the pubescence fuscous, the venation pale brown; the marginal vein is as long as the marginal cell. In the ♂ the venation is darker; the petiole is twice as long as the metathorax or four times as long as thick, slightly narrowed basally and apically, smooth, shining, with a grooved line at sides and a few punctures above; the body of the abdomen is pear-shaped, smooth and shining, with 3 grooved lines at extreme base and pubescent at tip.

HABITAT.—Arlington, Va., and Cedar Point, Md.

Types ♂ ♀ in Coll. Ashmead.

Described from 4 ♀ and 1 ♂.

Leptorhaptus rufus, sp. nov.

♂ ♀. Length, 4 to 4.5ᵐᵐ. The ♂ is entirely rufous; the antennæ, mandibles, palpi, tegulæ and legs pale brownish-yellow, the antennæ toward tips fuscous. First flagellar joint as long as the scape and pedicel together, excised at base, the following joints to the last, long, cylindrical, subequal. Mandibles bifid at tips with a large tooth within. Thorax with two furrows, the scutellum with a large fovea at base which is itself bifoveated at bottom, the metathorax with three keels. Body of abdomen long, pear-shaped, the petiole very long, four times as long as the metathorax or six or more times longer than thick, smooth and polished, but with grooved lines along the sides; second segment without grooved lines at base. Wings hyaline, the marginal vein a little longer than the marginal cell.

The ♀ has the head, thorax, and petiole black, the rest of the abdomen rufous; the abdomen is conic-ovate, the tip curving upwards; the petiole twice as long as the metathorax, fluted, gibbous towards the base; the antennæ, palpi, tegulæ, and legs brownish-yellow; the first flagellar joint much the longest, twice as long as the second.

HABITAT.—Jacksonville, Fla., and Washington, D. C.

Types ♂ ♀ in Coll. Ashmead.

The ♂ was captured in Florida, the ♀ near Washington.

<h2 style="text-align:center">MIOTA Förster.</h2>

Hym. Stud., II, p. 131 (1856).

Head subglobose, the occiput slightly emarginate, delicately margined, ocelli 3, rather close together in a triangle; eyes oval, pubescent.

Antennæ inserted on a frontal prominence, in ♀ long, filiform or sub-filiform, 15-jointed, the scape thicker than the flagellum and a little longer than the first flagellar joint, the second subequal with the first, the joints beyond to the last gradually becoming shorter and shorter, the penultimate joint not longer than thick, the last longer and a little stouter; in ♂ 14-jointed, the scape shorter, not reaching beyond the ocelli, the first flagellar joint emarginate at base, the following long, cylindrical.

Maxillary palpi 5-jointed, labial palpi 3-jointed.

Mandibles acute at tips.

Thorax ovate, the angles of the collar acute, the mesonotum with 2 deep furrows, the scutellum convex, with a deep fovea at base, the metathorax not longer than high, truncate at apex, the truncature margined and connected with a longitudinal carina on the dorsum.

Front wings with venation as in *Leptorhapta*, the marginal vein being rarely longer than the marginal cell.

The type of this genus was not mentioned or described by Förster.

TABLE OF SPECIES.

Abdomen rufo- or brownish-piceous 2
Apex of abdomen yellow or pale rufous, otherwise black.
  Antennæ thickened toward tips, the three penultimate joints about twice as long as thick, the scape only slightly longer than the first flagellar joint ♀ ....M. GLABRA Ashm.
  Antennæ not thickened toward tips, the joints very gradually shortening, the scape longer than the first flagellar joint ♀ ......M. ANALIS, sp. nov.
2. Abdomen brownish-piceous.
  Scape nearly twice as long as the first flagellar joint ♀ ..M. COLORADENSIS Ashm.
  Abdomen rufo-piceous.
  Scape shorter than the first flagellar joint, the latter emarginate at base ♂ .
                                        M. AMERICANA Ashm.

Miota glabra Ashm.

(Plate XIV, Fig. 2. ♀.)

Bull. No. 1, Col. Biol. Assoc., p. 12.

♀. Length, 2.6ᵐᵐ. Black, shining, pubescent. Mandibles pale. Antennæ 15-jointed, cylindrical, a little thickened toward tips, brown, darker at tips; the scape is only slightly longer than the first flagellar joint; pedicel long-oval; the joints after the first become gradually shorter, the three preceding the last about twice as long as thick, the last fusiform and about twice as long as the penultimate. Thorax with two grooves. Mesopleura with a transverse groove below the middle. Scutellum smooth, polished, with a large, deep fovea at base. Metathorax with 3 delicate keels, the posterior angles a little prominent. Legs brownish-yellow, the posterior pair somewhat rufous. Wings hyaline, pubescent, the venation brown; the marginal vein is about as long as the closed triangular marginal cell; the stigmal nervure is straight and less than half the length of the marginal. Abdomen black, polished, the apex rufous, the petiole subopaque, fluted, pubescent beneath.

HABITAT.—West Cliff, Colo.

Type ♀ in Coll. Ashmead.

Described from a single specimen, taken by T. D. A. Cockerell.

Miota analis, sp. nov.

♀. Length, 4ᵐᵐ. Black, shining, sparsely pilose. Antennæ and legs brownish-yellow, the antennæ fuscous beyond the middle. Abdomen piceous-black, the apex yellow. The antennæ are 15-jointed, long, filiform; the scape is very long, reaching far beyond the ocelli, and longer than the first flagellar joint; the pedicel small; first flagellar joint about 6 times as long as thick, the following joints gradually shortening, the last longer than the penultimate. Mesopleura with a deep oblique groove at the middle. Scutellum convex posteriorly, with a deep fovea at base. Metathorax with 3 keels, the sides hairy. Wings hyaline, the marginal vein as long as the marginal cell; the first branch of the stigmal nervure is a little oblique. The petiole of the abdomen is not quite twice as long as the metathorax, striated; the second abdominal segment has a long sulcus at base.

HABITAT.—Carolina.

Type in Royal Berlin Museum.

Described from a single specimen, labeled "Carolina, Zimmerman."

Miota coloradensis Ashm.

Psilomma coloradense Ashm., Bull. No. 1, Col. Biol. Assoc., p. 11.

♀. Length, 3ᵐᵐ. Polished black, pubescent. Eyes almost bare. Mandibles brown. Antennæ filiform, broken at tips, dark-brown above, yellowish beneath; scape very long, cylindrical, nearly twice as long as the first flagellar joint; pedicel not much longer than wide, narrowed at base; first flagellar joint about half the length of the scape, the following joints shorter. Thorax with 2 nearly parallel grooves, and between them anteriorly are 2 abbreviated grooves. Mesopleura smooth, polished, with a deep depression at the middle, terminating in a large fovea posteriorly. Scutellum smooth, rounded off posteriorly, and with a large quadrate fovea across the base. Metathorax smooth, tricarinated. Wings hyaline; veins brown; the marginal vein about as long as the short, triangular, closed marginal cell; the first branch of the radius is about one-third as long as the marginal vein, straight. Legs, including the coxæ, brownish-yellow, the posterior pair slightly obfuscated. Abdomen conic-ovate, brownish-piceous, shining; the petiole is about two and a half times as long as the metathorax, black, opaque, and fluted.

HABITAT.—West Cliff, Colo.

Type in Coll. Ashmead.

Described from a single specimen captured by T. D. A. Cockerell.

Miota americana Ashm.

Psilomma americana Ashm., Can. Ent., XIX, p. 197.

♂. Length, 3ᵐᵐ. Slender, polished black, pubescent. Antennæ and legs pale brownish-yellow, or honey-yellow, the former dusky at tips. Collar above rufous. Eyes nearly bare. Antennæ 14-jointed, setaceous, as long as the body, pubescent; the first flagellar joint is fully as long as the scape, excised at basal half, the following joints all long, but subequal with the first. Metathorax carinated, pubescent. Wings subhyaline, the venation pale brown, the marginal cell shorter than the marginal vein, the first branch of the radius straight, one-third the length of the marginal vein. Abdomen rufo-piceous, the petiole very long, fully thrice as long as the metathorax or nearly 5 times as long as thick, shining but with grooved lines; body of the abdomen pear-shaped, very little longer than the petiole.

HABITAT.—Jacksonville, Fla.

Type in Coll. Ashmead.

ACROPIESTA Förster.

Hym. Stud. II, p. 131 (1856).

(Type A. collaris Först.)

Head transverse, or subglobose, the vertex convex, the occiput straight, margined; ocelli 3, very small, arranged in a triangle; eyes oval, sparsely pubescent.

Antennæ inserted on a frontal prominence; in ♀ 15-jointed, subclavate, submoniliform, the scape slightly bent, extending beyond the ocelli, pedicel rounded, first funicular joint about twice as long as thick, subcylindric, narrowed at base, truncate at tip, the joints beyond all submoniliform, the last very large, fusiform, as long as the two preceding joints together; in ♂ 14-jointed, filiform, the first flagellar joint not, or scarcely, emarginate at base.

Maxillary palpi, 4-jointed; labial palpi, 3-jointed.

Mandibles short, acute, with a small tooth within.

Thorax subovoid, the prothorax visible above as a transverse ridge, produced anteriorly into a slight neck, angles straight; mesonotum trapezoidal with 2 deep furrows; scutellum convex with a deep fovea at base; metathorax carinated, with the posterior angles acute.

Front wings pubescent with a basal and a closed marginal cell; the marginal vein not as long as the marginal cell; the first branch of radius straight or but slightly oblique; hind wings with a single cell.

Abdomen conic-ovate, the petiole a little longer than thick, fluted; in ♀ 7-jointed, the second segment very large, third to sixth very short, the last acute, conical, longer than segments 3 to 6 united.

Legs clavate, pubescent; the middle and hind tibial spurs short and weak; the tarsi long, slender, the basal joint more than twice the length of the second.

Acropiesta flavicauda, sp. nov.

(Pl. XIV, Fig. 3, ♀.)

♀. Length, 3^{mm}. Black, shining, the last abdominal segment yellow. Antennæ 15-jointed, extending to apex of metathorax, incrassated toward tips, rufous, the last joint large, oblong, dark fuscous; first flagellar joint twice as long as the second, the joints beyond moniliform. Thorax with two furrows, the scutellum with a large deep fovea. Metathorax carinated, the sides covered with a fuscous pubescence. Tegulæ yellowish. Wings subhyaline, the venation fuscous, the marginal vein two-thirds the length of the closed marginal cell, the first branch of the radius slightly oblique. Legs rufous. Abdomen conic-ovate, black, polished, the petiole not longer than the metathorax, fluted; the third, fourth, fifth, and sixth segments very short, the last longer than these united, conical and yellow.

HABITAT.—Ottawa, Canada.

Type in Coll. Ashmead.

Described from a single specimen received from Mr. W. Hague Harrington.

Acropiesta subaptera, sp. nov.

♀. Length, 2.2^{mm}. Head globose, black; thorax and abdomen brownish-piceous; scape, pedicel, and legs brownish yellow. The whole body is polished, impunctured, and pubescent; the frontal prominence large; antennæ 15-jointed, moniliform, the first flagellar joint twice as long as the second, cylindric, joints 2 to 5 round, from here to the last transverse-moniliform. Thorax with two almost parallel furrows, very slightly converging toward each other posteriorly; scutellum convex, with a deep quadrate fovea at base; mesopleura with a smooth cross furrow on the disk; metathorax carinated. Abdomen longer than the head and thorax together, conic-ovate, the last segment conical, as long as the third, fourth, fifth, and sixth segments united, and yellowish, the ovipositor issuing from its tip; petiole scarcely twice as long as wide, fluted. Wings not fully developed, extending only to the base of the second abdominal segment, subfuscous and very pubescent.

HABITAT.—Marquette, Mich.

Type in Coll. Ashmead.

Described from a single specimen, taken at the above place by Mr. E. A. Schwarz.

BELYTA Jurine.

Hym., p. 311 (1807); Förster, Hym. Stud., II. p. 130, 133 (1856).

(Type *B. bicolor* Jur.)

Head su globose, the occiput narrowed, rounded; ocelli small, in a triangle; eyes rounded, hairy.

Antennæ inserted on a very prominent frontal projection; in ♀ 15-jointed, stout, moniliform or submoniliform, the scape stout, reaching far above the head, the pedicel rounded, the first flagellar joint obconic, the following joints to last moniliform or transverse-moniliform, usually increasing in size toward apex, the last oblong or ovate; in ♂ 14-jointed, long, filiform, the scape extending beyond the ocelli, pedicel rounded, the first flagellar joint about two-thirds the length of the scape, profoundly excised at base, the following shorter, cylindrical, 3 or 4 times larger than thick.

Maxillary palpi 5-jointed, long; labial palpi short, 3-jointed.

Thorax subovoid, depressed, the prothorax distinct from above, rounded before, the angles obtuse: mesonotum with 2 deep furrows; scutellum not very prominent, subconvex, with a profound fovea at base, and without lateral grooved lines; metathorax with the middle carina not extending to the apex, interrupted or broken, sometimes wanting, posterior angles prominent, acute.

Front wings pubescent, with a basal cell, and usually, but not always, with an open or imperfectly formed marginal cell; the marginal vein short: the first branch of radius short, oblique, with a hook or slight branch at tip; hind wings with one cell.

Abdomen ovate or oblong oval, depressed, 8-jointed, the petiole stout, seldom twice as long as thick, fluted, the second segment very large, with a longitudinal sulcus at base; the following segments all short, the last subtriangular.

Legs rather stout, clavate, pilose, the middle and hind tibiæ subclavate; tibial spurs 1, 2, 2, the last two short, but stout; tarsi 5-jointed, the basal joint nearly 3 times as long as the second.

TABLE OF SPECIES.

FEMALES.

Marginal cell closed.
  Wings fuscous or subfuscous.
    Coxæ black, legs honey-yellow.
      Flagellum fuscous............................B. MONILICORNIS Ashm.
    Coxæ and legs brownish-yellow.
      Flagellum brownish-yellow, the three apical joints dusky.
        Metathorax without a median carina.................B. FRONTALIS sp. nov.
    Coxæ and legs rufous. .
      Metathorax with a forked median carina............B. ERYTHROPUS sp. nov.
Marginal cell open.
  Wings hyaline.
    Metathorax with a forked median keel......................B. TEXANA sp. nov.

Belyta monilicornis Ashm.

Bull. No. 1. Col. Biol. Assoc., p. 12.

♀. Length, 3mm. Robust, black, shining, covered with a fine fuscous pubescence. Eyes bristly. Mandibles piceous. Antennæ 15-jointed, the flagellar joints after the first moniliform, the first joint about twice as long as the pedicel; the scape reddish brown, the rest of the antennæ dark fuscous. Mesonotal furrows distinct, but not deeply impressed. Scutellum with a large fovea at base. Mesopleura deeply impressed posteriorly. Metathorax truncate and squared off at apex. Legs dark honey yellow, the coxæ black, the hind femora slightly dusky above in the middle. Abdomen about as long as the head and thorax together, smooth and polished; the petiole stout, a little longer than thick, grooved and hairy above; the second segment, which occupies the larger portion of the body of the abdomen, has a median longitudinal furrow at base; apex surrounded by sparse, whitish hairs. Wings subhyaline, pubescent, the marginal cell closed, about twice as long as the marginal vein, the first branch of the radius or stigmal vein oblique.

HABITAT.—West Cliff, Col.

Type ♀ in Coll. Ashmead.

Described from a specimen received from T. D. A. Cockerell.

Belyta frontalis, sp. nov.

(Pl. XIV, Fig. 4, ♀.)

♀. Length, 3mm. Polished black, the abdomen brownish; frontal prominence very large, half the length of the head. Antennæ 15-jointed, incrassated toward tips, moniliform, brownish yellow, the two or three apical joints fuscous; scape very long, stout; first flagellar joint twice the length of the pedicel, the following joints to the last moniliform, the last oblong. Mandibles, palpi, and legs pale brownish yellow. Metathorax without a median carina, but with delicate lateral carinæ, the disk polished, the apex margined, the lateral angles acute. Wings subhyaline, the venation pale, the marginal vein one-third longer than the oblique first branch of the radius, the marginal cell closed, a little longer than the marginal vein. Petiole stout, scarcely longer than the metathorax, fluted.

HABITAT.—Delaware.

Type ♀ in Coll. American Entomological Society.

Described from a single specimen.

Belyta erythropus sp. nov.

♀. Length. 3.4ᵐᵐ. Black, shining, sparsely pilose; antennæ and legs rufous or reddish-yellow, the flagellum infuscated, the hind coxæ black basally. Antennæ 15-jointed, moniliform, the first flagellar joint one-third longer than the pedicel, the joints beyond transverse-moniliform, very slightly increasing in size toward the apex, the last conic. Scutellum with a transverse fovea at base. Metathorax with the middle carina forked at the middle; angles bluntly toothed. Wings subfuscous; tegulæ rufous; nervures pale brown; the marginal vein is not longer than the short, closed marginal cell; the first branch of the radius or stigmal vein oblique. Abdomen not longer than the thorax, the sides, apex and beneath, pilose; the petiole is stout, striate, and finely rugose, a little longer than thick; the second segment with some grooved lines at base.

HABITAT.—Wisconsin.

Type ♀ in Royal Berlin Museum.

Described from a single specimen labeled simply, "Wisconsin, Kumlin."

Belyta texana, sp. nov.

♀. Length, 3.4ᵐᵐ. Polished black; antennæ rufous; legs reddish-yellow. Antennæ 15-jointed, stout, the first flagellar joint only a little longer than the pedicel, the joints beyond transverse-moniliform, the last oval. Scutellum flattened, with a deep fovea at base. Metathorax finely rugose, the middle keel forked before the middle, the posterior angles produced. Tegulæ rufous. Wings hyaline, the marginal cell open, the marginal vein longer than the stigmal, the latter with a hook.

Abdomen scarcely as long as the head and thorax together; the petiole short, stout, rugose, scarcely longer than thick, and without any raised lines; base of second segment striated.

HABITAT.—Texas.

Type ♀ in Coll. American Entomological Society.

Described from a single specimen.

OXYLABIS Förster.
Hym. Stud., II, p. 130 (1856).

Lyteba Thoms. Öfv., 1858, p. 180.

(Type O. bisulca Nees.)

Head transverse, a little wider than the thorax, the occiput straight, not margined; ocelli 3, prominent, subtriangularly arranged; eyes oval, hairy.

Antennæ inserted on a frontal prominence; in ♀ 15-jointed, submoniliform; the scape subrobust, extending slightly beyond the ocelli, cylindric; pedicel oblong; first flagellar joint longer, obconic; the joints after the fourth, moniliform or submoniliform, the last ovate; in ♂ 14-jointed, long, filiform, the first flagellar joint about two-thirds the length of the scape, the following a little shorter, all covered with a short pubescence.

Maxillary palpi long, 5-jointed, the last joint the longest; labial palpi short, 3-jointed.

Mandibles short, curved, acute at tip, with a small tooth within.

Thorax as in *Belyta*, but with a large, acute spine at base of metathorax, the scutellum highly convex, with a furrow at sides, the posterior angles of the metathorax acute.

Front wings pubescent, with a basal cell and a closed marginal cell, rarely a little open toward apex. The marginal cell is always distinctly longer than the marginal vein, the stigmal being more or less oblique; hind wings with one cell.

Abdomen oval or oblong-oval, the petiole stout, fluted, a little longer than thick (longer and more slender in the ♂), the second segment very large, occupying most of the surface, sulcate at base, the following segments very short.

Legs as in *Belyta*, pilose, the tibial spurs distinct, basal joint of hind tarsi twice as long as the second.

Oxylabis spinosus Prov.

(Pl. XIV, Fig. 5, ♀.)

Aneurrhynchus spinosus Prov., Faun. Hym., II, p. 560.
Oxylabis spinosus Prov., Add., p. 405.

♀. Length, 2.5ᵐᵐ. Robust, polished black; legs and antennæ rufous. Antennæ 15-jointed, submoniliform; the pedicel is about half as long as the first flagellar joint, which is the longest joint; the three following joints longer than thick, the remaining joints moniliform. Mesonotum with two broad, deep furrows. Scutellum with a deep fovea at base, posteriorly elevated, convex. Postscutellum with a large, acute, rufous spine. Metathorax with a central groove, posteriorly margined, the angles acute. Tegulæ rufous. Wings fuscous, the marginal vein long, the marginal cell almost closed. Abdomen oblong-oval, covered with a grayish pubescence; the petiole broad and stout, hardly longer than thick, fluted; the second segment very large, with sparse punctures and channeled or grooved at base, the following segments very short.

HABITAT.—Cap Rouge, Canada, and Arlington, Va.

Type ♀ in Coll. Provancher.

The above description is drawn up from a specimen, agreeing in all particulars with Provancher's description, and collected by myself in Virginia.

CINETUS Jurine.

Hym., p. 310 (1807); Förster, Hym. Stud., II, pp. 130, 138 (1856).

Head transverse, or subglobose, the occiput rounded: ocelli small, triangularly arranged on the vertex: eyes oval, pubescent.

Antennæ inserted on a frontal prominence; in ♀ 15-jointed, filiform, or at the most subfiliform, rarely slightly thickened at tips, the scape long, cylindrical, reaching considerably beyond the ocelli; pedicel small, rounded, the first flagellar joint from one-half to two-thirds the length of the scape; the following joints cylindrical, becoming quite short before the last, the last a little longer than the penultimate: it is rare that the joints are moniliform; in ♂ 14-jointed, very long, cylindrical, sub-setaceous, pubescent, the scape slender, a little longer than the first flagellar joint; pedicel rounded; first flagellar joint excised at base, the following joints five or more times longer than thick.

Maxillary palpi 5-jointed; labial palpi 3-jointed.

Mandibles short, curved, with a small tooth within.

Thorax subovoid, the pronotum produced anteriorly into a small, rounded neck; mesonotum longer than wide, with two furrows; scutellum convex, with a profound fovea at base; metathorax short, carinated, the posterior angles not prominent or produced into a tooth.

Front wings broad, pubescent, with a basal cell, a closed marginal cell about as long as the marginal vein or longer, the first branch of the radius slightly oblique and the postmarginal nervure well developed.

Abdomen conic-ovate, with eight segments, the petiole in both sexes very long, four or more times longer than thick, the third segment longer than the fourth.

Legs slender, pilose, the femora subclavate, the tarsi very long, slender, the posterior tarsi longer than their tibiæ, the basal joint 2½ times as long as the second, the tibial spurs weak, scarcely discernible.

TABLE OF SPECIES.

FEMALES.

Polished black; abdomen sometimes rufous.
    Antennæ somewhat thickened toward tips, rufous, joints 8 to 14 transverse-quadrate.
        Marginal cell a little more than twice the length of the marginal vein.
                                            C. RUFICORNIS sp. nov.
    Antennæ filiform, the flagellum fuscous.
        Second flagellar joint two-thirds the length of the first; marginal cell 3 times the length of the marginal vein; body of abdomen dull rufous basally......................................C. MACRODYCTIUM Ashm.
    Second and first flagellar joints equal; legs honey-yellow; abdomen black.
        Marginal cell twice as long as the marginal vein; flagellar joints 13 and 14 equal..C. MELLIPES Say
        Marginal cell thrice as long as the marginal vein; flagellar joint 13 distinctly longer than 11........................C. SIMILIS sp. nov.

MALES.

Antennæ rufous, fuscous toward tips; legs rufous.

Marginal cell thrice as long as the marginal vein; abdomen rufous, the third segment not more than twice as long as the fourth.

C. MACRODYCTIUM Ashm.

Marginal cell not thrice the length of the marginal vein; abdomen black, the third segment about 4 times as long as the fourth.

C. CALIFORNICUS sp. nov.

Cinetus ruficornis, sp. nov.

♀. Length, 3ᵐᵐ. Polished, black; antennæ rufous; tegulæ and legs brownish-yellow. Antennæ 15-jointed, subclavate, the first flagellar joint a little more than twice as long as the pedicel, the joints beyond to the eighth gradually subequal, the eighth moniliform, the ninth, tenth, eleventh, and twelfth, transverse, the last cone-shaped. Metathorax carinated, the space between the keels smooth, shining. Wings hyaline, the venation brown, the marginal cell twice as long as the marginal vein, the first branch of the radial vein slightly oblique, a little shorter than the marginal. Abdomen conic-ovate, the petiole 1½ times as long as the metathorax, fluted, the second segment with a longitudinal sulcus at base, the third segment twice as long as the fourth, the seventh as long as the fourth, fifth, and sixth together, piceous, eighth very short, yellowish.

HABITAT.—Arlington, Va.

Type in Coll. Ashmead.

Cinetus macrodyctium, sp. nov.

♂ ♀. Length, 3 to 3½ᵐᵐ. Black, polished, pubescent; legs rufous; abdomen dull rufous, in ♀ with the petiole black. Antennæ in ♀ about as long as the body, filiform, 15-jointed, dark fuscous, pubescent, the first two joints rufous; first flagellar joint about one-third longer than the second, the joints beyond to eighth gradually shortening, the eighth being about half the length of the second; the joints from 8 to the last short, all, however, longer than thick, the last longer than the penultimate; in ♂ 14-jointed, the joints all long, cylindrical; the first three rufous, the following fuscous; the first flagellar joint is as long as the scape, and a little longer than the second, excised at base. Wings hyaline, pubescent, the marginal cell thrice as long as the marginal vein; the stigmal vein almost straight, about half the length of the marginal, or a little longer in the ♂. Abdomen a little longer than the head and thorax together, the petiole in ♂ strongly fluted, about twice the length of the metathorax, in ♀ a little shorter; rest of the body polished, pubescent; the second segment at base with a longitudinal sulcus.

HABITAT.—Arlington, Va., and West Cliff, Colo.

Types in Coll. Ashmead.

The ♀ of this species was received from Mr. T. D. A. Cockerell, from Colorado, but I have no doubt it is properly correlated with a ♂ taken in Virginia, as it agrees in venation and general appearance, although the color of the abdomen is slightly different.

Cinetus mellipes Say.

(Pl. xiv, Fig. 6, ♀.)

Belyta mellipes Say Lec., Ed. Say's Works, ii, p. 726.
Xenotoma mellipes Ashm. Can. Ent., xix, p. 199.
Cinetus mellipes Cr. Syn. Hym., p. 250.

♀. Length, 2.5ᵐᵐ. Polished black, shining, pubescent; antennæ fuscous, the first two joints honey-yellow; legs honey-yellow. Antennæ 15-jointed, filiform, reaching to the middle of the abdomen; scape more than twice as long as the first flagellar joint; first and second flagellar joints equal, 3 times as long as the pedicel; the following joints to eighth gradually shortening; from here to the last much shorter, the thirteenth and fourteenth being only a little longer than thick, equal, and shorter than the last which is conic. Wings hyaline, the marginal cell only twice the length of the marginal vein, the venation fuscous. Abdomen one-third longer than the head and thorax together, the petiole being more than twice as long as the metathorax, fluted; the body of abdomen is smooth, shining, covered with sparse white hairs along the sides, at tip and beneath, the first segment occupying nearly the whole surface, with a longitudinal sulcus at base; the second is very short, but twice the length of the third; the last conic, piceous.

HABITAT.—Indiana and Arlington, Va.

Specimens in Coll. Ashmead.

Cinetus similis, sp. nov.

♀. Length, 2.8ᵐᵐ. Polished black, pubescent; scape, pedicel, and legs honey-yellow; rest of antennæ subfuscous. Head transverse, wider than the thorax. Antennæ 15-jointed, filiform; the scape fully 4 times as long as the first flagellar joint, obclavate, first and second flagellar joints equal, twice as long as the pedicel or a little more than 3 times as long as thick, the following joints very gradually subequal, the last conical, one-third longer than the penultimate. Wings hyaline, pubescent, the marginal cell about 3 times as long as the marginal vein. Abdomen conic-ovate, black, polished, the petiole opaque, fluted, nearly twice the length of the metanotum, the third segment short but about twice the length of the fourth.

HABITAT.—Arlington, Va.

Type in Coll. Ashmead.

Allied to *C. mellipes* Say, but differs in the relative length of the antennal joints and in having a longer marginal cell.

Cinetus californicus, sp. nov.

♂. Length, 2.5ᵐᵐ. Polished black, nearly devoid of pubescence, the

petiole beneath with long, sparse hairs. Antennae fuscous, the scape
and pedicel rufous, yellowish at base, the joints long, cylindrical, the
first flagellar joint the longest. Legs reddish-yellow, paler at articulations. Tegulae yellowish. Wings hyaline, the venation dark brown.
Metathorax smooth, shining, carinated. Abdomen black, shining,
depressed, the petiole 2½ times as long as the metathorax, striated, the
second segment striated at the extreme base, with a central longitudinal
sulcus, the third segment nearly as long as all the following together.

HABITAT.—Santa Cruz Mountains, California.

Type in Coll. Ashmead.

Described from a single specimen. The length of the third abdominal segment renders this species easy of identification.

Cinetus nasutus Prov.

Add. et Corr., p. 178; Cress., Syn. Hym., p. 251.

♀. Long. 15 pce. Noir avec les pattes roux-jaunâtre. La face prolongée en
museau dans sa partie inférieure. Tête subglobuleuse; les antennes fortes, de 15-
articles, le 1er très long, les derniers granuleux, s'épaississant insensiblement en
massue, quelque peu brunâtre. Thorax rétréci en avant, les sillons parapsidaux du
mésonotum profonds, le divisant en 3 lobes. Ailes hyalines, velues, à radiale très
petite, la nervure stigmatique en voyant un rayon vers la base de l'aile. Pattes d'un
beau jaune-miel, sans aucune tache. Abdomen à pédicule fort et assez allongé,
scabre, le reste en oval, déprimé, poli, le 2e segment très grand avec une petite
fossette à la base. (Provancher.)

HABITAT.—Ottawa, Canada.

Type in Coll. Provancher.

Not recognized, and probably not a true Cinetus.

XENOTOMA Förster.

Hym. Stud., ii, p. 129 (1856).

Acoretus Hal., Nat. Hist. Rev., 1857, p. 169.

(Type X. bicolor Nees nec Jurine).

Head transverse, the occiput delicately margined behind; ocelli 3,
triangularly arranged; eyes broadly oval, pubescent.

Mandibles large, acute, sickle-shaped, crossing each other at tips.

Antennae inserted on a frontal prominence, filiform or subfiliform, in
♀ 15-jointed, the 5 or 6 terminal joints very short, the last conic or
ovate, the scape cylindric, a little curved, and swollen at the middle;
the first flagellar joint is about twice as long as thick and a little longer
than the second; in ♂ 14-jointed, the scape not, or scarcely, reaching
beyond the ocelli, the first flagellar joint as long, or nearly as long, as
the scape, excised at base.

Thorax subovoid, the anterior angles of pronotum straight; mesonotum with 2 furrows; scutellum convex, depressed, and with a deep
fovea at base; metathorax as long as high, carinated, truncate, and
margined posteriorly.

Front wings pubescent, the marginal fringe short, a basal and a closed
marginal cell, the marginal vein never as long as the marginal cell,

usually about half as long; stigmal vein or first branch of the radius slightly oblique.

Abdomen conic-ovate, the tip with a slight upward curve, the petiole as long as, or a little longer than, the metathorax, distinctly fluted or striated; the third segment is short, but usually longer than the fourth.

Legs much as in *Cinctus*, but with the tibial spurs distinct.

This genus is at once recognized by the long falcate mandibles.

The species known to me may be separated as follows:

Head, thorax, and petiole black.
> Scape, pedicel, and legs yellow ♂ ....X. XANTHOPUS sp. nov.
Head, except clypeus, black; thorax and abdomen dull rufous.
> Antennæ and legs brownish-yellow ♂ ....X. MANDIBULARIS sp. nov.
Brownish; vertex and disk of thorax blackish ♀ ....X. XANTHOPUS.

Xenotoma xanthopus, sp. nov.

♂. Length, 2.1ᵐᵐ. Polished black, pubescent; the abdomen reddish-brown; mandibles, palpi, two basal joints of antennæ, and legs bright yellow; flagellum fuscous. Antennæ 15-jointed, setaceous, the scape very little longer than the first flagellar joint; the first flagellar joint the longest, slightly excised at base, the joints beyond very slightly shortening. Metathorax smooth, with a middle carina, the lateral carinæ irregular and broken, descending obliquely toward the posterior coxæ. Wings hyaline, the marginal cell fully two and a half times as long as the marginal vein. Abdomen as long as the head and thorax together, pilose beneath, the petiole a little longer than the metathorax with raised lines.

♀. Length, 3ᵐᵐ. Brownish, with a fuscous pubescence; vertex of head and disk of thorax blackish; antennæ, mandibles, palpi, and legs yellow. Antennæ 15-jointed, filiform, a little shorter than the body; scape more than thrice as long as the first flagellar joint; pedicel rounded; first flagellar joint nearly thrice as long as the pedicel, the joints beyond to sixth very gradually subequal, joints from sixth to last hardly longer than thick. Abdomen a little longer than the head and thorax together, the petiole twice as long as the metathorax, fluted, the third segment one-third longer than the fourth, the second, with faint punctures laterally otherwise it agrees with ♂.

HABITAT.—New Jersey.

Types in Coll. Ashmead.

Described from 1 ♂ and 1 ♀ specimen.

Xenotoma mandibularis, sp. nov.

(Pl. XIV, Fig. 7, ♀.)

♂ ♀. Length, 2.5ᵐᵐ. Polished, pubescent; head black, the thorax and abdomen varies from a blackish or brown to dull rufous, more especially rufous in the ♂. Antennæ and legs pale brownish-yellow. Mandibles long, sickle-shaped, crossing and extending far beyond each other at tips. Eyes oval, vertical. Antennæ in ♀ 15-jointed, filiform, the first flagellar joint thrice as long as the pedicel, the joints beyond to the ninth gradually shortened, joints tenth to thirteenth about equal, a little longer than thick; in ♂ 14-jointed, filiform, the first flagellar joint excised at base, about two-thirds as long as the scape, all the joints beyond long, cylindric, subequal. Metathorax short, abrupt, and keeled, the posterior angles subacute. Wings hyaline, the marginal cell scarcely twice as long as the marginal vein, the stigmal slightly shorter than the marginal. Abdomen ovate, depressed, the petiole slightly more than twice as long as the metathorax, in ♂ fluted, in ♀ with a short striate sculpture but not distinctly fluted or grooved.

HABITAT.—Texas.

Types in Coll. Ashmead.

Described from two specimens.

ZELOTYPA Förster.

Hym. Stud., II, p. 130 (1856).

Head transverse, the ocelli in a triangle, the eyes oblong-oval, pubescent.

Antennæ inserted on a frontal prominence, in ♀ 15-jointed, filiform, cylindrical, the last joint enlarged, fusiform, the scape long, slender, extending far above the ocelli; in ♂ 14-jointed, the scape extending scarcely beyond the ocelli, the first flagellar joint about two-thirds the length of the scape, excised at base.

Mandibles meeting at tips, bifid.

Thorax subovoid, the anterior angles of pronotum straight; mesonotum with 2 furrows; scutellum convex, with a deep fovea at base; metathorax carinated, the posterior angles not prominent.

Front wings with a basal and a closed marginal cell, the latter large, much larger than the marginal vein, the marginal vein being usually shorter than half the length of the marginal cell; stigmal or first abscissa of radius straight, at a right angle with the margin.

Abdomen ovate, the petiole not or scarcely longer than the metanotum, strongly fluted or striated.

Legs rather long, pubescent; the anterior tibiæ clavate, their tarsi twice as long as the tibiæ; middle and hind tibiæ long, slender, very little shorter than their tarsi; basal joint of hind tarsi about 3 times as long as the second; tibial spurs short.

The type of this genus has never been described. Our species may be thus tabulated:

TABLE OF SPECIES.

MALES.

Antennæ rufous... 3
Antennæ blackish or fuscous, the three or four basal joints pale rufous.
  Marginal vein shorter than the marginal cell and less than twice as long as the first branch of radius 2
  Marginal vein as long as the marginal cell, or 2½ times as long as the first branch of radius.
    Legs reddish-yellow.................................Z. TEXANA, sp. nov.

2. Legs brownish-yellow.

Marginal cell 2½ times as long as the marginal vein; abdomen brownish or rufous basally .... Z. LONGICORNIS, sp. nov.

Legs yellow.

Marginal cell hardly twice as long as the marginal vein; abdomen entirely black .... Z. FLAVIPES, sp. nov.

3. Marginal cell and the marginal vein long, about equal in length.

Legs reddish-yellow .... Z. RUFICORNIS, sp. nov.

Zelotypa texana, sp. nov.

♂. Length, 2.5ᵐᵐ. Polished black, with a glittering pubescence; antennæ fuscous, the three basal joints yellow; legs reddish-yellow. Face with white pile. Angles of collar acute. Antennæ 14-jointed, filiform, the first flagellar joint thrice as long as the pedicel, very slightly excised at the base, the following joints shorter, subequal. Metathorax carinated, pubescent. Wings hyaline, the venation brown, the marginal cell about as long as the marginal vein, the latter fully 2½ times as long as the straight first branch of radius. Abdomen long, oval, the petiole 2½ times as long as thick, fluted.

HABITAT.—Texas.

Type in Coll. Ashmead.

Described from a single specimen.

Zelotypa longicornis, sp. nov.

(Pl. xv, Fig. 1, ♂.)

♂. Length, 3.2ᵐᵐ. Polished black, pubescent; antennæ fuscous, basally yellowish; tegulæ and legs brownish yellow. Antennæ 14-jointed, filiform, and pubescent, longer than the body, the first flagellar joint as long as the scape, excised one-third its length at base, the joints beyond all long, very gradually shortened, the penultimate nearly 5 times as long as thick. Metathorax carinated. Wings subfuscous, the marginal cell long, 2½ times as long as the marginal vein, the latter ½ longer than the stigmal. Abdomen ovate, black, shining, the second segment rufous basally, the petiole 1½ times as long as the metathorax, or 2½ times as long as thick, fluted.

HABITAT.—Arlington, Va.

Type in Coll. Ashmead.

Zelotypa flavipes, sp. nov.

♂. Length, 2½ᵐᵐ. Polished black, sparsely pubescent, more densely pubescent on the abdomen beneath; antennæ dark fuscous, almost black, the three basal joints rufous; legs pale yellow. Antennæ 14-jointed, as long as the body, the first flagellar joint two-thirds the length of the scape, excised for nearly half its length basally, the joints beyond gradually decreasing, the penultimate scarcely thrice as long as thick, the last a little longer. Metathorax carinated. Wings subfuscous, the marginal cell not more than twice as long as the marginal vein, the latter about twice as long as the straight first branch of radius. Abdomen oblong-oval, black, polished, with a glittering white pile beneath, the petiole more than twice as long as thick, fluted.

HABITAT.—Fort George Island, Fla.

Type in Coll. Ashmead.

Zelotypa ruficornis, sp. nov.

♂. Length, 3ᵐᵐ. Polished black, covered with a sparse white pubescence; antennæ pale rufous: legs reddish-yellow, the tarsi pale. Antennæ 14-jointed, slender, filiform, as long as the body, the first flagellar joint a little shorter than the scape, strongly excised half its length basally, the following joints all long, slightly subequal, about 5 times as long as thick. Metathorax carinated. Wings subhyaline, the marginal cell as long as the marginal vein, the latter nearly twice as long as the stigmal. Abdomen oblong-oval, black, shining, the tip and beneath with white hairs, the petiole very long, more than twice as long as the metathorax, or about 3½ times as long as thick, fluted.

HABITAT.—Arlington, Va.

Type in Coll. Ashmead.

PANTOCLIS Förster.

Hym. Stud., II, p. 131 (1856).

(Type ? Balyta brevis Nees.)

Head subglobose or transverse, the occiput rounded; ocelli 3, close together in a triangle; eyes broadly oval or rounded, pubescent. Antennæ inserted on a frontal prominence: in ♀ 15-jointed, gradually thickened toward tips, submoniliform or moniliform; the scape is stout and extends to or a little beyond the ocelli, or about as long as the first four or five flagellar joints united; pedicel rounded; the first flagellar joint a little longer than thick, obconic; the second and third short, the following gradually enlarged, moniliform, submoniliform, or transverse; in ♂ 14-jointed, filiform or setaceous, densely pubescent; first flagellar joint about two-thirds as long as the scape, excised basally, the following joints shorter, the three or four joints before the last less than three times as long as thick, rarely longer.

Mandibles acute with a tooth within, slightly crossing each other at tips. Thorax as in Zelotypa, the posterior angles of metathorax often prominent.

Front wings pubescent, with a basal cell and a small triangular closed marginal cell, the marginal vein short, rarely much longer than the first abscissa of the radius, the latter oblique.

Abdomen ovate, more rarely oblong-oval, the petiole short, the second segmen toccupying most of the surface, with a median longitudinal sulcus at base, the following segments short, about equal; in the ♂ the petiole is longer, about three times as long as thick.

Legs pubescent, the tibial spurs short but distinct.

TABLE OF SPECIES.

Pantoclis montana, sp. nov.

♂ ♀. Length, 2.5 to 3.2ᵐᵐ. Polished black, pubescent; legs rufous, the hind coxae black. Antennae in ♀ 15-jointed, subclavate, the joints rounded, the first flagellar joint twice as long as the pedicel, slender towards the base; the second about half the length of the first, the joints from the fifth quadrate-moniliform, slightly pedicellate, the last one-shaped; in ♂ 14-jointed, long, setaceous, the first flagellar joint two-thirds the length of the scape, excised towards base, the following joints subequal, the penultimate twice as long as thick, the last one-third longer. Metathorax with 5 keels, a central, and two parallel lateral keels; the lateral angles prominent. Abdomen conic-ovate, a little depressed; the petiole as long as the metathorax, strongly fluted; the third, fourth, and fifth segments very short, equal; the sixth slightly longer; the seventh more than twice as long as the sixth and slightly longer than the eighth.

HABITAT.—Santa Cruz Mountains, California and Colorado.

Types in Coll. Ashmead.

Described from several specimens.

Pantoclis picipes, sp. nov.

♂. Length, 2.6ᵐᵐ. Polished black; antennae piceous-black, the base of the first flagellar joint pale brown; legs piceous, trochanters, tips of femora, and tibiae and tarsi, brownish-yellow. Antennae 14-jointed, setaceous, as long as the body, the first flagellar joint two-thirds the length of the scape, the joints beyond shorter, very gradually shortened to the last, which is a little longer, the penultimate being about thrice as long as thick. Tegulae piceous. Wings subfuscous, the venation brown, the marginal cell less than twice as long as the marginal vein, the first branch of radius very oblique and very little shorter than the marginal. Metathorax with 5 keels. Abdomen smooth, shining, with a sparse white pile along the sides at tip and beneath, the petiole not quite as long as the metathorax, channelled.

HABITAT.—Washington, D. C.

Type in Coll. Ashmead.

Pantoclis megaplasta, sp. nov.

♀. Length, 2ᵐᵐ. Head and thorax black; metathorax and abdomen brownish-piceous; antennae fuscous; the scape, pedicel, and legs pale brownish-yellow. Antennae 15-jointed, filiform, the last joint enlarged, oblong, about 5 times as large as the penultimate joint, the first flagellar joint very slightly longer than the pedicel, but slenderer, the following joints subequal, the 2 penultimates scarcely longer than thick. Tegulae yellow. Wings hyaline, the venation pale brown, the marginal cell twice as long as the marginal vein, the first branch of the radius oblique, shorter than the marginal. Metathorax keeled, brownish. Abdomen brownish-piceous, the tip slightly turned upwards, the petiole a little more than twice as long as thick, fluted.

HABITAT.—Manhattan, Kans.

Type in Coll. Ashmead.

Described from a single specimen received from Prof. E. A. Popenoe. The large terminal antennal joint and the color of the abdomen, the tip curving upwards, readily distinguish the species.

Pantoclis insularis, sp. nov.

♂ ♀. Length, 2 to 2.6^{mm}. Polished black, pubescent; antennæ and legs brownish-yellow. Antennæ in ♀ 15-jointed, gradually thickened toward tips, the last joint cone-shaped, twice as long as the penultimate, scape more than thrice as long as the first flagellar joint, the latter twice as long as the pedicel, the joints beyond to the seventh shortening, from the seventh to the last moniliform, wider than long; in ♂ 14-jointed, setaceous, fuscous, the first flagellar joint excised at base, a little shorter than the scape, the latter yellowish, the joints beyond subequal. Tegulæ yellow. Wings hyaline, the venation yellowish, the marginal cell twice as long as the marginal vein, the first branch of the radius two-thirds the length of the marginal. Metathorax with 3 keels, smooth on disk. Abdomen black, the tip turned upwards, the second segment basally and the sixth rufous, venter pilose, the petiole 1½ times as long as thick, fluted.

HABITAT.—Fort George Island, Florida.

Types in Coll. Ashmead.

Described from several specimens.

Pantoclis crassicornis, sp. nov.

♂ ♀. Length, 3 to 4^{mm}. Polished black, and covered with a fuscous pubescence; antennæ, mandibles, and legs rufous. Head in ♀ longer than wide, the face with two furrows extending to the clypeus. Antennæ 15-jointed, much incrassated, the scape very stout, as long as the pedicel and first four flagellar joints united, the first flagellar joint about twice as long as the pedicel, the joints beyond moniliform, after the sixth transverse-moniliform and slightly pedicellate. Tegulæ rufous. Wings, subfuscous, the venation brown, the marginal cell not longer than the marginal vein, the first branch of the radius oblique, about half the length of the marginal. Metathorax flat, keeled, with the posterior angles prominent. Abdomen oblong ovate, depressed, black, pubescent, as long as the head and thorax together, the second segment rufo-piceous basally, the petiole stout, about as long as the metathorax.

The ♂ is the smaller, more slender, the head transverse, the antennæ 14-jointed, filiform, brown; the scape and pedicel yellow; the flagellar joints nearly equal in length; the first emarginated at base; none of the joints are less than 4 times as long as thick.

HABITAT.—Arlington, Va.

Types in Coll. Ashmead.

Pantoclis analis, sp. nov.

(Pl. xv. Fig. 2, ♀.)

♀. Length, 3.2ᵐᵐ. Polished black, sparsely pubescent; antennæ rufous, blackish toward tips; legs brownish-yellow. The frontal prominence beneath is rugoso-punctate. Antennæ 15-jointed, thickened toward tips, moniliform; the scape rather stout, twice as long as the pedicel, the joints beyond moniliform, after the second, transverse-moniliform, the last conic, a little longer than the penultimate. Mesopleura impressed with a large fovea. Tegulæ yellowish. Wings subhyaline, the marginal cell about five times as long as the marginal vein, the latter shorter than the oblique first branch of the radius. Metathorax carinated. Abdomen rufo-piceous, the two last segments reddish-yellow, the seventh and eighth segments about equal, as long as the third, fourth, and fifth united; the spiracles distinct; the petiole stout, as long as the metathorax, fluted.

HABITAT.—Washington, D. C.

Type in Coll. Ashmead.

Described from a single specimen captured by E. A. Schwarz.

Pantoclis ruficauda, sp. nov.

♀. Length, 3ᵐᵐ. Polished black; the apex of abdomen with long white hairs, the last segment conically pointed, reddish; antennæ and legs brownish-yellow, the former dusky toward tips. Antennæ 15-jointed, filiform-moniliform, the first flagellar joint twice as long as the pedicel, the joints beyond to the last, equal, moniliform, scarcely wider than long, the last cone-shaped, a little longer than the penultimate; metathorax carinated. Tegulæ yellow. Wings subhyaline, the marginal cell hardly 5 times as long as the marginal vein, the latter very slightly shorter than the oblique first branch of the radius. Abdomen conic-ovate, black, highly polished, the last segment conically pointed, twice as long as the seventh, the sixth a little shorter than the seventh, but twice as long as the fifth, the petiole as long as the metathorax, not longer than thick, fluted.

HABITAT.—Arlington, Va.

Type in Coll. Ashmead.

Described from one specimen.

Pantoclis flavipes, sp. nov.

♀. Length, 2.5ᵐᵐ. Polished black, pubescent; antennæ and legs bright yellow, the former after the first flagellar joint slightly fuscous; mandibles and clypeus rufous. Antennæ 15-jointed, subfiliform, the first flagellar joint two-thirds the length of the scape, or thrice as long as the pedicel, the joints beyond to the seventh subequal, beyond short, not, or scarcely, longer than thick. Metathorax carinated. Tegulæ yellow. Wings subhyaline, the venation brown, the marginal cell 4 times as long as the marginal vein, the latter slightly shorter than the oblique first branch of the radius. Abdomen black, pubescent beneath and at tip, the second segment rufous toward base, both above and beneath, the petiole shorter than the metathorax, not longer than thick, striated, segments third, fourth, fifth, and sixth short, equal, seventh twice as long as the sixth and a little longer than the eighth.

HABITAT.—Arlington, Va.

Types in Coll. Ashmead.

Pantoclis coloradensis Ashm.

Zelotypa coloradensis Ashm., Bull. No. 1, Col. Biol. Assoc., p. 12.

♀. Length, 3ᵐᵐ. Subrobust, polished black, pubescent. Face with a deep impression just above the clypeus. Eyes pubescent. Antennæ brownish-yellow, the flagellum fuscous; the scape is nearly twice as long as the first flagellar joint; pedicel globular; first flagellar excised for one-third its length basally; the following joints shorter, about thrice as long as thick. Thorax with 2 furrows, converging posteriorly; mesopleura impressed across the disk; scutellum convex, with a deep fovea across the base; metathorax short, carinated. Legs brownish-yellow, the posterior coxæ dusky at base. Wings hyaline, pubescent, the venation pale brown; the marginal vein is only one-third the length of the marginal cell, the first branch of the radius about as long as the marginal, oblique.

HABITAT.—West Cliff, Colo.

Type in Coll. Ashmead.

Described from one specimen received from T. D. A. Cockerell.

Pantoclis californica, sp. nov.

(Pl. xv, Fig. 2a, ♂ antenna.)

♂ ♀. Length, 2 to 2.2ᵐᵐ. Polished black; antennæ, mandibles, tegulæ, and legs rufous, the former dusky toward tips. Antennæ in ♀ 15-jointed, clavate, moniliform, the first flagellar joint twice as long as the pedicel, the joints beyond moniliform, after the fourth, transverse-moniliform; in ♂ 14-jointed, filiform, pale brownish; the first flagellar joint about as long as the scape, excised at base, the joints beyond very gradually shortening, the penultimate a little shorter than the last, 2½ times as long as thick. Metathorax carinated, the posterior angles projecting. Wings subfuscous; the marginal cell about thrice as long as the marginal vein, the latter two-thirds the length of the oblique first branch of the radius. Abdomen conic-ovate, polished black; the petiole as long as the metathorax, fluted; segments third, fourth, and fifth, very short, equal; sixth a little longer; seventh and eighth about equal, slightly more than twice as long as the sixth; the eighth rufous.

HABITAT.—Santa Cruz Mountains, California.

Types in Coll. Ashmead.

Described from 1 ♂. 2 ♀ specimens. Comes near to *P. montana*, but in that species the antennæ and hind legs are black, the seventh abdominal segment is a little longer than the eighth, and the latter is black, not rufous.

Pantoclis rufescens, sp. nov.

♀. Length, 3ᵐᵐ. Entirely brownish-piceous; antennæ and legs, brownish-yellow. Antennæ 15-jointed, subfiliform-moniliform; the first flagellar joint twice as long as the pedicel, the joints beyond moniliform; after the third transverse-moniliform, the last one-third longer than the penultimate. Metathorax short, carinated, the posterior angles slightly prominent. Wings subfuscous, the marginal cell 2½ times as long as the marginal vein, the latter slightly shorter than the oblique first branch of the radius. Abdomen conic-ovate, the petiole 1½ times as long as thick, fluted; the segments, 3, 4, and 5 short, equal; the sixth a little longer, the seventh thrice as long as the sixth and a little shorter than the eighth.

HABITAT.—Texas.

Type in Coll. Ashmead.

Described from a single specimen.

Pantoclis floridana, sp. nov.

♀. Length, 2.5ᵐᵐ. Polished black, pubescent; antennæ, mandibles, tegulæ, and legs brownish-yellow, the former dusky toward tips. Antennæ 15-jointed, filiform-moniliform; the first flagellar joint slightly more than twice as long as the pedicel; the joints after the third equal, moniliform. Metathorax with 5 keels, the posterior angles not prominent. Wings subfuscous, the marginal cell thrice as long as the short marginal vein, the latter shorter than the oblique first branch of the radius. Abdomen oval; the petiole about 1½ times as long as thick, striated; the segments 3, 4, 5, and 6 very short, about equal; the seventh longer, rufous.

HABITAT.—Jacksonville, Fla.

Type in Coll. Ashmead.

ZYGOTA Förster.

Hym. Stud., ii, p. 131 (1856).

(Type *Belyta abdominalis* Nees.)

Head transverse or subglobose; the occiput straight, delicately margined behind; ocelli 3, arranged in a curved line, or subtriangularly arranged; eyes oval, hairy.

Antennæ inserted on a frontal prominence; in ♂ 14-jointed, filiform; the scape rather short, extending scarcely beyond the ocelli; the first flagellar joint as long, or nearly as long, as the scape, deeply excised at base; in ♀ 15-jointed, thickened toward tips; the scape long and stout, reaching beyond the ocelli; the first flagellar joint ob conic, longer than the pedicel; the following joints transverse, sub moniliform or perfoliated; the last ovate, conical, or oblong.

Maxillary palpi with joints 2 and 3 dilated.

Thorax as in *Zelotypa*, the posterior angles of metathorax promi nent, acute, or toothed.

Front wings pubescent, with a basal cell and a very long, lanceolate marginal cell, which is usually open at tip: the marginal vein is short, not or scarcely longer than the stigmal or the first abscissa of the radius.

Abdomen ovate, slightly pointed at tip; the petiole usually stout, a little longer than thick, fluted.

Legs similar to *Pantoclis*, but the anterior tibiæ in the ♂ is twisted and armed at the middle with a spine or tooth.

TABLE OF SPECIES.

MALES.

Marginal cell not especially long, closed; the first branch of radius oblique; antennæ dark fuscous.............................Z. CALIFORNICA, sp. nov.
Marginal cell long, open; the first branch of radius straight ....Z. AMERICANA Ashm.

FEMALES.

First flagellar joint not twice as long as the pedicel.............Z. AMERICANA Ashm.
First flagellar joint about twice as long as the pedicel..........Z. TEXANA, sp. nov.

Zygota californica, sp. nov.

♂. Length, 3.5ᵐᵐ. Black, shining, covered with a fuscous pubescence. Antennæ 14-jointed, brown-black, extending to the middle of the abdo men; the first flagellar joint slightly shorter than the scape, strongly excised at base; the joints beyond from 3½ to 4 times as long as thick, the last a little longer than the penultimate. Metathorax smooth, shining between the central and lateral carinæ. Tegulæ rufo-piceous. Wings subfuscous, the venation piceous; the marginal vein is slightly longer than the oblique first branch of the radius, the marginal cell closed.

Legs brownish-yellow, the hind coxæ slightly dusky basally, the an terior tibiæ with a strong spine at about two-thirds its length. Ab domen black, polished, with a greyish pubescence beneath; the petiole twice as long as the metathorax.

HABITAT.—Placer County, Cal.

Type in National Museum.

Described from a single specimen collected by Mr. A. Koebele, in September.

Zygota americana Ashm.

(Pl. XV, Fig. 3, ♂.)

Can. Ent., XX, p. 51.

♂ ♀. Length, 3 to 3.5ᵐᵐ. Polished black, covered with a fuscous pubescence. Antennæ 15-jointed, filiform-moniliform; the first flagel-

lar joint not twice as long as the pedicel, the three following joints
nearly round, the remaining to the last, transverse-moniliform, subpedicellate. the last cone-shaped. Mesonotal furrows broad, distinct.
Scutellum with a deep depression at base. Metathorax carinated.
Legs, including coxae, honey-yellow, the first joint of anterior tarsi
long, deeply emarginate at base. Wings fusco-hyaline, pubescent, the
venation brown, the marginal cell very long. not quite closed, the marginal vein very little longer than the first abscissa of radius. Abdomen ovate, black, slightly rufous basally, the venter densely pubescent,
the petiole scarcely twice as long as wide, fluted.

The ♂ antennae are 14-jointed, long, filiform, pubescent, the pedicel
rounded, the first flagellar joint about 5 times as long as thick, excised
basally, the following joints shorter. The anterior tibiae is twisted and
produced outwardly into a strong tooth or spine, the apical spine long,
curved.

HABITAT.—Ottawa, Canada.

Types in Coll. Ashmead.

Zygota texana. sp. nov.

♀. Length, 3ᵐᵐ. Black, shining, somewhat densely covered with a
fine fuscous pubescence; the apex of the seventh dorsal abdominal
segment and the last ventral segment, rufous; antennae rufous: tegulae
and legs brownish-yellow; palpi and mandibles pale.

Antennae 15-jointed, moniliform, the first flagellar joint about twice
as long as the pedicel, the last oblong, the five preceding transverse-
moniliform. Wings subfuscous. the venation brown, the marginal cell
very long. more than 5 times as long as the marginal vein, the latter
scarcely longer than the first abscissa of radius, the latter almost
straight. Metathorax carinated. Abdomen about as long as the head
and thorax together. the petiole along the sides and at apex, and the
venter, densely pubescent, the second segment laterally very finely
punctate.

HABITAT.—Texas.

Type in Coll. Ashmead.

ACLISTA Förster.

Hym. Stud., II, p. 128 (1856).

Head globose or subglobose, the cheeks usually full; ocelli 3, small,
in a triangle. not prominent; eyes rounded, pubescent.

Antennae inserted usually on a very prominent frontal projection:
in ♀ 15-jointed, incrassated or clavate-moniliform; scape stout, reaching far above the ocelli; pedicel smaller than the first flagellar joint,
the latter usually only a little longer than thick; the following joints
all moniliform or transverse-submoniliform, sometimes slightly pedicellate; in ♂ long, filiform, 14-jointed, covered with a short, dense pubes-

cence, the scape reaching beyond the ocelli, the pedicel small, annular, or rounded, the first flagellar joint lengthened, excised one-third its length at base, the following joints shorter, the terminal joint usually shorter than the second.

Mandibles short, arcuated.

Thorax subovoid, the pronotum distinctly visible from above, sometimes lengthened, narrowed before; mesonotum with 2 deep furrows; scutellum subconvex or flattened, with a large fovea at base; metathorax short, carinated, the posterior angles prominent or acute.

Front wings pubescent, with a distinct basal cell and a short, open marginal cell; the marginal vein is usually very short; the first branch of radius short, oblique, with a slight uncus from the tip; ♀ sometimes apterous or subapterous.

Abdomen ovate or conic, the petiole longer than thick, a little swollen in the middle.

Legs pubescent or pilose, the femora clavate, the tibiae clavate or subclavate, tibial spurs short but distinct, the basal joint of hind tarsi 3 or more times longer than the second.

TABLE OF SPECIES.

First flagellar joint one-third longer than the pedicel, the joints beyond the third moniliform, equal, not wider than long, the last one-third longer than the penultimate.

Abdomen pointed ovate.............................A. BOREALIS, sp. nov.

Antennæ filiform, the joints long, cylindrical.

First flagellar joint one-third longer than the pedicel, excised at base, the joints beyond subequal, the penultimate at least thrice as long as thick.

A. FLORIDANA, sp. nov.

Aclista rufa, sp. nov.

♀. Length, 2.5ᵐᵐ. Rufo-piceous, the head dusky or blackish; antennæ and legs brownish yellow. Antennæ 15-jointed, clavate-moniliform, the first flagellar joint slender, not quite as long as the pedicel, which is much stouter, the joints beyond all moniliform and much enlarged after the seventh, the last oblong. Wings subhyaline, the marginal cell entirely wanting, the marginal vein about twice as long as thick, the stigmal represented only by a stump of a vein. Abdomen long, conic-ovate, the petiole short, not longer than thick, almost smooth, the seventh segment as long as the fourth, fifth, and sixth united and slightly longer than the eighth.

HABITAT.—Texas.

Type in Coll. Ashmead.

Aclista rufescens, sp. nov.

♀. Length, 3ᵐᵐ. Dark brownish piceous, the head almost black; antennæ and legs brownish-yellow, the three or four terminal joints of the antennæ black.

It is close to *rufa*, but differs in having longer and more slender antennæ, the scape longer and slenderer, the pedicel longer than the first flagellar joint, while the flagellum is longer, the joints towards apex not so large or transverse, the four or five terminal joints being black; the marginal cell is entirely wanting, the marginal vein being punctiform, not longer than thick; the abdomen is longer, stouter, and more pointed, the seventh and eighth segments being equal.

HABITAT.—Santa Cruz Mountains, California.

Type in Coll. Ashmead.

Described from a single specimen.

Aclista rufopetiolata, sp. nov.

(Pl. XV, Fig. 4, ♀.)

♀. Length, 2.5ᵐᵐ. Black, shining, pubescent, the metathorax petiole apex of abdomen rufous, basal joint of antennæ and legs pale rufous.

Antennæ 15-jointed, clavate-moniliform, the first flagellar joint is twice as long as the pedicel, the joints beyond shorter, all moniliform.

Wings fuscous, the marginal cell open, the marginal vein half the length of the first branch of the radius being only a little longer than thick, the second branch of radius not developed. Abdomen pointed

ovate, the petiole stout, not longer than thick, fluted, the sixth and
seventh segments about equal, the eighth shorter.

HABITAT.—Washington, D. C.

Type in Coll. Ashmead.

Aclista missouriensis, sp. nov.

♂. Length, 2.8ᵐᵐ. Polished black, sparsely pubescent; antennæ,
tegulæ, and legs brownish-yellow. Antennæ 14-jointed, rather stout,
cylindrical, the first flagellar joint as long as the scape, the basal two-
thirds excised, the following joints very slightly shortened, about thrice
as long as thick. Metathorax carinated. Wings subhyaline, the marginal cell partially present, the marginal vein thrice as long as thick,
the first branch of radius oblique and short, the second branch of the
radius present being two-thirds the length of the marginal. Abdomen
black, the second segment toward the base more or less rufous, the petiole
stout, as long as the metathorax, rugose, with a furrow along the sides.

HABITAT.—St. Louis, Mo.

Type in Coll. Ashmead.

Aclista conica, sp. nov.

♀. Length, 2ᵐᵐ. Polished black, nearly devoid of pubescence;
antennæ and legs rufous. Antennæ 15-jointed, clavate-moniliform, the
first flagellar joint much more slender but not longer than the pedicel,
the joints beyond sub-moniliform, transverse after the third, the terminal joint oval, longer than the penultimate. Metathorax carinated.

Wings subhyaline, the marginal cell entirely wanting, the marginal
vein more than thrice as long as thick, the first branch of radius short.
Abdomen conic-ovate, the petiole short, not longer than thick, with
faint raised lines but not distinctly fluted, the sixth and seventh segments about equal, longer than the preceding segment.

HABITAT.—Arlington, Va.

Type in Coll. Ashmead.

Aclista rugosopetiolata, sp. nov.

♀. Length, 2.6ᵐᵐ. Polished black, pubescent; the thorax rather flat,
the scutellum with two foveæ at base connected by a grooved line.
Antennæ 15-jointed, very stout and incrassated, the scape nearly half
the length of the flagellum, the first flagellar joint a little longer than
the pedicel and as stout, the remaining joints all transverse-moniliform,
the last oval. Metathorax rugose, carinated.

Wings subhyaline, the marginal vein about thrice as long as thick,
the stigmal vein short, oblique, ending in a hook. Abdomen long-oval,
pilose toward tip and beneath, the petiole stout, longer than thick,
rugose and opaque, with a channel along the sides.

HABITAT.—Arlington, Va.

Type in Coll. Ashmead.

Aclista californica, sp. nov.

♀. Length, 2.5ᵐᵐ. Polished black, pubescent; antennæ black, the basal 5 joints rufous; legs rufous, the hind coxæ black basally. Antennæ 15-jointed, moniliform, the first flagellar joint twice the length of the pedicel, the second shorter than the pedicel, the joints beyond moniliform. Metathorax carinated, very pubescent. Wings subfuscous, the marginal vein about twice as long as thick, the stigmal short oblique with a hook at tip (all that remains of the radius.) Abdomen oblong-oval, black, the tip and beneath sparsely pubescent, the petiole as long as the metathorax, strongly fluted.

HABITAT.—Santa Cruz Mountains, California.

Type in Coll. Ashmead.

Described from a single specimen.

Aclista borealis, sp. nov.

♀. Length, 2.6ᵐᵐ. Polished black, pubescent; antennæ and legs brownish-yellow. Antennæ 15-jointed, filiform-moniliform, the first flagellar joint one-third longer than the pedicel, the joints beyond the third, except the last, all of an equal size, moniliform, very slightly wider than long, the last one-third longer than the penultimate. Wings subfuscous, the marginal vein about as long as the stigmal, the latter oblique with a slight branch at tip. Abdomen pointed, ovate, black, shining and pubescent, the petiole as long as the metathorax, strongly fluted.

HABITAT.—Ottawa, Canada.

Type in Coll. Ashmead.

Described from a single specimen.

Aclista floridana, sp. nov.

♂. Length, 2.6ᵐᵐ. Polished black, pubescent; antennæ and legs brownish-yellow. Antennæ 14-jointed, setaceous, the first flagellar joint a little longer than the scape, excised at base, the following joints gradually subequal to the last, the penultimate joint being about thrice as long as thick and shorter than the last. Metathorax smooth, shining, carinated, the central carina forked from the middle. Wings subfuscous, the marginal vein a little longer than the oblique stigmal, the radius (or second branch) partially developed, as long as the stigmal, and forming an open marginal cell. Abdomen oblong-oval, black, shining, the tip and beneath with whitish pile, the petiole 1½ times as long as thick, delicately keeled above with a deep channel along the sides, the seventh and eighth segments of an equal length, longer than the sixth.

HABITAT.—Jacksonville, Fla.

Type in Coll. Ashmead.

PSILOMMA Förster.

Hym. Stud., II, p. 132 (1856).

Opazon Hal., Nat. Hist. Rev., IV, p. 170 (1857).

Head subglobose, the occiput rounded, the ocelli 3 in a triangle; eyes oval, bare.

Antennæ in ♀ incrassated toward tips, submoniliform, the scape long, the pedicel shorter than the first flagellar joint, the joints beyond transverse-moniliform, the last oblong; in ♂ 14-jointed, filiform, the pedicel rounded, the first flagellar joint elongate, cylindric, the second shorter, slightly excised outwardly at base, the joints beyond to the last short, scarcely twice as long as thick, the last fusiform, longer than the penultimate.

Mandibles short, pointed, but bifid at tips.

Thorax subovoid, the pronotum anteriorly produced into a slight neck; mesonotum with two furrows; scutellum convex, with a profound fovea at base and a transverse, punctate line before the apex; postscutellum flat, rugose; metathorax carinated.

Front wings pubescent, fringed, the marginal vein punctiform, the stigmal short, with a long radius from its tip, forming a long, lanceolate marginal cell, which is open at tip.

Abdomen conic-ovate, subdepressed, the petiole long.

Legs long, the femora swollen, clavate, the tibiæ subclavate, the tarsi shorter than their tibiæ (except the anterior pair), the basal joint short, not twice as long as the second.

Förster indicated no type.

Psilomma columbianum, sp. nov.

(Pl. XV, Fig. 5, ♀; *a*, ♂ antenna.)

♂ ♀. Length, 3.6 to 4ᵐᵐ. Polished black; antennæ and legs rufous; tegulæ yellow. Antennae in ♀ 15-jointed, clavate-submoniliform, the first flagellar joint a little longer than the pedicel, the following to the last transverse-submoniliform, the last the stoutest, conic; in ♂ 14-jointed, filiform, the pedicel small, round, the first flagellar joint the longest, about 3½ times as long as the pedicel, the second slightly shorter, the remaining cup-shaped, loosely articulated. Thorax narrowed before, with two broad furrows, the scutellum foveated at base, the metathorax subquadrate, rugose. Wings subfuscous, the marginal vein punctiform, the first branch of radius very short, the second branch visible as a long fuscous streak, forming a very long, lanceolate, open marginal cell. Abdomen conic-ovate, pilose beneath and toward tip, black, smooth, and shining; the petiole long, finely roughened, with a channel at the sides, the anal segment punctate.

HABITAT.—Washington. D. C.

Types in Coll. Ashmead and National Museum.

Described from 1 ♀ and 2 ♂ specimens.

ISMARUS Haliday.

Nat. Hist. Rev., IV, p. 169 (1857).

Entomius H. Schf., Nom. Ent., II.

(Type *I. dorsiger* Curtis.)

Head transverse, much wider than the thorax, the occiput rounded; ocelli 3, triangularly arranged; eyes rounded, bare.

Antennae in ♀ 15-jointed, submoniliform; in ♂ 14-jointed, filiform, the fourth joint excised at base.

Maxillary palpi 4-jointed; labial palpi 3-jointed.

Mandibles short, arcuate, bidentate.

Thorax subovate, the pronotum visible from above, the anterior angles straight; mesonotum without furrows; scutellum convex, margined at sides and with a fovea at base; metathorax short, rugulose, the angles unarmed.

Front wings pubescent, with a small, triangular, closed marginal cell and a basal cell; the marginal vein is longer than the marginal cell.

Abdomen ovate or oval, the petiole short, only a little longer than thick, the second segment large, occupying about half of the remaining surface, with a median sulcus at base, the following segments short, about equal.

Legs moderate, pilose or pubescent, the hind tarsi not longer than their tibiae, the basal joint three times as long as the second.

This genus is not found in our fauna. On Pl. XV, Fig. 6, I figure the male of *Ismarus rugulosus* Först., to give our students some idea of the genus.

ANOMMATIUM Förster.

Hym. Stud., II, pp. 130 et 140 (1856).

Head subglobose, without ocelli; the eyes rounded, pubescent.

Antennae inserted on a frontal prominence, in ♀ 14-jointed, incrassated toward tips, the last joint strongly developed, nearly round, and much thicker than the penultimate.

Thorax as in *Aclista*.

Wings entirely wanting.

Abdomen ovate or conic, the petiole a little longer than thick.

Legs of moderate length, the femora clavate, the tarsi not especially long, joints 2 to 4 gradually subequal, the last joint as long as the two preceding; claws stout.

The type of this genus is not mentioned by Förster and the genus is not yet recognized in our fauna.

ANECTATA Förster.

Hym. Stud., II, p. 130 (1856).

Head transverse or subglobose; ocelli 3, close together, in a triangle; eyes oval, pubescent.

Antennæ inserted on a frontal prominence, 14-jointed, filiform in both sexes, or in ♀ sometimes slightly incrassated toward tips, the joints, for the most part, moniliform or submoniliform, the last conic or ovate; in the ♂ the joints are cylindrical, the first flagellar joint the longest and usually excised at base.

Thorax as in *Cinetus*.

Wings pubescent, with a basal cell and a closed marginal cell, the stigmal vein usually oblique (rarely straight), the marginal vein very short, seldom much longer than the stigmal.

Abdomen obovate or conic, the petiole two or three times as long as thick, the second segment with a longitudinal sulcus at base.

Legs as in *Aclista*, the basal joint of hind tarsi less than three times as long as the second; tibial spurs short, but distinct.

No type seems to have been described or indicated by Förster, but the genus is a valid one, readily distinguished by the 14-jointed antennæ in female, and the closed marginal cell.

Only two other genera have 14-jointed antennæ in the ♀, viz, *Anommatium* and *Pantolyta*; the former is wingless and without ocelli, the latter winged but without a marginal cell.

TABLE OF SPECIES.

All coxæ pale 2
Hind coxæ black; legs reddish-yellow.
  Antennæ clavate-moniliform, black; the tip of scape and pedicel rufous, ♀.
    Marginal cell not twice as long as the marginal vein...A. CALIFORNICA, sp. nov.
  Antennæ filiform, fuscous, base and tip of scape, pedicel, and base of first flag-
    ellar joint yellowish, ♂.
  Marginal cell more than twice as long as the marginal vein.

                                         A. POLITA, sp. nov.
2. Antennæ rufous, fuscous toward tips.
  Antennæ filiform, submoniliform, the first flagellar joint thrice as long as the
    pedicel, the following subequal, the three preceding the last monili-
    form, ♀.
  Marginal cell twice as long as the marginal vein.......A. RUFIFRONS Ashm.
  Antennæ setaceous, fuscous, yellowish basally, the first flagellar joint a little
    shorter than the scape, excised at base, the joints beyond gradually
    shortening, the penultimate 2¼ times as long as thick, ♂.
  Marginal cell more than twice as long as the marginal vein.

                                         A. GEORGICA, sp. nov.

Anectata californica. sp. nov.

♀. Length, 2ᵐᵐ. Polished black; tip of scape, pedicel and legs, rufous or reddish-yellow. Antennæ 14-jointed, gradually incrassated toward tips, moniliform; the pedicel and first flagellar joint of an equal length; the joints beyond to fifth subequal; joints 5 to 8 rounded; 8 to 11 transverse-moniliform; the last cone-shaped. twice as long as the penultimate. Wings subhyaline, the venation brown, the marginal cell less than twice as long as the marginal vein; the first branch of radius oblique, about half the length of the marginal. Metathorax carinated.

Abdomen shining black, impunctured, with some sparse pile toward the apex and beneath; the petiole slightly longer than the metathorax, strongly fluted.

HABITAT.—Santa Cruz Mountains, California.

Type in Coll. Ashmead.

Anectata polita, sp. nov.

♂. Length, 3ᵐᵐ. Polished black, pubescent; antennæ fuscous, tip of scape, pedicel, and base of first flagellar joint yellowish; legs brownish-yellow; the middle and posterior coxæ black basally; mandibles rufous.

Antennæ broken off at tips, the scape a little longer than the pedicel and first funicular joint together, the first funicular joint more than one-third longer than the second, excised at basal half.

Wings hyaline, the marginal cell more than twice as long as the marginal vein; the first branch of radius nearly straight, one-third shorter than the marginal. Abdomen as long as the head and thorax together; the petiole strongly fluted, a little longer than the metathorax, the second ventral segment with distinct punctures toward the base.

HABITAT.—West Cliff, Colo.

Type in Coll. Ashmead.

Described from a single specimen received from Mr. T. D. A. Cockerell.

Anectata hirtifrons Ashm.

(Pl. xv, Fig. 8, ♀.)

Can. Ent., xix, p. 198.

♀. Length, 3ᵐᵐ. Black, shining, pubescent; the face with rather dense whitish pubescence; antennæ rufous, a little dusky at tips; legs pale rufous or brownish-yellow. Antennæ subfiliform, 14-jointed; scape slightly bent, nearly thrice as long as the first flagellar joint; pedicel not longer than thick; first flagellar joint a little longer than the second, the following gradually shortened to eighth, the three preceding the last very slightly wider than long, the last oblong. Wings hyaline, the pubescence fuscous, the marginal vein one-half the length of the marginal cell, the first branch of the radius straight, one-half shorter than the marginal vein. Abdomen a little shorter than the head and thorax together, with the tip turned upward, covered beneath and at apex with a whitish pubescence; petiole as long as the metathorax, strongly fluted.

HABITAT.—Ottawa, Canada.

Type in Coll. Ashmead.

The only specimen I possess or have seen of this species was received some years ago from Mr. W. Hague Harrington.

Anectata georgica, sp. nov.

♂. Length, 3½ᵐᵐ. Polished black, with a sparse white pile; antennæ rufous; legs and tegulæ brownish-yellow. Antennæ 15-jointed, long, setaceous; the first flagellar joint nearly as long as the scape, strongly excised at base; the joints beyond shorter, very gradually shortened; the penultimate joint thrice as long as thick, the last slightly longer. Metathorax smooth, shining, carinated. Wings subhyaline, the venation pale brown, the marginal cell $2\frac{1}{2}$ times as long as the marginal vein, the first branch of the radius very oblique, a little shorter than the marginal vein. Abdomen black, polished, pilose; the petiole longer than the metathorax or a little more than twice as long as thick, strongly fluted.

HABITAT.—Georgia.

Type in Coll. Ashmead.

PANTOLYTA Förster.

Hym. Stud., ii, p. 130 (1856).

(Type *Belyta heterocera* Hal.)

Head subglobose, the face produced into a ledge for the insertion of antennæ; ocelli minute; eyes rounded.

Antennæ inserted on a frontal prominence; in ♀ 14-jointed submoniliform; the pedicel larger and stouter than the first flagellar joint; in ♂, filiform.

Maxillary palpi, 5-jointed; labial palpi, 3-jointed.

Mandibles bifid at tips, subrostriform.

Thorax as in *Aneclata*.

Front wings pubescent, with a distinct basal cell, but without a marginal cell; the marginal vein variable, punctiform, or as long as the basal nervure.

The radius always short, but slightly developed.

Abdomen conic-ovate, the petiole short.

Legs as in *Aneclata*, the femora and tibiæ clavate, pilose.

This genus very closely resembles *Polypeza* Förster in the *Diapriinæ*, but that genus has no basal cell in the hind wings.

The following species is the only one as yet detected in our fauna:

Pantolyta brunnea, sp. nov.

(Pl. XVI, Fig. 1, ♀.)

♀. Length, 2^{mm}. Light brown to brownish-yellow; antennæ and legs pale brownish-yellow; eyes dark brown, rounded. Head globose, the face with a prominent frontal ledge for the insertion of antennæ, the occiput rounded; mandibles subrostriform, conical, bifid at tips. Antennæ four-fifths the length of the body, clavate, gradually incrassated towards tips, pubescent, the joints, after the sixth, moniliform; scape long, cylindrical, slightly bent, and nearly half the length of the flagellum, the apical margin beneath produced into a minute spine; pedicle oblong, stout; first funiclar joint the slenderest and longest joint, obconical, about three times as long as thick at tip; second funiclar joint about two-thirds the length of the first; the third still shorter but stouter; all joints beyond moniliform, gradually increasing in size. Thorax subovate, smooth, shining, trapezoidally narrowed anteriorly from the tegulae; mesonotum about as long as wide across the base, the parapsidal furrows being distinct, the middle lobe at base being one-third shorter than at apex; scutellum deeply foveated at base; postscutellum short with a median carina; metanotum as long as the scutellum, areolated. Wings hyaline, covered with a long pubescence, the venation pale; the marginal vein is as long as the basal; the postmarginal not at all developed; the radial vein small, the first abscissa being scarcely twice as long as thick, while the second abscissa is much longer, and extends forward parallel with the outer margin of the wing, leaving very little space, so that there is scarcely any trace of the marginal cell. Legs pilose, the femora and tibiae strongly clavate. Abdomen conic-ovate, the tip conically pointed and curving downwards; petiole scarcely twice as long as thick.

HABITAT.—Arlington, Va.

Type in Coll. Ashmead.

It bears a close resemblance to *Polypeza Pergandei*, and without a careful examination for the cell in the hind wing could easily be confused with that species.

Subfamily X.—DIAPRIINÆ.

Head globose or subglobose, rarely oblong. Ocelli 3, close together, and triangularly arranged. Mandibles most frequently short, obtuse, bidentate at apex, rarely rostriform. Maxillary palpi 4- or 5-jointed; labial palpi 2-jointed. Antennae porrect, inserted on a frontal prominence, in males 13- or 14-jointed, filiform, nodose verticillate or moniliform; in females 12- or 13-jointed, and in a single case 14-jointed, clavate. Pronotum scarcely visible from above; mesonotum rarely longer than wide, and with or without furrows; scutellum most frequently convex, and foveated at base; metathorax short, emarginate. Front wings strongly fringed, most frequently without distinct cellules, the marginal nervure punctiform or wanting; hind wings never with a basal cell, veinless. Apterous forms frequent. Abdomen subpetiolate, rarely with a very long petiole, ovate, conic-ovate, or oblong-oval, composed of 7 or 8 segments, the second always large. Legs clavate, the tibial spurs. 1, 2, 2, the middle and posterior pairs usually weak; tarsi long, slender, 5-jointed, claws simple.

This subfamily could only be confounded with the *Belytinæ*, but the 2-jointed labial palpi, the absence of a cell in the hind wings, and the less distinctly veined front wings readily separate it. The habitus, except in a few genera, is also quite different; the frontal prominence is rarely so distinct, the antennae in the female are shorter, more distinctly clavate, the venation in front wings quite different, while the abdomen is, as a rule, less distinctly pointed, and most frequently with a much shorter petiole.

The only genera, probably, that would give the student trouble in placing, are *Synacra* and *Polypeza*. The first mentioned was originally placed by Förster in the *Belytinæ*, as the venation is more strongly developed than in any other Diapriid, and the abdomen is shaped as in many of the Belytids, but there is *no basal cell in the hind wings*.

Loboscelida Westw., Thes. Ent. Oxon., p. 171, is not included in the table of genera, as I believe it to be a Cynipid genus.

While comparatively little is yet known of the habits of the greater number of the *Diaprinæ*, from such as have been reared we are justified in stating the group is parasitic only on Dipterous larvæ, as all carefully bred specimens, both in Europe and America, have been reared only from these insects. *Diapria melanocorypha* Ratz. is recorded as having been bred from *Cryptorhynchus lapathi*, and *Entomacis cordipennis* Förster from a spider's nest; but we know now that various Diptera are found in the burrows of Rhynchophora and other wood-boring Coleoptera and that certain Diptera, *Leucopis*, etc., are parasites of spiders, and it is safe to infer that these Diapriids came from Dipterous parasites overlooked by Ratzeburg and Förster.

The group may be divided conveniently into two tribes, as follows:

Submarginal vein reaching the costa at about half the length of the wing or a little before; if it does not reach the costa it attains nearly half the length of wing and ends in a stigma; costal cell most frequently closed.
Tribe I—SPILOMICRINI

Submarginal vein never reaching the costa beyond one-third the length of the wing; costal cell most frequently open, or the subcostal and costal nervures are confluent .... Tribe II—DIAPRIINI

Tribe I—SPILOMICRINI.

The longer submarginal vein, that almost invariably reaches the costa at about half its length, and the closed (rarely open) costal cell can be depended upon to distinguish the genera in this tribe; it is only in two genera, *Aneurhynchus* and *Labolips*, that the submarginal vein does not reach the costa, but even in these genera it is long and terminates in a stigma.

The genera are not numerous and may be synoptically represented as follows:

TABLE OF GENERA.

FEMALES.

1. Antennæ 14 jointed; mesonotum with 2 furrows .... POLYPEZA Förster
   Antennæ 12-jointed... 3
   Antennæ 13-jointed.
   Mesonotum without furrows (rarely slightly developed posteriorly) 2
   Mesonotum with 2 distinct furrows.
   Metathorax armed with a long curved spine at base .... HOPLOPRIA Ashm.
   Metathorax not armed at base with a curved spine.
   Basal nervure absent or subobsolete.
   Abdomen rounded or truncate at tip; costal cell open; stigmal vein often with a backward directed branch................. HEMILEXIS Först.
   Abdomen conically pointed; costal cell closed; stigmal vein simple.
   PARAMESIUS Westw.

Basal nervure present.

    Abdomen conically pointed, the second segment without sulci at base, overlapping the apex of petiole; marginal nervure distinct; costal cell closed... SPILOMICRUS Westw.

    Abdomen rounded or truncate at apex, the second segment with sulci at base; marginal nervure very short; costal cell usually open.

                                                     HEMILEXIS Förster

2. Basal nervure wanting; stigmal nervure longer than the marginal.

                                       HEMILEXODES Ashm., gen. nov.

3. Submarginal nervure not attaining the costa, ending in a stigma 5

Submarginal nervure attaining the costa.

    Wings with a basal nervure 4

    Wings without a basal nervure.

        Mesonotum with 2 furrows.

            Club 5-jointed .... SYNACRA Förster

            Club 4-jointed .... GLYPTONOTA Förster

4. Mesonotum without furrows.

    Club abrupt, 3-jointed; face keeled at the sides..TROPIDOPSIS Ashm., gen. nov.

    Mesonotum with 2 furrows.

        Club 5-jointed .... IDIOTYPA Förster

5. Mesonotum with 2 furrows.

    Stigma with a stigmal vein; abdomen with 3 sulci at base.

                                     ANEURHYNCHUS Westw.

    Stigma without a stigmal vein; abdomen with a single sulcus at base.

                                     LABOLIPS Haliday

MALES.

Antennæ 14-jointed 3

Antennæ 13-jointed.

    Mesonotum without furrows 2

    Mesonotum with furrows.

        Metathorax armed at base with a curved spine.

            Flagellar joints very long, cylindrical........HOPLOPRIA, Ashm., gen. nov.

        Metathorax not armed at base with a curved spine.

            First flagellar joint not half as long as the second.

                Basal vein absent; costal cell closed...............PARAMESIUS Westw.

            First flagellar joint as long or longer than the second.

                Mesonotum not longer than wide.

                    Costal cell open; basal nervure present.

                        Stigmal vein much longer than the marginal.....HEMILEXIS Förster.

                Mesonotum longer than wide.

                  Costal cell closed; basal nervure present.

                      Abdomen with sulci at base ....IDIOTYPA Förster.

                      Abdomen without sulci at base ....SPILOMICRUS Westw.

2. Basal nervure present; costal cell closed.

    Base of second segment overlapping the apex of petiole ....(SPILOMICRUS.)

    Basal nervure absent; costal cell open or at least open at base.

        Flagellar joints long, cylindrical, pilose, the first longer than the second.

                        HEMILEXODES Ashm., gen. nov.

3. Submarginal vein not reaching the costa, ending in a stigma.................. 4

Submarginal vein reaching the costa.

    Basal nervure wanting.

        Mesonotum with 2 furrows.

            Eyes hairy ....SYNACRA Förster.

        Mesonotum without furrows.

            Flagellar joints elliptic-oval................TROPIDOPSIS Ashm., gen. nov.

4. Stigma with a stigmal vein ......................ANEURHYNCHUS Westw.

POLYPEZA Förster.

Hym. Stud., II, p. 123 (1856).

Head globose, rounded before and behind, the occiput not margined; ocelli 3, triangularly arranged; eyes rounded, pubescent.

Antennæ 14-jointed, inserted on a frontal prominence; in ♀ submoniliform, gradually incrassated toward tips; in ♂ filiform, the joints cylindrical.

Mandibles subrostriform, bidentate.

Maxillary palpi 3-jointed: labial palpi 2-jointed.

Thorax ovate, narrowed into a short rounded collar anteriorly; mesonotum with two furrows; scutellum rounded, convex, with a deep fovea at base; metathorax with a central longitudinal carina.

Front wings pubescent, ciliated, the submarginal nervure reaching the costa at about half its length, the marginal nervure distinct, postmarginal and stigmal nervures very minute, scarcely developed; basal nervure distinct.

Abdomen conic-ovate, petiolated, the petiole as long as the metanotum: ovipositor prominent.

Legs clavate, pilose, the tarsi fully as long or longer than the tibiæ, 5-jointed, the basal joint of hind tarsi fully 3 times as long as the second.

This genus is the only one in the tribe with 14-jointed antennæ in the ♀ and closely resembles a belytid, without a marginal cell.

Polypeza Pergandei, sp. nov.

(Pl. XVI, Fig. 2, ♀.)

♂ ♀. Length, 1 to 1.2mm. Polished black; metathorax, petiole, and base and tip of abdomen dull rufous, or piceous. Antennæ in ♀ 14-jointed, submoniliform, gradually incrassated toward tips; pedicel longer and thicker than the first flagellar joint: second and third flagellar joints shorter than the first, the joints beyond moniliform; in ♂ 14-jointed, long, filiform, fuscous, the first flagellar joint thrice as long as the pedicel.

Thorax with 2 distinct furrows: scutellum convex, with a deep depression at base: metathorax short, smooth, but bounded by a carina behind. Tegulæ yellowish. Wings hyaline, pubescent, very broad, and with a long fringe at the margins: the submarginal vein reaches the costa a little before the middle of the wing; the marginal vein is rather long, linear; stigmal vein short, with an uncus at tip and a slight indication of a backward directed branch; basal cell distinct. Legs honey-yellow, the femora and tibiæ slightly infuscated. Abdomen conic-ovate, petiolated: the petiole twice as long as thick, striated; the second segment occupies most of the remaining surface: segments 3 to 6 very short, the seventh conically pointed, longer than segments 3 to 6 united.

HABITAT.—Arlington, Va.

Types in Coll. Ashmead.

Described from four specimens.

This interesting species is dedicated to my friend Mr. Theo. Pergande, to whom I am indebted for two of the four specimens.

HOPLOPRIA Ashm., gen. nov.

(Type *H. pulchripennis* Ashm.)

This genus is founded upon a South American species in the Berlin Museum, and is clearly allied to *Paramesius* Westwood, but differs in having a curved acute spine at the base of the metathorax and in having only 4-jointed maxillary palpi. The joints of the antennae are also much longer than any other Diapriid genus.

H. pulchripennis, sp. nov.

♂. Length, about 4ᵐᵐ. Black; antennae, legs, and abdomen ferruginous, the latter dusky at apex. Head opaque, finely shagreened; cheeks, prothorax, metathorax, and the petiole beneath, woolly. Antennae 13-jointed, longer than the body; the first flagellar joint about as long as the scape, the second very slightly shorter, the following about equal, all cylindrical. Thorax smooth, shining, bisulcate; scutellum with two large deep foveae at base; metathorax armed at base with a curved acute spine. Wings hyaline, with a large smoky blotch extending from near the base to one-third the length of the wing, having 4 rays; one ray extends to the stigma and another beyond this to the costal margin, and opposite these are the other two which extend to the hind margin; there is also a smoky spot at the apex of the wing.

HABITAT.—Bogota.

Type in Berlin Museum.

HEMILEXIS Förster.

Hym. Stud., II, p. 123 (1856).

Entomacis Först. loc. cit., p. 121.
Glyphidopria Hal. Nat. Hist. Rev., IV. p. 172 (1857).

Head transverse or subglobose, the occiput rounded, the cheeks bare; ocelli 3 small, in a triangle; eyes bare.

Antennae inserted on a slight frontal prominence, 13-jointed in both sexes; in ♀ submoniliform, incrassated toward tips or clavate; in ♂ filiform, the first and second funiclar joints elongate, the second the shorter and thicker, slightly curved, the following oval or oblong-oval.

Maxillary palpi 5-jointed.

Thorax ovoid, the prothorax produced into a little neck anteriorly, scarcely visible from above; mesonotum short, rounded before with two furrows, the furrows sometimes abbreviated anteriorly, but rarely entirely wanting; scutellum with a moderately deep fovea at base; metathorax short, submarginate, the angles not acute.

Front wings pubescent, strongly ciliated and often emarginate at apex;

the submarginal vein terminates in a short or punctiform marginal vein just before the middle of the wing; and there is a distinct oblique stigmal vein, usually three times as long as the marginal, with sometimes a distinct backward directed branch: the costal cell is open and the basal nervure is present, although faint or subobsolete.

Abdomen long-oval, subtruncate or but slightly pointed at apex, the petiole in the ♀ is usually short, in the ♂ long, woolly, or pubescent; the second segment occupies most of the remaining surface and has sulci at base above.

Legs clavate, pubescent, or pilose, the basal joint of posterior tarsi twice as long as the second.

Dr. Förster erected the genus *Entomacis* upon specimens with a "heart-shaped piece" cut out of the apex of the front wing, but this character in itself will not hold to found a genus, and I have here joined it to *Hemilexis*, as I can find no other character to separate them. I have species with emarginate, slightly emarginate, and entire wings, but in all other essential characters, viz, the 13-jointed antennæ and two furrows on the mesonotum, they agree with *Hemilexis*, so that if I have properly identified these genera they should be united.

Nothing is known of the parasitism of the genus, except what Förster tells us: that he bred specimens of an *Entomacis* from a spider's nest. In all probability the spider was already infested with a Dipterous parasite, *Leucopis* sp., and the Diapriid came from the Dipteron.

Three species are known to me, separated as follows:

Apex of wings more or less emarginated; stigmal vein longer than marginal.
  Antennæ yellow, gradually incrassated, moniliform, the last joint large, fusiform, fuscous. ♀.
    Legs reddish-yellow; petiole black ....H. SUBEMARGINATA, sp. nov.
  Antennæ subclavate, submoniliform, the flagellum fuscous. ♀.
    Legs and petiole, honey-yellow.....................H. MELLIPETIOLATA, Ashm.
Apex of wings entire, stigmal vein more than twice as long as the marginal.
  Antennæ filiform, fuscous, the first and second flagellar joints elongate, the second the shorter and produced toward one side at apex, the following joints submoniliform. ♂.
    Legs rufous; trochanters, tibiæ, and tarsi yellow .....H. CALIFORNICA, sp. nov.
  Antennæ filiform, brownish-yellow, the second flagellar joint a little longer than the first and third. ♂.
    Legs honey-yellow...............................H. SUBEMARGINATA, sp. nov.

Hemilexis subemarginata, sp. nov.

(Pl. XVI, Fig. 4. ♀.)

♀. Length, 1.4^{mm}. Polished black, impunctured; antennæ and legs reddish-yellow. Head subglobose, well rounded before and behind. Mandibles rufous, projecting, bidentate at tips. Antennæ 13-jointed, gradually thickened toward tips, the last joint large, fusiform, larger than the two preceding united; flagellar joints one to four cylindric, subequal, the joints beyond to last moniliform, loosely joined. Thorax short, rounded before, with two short, abbreviated impressed lines behind, not extending forward to the middle of the mesonotum; scutellum with a transverse fovea at base; metanotum tricarinate, the truncature bounded by a carina above, slightly produced at the lateral angles. Wings hyaline, fringed, slightly emarginated at apex; stigmal nervure oblique, three times as long as the short marginal; basal nervure subobsolete. Abdomen oblong-oval, truncate at apex, polished, shining; the petiole a little longer than thick, finely striated: the extreme base of second segment with some striæ.

The ♂ agrees well with the ♀ except the mesonotal furrows are a little longer, the head larger, antennæ filiform, the second flagellar joint a little longer and stouter than the others, the others about three times as long as thick, while the wing venation is darker, more distinct, the stigmal vein shorter, with a slight backward directed branch from its tip.

HABITAT.—District of Columbia.

Types in Coll. Ashmead.

Hemilexis mellipetiolata Ashm.

Can. Ent., XIX, p. 196; Cres. Syn. Hym., 251.

♀. Length, 1ᵐᵐ. Polished black, impunctured; scape, legs, and abdominal petiole, honey-yellow; flagellum rust-brown. Antennæ 13-jointed, long, gradually incrassated toward tips; flagellar joints 1 to 5 subcylindric, the following submoniliform. Thorax without furrows, scutellum foveated at base; metapleura pubescent. Wings hyaline, strongly fringed, the apex slightly emarginated, the stigma and nervures pale yellow, the stigmal nervure much longer than the short stigma. Abdomen oval, truncate at apex, black, shining, the petiole long, yellow, about three times as long as thick, finely striated.

HABITAT.—Jacksonville, Fla.

Type in Coll. Ashmead.

Quite distinct from the other species in the long yellow petiole, absence of mesonotal furrows, pubescent metapleura and in the relative length of antennal joints.

Hemilexis californica, sp. nov.

♂. Length, 1.5ᵐᵐ. Polished black, impunctate; antennæ brown, the scape and legs yellowish, the hind coxæ and thickened parts of the femora and tibiæ fuscous or piceous. Head more transverse than in the preceding species. Mandibles rufous. Antennæ 13-jointed, filiform, the first flagellar joint long, three times as long as the pedicel, cylindrical, the second stouter and slightly curved and dilated toward apex, the following joints still shorter, loosely joined, from an elliptic-oval to moniliform, the last ovate. Thorax with the parapsidal furrows wanting anteriorly; scutellum with a deep fovea all across the base, metapleura finely rugose. Wings hyaline, fringed, the nervures pale yellow, the stigmal vein oblique, three times as long as the marginal; basal nervure distinct. Abdomen oblong-oval, truncate posteriorly, the petiole a little more than twice as long as thick, faintly striated.

HABITAT.—Santa Cruz Mountains, California.

Types in Coll. Ashmead.

Hemilexis brevicornis Say.

Psilus brevicornis Say., Lec. Ed. Say, I. p. 221.
Diapria brevicornis Ashm., Can. Ent., XIX, p. 396; Cress. Syn. Hym., p. 251.

Black, polished, immaculate; tibiæ and tarsi piceous.
Inhabits St. Peters river.

Antennæ short, first joint much elongated, second joint longer than the remaining ones, which are subcylindric-quadrate; mandibles pale, testaceous; thorax convex, rounded, two faint impressed lines each side converging to the scutel and on the posterior margin two indistinct dull whitish spots; scutel elevated, convex, feet dark, piceous; thighs nearly black; wings, costal nervure indistinct; abdomen depressed, fusiform, acute at tip. Length more than one-twentieth of an inch. (Say.)

This species is unknown to me, but evidently belongs here.

PARAMESIUS Westwood.

Phil. Mag., I, p. 129 (1832); Först. Hym. Stud., II. p. 123 (1856).

Cerapsilon Hal. (?)

(Type *P. rufipes* Westw.)

Head transverse, the occiput rounded, the cheeks bare; ocelli 3, distinct in a triangle; eyes oval.

Antennæ inserted on a frontal prominence, 13-jointed in both sexes, in ♀ gradually incrassated toward tips, clavate, the scape reaching far beyond the ocelli, slenderer toward the base; pedicel stouter and longer than the first funiclar joint, narrowed basally; first funiclar joint only a little longer than thick, a little shorter than the second; seventh to twelfth wider than long, the thirteenth fusiform or ovate, more than twice as large as the preceding; in ♂ longer than the body, filiform, the pedicel and the first funiclar joints about equal, the second nearly thrice as long as the first, slightly emarginate at base, the following joints long, cylindrical, four or more times longer than thick.

Maxillary palpi 5-jointed.

Mandibles bifid at tips.

Thorax oblong-oval, narrowed in front, the prothorax transverse, woolly at the sides; mesonotum with 2 distinct furrows; scutellum deeply bifoveated at base, the foveæ often confluent, forming one deep fovea, or separated only by a slight carina; metathorax short, emarginated posteriorly, a short conic keel at base, and with the posterior angles subacute.

Front wings pubescent, ciliated, the costal cell completely closed; the submarginal vein joins the costa at or a little before the middle of the wing; the marginal vein is about thrice as long as thick, with a stump of a stigmal vein; basal vein absent or indistinct.

Abdomen pointed-ovate, the petiole rather long, fluted; the second segment occupies the greater part of its surface, with sulci at base; the apical joint pointed, stylus-like.

Legs rather long, pilose, the femora clavate, the posterior tarsi a little longer than the tibiae, or fully as long, the basal joint one-third longer than the second.

The genus is sufficiently distinct from the other genera with 13-jointed antennae by the length of the submarginal vein, the absence of a basal nervure, the more pointed abdomen, the second segment with sulci at base, and by the antennal characters of the male.

I am not aware that anything has been published respecting the habits of the genus, but if my memory serves me rightly I saw specimens of *P. rufipes* Westw. in the Berlin Museum, reared from *Eristalis tenax*, along with *Diapria conica*.

Our species seem to be quite distinct from the described European species, and may be recognized by the aid of the following table:

TABLE OF SPECIES.

FEMALES.

Wings subhyaline or tinged with fuscous 2
Wings hyaline.
  Antennae honey-yellow or brownish-yellow, clavate-moniliform.
    Flagellum twice as long as the scape, the 5 or 6 terminal joints fuscous or black, the last enlarged, fusiform, more than thrice as large as as the penultimate; legs brownish-yellow; abdomen entirely black ....P. PARVULUS, sp. nov.
    Flagellum elongated, the last joint only black, globose-ovate; legs honey-yellow.
      Mesonotal furrows complete; abdomen pale at base...P. TERMINATUS Say.
      Mesonotal furrows wanting; abdomen not pale at base..P. CLAVIPES Ashm.
  Antennae fuscous or black.
    Flagellum 2½ times as long as the scape, the last joint fusiform, about thrice as large as the penultimate; legs pale reddish, all coxae black, the femora dusky ....P. UTAHENSIS, sp. nov.
2. Antennae and legs reddish-yellow.
  Last four joints of antennae black.
    Scutellum with two foveae at base; scapulae with a grooved line.
                                          P. SPINOSUS, sp. nov.
    Scutellum unifoveate, the fovea crenulate; scapulae without a groove.
                                          P. PALLIDIPES, sp. nov.

MALES.

Wings fusco-hyaline.
  Legs, including coxae, pale brownish-yellow; antennae pale ferruginous.
    Pedicel and the first funiclar joints of an equal length, the joints beyond slightly more than 4 times as long as thick.
                                          P. PALLIDIPES, sp. nov.
    Pedicel a little shorter but stouter than the first funiclar joint, the joints beyond very long, several times longer than thick.
                                          P. SPINOSUS, sp. nov.
Wings dark, fuscous.
  Legs ferruginous, all coxae and the swollen portion of femora black, the tibiae toward apex more or less fuscous; antennae black, the pedicel shorter than the first funiclar joint..P. OREGONENSIS, sp. nov.

Paramesius parvulus. sp. nov.

♀. Length, 1.5ᵐᵐ. Polished black, impunctured, with some sparse hairs; antennæ, except the five or six terminal joints, and legs, honey-yellow or brownish-yellow; mandibles reddish: palpi white. Antennæ 13-jointed, about as long as the body; the flagellum clavate, twice as long as the scape, the 5 or 6 terminal joints black or blackish, the last joint being very large, fusiform, as long as the three preceding joints united and much stouter; the pedicel is longer and stouter than the first flagellar joint, the joints beyond submoniliform, gradually widening toward tips; the penultimate quadrate. Thorax with two furrows; the scutellum foveated at base and grooved at the sides; posterior margin of prothorax at the sides with a row of punctures, otherwise smooth and polished; metathorax finely rugose, pubescent, with a prominent acute median carina above. Wings clear, hyaline, fringed, the submarginal vein pale yellow, reaching the costa at half the length of the wing; the marginal vein fuscous or brown, twice as long as thick at tip. Abdomen petiolated, conic-ovate, longer than the head and thorax united, polished black, the petiole more than twice as long as thick, coarsely fluted, pubescent beneath, the second segment fully twice as long as the petiole, apex with sparse, long, white hairs.

HABITAT.—District of Columbia and Arlington, Va.

Type in Coll. Ashmead.

Paramesius terminatus Say.

Psilus terminatus Say, Leconte Ed. Say, II, p. 778.
Paramesius terminatus Ashm., Can. Ent., XIX, p. 196.
Diapria terminata Cr. Syn. Hym., p. 251.

♀. Length, 2.5ᵐᵐ. Polished black, impunctured, sparsely pubescent; antennæ, except toward apex, and legs honey-yellow; petiole obscurely rufous. Antennæ 13-jointed, the pedicel longer than the first flagellar joint, funiclar joints 1 to 5 subequal, the first about thrice as long as thick at apex, the fifth scarcely twice as long as thick, the joints from here moniliform, the last fusiform, longer than the two preceding joints united. Thorax with 2 furrows; the scutellum truncate behind, with a quadrate, crenate fovea at base, the mesopleura below with a few longitudinal striæ; metathorax rugulose, pubescent, with a prominent acute carina above. Abdomen conic-ovate, highly polished, the tip pointed, piceous, the petiole about twice as long as thick, striated.

HABITAT.—Bladensburg, Md.

Specimens in Coll. Ashmead.

Paramesius clavipes Ashm.

Can. Ent., XX, p. 53.

♀. Length, 2ᵐᵐ. Polished black, impunctured, covered with some long, sparse hairs on the head, thorax, and surrounding the apex of abdomen; the metathorax and petiole beneath rather densely pubescent; antennæ, except the last joint, and the legs honey—or reddish— yellow. Antennæ 13-jointed, subclavate, the first flagellar joint slightly shorter than the pedicel, the following joints to the fifth subequal, from thence moniliform, loosely joined, the last joint fusiform, more than twice the length of the preceding, fuscous or black. Thorax without furrows, somewhat flat, the sides compressed; collar reddish laterally; mesopleura entirely smooth; metapleura rugose, the mesonotal median keel small. Wings hyaline, pubescent, the venation pale yellow, the marginal nervure short, about twice as long as thick at apex. Legs with the femora strongly clavate, the tarsi very long, the anterior and middle pairs being much longer than their tibiæ. Abdomen pointed or conic-ovate, smooth and polished, the petiole a little longer than thick, strongly fluted, the second segment with two lateral impressions at extreme base.

HABITAT.—Ottawa, Canada.

Type in Coll. Ashmead.

Described from a single specimen received from Mr. W. Hague Harrington.

Paramesius utahensis, sp. nov.

♀. Length, 2.6ᵐᵐ. Polished black, impunctured, with a sparse cinereous pubescence. Antennæ two-thirds the length of the body, subclavate, black or fuscous, the basal flagellar joints a little reddish, the pedicel a little thicker but not, or scarcely, longer than the first flagellar joint, flagellar joints 6 to 10 quadrate, the last fusiform, about thrice as long as the penultimate. Thorax with 2 furrows, abbreviated anteriorly, the scutellum truncate behind with a large crenate furrow across the base; mesopleura along the mesopectus longitudinally striate; metapleura rugulose. Wings hyaline, the marginal nervure brown, only twice as long as the short stigmal. Legs pale brown, the coxæ black, the thickened part of the femora piceous, sometimes with the posterior tibiæ fuscous. Abdomen conic-ovate, polished, black, the petiole twice as long as thick, fluted and pubescent beneath.

HABITAT.—Utah Lake, Utah.

Types, two ♀ specimens, in Coll. Ashmead.

Taken by Mr. E. A. Schwarz, in Utah.

Paramesius spinosus, sp. nov.

♀. Length, 3ᵐᵐ. Polished black, impunctured, with a sparse cinereous pile; antennæ, except the 3 terminal joints, and legs, reddish-yellow. Antennæ 13-jointed, about as long as the body, subclavate, pubescent, the scape very long, first flagellar joint a little longer than the pedicel, joints 2 to 5 subequal, the fifth thicker than the preceding joint, the following joints to the last submoniliform, the last fusiform, a little thicker and longer than the two preceding joints together. Thorax with 2 complete furrows, the scutellum truncate at apex with a grooved line at the sides and bifoveated at base; mesopleura along the mesopectus rugose or coarsely punctate; metapleura rugulose, pubescent, metanotum with an acute triangular keel just behind the postscutellum, the posterior angles acute. Wings subhyaline, pubescent, the marginal nervure at least four times as long as thick at tip, the stigmal vein not longer than wide. Abdomen long, conically pointed, polished, black, the petiole stout, fluted, 2½ times as long as thick, pubescent beneath.

The ♂ measures but 2ᵐᵐ, the antennæ being nearly twice as long as the body, filiform; the pedicel and first funiclar joint both short, the latter a little longer, the following joints all very long, cylindrical, several times longer than thick, as long as the scape, the abdomen pear-shaped, the petiole 3½ times as long as thick, fluted.

HABITAT.—Washington, D. C.

Types in Coll. Ashmead.

Described from two specimens. The species is readily distinguished by the bifoveolated scutellum and the rugoso-punctate surface of the lower part of the mesopleura.

Paramesius oregonensis, sp. nov.

♂. Length, 4ᵐᵐ. Polished black, sparsely pilose, the hairs fuscous. Head subglobose, the frontal ridge beneath foveated at the middle; clypeus separated, convex, with a row of punctures at base; mandibles ferruginous, with some golden hairs and several punctures toward base. Antennæ 13-jointed, filiform, longer than the body, black; the pedicel is a little shorter than the first funiclar joint; the second funiclar joint is the longest joint, fully as long as the pedicel and first joint united, and excised at base, the following joints all long, cylindrical, very slightly shorter than the second. Collar at sides pubescent, mesonotum with 2 furrows, scutellum with a crenate furrow at base, while the metathorax is rugose pubescent. Wings fuscous, the submarginal and marginal nervures black, the stigmal about as long as the marginal.

Legs black, the base of tibiæ and tarsi, piceous. Abdomen oblong-oval, polished, black, the petiole 3 times as long as thick, strongly fluted.

HABITAT.—Portland, Oregon.

Type in Coll. Ashmead.

Described from a single specimen, collected by H. F. Wickham.

Paramesius pallidipes, sp. nov.

(Pl. XVI, Fig. 5, ♀.)

♂. Length, 2.5ᵐᵐ. Polished black, covered with some sparse, long hairs. Antennæ 13-jointed, longer than the body, cylindrical, pale ferruginous; pedicel and first flagellar joint of an equal length, not half as long as the second, the joints after the first about equal, slightly more than 4 times as long as thick, the second excised at base. Mesonotum with two furrows; scutellum with a deep fovea across the base, crenulate at bottom; middle carina of metathorax acutely prominent at base. Legs pale, brownish-yellow. Wings fusco-hyaline, the venation brown, the stigmal about as long as the marginal vein, short; basal vein indistinct. Abdomen black, smooth, shining, the petiole 2½ times as long as thick, coarsely fluted, pubescent beneath.

♀. Length, 3^{mm}. Agrees well with the ♂, except the antennæ are gradually incrassated toward tips, flagellar joints 1 to 6 gradually shortening, the basal joints rather long, the first about as long, but not so thick, as the pedicel, joints 7 to 9 very little longer than thick, the last fusiform, stouter and a little longer than the two preceding joints united.

HABITAT.—Carolina and District of Columbia.

Type ♂ in Berlin Museum, the ♀ in Coll. Ashm.

Described from a single ♂ specimen in Berlin Museum, labeled as having come from Dr. Zimmermann, and 2 ♀ specimens in my collection taken near Washington.

SPILOMICRUS Westwood.

Phil. Mag., I, p. 129 (1832); Förster Hym. Stud., II. p. 123.

(Type S. *stigmaticalis* Westw.)

Head transverse or subglobose, the occiput rounded, the cheeks slightly pubescent; ocelli 3 triangularly arranged; eyes broadly oval.

Antennæ inserted on a frontal prominence, 13-jointed in both sexes, in ♀ incrassated toward tips, submoniliform, the scape subclavate extending beyond the ocelli; pedicel stouter and longer than the first funiclar joint; in ♂ filiform long, covered with a short, dense pubescence, the pedicel small, rounded, the second funiclar joint usually as long or longer than the first, the first slightly emarginate at base, the joints beyond long, cylindrical.

Maxillary palpi 5-jointed.

Mandibles short, bifid.

Thorax long ovoid, the prothorax visible from above, more or less woolly, especially at the sides; mesonotum with two furrows often abbreviated anteriorly, seldom entirely wanting; scutellum with two oblong fovea at base, metathorax short, emarginate posteriorly, tridentate.

Front wings pubescent, ciliated, the submarginal vein terminating in a punctiform marginal vein before the middle of the wing, the costal cell distinct, closed, the basal vein usually quite distinct; the short stigmal vein often has a backward-directed branch.

Abdomen oblong-oval, composed of seven segments in the ♀ and eight in the ♂, the petiole long, fluted, its apex covered by the projecting base of the second segment, the second segment occupying most of the surface and without sulci at base.

Legs clavate, pilose, the tarsi long, slender, the posterior tarsi with the basal joint almost twice as long as the second.

Resembles both *Paramesius* and *Basalys;* from the former it is distinguished by the distinct basal nervure, usually a less distinctly pointed abdomen and the structure of the second segment, which overlaps the apex of the petiole and is without sulci at base, while the distinct mesonotal furrows, longer submarginal vein, and the wholly different antennæ in the ♂ separate it from the latter.

TABLE OF SPECIES.

Scutellum with 2 foveæ at base.
  Scapulæ not separated... 2
  Scapulæ separated posteriorly.
    ♀. Five terminal joints of antennæ black, rest of antennæ brownish-yellow
      or rufous.
      Scutellum with a row of 6 large punctures at apex, the shoulders with a
        grooved line .... S. ARMATUS Ashm.
      Scutellum with a row of 6 small punctures at apex, the shoulders without
        a grooved line .... S. ATRICLAVUS, sp. nov.
    ♂. Antennæ and legs brownish-yellow, the flagellar joints long, nearly
      equal, the first two a little longer, about five times as long as thick
                                S. FLAVICORNIS, sp. nov.
2. Coxæ and femora black.
    ♂. Antennæ black, first three flagellar joints about equal, the second excised at
      base, about four times as long as thick, the following three times as
      long as thick........................... S. ATROPETIOLATUS Ashm.
                                    ?S. FOVEATUS Prov.
                                  ?S. LONGICORNIS Prov.

Spilomicrus armatus, sp. nov.

(Pl. xvi, Fig. 7, ♀.)

♀. Length, 3.1ᵐᵐ. Polished black, impunctured, with long, sparse hairs; a tuft of pubescence on cheeks behind, collar at sides and petiole beneath, griseous. Antennæ 13-jointed, brownish-yellow, the 5 terminal joints black or fuscous; the flagellum is about 2½ times as long as the scape, thickened toward tip, the joints 6 to 10 quadrate, the last oblong. Thorax with two furrows, the scapulæ with a long grooved furrow; scutellum rather flat, with 2 large foveæ at base, grooved lines at sides, at apex truncate and with a row of 6 large punctures; metathorax with an acute, triangular central carina, the sides pubescent. The mesopectus has 3 or 4 distinct striæ just above the middle coxæ. Wings subhyaline, the stigma rufo-piceous, triangular; basal nervure wanting, indicated only by a wrinkle in the wing surface. Legs brownish-yellow, the first joint of trochanters long, clavate. Abdomen oblong-oval, polished, black, the petiole a little more than twice as long as thick, strongly fluted, with long, sparse pubescence above and woolly beneath, the wool extending on to the second ventral segment.

HABITAT.—District of Columbia.

Type in Coll. Ashmead.

Spilomicrus atriclavus, sp. nov.

♀. Length, 3mm. Polished, black, impunctured, and agreeing closely with *A. armatus* but differs as follows: the five terminal joints of antennæ are black, and more abruptly separated from the funicle, the first four being transverse-quadrate, about twice as broad as long; the scapulæ are without a grooved line; there is a row of 6 small punctures at apex of scutellum; the wings are subfuscous, while the legs are more distinctly rufous.

HABITAT.—Ottawa, Canada.

Type in Coll. Ashmead.

Described from a single specimen received from Mr. W. Hague Harrington.

Spilomicrus flavicornis, sp. nov.

(Pl. XVI, Fig. 7 *a*, ♂ antenna.)

♂. Length, 1.4mm. Polished black; head twice as wide as thick through antero-posteriorly, the cheeks oblique. Mandibles and palpi pale. Antennæ 14-jointed, honey-yellow, about as long as the body: scape as long as the anterior femora, excluding the trochanters; pedicel longer than thick; first and second flagellar joints equal, longer than any of the following; joints beyond to the last moniliform, only slightly longer than thick, the last joint very little longer than the penultimate. Thorax smooth, with distinct but delicate mesonotal furrows. Pronotum with an impression at sides just in front of the tegulæ. Scutellum with the fovea at base separated into two parts by a delicate carina. Metapleura finely rugose, scarcely pubescent. Tegulæ rufous. Wings hyaline, pubescent, ciliated, the veins yellowish. Legs, including coxæ, honey-yellow. Abdomen oval, polished, the petiole nearly thrice as long as thick, finely sculptured, with a longitudinal furrow above toward the sides, and very finely, sparsely pubescent.

HABITAT.—Arlington, Va.

Types in National Museum and Coll. Ashmead.

Described from specimens taken by Mr. Theo. Pergande, July 13, 1884.

Spilomicrus atropetiolatus Ashm.

Ismarus atropetiolatus Ashm., Bull. No. 1 Col. Biol. Assoc., p. 11.

♀. Length, 2mm. Black polished, with some long, scattered, sparse, grayish pubescence, the face, as well as the metapleura and petiole, more densely pubescent. Face prominent. Eyes bare. Antennæ 13-jointed, black, pubescent, the scape very long, slender, a little more than twice as long as the third joint; the pedicel short, much less than half as long as the third, narrowed at base, the third and fourth joints of about an equal length, the fifth joint and joints beyond slightly shorter than the third, and all narrowed at base and of about an equal length. Thorax smooth, without grooves. Mesopleura smooth, bare, with a slight furrow a little below the middle, more distinct anteriorly and almost obliterated posteriorly. Scutellum smooth, quadrate behind, the margins delicately keeled, and with two rather deep oblong foveæ at base. Metathorax and metapleura delicately sculptured; the metathorax has a delicate median longitudinal keel, which becomes forked posteriorly. Legs dark red, pubescent. Abdomen polished, bare, except a few hairs surrounding the apex; the petiole is about half the length of the abdomen, opaque, fluted, and hairy. Wings subhyaline, pubescent; veins pale brown, the marginal vein black or piceous.

HABITAT.—West Cliff, Colo.

Type in Coll. Ashmead.

This species was received from Mr. T. D. A. Cockerell.

Spilomicrus longicornis Prov.

Nat. Can., xii, p. 262; Faun. Ent. Can., ii, p. 561. ♀ ; Cress., Syn. Hym., p. 251.

♀.—Long. .11 pce. Noir, poli, brillant; les mandibules, la base des antennes, les écailles alaires avec les pattes, jaune-roussâtre. Antennes longues, un peu plus épaisses à l'extrémité, noires, roussâtres à la base, insérées sur un tubercule frontal. Thorax plus épais en avant; écusson proéminent avec une petite fossette à la base. Ailes hyalines, fringées, velues, la nervure formant la petite cellule radiale, noire, bien distincte, se prolongeant inférieurement jusqu'à la rencontre ou peu s'en faut du cubitus, point d'autres nervures distinctes à part celles de la base. Pattes longues, grêles, les cuisses et les jambes légèrement renflées, les hanches noires. Abdomen à pédicule strié, de la moitié de sa longueur environ, le reste en forme de losange, terminé en pointe, déprimé, poli, brillant. Tarière non apparente. (Provancher.)

HABITAT.—Canada.

Unknown to me. Judging from the description of the venation, the species is a Belytid and not a Diapriid.

Spilomicrus foveatus Prov.

Add. et Corr., p. 176; Cress. Syn. Hym. N. A., p. 251.

♀.—Long. .13 pce. Noir, les pattes et les antennes jaunes-miel. La tête en carré avec la face renflée jusqu'au milieu, formant une protubérance sur laquelle sont insérées les antennes. Antennes de 13 articles, jaunes, les articles terminaux bruns, le premier article fort long, les suivants très courts, poilus, les 4 à 5 terminaux épaissis graduellement en massue. Thorax poli, brillant, beaucoup large en avant, l'écusson brillant avec une petite fovéole transverse en avant. Ailes velues, fringées, sans taches n'ayant de bien apparente qu'une petite nervule stigmatique un peu courbé, d'autres nervures moins distinctes formant une cellule brachiale triangulaire. Pattes d'un beau jaune-miel, avec les cuisses et les jambes renflées en massue à l'extrémité. Abdomen à pédicule court, mais distinct, déprimé à 2e segment très grand, rétréci en pointe et poilu à l'extrémité; tarière droite, très courte.

Se distingue surtout du *longicornis*, par son abdomen à pédicule plus court et ses antennes plus distinctement en massue. (Provancher.)

HABITAT.—Cap Rouge Canada.

Unknown to me, but evidently not a true *Spilomicrus*.

HEMILEXODES Ashm., gen. nov.

(Type *H. floridana*.)

Head subglobose, the cheeks rounded; ocelli 3 small, triangularly arranged; eyes oval.

Antennæ 13-jointed in both sexes, inserted on a slight frontal prominence; in ♀ incrassated toward tips, submoniliform; in ♂, long filiform, the joints long, cylindrical, pilose.

Maxillary palpi 4-jointed.

Thorax subovoid, the mesonotum broader than long, without furrows, the scutellum subconvex with grooved lines at the sides and a fovea across the base, the metanotum very short, the posterior face carinated above.

Front wings pubescent, ciliated, the submarginal nervure reaching the costa just before its middle, marginal vein punctiform, stigmal vein distinct, oblique, two or more times longer than the marginal nervure; basal nervure wanting the postmarginal nervure distinct.

Abdomen oblong-oval, petiolated, the second segment very long, with sulci at base.

Legs clavate.

Distinguished from *Hemilexis* in having no mesonotal furrows and no basal nervure; also by the difference in the male antennæ, all the flagellar joints being long and cylindrical, pilose. ·

Hemilexodes floridana, sp. nov.

(Pl. xvi, Fig. 8, ♀.)

♀. Length, 1.5ᵐᵐ. Entirely honey-yellow, except the head, which is brownish. Antennæ 13-jointed, the flagellar joints, after the fourth, round-moniliform. Thorax without furrows; scutellum with a transverse fovea at base; metathorax carinated. Wings hyaline, strongly fringed, entire, the nervures pale yellow, the stigmal vein long, oblique, more than 3 times as long as the short marginal. Abdomen oblong oval, truncate at tip, smooth, polished, the petiole 1½ times as long as thick with raised lines toward the sides.

HABITAT.—Jacksonville, Fla.

Types in Coll. Ashmead.

Other species in this genus are known to me from the West Indies.

SYNACRA Förster.

Hym. Stud., II, p. 128 (1856).

Chlidonia H. Sch., Forts. Faun. Ins. Germ. Cl. vii.
Artibolus Hal., Nat. Hist. Rev., iv, p. 173 (1857).

(Type *Diapria brachialis* Nees).

Head subglobose, the cheeks rounded, the occiput scarcely margined; ocelli 3, triangularly arranged, rather close together on the crown; eyes oval.

Antennæ inserted on a frontal prominence; in ♀ 12-jointed, terminating in a 5-jointed club; in ♂ 14-jointed, filiform, the flagellar joints long, cylindrical, pilose, the first excised.

Mandibles subrostriform, bidentate at tips.

Thorax ovate, the mesonotum with 2 distinct furrows, the scutellum convex, with a fovea at base; metathorax smooth.

Front wings pubescent, the submarginal nervure reaching the costa near the middle, the marginal vein short, the stigmal vein distinct with an appendage, basal nervure distinct.

Abdomen petiolate, oblong, oval or obovate, the petiole short, rugulose, the second segment very long.

Legs clavate.

No species in this genus is known out of the European fauna.

GLYPTONOTA Förster.

Hym. Stud., II. p. 122 (1856).

(No type mentioned.)

Head transverse, the frons rounded, not impressed, occiput rounded; ocelli 3, in a triangle, eyes rounded.

Antennæ inserted on a slight frontal prominence; in ♀ 12-jointed, clavate, the 4 or 5 terminal joints enlarged, the last being twice as large as the penultimate; scape cylindrical, slightly curved, reaching a little beyond the ocelli; pedicel oval, shorter than the first funiclar joint; in ♂ 13-jointed, filiform.

Maxillary palpi 5-jointed.

Mandibles short, bidentate at tips.

Thorax oblong-oval; prothorax visible from above, impressed laterally; mesonotum with 2 distinct furrows; scutellum large, subconvex, foveated at base and with a distinct frenum posteriorly; metathorax emarginated, the posterior angles acute.

Front wings pubescent, the submarginal vein reaching the costa at about half the length of the wing, the marginal vein punctiform, with a short oblique stigmal vein; basal vein wanting.

Abdomen ovate, 7- or 8- segmented, the petiole longer than thick, faintly strigose, beneath towards apex angulately produced; second segment very large, occupying more than two-thirds of the body of the abdomen, with 3 sulci at base.

Legs moderate, sparsely pilose, the femora and tibiæ clavate, tibial spurs weak, the posterior tarsi short, the basal joint more than twice as long as the second.

The genus seems to be a valid one, although Förster mentions no type, and I can find no described European species.

The following, therefore, appears to be the first species described:

Glyptonota nigriclavata, sp. nov.

(Pl. XVII. Fig. 1, ♀.)

♀. Length, 2.8^{mm}. Polished black, sparsely pilose; antennæ, except the 4 terminal joints (or club) which are black, mandibles, palpi, tegulæ, and legs, pale brownish-yellow. Antennæ 12-jointed, clavate; pedicel stouter and a little longer than the first funiclar joint; the funiclar joints to the club are very gradually shortened, but gradually widened, the last two, or the fifth and sixth, being wider than long, the club joints are slightly pedicellate, transverse, increasing in size, the last very large cone-shaped. Mesonotal furrows deep and broad. Scutellum with a punctate frenum and two broad foveæ at base. Metathorax rugoso-punctate, pubescent. Wings hyaline, pubescent, fringed. Abdomen smooth, polished, impunctured, the petiole longitudinally striated.

HABITAT.—Washington, D. C.

Type in Coll. Ashmead.

Described from a single specimen, taken by myself along the canal at Georgetown.

TROPIDOPSIS Ashm., gen. nov.

(Type *T. clavata* Ashm.)

Head subglobose, the face in the female flat, with delicate side keels; ocelli 3, triangularly arranged; eyes large, rounded.

Antennæ in ♀ 12-jointed, terminating in a large, abrupt, 3-jointed club, the funicle slender; in ♂ 14-jointed, filiform, longer than the body, the flagellar joints, except the first and last, elliptic-oval, pilose.

Mandibles short, bifid.

Thorax as in *Loxotropa*, the metathorax emarginated posteriorly, the angles a little prominent, centrally with an acute carina.

Front wings pubescent, the submarginal nervure reaching the costa at about the middle, the marginal nervure about twice as long as thick, the stigmal vein not developed, the basal nervure distinct.

Abdomen petiolate, oblong-oval, the second segment very long.

Legs clavate.

Allied to *Loxotropa*, but readily distinguished by the much longer submarginal nervure, presence of facial keels, and the difference in the male antennæ.

Tropidopsis clavata, sp. nov.

(Pl. XVII, Fig. 2. ♀; a, ♂ antenna.)

♀. Length, 1.4 ᵐᵐ. Brownish-red or ferruginous, smooth, polished; antennæ, except club and legs, pale, or yellowish. Antennæ 12-jointed, terminating in an abrupt 3-jointed black club, the first two joints of which are quadrate, the last oblong; funicle 7-jointed, slender, the first joint about twice as long as the second, the following not longer than thick, the last two or three slightly transverse; pedicel obconic, much larger and stouter than the first funiclar joint. Head globose, the face flat with delicate side keels. Eyes large, rounded. Scutellum with a single fovea at base. Metanotum emarginate, the angles a little promi-nent. Abdomen oblong-oval, the petiole a little longer than thick, pubescent. Wings hyaline, fringed, the submarginal vein reaching the costa at about the middle of the wing, ending in a subtriangular marginal vein; basal nervure straight.

♂. Length, 1.3 ᵐᵐ. Antennæ longer than the body; 14-jointed, filiform, the flagellar joints, except the first three and the last, elliptic-oval, the fourth and following stouter than the three preceding. Abdomen oblong-oval, the petiole about twice as long as thick, cylindric, striated, subpubescent; otherwise color, etc., as in female.

HABITAT.—St. Vincent, West Indies.

Types in National Museum.

IDIOTYPA Förster.

Hym. Stud., II. p. 122 (1856).

Mionopria Hal., Nat. Hist. Rev., IV. p. 172.

(Type *M. maritima* Hal.)

Head subrotund, the occiput rounded; ocelli 3, in a triangle; eyes large, rounded or oval.

Antennæ in ♀ 12-jointed, clavate, the pedicel shorter than the first funiclar joint; club 5-jointed; in ♂ 13-jointed, filiform, pubescent, the scape reaching beyond the ocelli, cylindric; pedicel small; first flagellar joint as long or slightly longer than the second; second, more or less excised at base; the following joints all cylindrical, equal or subequal.

Thorax ovoid, the prothorax distinctly visible from above, transverse, the sides bare; mesonotum wider than long, with two distinct furrows, slightly diverging anteriorly; scutellum subconvex, with an impressed line at sides and unifoveated at base; metathorax short, emarginated posteriorly, a conic prominence at base above, and with the angles acute.

Front wings pubescent, the costal cell complete; the submarginal vein reaches the costa at about half its length, terminating in a distinct marginal vein; the postmarginal vein sometimes slightly developed; the stigmal vein usually short, but distinct, with a spurious vein from its tip, directed backwards; the basal cell distinct, although the median nervure is more or less hyaline before it joins the basal nervure.

Abdomen subovate or oblong-oval; in ♀ 8-jointed, the petiole distinct, two or three times as long as thick; second segment very large, with a median sulcus at base; following segments short.

Legs rather long, the femora and tibiæ clavate, basally very slender, pilose, the tibial spurs weak but distinct, the posterior tarsi longer than their tibiæ, the basal joint more than thrice the length of the second, the terminal joint also longer than usual; claws small, simple.

Not found in our fauna. On Pl. XVII, Fig. 3, I figure the female of *Idiotypa pallida* Ashm. from St. Vincent, West Indies.

ANEURHYNCHUS Westw.

Phil. Mag., I. p. 129, 1832.

Syn. *Mythras* Hal.

(Type *A. galesiformis* Westw.)

Head transverse, with a short frontal sulcus, the occiput not emarginated, rounded. Eyes rounded. Ocelli 3, in a triangle,

Antennae inserted on a frontal prominence; in ♀ 12-jointed, clavate or strongly incrassated toward tips, the scape reaching beyond the ocelli, thickened apically; pedicel rounded; in ♂ 14-jointed, long, filiform; the scape cylindrical.

Maxillary palpi 5-jointed.

Mandibles short, bidentate.

Thorax oblong; prothorax visible from above, the angles rounded; mesonotum with two distinct furrows; scutellum subconvex, bifoveated at base; metathorax short, submarginated, the posterior lateral angles acute.

Front wings with the submarginal vein terminating in a stigma before attaining the costa, and from which issues a short stigmal vein knobbed at tip; basal vein absent.

Hind wings veinless.

Abdomen long, oval, 8-jointed, the petiole strigose, hairy.

Legs moderate, the femora clavate, the tibiae subclavate, pilose; basal joint of posterior tarsi three or four times as long as the second; claws small, simple.

This genus comes nearest to *Labolips*, but is readily distinguished by the maxillary palpi and the knobbed stigmal vein.

Abbé Provancher described three species under this genus, none of which belong here. *Aneurhynchus aneurus* Prov. is a ♂ Platygasterid belonging to the genus *Isocybus*, while *A. spinosus* and *A. inermis* Prov. are Belytids.

I have recognized three species in our fauna, which may be tabulated as follows:

FEMALES.

Frontal ridge pale rufous.
　Legs, including coxae, honey-yellow; ventral segments punctate, the last very
　　closely punctate...A. MELLIPES Ashm.

MALES.

Frontal ridge pale rufous.
　Legs and antennae pale brownish-yellow; ventral segments smooth, impunctured.
　　　　　　　　　　　　　　　　　　　　　　　A. FLORIDANUS, sp. nov.
Frontal ridge black.
　Legs and antennae brownish-yellow; ventral segments finely, sparsely punctate.
　　　　　　　　　　　　　　　　　　　　　　　A. VIRGINICUS, sp. nov.

Aneurhynchus mellipes Ashm.

Can. Ent. xx, p. 52.

♀. Length, 2.5ᵐᵐ. Polished black, sparsely pubescent. Antennae 12-jointed, stout, clavate, rufous; the scape is very thick, a little shorter than half the length of the flagellum; pedicel stouter, but not as long as the first flagellar joint; second flagellar joint shorter than the first; third, shorter than the second; from thence the joints are transverse, increase in size and width, and are well separated. Mesonotal furrows distinct. Scutellum with two transverse foveae almost meeting at base. Metathorax closely punctate, pubescent, and with some irregular carinæ. Legs, including the coxæ, honey-yellow. Abdomen polished black, the petiole rugose with some longitudinal sulci. Wings subhyaline, pubescent: the submarginal vein ends in a callosity and a short stigmal vein that also terminates in a knob; from this knob there is a slight trace of a radial vein; the submarginal vein is very pale.

HABITAT.—Ottawa, Canada.

Type in Coll. Ashmead.

Described from a single specimen received from Mr. W. Hague Harrington.

Aneurhynchus floridanus, sp. nov.

(Pl. XVII, Fig. 4, ♀ ; *a*, ♂ antenna.)

♂. Length, 3ᵐᵐ. Polished black, sparsely pubescent; antennæ, mandibles, palpi, tegulæ, and legs, including coxæ, pale brownish-yellow. Antennæ 14-jointed, hairy, extending to the middle of the abdomen; scape as long as the first and second flagellar joints united; pedicel rounded; first funicular joint about twice the length of the pedicel; the second emarginate at base with a slight tooth beneath, the joints beyond nearly of an equal length, nearly four times as long as thick. Thorax with two deep, broad furrows; the collar with a deep, smooth impression at sides; scutellum with lateral ridges and two large foveæ at base, separated only by a slight carina, the bottom slightly crenate; metathorax rugose with a slight median carina. Wings hyaline, very pubescent, the pubescence dusky; the submarginal vein does not reach the costal edge and terminates in a stigma, the stigmal vein very oblique; the radius or second branch of the stigmal is visible as a long dusky streak, thus forming a very long, open radial cell; basal vein present. Abdomen smooth, polished black, impunctured, sparsely covered with long hairs, the anus pale rufous; the petiole is about twice as long as thick, with irregular grooved lines above, and a deep grooved channel at the sides.

HABITAT.—Fort George Island, Florida.

Type in Coll. Ashmead.

Described from a single specimen.

Aneurhynchus virginicus, sp. nov.

♂. Length, 2.8ᵐᵐ. Very closely allied to *A. floridanus* in color and sculpture, but at once separated by the hind coxæ being black and by the sparsely, but distinctly, punctate ventral segments; while the anus is black and the antennæ are slightly stouter, the flagellar joints being in consequence relatively shorter.

HABITAT.—Harpers Ferry, W. Va.

Type in Coll. Ashmead.

Described from a single specimen taken by Mr. E. A. Schwarz.

LABOLIPS Haliday.

Nat. Hist. Rev., 1857, p. 173.

(Type *L. innupta* Hal.).

(Pl. xvii, Fig. 5, ♀.)

Head rounded. Eyes small, protuberant, hairy. Ocelli 3, in a triangle.

Antennæ, in ♀, 12-jointed, clavate, the scape cylindrical; the pedicel longer than the first flagellar joint, the second to the last transverse-moniliform, slightly pedicellate, the last oval, twice as long as the preceding; in ♂, 14-jointed, filiform.

Maxillary palpi obsolete; labial palpi not jointed.

Mandibles bidentate.

Thorax oblong, depressed; prothorax attenuated into a neck; mesonotum with two furrows; scutellum transverse, subquadrate; metathorax short, truncate at apex, with the angles a little prominent.

Front wings with the submarginal vein not reaching the costa, terminating in a stigma but without a stigmal vein; basal vein subobsolete. Hind wings veinless.

Abdomen oblong, oval, petiolate.

Legs normal, the femora stout, clavate; the tibiæ subclavate.

This genus has not yet been discovered in our fauna, and the single species described by Haliday, from England, is the only one known.

Tribe II.—DIAPRIINI.

In this tribe, the furthest removed from the *Belytinæ*, the submarginal nervure reaches the costa at one-third its length or a little before, the costal cell being most frequently open; or the costal and subcostal nervures are confluent as in the *Braconidæ* (*Loxotropa*, *Basalys*, etc.); or it is not, or but slightly, developed (*Galesus*). Eleven genera having these characteristics in common are here brought together. All those of which the parasitism is known, are found living upon Dipterous larvæ.

TABLE OF GENERA.

FEMALES.

Antennæ 13-jointed (except in *Myrmecopria*)..............................4
Antennæ 12-jointed.
  Face not greatly lengthened..2
  Face greatly lengthened; mandibles rostriform.
    Mesonotum with 2 furrows ....GALESUS Curt.
2. Head transverse or subglobose, rarely a little longer than wide...............3
  Head large and flat; ocelli wanting; mesonotum without furrows; legs short and
    stout......................................PLATYMISCHUS Westw.
3. Wings with a basal nervure.
  Mesonotum without furrows; club 3 or 4 jointed...........LOXOTROPA Först.
  Wings without a basal nervure.
  Mesonotum without furrows.
    Scutellum foveated at base.
      Tip of scutellum compressed from the sides, the disk or apex with a median
        carina; abdomen usually conically pointed.
                              TROPIDOPRIA Ashm., gen. nov.

Tip of scutellum not compressed from the sides, rounded or truncate, without a carina; abdomen ovate or oblong-oval, often truncate at tip. Petiole much longer than thick; metathorax always with a distinct ridge or conic prominence at base..........................DIAPRIA Latr.

Petiole not longer than thick, densely woolly: metathorax most frequently without a conic prominence at base, usually arcolated.

 Club 3-jointed..........................CERATOPRIA Ashm., gen. nov.

 Club 4 or 5 jointed.....................TRICHOPRIA Ashm., gen. nov.

Scutellum not foveated at base.

 Axillæ not separated..........................PILENOPRIA Ashm., gen. nov.

4. Scutellum not foveated; club consisting of one enlarged joint.

Mesonotum without furrows.................................MONELATA Först.

Scutellum foveated at base, club with more than one joint.

Mesonotum without furrows, or only slightly indicated posteriorly.

 Basal nervure present; antennal club abrupt, 4-jointed......BASALYS Westw.

 Basal nervure absent; antennal club not abrupt, the flagellum subclavate; antennæ 14-jointed......................MYRMECOPRIA Ashm., gen. nov.

MALES.

Antennæ 13-jointed 4

Antennæ 14-jointed.

 Scape not especially developed... 2

 Scape abnormally developed.

  Apterous; mesonotum without furrows..............PLATYMISCHUS Westw.

2. Face not lengthened; mandibles not rostriform............................... 3

Face lengthened; mandibles rostriform.

 Mesonotum with 2 furrows.................................GALESUS Curt.

3. Wings with a basal nervure.

 First flagellar joint much shorter than the second ....BASALIS Westw.

 First flagellar joint not shorter than the second.

  Mesonotum without furrows............................LOXOTROPA Först.

Wings without a basal nervure; mesonotum without furrows.

Scutellum foveated at base.

 Tip of scutellum compressed from the sides; the disk or apex with a median carina ....TROPIDOPRIA Ashm., gen. nov.

 Tip of scutellum not compressed from the sides; rounded or truncate, without a carina.

  Antennæ nodose-verticillate, the joints pedunculated........DIAPRIA Latr.

  Antennæ filiform or moniliform, not nodose-verticillate.

   Antennæ filiform, the joints from the fourth oval or moniliform, but not thickened toward tips.

    Second flagellar joint longer and thicker than the first, usually curved or angulate towards one side, the joints beyond rounded, with long bristles ....CERATOPRIA Ashm., gen. nov.

    Second flagellar joint shorter than the first; the first four or five joints twice longer than thick; the joints beyond long-oval or moniliform with short hairs, or joints after the second moniliform, pilose.

           TRICHOPRIA Ashm., gen. nov.

Scutellum not foveated at base.

 Second flagellar joint about as long as the first, the joints beyond long-oval or moniliform, hairy ....PILENOPRIA Ashm., gen. nov.

4. Scutellum not foveated at base.

Mesonotum without furrows.

 First flagellar joint not half as long as the second........;....MONELATA Först.

GALESUS Curtis.

Brit. Ent., p. 311 (1831).

Coptera Say, Bost. Jour., I. p. 282.
Anisoptera Her.-Schaef.

(Type *G. cornutus* Panz.)

Head oblong or horizontal, with a profound frontal sulcus, the angles of which are acute; occiput broad, rounded; ocelli 3.

Antennae, in ♀, 12-jointed, clavate-submoniliform, the pedicel very small, rounded; the scape short, stout; in ♂, 14-jointed, filiform; the scape usually angulated a little beyond the middle; the pedicel rounded but larger than the first flagellar joint.

Maxillary palpi 5-jointed.

Mandibles prolonged, rostriform, dentate.

Thorax: Prothorax visible from above; mesonotum with two furrows, slightly converging posteriorly; scutellum rather large, somewhat flat or subconvex, with a grooved line along the sides and bifoveated at base; metathorax emarginated at apex, the lateral angles acute.

Front wings folded, often emarginate or with a slit at apex, pubescent, the submarginal vein not developed or terminating before attaining the costa; the basal cell incomplete; the basal vein obsolete or at least water-lined.

Hind wings veinless.

Abdomen petiolated; in ♀ ovate, in ♂ oblong-oval, composed of apparently but 2 or 3 segments, the second occupying most of the surface and generally inclosing the apical segments, with a sulcus at base; petiole longer than thick, grooved.

Legs rather short, the femora stout, the tibiae clavate, their spurs weak; basal joint of hind tarsi about one-third longer than the second; claws small, simple.

In the oblong horned head, the rostrate mandibles, and the shape of the scape, this genus is quite distinct from all others in the group.

The genus *Coptera* Say, is without doubt identical, and was recognized as synonymous by Mr. A. H. Haliday as early as 1857; besides I have identified his type, *Coptera polita*, which proves to be nothing but a small *Galesus* with emarginated and folded wings.

On Pl. XVII, Fig. 6, I figure *Galesus 6-punctatus* Ashm. from St. Vincent.

Our species may be thus tabulated:

TABLE OF SPECIES.

FEMALES.

Frontal prominence with a large, broad, diamond-shaped fovea; its apex not emarginated between the antennae.
  Antennae brown, the legs rufous...............................G. QUEBECENSIS Prov.
Frontal prominence with a less distinct fovea, and distinctly emarginate between, the antennae.
  Antennae entirely black; the legs piceous.................G. ATRICORNIS, sp. nov.
  Scape black, the flagellum piceous; legs honey-yellow...........G. POLITUS Say.

Coxæ pale.

Antennæ not longer than the body, brown; legs honey-yellow....G. POLITUS Say.

Coxæ black.

Antennæ longer than the body, piceous; legs piceous.......G. FLORIDANUS Ashm.

Antennæ shorter than the body; scape black, the flagellum brown.

   Flagellar joints twice as long as thick, except the first, which is thrice as long as thick; last two apical abdominal segments rugoso-punctate.

                                      G. TEXANUS, sp. nov.

   Flagellar joints not longer than thick, except the first, which is one and a half times as long as thick; last two apical abdominal segments smooth.

                                        G. PILOSUS, sp. nov.

Galesus quebecensis Prov.

Pet. Faun. Can., II, p. 559; Ashm. Can. Ent., XIX, p. 195; Cress. Syn. Hym., p. 251.

♀. Length, 3.2ᵐᵐ. Black, polished, shining; head twice as long as wide, pilose, the cheeks with a tuft of griseous wool; anteriorly profoundly sulcate and exhibiting 4 acute teeth; frontal prominence with a large, broad, diamond-shaped fovea, its apex not distinctly emarginated between the antennæ. Antennæ 12-jointed, extending to the apex of metathorax, brown; the first flagellar joint thrice as long as thick, the second twice as long as thick; the joints from here gradually incrassated, shorter but broader, submoniliform; the last very large; cone-shaped, as long as the first. Thorax with two deep furrows, the scapulæ with a deep sulcus near the tegulæ; tegulæ large, black; metathorax rugose, opaque, pubescent. Wings subfuscous, hyaline basally, pubescent; folded, but not fissured. Legs, including coxæ, rufous. Abdomen a little longer than the head and thorax together, polished black; the petiole stout, with deep longitudinal sulci, sparsely pilose.

♂. Length, 4ᵐᵐ. Black, polished, shining, with the legs red. Antennæ hairy, the joints separated at the sutures; front excavated before for the insertion of the antennæ, with a small point at the middle beneath and another on each side. Thorax elongated, depressed; the metathorax strongly punctured with two carinæ united at the base and diverging toward the summit; the mesothorax tuberculous. Tegulæ large, red, black at base. Legs of a beautiful red, the thighs swollen, and more or less black. Abdomen pedunculate; the petiole grooved above and hairy at the sides, the rest forming a rather short oval; the second segment with small foveæ at base.

HABITAT.—Canada and District of Columbia.

Type ♂ in Coll. Provancher; ♀ in Coll. Ashmead.

The ♂ of this species is unknown to me, and the above description is taken from Abbé Provancher; the ♀ is described from a specimen captured at Washington, D. C., which could only be correlated with this species.

Galesus atricornis, sp. nov.

♀. Length, 2.6ᵐᵐ. Black, polished. Antennæ 12-jointed, clavate, black, reaching only to the tegulæ; first flagellar joint only a little more than twice as long as thick, joints beyond to sixth shorter, subequal;

from here to the last they are longer, quadrate, the last large, cone-shaped. Head oblong, smooth, anteriorly just above the frontal excavation roughened, the angles acute; the frontal prominence emarginated, punctate, with a small discal fovea; beneath the eye is a large, long sulcus. Mandibles black. Thorax with two furrows that terminate anteriorly just before reaching the collar. Scutellum with two large, oblique foveae. Metathorax pubescent. Wings fusco-hyaline, pubescent, folded, with a deep fissure at apex. Legs piceous, the trochanters, the slender portion of the tibiae, and the tarsi honey-yellow. Abdomen as long as the head and thorax together, smooth, polished, the apex with some sparse white hairs, the petiole longitudinally grooved, with some sparse pubescence at the apex above and beneath.

HABITAT.—Ottawa, Canada.

Type in Coll. Ashmead.

Described from a single specimen received from Mr. W. Hague Harrington.

Galesus politus Say.

Coptera polita Say, Lec. Ed. Say's Works, II. p. 727; Ashm., Can. Ent., XIX, p. 195; Cress. Syn. Hym. N. A., p. 250.

♂ ♀. Length, 2.2 to 2.6ᵐᵐ. Polished black. Antennæ 12-jointed, clavate, reaching slightly back of the tegulæ; scape black, angulated within before the tip; pedicel and the 4 or 5 following joints honey-yellow, the joints beyond piceous, submoniliform, the last the largest, oblong, sometimes wholly piceous. Antennæ in ♂ 14-jointed, filiform, a little shorter than the body, pale brown, the scape black, the flagellar joints about 2½ times as long as thick. Head nearly twice as long as wide, smooth, polished, the anterior angles acute; frontal prominence with a discal fovea and an emargination at apex. Mandibles rufous or pale. Thorax smooth, polished, with two distinct furrows, the middle lobe with a fovea posteriorly. Scutellum with two oblong foveae at base. Metathorax pubescent. Wings subhyaline, folded, pubescent, with a deep fissure at apex. Legs, including coxae, honey-yellow or pale rufous. Abdomen oblong oval, polished, not as long as the head and thorax together.

HABITAT.—Indiana, District of Columbia, Oakland, Md., and Fort George, Fla.

Specimens in Coll. Ashmead.

This species is not rare and is widely distributed, occurring from Florida to Canada.

Galesus floridanus Ashm.

Can. Ent., III, p. 195.

♂. Length, 1.5ᵐᵐ. Black, polished, covered with sparse scattered pile. Head only a little longer than wide, smooth, polished, scarcely emarginated anteriorly, not cornuted. Antennæ 14-jointed, longer than the body, brownish piceous, the flagellar joints with long whitish hairs; scape swollen at the middle but not angulated. Legs honey-yellow or pale rufous, coxæ black, the femora piceous. Wings hyaline, pubescent.

HABITAT.—Fort George, Fla.

Type in Coll. Ashmead.

This species on account of the non-cornuted head may not be a genuine *Galesus*.

Galesus texanus, sp. nov.

♂. Length, 3.5ᵐᵐ. Black, shining; flagellum dark brown; scape and coxæ black; legs honey-yellow, pilose; palpi pale; mandibles blackish. Head twice as long as wide, smooth, except the stemmaticum, which is roughened. Antennæ extending to apex of the petiole, the flagellar joints, except the first, which is longer, fully twice as long as thick. Thorax with two deep, wide furrows, converging and meeting at base of the scutellum; metathorax carinated. Wings subhyaline, pubescent, slightly emarginated at apex, and folded. Abdomen nearly as long as the head and thorax together, polished, the petiole stout, twice as long as thick, fluted, the second segment occupying nearly the whole of the remaining surface, with a long median impressed line, and a fovea on each side of the line at base, apex with some sparse white hairs.

HABITAT.—Texas.

Type in Coll. Ashmead.

Galesus pilosus, sp. nov.

♂. Length, 2ᵐᵐ. Black, polished, pilose; the head only a little longer than wide, cornuted anteriorly, the vertex behind with a row of close punctures; flagellum brown, the joints, except the first and last, moniliform, not longer than wide; scape and coxæ black, the legs distinctly rufous. Collar fluted. Scutellum with 2 large foveæ at base. Metathorax rugose. Wings subhyaline, pubescent. Abdomen a little longer than the thorax, the petiole 1½ times as long as thick, striated, the second segment sulcate at base.

HABITAT.—Texas.

Type in Coll. Ashmead.

Described from a single specimen. The difference in the head, antennæ, petiole, and the sculpture of the second abdominal segment readily distinguish the species.

PLATYMISCHUS Westw.

Phil. Mag., i, p. 128.

(Type *P. dilatatus* Westw.)

(Pl. XVII. Fig. 7, ♂.)

Head oblong, flattened, the occiput subemarginated. Eyes rounded. Ocelli in ♀ wanting, in ♂ very small.

Antennae in ♂ 14-jointed, the scape abnormally enlarged, dilated, with a deep emargination within; the flagellum slightly incrassated toward apex, the joints after the first moniliform; in ♀ 12-jointed, clavate, the scape linear, the pedicel longer than the first flagellar joint, joints 2 to 5 subequal, subglobular, the following forming the club.

Mandibles bidentate.

Thorax oblong-quadrate, the prothorax not visible from above; mesonotum convex, without furrows; metathorax very short, not emarginated, the angles not acute, in ♂ sparsely, in ♀ densely woolly.

Wings wanting in both sexes.

Abdomen oval, truncate behind. the petiole very short, stout, woolly, the second segment occupying nearly the whole surface.

Legs short, stout, compressed, clavate, the tarsi short.

This curious genus, known only in a single species, has not yet been recognized in America, the species described as such by Abbé Provancher under the name *Platymischus torquatus* (Add. à la Faune Hym., p. 182), being a Diapriid belonging to the genus *Tropidopria* and allied to *Diapria conica* Latr. and *D. carinata* Thomson.

The genus *Platymischus* was originally characterized from the male; and to the Rev. T. A. Marshall we are indebted for perfecting the generic diagnosis by the discovery of the female. Mr. Marshall believes the genus is parasitic on Dipterous larvae inhabiting low marshes.

LOXOTROPA Förster.

Hym. Stud., II, p. 122, 1856.

(Type *L. acolutha* Först.)

Head subglobose, often oblong. with a frontal sulcus, the occiput rounded; cheeks woolly: eyes rounded: ocelli 3, minute. Antennae in ♀ 12-jointed, clavate; the three or four terminal joints most frequently abruptly, enormously enlarged; pedicel usually much larger than the first funicular joint; in ♂ 14-jointed, filiform; scape subclavate; pedicel short: first and second funicular joints lengthened, the first slightly the longer, the second thicker and usually curved, the joints beyond oval or cup-shaped, seldom much longer than wide.

Mandibles short, bifid.

Maxillary palpi, 5-jointed.

Thorax ovoid: prothorax slightly visible from above, usually covered with a whitish wool. always present at sides; mesonotum smooth without furrows; scutellum subconvex, or flat, smooth, truncate at apex, with distinct grooved lines along the sides. and a fovea at base; metathorax short, emarginated.

Front wings pubescent, the submarginal vein terminating in a small stigmated marginal vein before attaining half the length of the wing; basal vein always present.

Abdomen oblong-oval, in ♀ showing 7 segments; the petiole short, woolly; the second segment occupying nearly the whole surface; in ♂ 8-segmented, the petiole longer.

Legs moderate, the femora and tibiæ clavate, the tibial spurs weak, the basal joint of hind tarsi fully twice as long as the second, in the ♂ somewhat shorter.

This genus was originally confused with *Diapria*, but is at once distinguished from it and closely allied genera by the frontal impression, the shape of the scutellum, and the distinct basal nervure. In the shape of the head the female, in many species, recalls somewhat *Galesus*, but the wings and antennæ are quite different.

Our species may be distinguished by the following table:

TABLE OF SPECIES.

FEMALES.

Wingless or subapterous forms 3
Winged.
  Frons superiorly not angulated in front of the eyes........................ 2
  Frons superiorly angulated in front of the eyes.
    Antennæ and legs reddish-yellow or rufous; the club abrupt, 3-jointed, black.
    Fovea of scutellum small, rounded, not reaching entirely across the base.
      Metathorax and petiole yellowish .... L. COLUMBIANA, sp. nov.
      Metathorax and petiole black .... L. RUFICORNIS, sp nov.
2. Scutellum with a single fovea at base.
    Antennæ and legs rufous, the club abrupt, 3-jointed.
      Fovea of scutellum large, transverse, deep, extending entirely across the base,
        the lateral grooves entire .... L. ABRUPTA, Thoms.
      Fovea of scutellum small, rounded, shallow, not extending entirely across
        the base, the lateral grooves obliterated basally.
                                        L. CALIFORNICA, sp. nov.
  Scutellum with two small confluent foveæ at base.
    Antennæ and legs yellow, the club abrupt, 3-jointed.
      Second funicular joint much shorter than the first........ L. FLAVIPES, sp. nov.
3. Wings deformed, not extending more than two-thirds the length of abdomen.
    Head and abdomen black, thorax rufous.
      Antennæ and legs yellow, the club abrupt, 3-jointed, black .. L. NANA, sp. nov.
  Apterous.
    Entirely black.
      Antennæ and legs rufous, the club abrupt, 3-jointed, black.
        Scutellum with a small rounded fovea .... (L. CALIFORNICA.)
      Antennæ and legs rufo-piceous, the antennæ gradually clavate.
        Scutellum with a small, shallow fovea........... L. PEZOMACHOIDES, Ashm.

Loxotropa columbiana, sp. nov.

♀. Length, 1.5ᵐᵐ. Polished black, impunctured; antennæ, except the large, abrupt, 3-jointed club, the legs, apex of metathorax, and petiole, reddish-yellow. Wings hyaline, pubescent, the stigma piceous, the submarginal and basal nervures pale yellowish. The antennæ are 12-jointed, reaching to the base of the abdomen; scape as long as the funicle, subclavate; pedicel obconic, longer and thicker than the first funicular joint; funicular joints 2 to 5 a little longer than thick; the sixth not longer than thick; the seventh transverse; club very large, abrupt, 3-jointed, black; the joints quadrate, the last rounded at apex.

Anterior angles of thorax acute, piceous; the scutellum with a small, shallow fovea at base, separated into two parts by a delicate carina, a slight groove at each side posteriorly, the apex truncate; metathorax short, yellowish, with a central carina. Abdomen oblong-oval, polished black, the petiole yellow.

HABITAT.—District of Columbia.

Type in Coll. Ashmead.

Loxotropa ruficornis, sp. nov.

(Pl. XVII, Fig. 8, ♀ ; a, head of ♂.)

♂. Length, 1.6ᵐᵐ. Polished black; head twice as wide as thick antero-posteriorly, obliquely rounded off posteriorly. Mandibles rufous. Palpi pale. Antennae 14-jointed, rufous, longer than the body; second flagellar joint longer than the first, swollen toward the tip, the joints beyond to the last, oval-moniliform, twice as long as thick, the last nearly twice as long as the penultimate. Thorax smooth with deep mesonotal furrows, the middle lobe prominent, convex. Pronotum not impressed at sides in front of the tegulae. Scutellum with a single large, smooth fovea at base. Metapleura finely sculptured, rufous toward the hind coxae. Tegulae rufous. Wings hyaline, pubescent, with moderately short cilia. Legs, including coxae, reddish-yellow. Abdomen oval, polished, the petiole hardly twice as long as thick, finely sculptured and pubescent.

♀. Length, 1.65ᵐᵐ. Antennae, except the abrupt, 3-jointed black; club, rufous, flagellar joints 5 to 7 a little transverse; frons with acute tubercles before the eyes; scutellum with a small shallow fovea at base, not divided at base by a carina; postscutellum opaque, shagreened; metanotum, with an acute prominence above, behind the postscutellum, and a deep, oblong fovea above the posterior angles. Wings clear, hyaline, pubescent, while the abdomen is oblong-oval, polished, the petiole finely striated and covered with a griseous wool.

HABITAT.—Arlington, Va.

Types in Coll. Ashmead.

Described from specimens taken by myself while beating in low places along the Potomac River.

Loxotropa abrupta Thoms.

Basalys abrupta Thoms., Öfv., p. 368, ♀ ; Ashm. Can. Ent., XX, p. 54.

♀. Length, 2ᵐᵐ. Polished black, impunctured; antennae, except the abrupt 3-jointed black club, and legs, rufous. Head subglobose, the frons somewhat flat, but not horned or tubercular. Antennae 12-jointed; scape rather stout, as long as the funicle; pedicel stout, about twice the length of the first funiclar joint, funiclar joints 2 to 5 not or scarcely longer than thick, 6th and 7th transverse. Scutellum with a fovea all across the base, connected with lateral grooved lines. Metapleura covered with a cinereous pubescence. Wings subhyaline, pubescent; stigma brown, nervures pallid. Abdomen oblong-oval, polished black, about as long as the thorax; the petiole rugose, pubescent.

HABITAT.—Europe, Canada.

A specimen of what is evidently this European species was sent to me, more than four years ago, by Mr. W. Hague Harrington.

Loxotropa californica, sp. nov.

♀. Length, 1.8ᵐᵐ. Allied to *L. abrupta*, but differs as follows: The first flagellar joint is two-thirds the length of the pedicel; the scutellum has a very shallow not large fovea at base, the lateral impressed lines only distinct posteriorly; wings hyaline or very slightly tinged with fuscous; sometimes wanting; the head is a little wider before than behind; cheeks behind, collar at sides, and petiole densely woolly.

HABITAT.—Santa Cruz Mountains, California.

Types in Coll. Ashmead.

Loxotropa flavipes, sp. nov.

♀. Length, 1.3 to 1.6ᵐᵐ. Polished black, impunctured; antennæ (except club) and legs brownish-yellow or reddish-yellow; wings hyaline, pubescent, the stigma piceous. The head is a little longer than wide and a little wider before than behind, the frons flattened. Antennæ 12-jointed, as in *L. columbiana*, the three last funiclar joints gradually widened, transverse; club abrupt, 3-jointed, joints 1 and 2 quadrate, joint 3 oblong. Scutellum with 2 small foveæ at base, usually more or less confluent, and with lateral grooved lines posteriorly. Metathorax roughened, pubescent, with an acute prominence above. Abdomen oblong-oval, black, shining, the petiole with a long, cinereous pubescence.

The ♂ agrees with the female structurally, except the foveæ at base of scutellum are deeper and always confluent, and in having 14-jointed, filiform-moniliform antennæ. Legs and antennæ honey-yellow. The pedicel and the second flagellar joint are about of an equal length, the latter a little curved or dilated toward one side at apex, the first more slender and shorter, the following, to the last, rounded-moniliform, the last ovate.

HABITAT.—Maryland, Virginia, and District of Columbia.

Types in Coll. Ashmead.

Loxotropa nana, sp. nov.

♀. Length, 1ᵐᵐ. Head and abdomen black, polished; thorax rufous; legs and antennæ, except the club, brownish-yellow, or honey-yellow; wings not fully developed, narrow, not reaching to the tip of the abdomen. The head is longer than wide, a little narrower behind than before, the frons flattened. Antennae, 12-jointed, with an abrupt 3-jointed, black club; pedicel much stouter and larger than the first funiclar joint; funiclar joints 2 to 5 a little wider than long, the following transverse and shorter. Scutellum with a small rounded fovea at base.

HABITAT.—Jacksonville and Fort George Island, Fla.

Types in Coll. Ashmead.

Loxotropa pezomachoides Ashm.

Can. Ent., xx, p. 53.

♀. Length, 1.2ᵐᵐ. A small apterous, highly polished, black species, with dark rufous-colored legs and antennae, and covered with long, sparse hairs; the cheeks and collar woolly. The antennae are long, gradually incrassated towards tips, the first flagellar joint a little longer but not quite so thick as the pedicel, the joints 2 to 5 longer than thick; joints 6 to 9 moniliform, the last conic nearly twice as long as the preceding. Thorax narrowed anteriorly, the scutellum with a very small, slightly impressed fovea at base; the metathorax rugose.

The abdomen is ovate, large, fully twice as wide as the thorax, highly polished, with some sparse, long hairs and a very short petiole.

HABITAT.—Ottawa, Canada.

Types in Coll. Ashmead.

Described from four specimens. Quite distinct from the other species of this genus, and probably not a genuine *Loxotropa*, being more closely allied structurally to the new genus *Trichopria*.

TROPIDOPRIA Ashmead, gen. nov.

(Type *Diapria conica* Fabr.)

Head rounded, smooth, the occiput rounded; cheeks woolly; ocelli 3, small, placed on the anterior part of the head; eyes round, usually with 3 or 4 bristles.

Antennae inserted on a frontal prominence; in ♀ 12-jointed, clavate, submoniliform; scape cylindric, reaching beyond the ocelli; pedicel oval, thicker than the first funiclar joint; the first funiclar joint cylindric, narrowed basally, at least 4 or 5 times as long as thick, the second, third, and fourth shorter, subequal, all narrowed basally; the joints beyond gradually increase in size and are more or less moniliform, the last the largest, fusiform; in the ♂ 14-jointed, filiform, simple, or pedicellate-nodose with whorls of hairs.

Mandibles short, arcuate, with 2 small teeth at apex.

Maxillary palpi 5-jointed.

Thorax ovoid, the sides flat; prothorax visible from above, densely woolly; mesonotum smooth, longer than wide, without furrows; scutellum convex, compressed from the sides, with a delicate median carina on the disk and a deep fovea at base; metascutellum conically prominent; metathorax short, woolly, with subacute angles.

Front wings ample, pubescent and fringed; the submarginal vein terminates in a short punctiform marginal vein, at about one-third the length of the wing; otherwise entirely veinless.

Abdomen long, conic-ovate, composed of 6 segments, the petiole longer than thick, fluted, pubescent or woolly; the second occupies about half the whole surface; the third and fourth very short; the fifth as long as the third and fourth together; the sixth still longer, pointed and stylus-like.

Legs long, pilose, especially the tibiæ, the femora and tibiæ clavate, the tibial spurs well developed, or at least not weak, posterior tarsi shorter than their tibiæ, the basal joint scarcely twice as long as the second, claws well developed, falcate.

This genus is founded upon *Diapria conica* Fabr., and is closely allied to *Diapria*; but the carinated scutellum and the difference in the abdomen readily distinguish it, not only from this genus but from all others in the group. In the carinated scutellum we see the first approach to the cupuliform scutellum in the group *Eucoilinæ*, in the family *Cynipidæ*, and like that group, the genus is parasitic, so far as we know, on Dipterous larvæ.

The following table will aid in determining our species:

TABLE OF SPECIES.

FEMALES.

Abdomen not produced into a long, conical point, although obtusely pointed .. 2
Abdomen produced into a long conical point.
    Scutellum with a large fovea at base, the bottom usually with 3 or 4 keels.
        Antennæ rufous, the five or six terminal joints black or fuscous; legs, including coxæ, rufous ....T. CONICA Fabr.
2. Scutellum with but a single keel at the bottom of the fovea, the continuation of the scutellar carina.
    Antennæ, coxæ, and middle of femora piceous or blackish.
        Pedicel longer than the first funiclar joint..............T. CARINATA Thoms.
        Pedicel much shorter than the first funiclar joint.........T. TORQUATA Prov.
    Scutellum with the bottom of the fovea slightly aciculated, but the scutellar carina ceases before reaching it.
    Antennæ, coxæ, and middle of femora rufo-piceous.
        Club 5-jointed, first flagellar joint nearly twice as long as the pedicel.
                        T. SIMULANS, sp. nov.
    Antennæ, except club, and legs brownish-yellow; club 4-jointed, the first three flagellar joints equal, very little longer than the pedicel, the second and third shorter than first.........T. TETRAPLASTA, sp. nov.

Tropidopria conica Fabr.

(Pl. xviii, Fig. 1, ♀.)

Cynips No. 33, Geoffr., Hist. des Ins., ii.
Cynips phragmitis Schrank, En. No. 617; Vill. Ent., iii, p. 76, No. 21 (?).
Ichneumon conicus Fabr., Ent. Sys., ii, 188; Vill. Linn. Ent., iii, 212.
Psilus conicus Jur., Hymn., p. 319; Spin. Ins. Lig. Fasc., iii, p. 166.
Diapria conica Latr., Hist. Nat., xiii, 231; Gen. Crust. et Ins., iv, 37, ♀; Nees,
   Mon., ii., 325, ♂ ♀; Steph. Ill. M., vii, Suppl. 10, Pl. xlvi, fig. 2,
   ♀; Ratz., Ichn. d. Forst. Ins., iii, 186 (Econ.); Thoms., Öfv., 1858,
   p. 360, ♂ ♀; Marsh., Cat. Brit. Oxy., p. 112.

♂ ♀. Length, 2.5 to 3ᵐᵐ. Polished black; cheeks, collar, metathorax and petiole woolly, the rest of the surface with some sparse
hairs; antennae, mandibles and legs rufous, the scape more or less rufo-
piceous, the five apical joints black, the first four funicular joints subcylindrical, a little thicker at tip than at base, the first the longest, one
and a half times longer than the pedicel, the three following joints
equal or very slightly shortened, the fifth much thickened at tip, the
sixth, seventh, eighth, and ninth, oval-moniliform, the last cone-shaped,
slightly longer than the preceding.

In the male the antennae are verticillate, the flagellar joints elon-
gate-cylindrical, the second outwardly emarginated at base. Scutellum with a large deep fovea at base, separated into two parts
by the medial carina that extends posteriorly to the tip of the scutellum, the bottom with 3 or 4 raised lines; laterally the scutellum is
densely pubescent and posteriorly there are some erect hairs. Metathorax densely pubescent or woolly, with a raised prominence at base.
Tegulae black or piceous. Wings hyaline, ciliated. Abdomen longer
than the head and thorax together, conically pointed, the petiole pubescent, finely rugose, with two short carinae at base above; second
segment very long, with two foveolae at base; third and fifth segments
about equal in length, twice as long as the fourth; sixth produced into
a conic point, about as long as the third, fourth, and fifth together, the
ovipositor slightly projecting from its tip.

Habitat.—England, Europe, and North America.

Specimens in National Museum and Coll. Ashmead.

This European species is now evidently widely distributed in North
America, and has probably been imported with its Syrphid host, *Eris-
talis tenax*. I have specimens from Long Island and Albany, N. Y.;
Ottawa, Canada; Arlington, Va.; Washington, D. C.; and Marquette,
Mich. It is recorded by Kirchner (Cat. Hym. Europae, p. 204) as para-
sitic on *Eristalis tenax*, and I have seen specimens in the Berlin Mu-
seum reared from this insect, while Mr. L. O. Howard informs me
Dr. Lintner has reared it from this same Dipteron at Albany, N. Y.

Tropidopria carinata Thoms.

Diapria carinata Thoms., Öfv., 1858, p. 361.

♀. Length, 2.5ᵐᵐ. Polished black; cheeks, collar, metapleura, and
petiole posteriorly, woolly; rest of its surface with long, sparse hairs;

antennæ, mandibles, coxæ, and the thickened portion of the femora and tibiæ piceous or brown-black, rest of the legs rufous; the pedicel is stouter and a little longer than the first funiclar joint, the joints beyond to the sixth very gradually decreasing, but slightly thickened; the seventh large, oval, the three terminal joints greatly enlarged, the last oblong. Fovea of scutellum smooth at base, separated into two parts by the scutellar carina, the sides not densely pubescent. Metathorax with a prominence at base and foveated at sides, nearly bare. Tegulæ black. Wings subhyaline, pubescent. Abdomen not longer than the head and thorax united, not conically pointed at apex, and depressed, the petiole not much longer than thick, fluted; the second segment occupies the greater portion of its surface, the third short, slightly longer than the fourth, the fifth longer than the third, the sixth not longer than the fifth.

HABITAT.—Canada and Europe.

Specimens in Coll. Ashmead.

A single specimen received from Mr. W. H. Harrington, taken at Ottawa, Canada, could not be separated from European specimens of this species in my collection, and, like *T. conica*, it has probably been imported into this country on some Dipterous host.

Tropidopria torquata Prov.

Platymischus torquatus Prov., Add. et Corr., p. 182.

♀. Length, 2.8^{mm}. Closely allied to *G. carinata*, but with the following distinct differences: The pedicel and two or three funiclar joints are more or less rufous, the pedicel much shorter than the first funiclar joint, and the club really begins with the fifth funiclar joint, the joints from the sixth to the last being quadrate, the last conic; scutellum with a large, smooth fovea at base, without raised lines at bottom and not separated into two parts by the forward extension of the median carina; tegulæ black; wings hyaline; the legs show more rufous than in *D. carinata*, the femora and tibiæ not being so dark; petiole of abdomen longer, more pubescent, and not distinctly fluted; the third segment is longer than either the fourth or the fifth, the latter very slightly longer than the fourth, while the sixth is very little longer than the third.

HABITAT.—Ottawa, Canada.

Type in Coll. Ashmead.

Described from Provancher's type, given me by Mr. W. H. Harrington.

Tripidopria simulans, sp. nov.

♀. Length, 2 to 2.5^{mm}. Approaches nearest to *T. torquata*, but differs as follows: Antennæ dark rufous, very gradually clavate, more slender than in *torquata*, the pedicel one-third shorter than the first funiclar joint, the funiclar joints to the fifth slender, decreasing in length, the fifth stouter, the joints beyond to the last oval-moniliform, the last cone-shaped. The scutellum has a single large fovea at base, the bottom with some raised lines; the median carina distinct only at the tip of the scutellum. Coxæ piceous-black, the legs rufous. Tegulæ black. Wings subhyaline, with a fuscous pubescence. The third abdominal segment is very little longer than the fourth, the fifth and sixth about equal, and longer than the third and fourth together.

HABITAT.—Ottawa, Canada.

Types in Coll. Ashmead.

Described from two ♀ specimens received from Mr. W. Hague Harrington.

Tropidopria tetraplasta, sp. nov.

♀. Length, not quite 2^{mm} Polished, black; cheeks bare; collar, metapleura and petiole woolly; dorsum of metathorax bare, smooth, with a median carina; mandibles piceous; antennæ, except the club, and the legs, including coxæ, pale brownish-yellow. Antennæ 12-jointed, terminating in a 4-jointed club; the pedicel is much stouter, but not quite as long as the first flagellar joint; the first, second, and third flagellar joints equal, the three following a little shorter and thicker, the club rather abrupt, the last joint the largest, oval, but scarcely longer than the penultimate. Scutellum with a large fovea at base, its bottom with about 5 raised lines; dorsum of scutellum carinated, but the carinæ not extending into the fovea. Wings large, ciliated, the marginal vein triangular, pale brown. Abdomen pointed-ovate, scarcely longer than the head and thorax together, the petiole short, not longer than thick, woolly, the apex pointed, pilose.

HABITAT.—Washington, D. C.

Type in Coll. Ashmead.

Described from a single specimen taken in October. It comes nearest to *T. simulans*, but the antennæ are quite different, having a distinct 4-jointed club, and the color of antennæ and legs are pale yellowish.

DIAPRIA Latreille

Préc., p. 110 (1796); Förster, Hym. Stud., II, p. 123 (1856).
Psilus Panz.

(Type *D. elegans* Jur.)

Head rounded or subglobose, without a frontal sulcus, the occiput rounded. Ocelli 3, small, placed anteriorly. Eyes rounded. Antennæ on a frontal prominence, in ♀ 12-jointed, clavate, or gradually incrassated toward tips, sometimes the last three or four joints much enlarged; the scape extends beyond the ocelli; the pedicel is usually longer and stouter than the first funiclar joint; in ♂ 14-jointed, nodose-pedicellate, with whorls of long hairs, the first funiclar joint longer than the second, the second curved or dilated toward one side.

Maxillary palpi rather short, 5-jointed.

Mandibles bifid at tips.

Thorax ovoid, the prothorax slightly visible from above, usually woolly; mesonotum a little longer than wide, smooth without furrows; the mesopleura smooth, not impressed; scutellum rather small, subconvex, rounded off posteriorly without a medial carina, and unifoveated at base; metathorax short, woolly, or pubescent, with an angulated prominence at base, and more or less acute lateral angles posteriorly.

Front wings pubescent, with submarginal vein terminating in a punctiform marginal vein at about one-third the length of the wing; otherwise entirely veinless.

Abdomen ovate or oval, 7 or 8 segmented; the petiole longer than thick, woolly, the second segment occupying most of its surface, the following segments being very short.

Legs rather long, pilose; the femora and tibiæ clavate; the tibial spurs distinct; posterior tarsi at least as long as the tibiæ; the basal joint one-third longer than the second.

The student will have no difficulty in recognizing the males in this genus, as here restricted; but with the females it is quite different, many females in the new genera *Ceratopria*, *Trichopria*, and *Phænopria* closely resembling those of *Diapria*. *Phænopria* can always be distinguished by the absence of a fovea at base of the scutellum; *Ceratopria* most frequently has two small foveæ at base of the scutellum; the head is usually longer than wide, with a frontal sulcus, and the club of antennæ is generally abrupt; while *Trichopria*, which is the most closely allied, may be distinguished by the less prominent ridge at base of the metathorax, which is areolated or bifovealated; the scutellum has one or two shallow foveæ at base, while the abdomen is oblong-oval and less pointed at apex.

The following table will aid in determining our species:

TABLE OF SPECIES.

FEMALES.

Club of antennæ 4-jointed 3
Club of antennæ 3-jointed.
  Antennæ and legs reddish-yellow, or yellow, sometimes more or less piceous,
    the club black 2
  Antennæ and legs black.
    Second funiclar joint two-thirds the length of the first, the following joints
      to the club cylindrical, at least twice as long as thick.
                            D. CALIFORNICA, sp. nov.
    Second funiclar joint half the length of the first and very little longer than
      thick, the following joints to club short, moniliform.
                            D. UTAHENSIS, sp. nov.
2. First funiclar joint thrice as long as thick, the following joints at least twice as
  long as thick.
  Legs uniformly reddish-yellow .... D. ERYTHROPUS, sp. nov.
  Legs piceous, the slender parts of femora, tibiæ, and the tarsi, honey-yellow.
                            D. AGROMYZÆ Fitch.
  First funiclar joint only twice as long as thick, the following joints scarcely
    longer than thick.
  Legs yellow .... D. TEXANA, sp. nov.

3. Head, thorax, and abdomen black.. 4
   Head and abdomen black, the thorax rufous..........D. ERYTHROTHORAX Ashm.
4. Club slender, the pedicel much shorter than the first funicular joint.
   Antennæ and legs long, yellow, the club black; funicular joints all very long, slender, the first the longest, twice as long as the pedicel, the others subequal..D. COLUMBIANA, sp. nov.
   Club incrassated, the pedicel longer than the first funicular joint.
   Antennæ and legs piceous or black, trochanters, base of tibiæ, and tarsi yellow.
   Last joint of club much enlarged, 4 times as large as the penultimate, the first joint distinctly larger than the last funicular joint ...D. ARMATA Ashm.
   Last joint of club not especially enlarged, only twice as large as the penultimate, the first joint not much longer than the last funicular joint.
                                                                D. COLON Say.
   Antennæ and legs rufous or reddish-yellow, the club, or the terminal joints of club alone, black.
   Three terminal joints black........................D. TETRAPLASTA, sp. nov.
   Club wholly black.
   Last four funicular joints not or scarcely longer than thick.
                                                                D. VIRGINICA, sp. nov.
   All funicular joints fully twice as long as thick ....D. MUSCÆ, sp. nov.

MALES.

Head and abdomen black, thorax rufous..............D. ERYTHROTHORAX Ashm.
Wholly black, antennæ and legs ringed with piceous.........D. MUSCÆ, sp. nov.

The following species described by Thomas Say under the old genus *Psilus*, probably belong in this genus: *Psilus abdominalis*, *P. obtusus*, *P. ciliatus*, and *P. apicalis*. They remain unknown to me.

Diapria californica, sp. nov.

(Pl. XVIII, Fig. 3, ♀: *a*, ♂ antenna.)

♀. Length, 1.4 to 1.5 ᵐᵐ. Polished black, impunctured, covered with sparse long hairs; sides of collar and petiole woolly; palpi brown black; legs black, pilose, the tarsi dark brown. Antennæ 12-jointed, black, sparsely pilose, terminating in a 3-jointed club; the first flagellar joint is longer than the pedicel, the latter much the stouter; flagellar joints 2 to 6 subequal, the seventh a little longer and stouter than the sixth; club 3-jointed, the last joint ovate, not quite twice as long as the preceding joint but wider at base. Scutellum with a deep fovea at base connected with lateral impressed lines. Metathorax opaquely sculptured, with a prominence above and covered with a fine pubescence. Wings hyaline or very faintly tinged, strongly fringed, the marginal nervure wedge-shaped, rust-brown; there is also a fuscous streak across the wing below the marginal nervure. Abdomen ovate, polished, the last segment produced into a little point, with long sparse hairs; petiole a little longer than thick, roughened and covered with a fine woolly pubescence.

The ♂ agrees with the female except the antennæ are 14-jointed, long filiform, with the scape black, the flagellum brown, pilose; the first flagellar joint is the longest joint, the second is a little shorter, clavate, a little curved, the following, very little shorter, fusiform.

HABITAT.—Santa Cruz Mountains, California.

Types in National Museum.

Diapria utahensis, sp. nov.

♀. Length, 1.4ᵐᵐ. Agrees in every respect with *D. californica*, except as follows: The antennæ are a little shorter; the first flagellar joint is scarcely longer than the pedicel; the second, about half the length of the first; the following to the club all short, not or scarcely longer than thick, very slightly widened, moniliform; club 3-jointed, the first joint moniliform, the last oblong. The fovea at base of scutellum is shallow as compared with *D. californica*, and not connected with the lateral impressed lines. The petiole is as thick as long; therefore a little shorter than in *D. californica*, while the body of the abdomen is a little longer.

HABITAT.—Salt Lake, Utah.

Types in Coll. Ashmead.

Two ♀ specimens received from Mr. E. A. Schwarz.

Diapria erythropus, sp. nov.

♀. Length, 1.6 to 1.8ᵐᵐ. Polished black, impunctured; antennæ, except club, and legs, reddish-yellow; metathorax and petiole piceous; wings hyaline, strongly fringed.

Antennæ long, 12-jointed, terminating in a 3-jointed, black club; the first flagellar joint is about as long as the pedicel, more slender, cylindrical, joints 2 to 6 subequal, seventh much longer and stouter than the sixth; club joints all longer than thick. Collar woolly at sides. Scutellum faintly foveated at base, very slightly ridged toward apex, the sides sloping and with a slight impressed line. Abdomen ovate, a little longer than the thorax, polished black, the petiole piceous or yellowish, scarcely longer than thick, pubescent above and beneath.

The ♂ differs in having long 14-jointed antennæ, rufo-piceous toward tips, the joints relatively as in *D. californica*, the thickened parts of the legs more or less piceous, while the abdomen is shorter and of an oval shape, colored as in the female.

HABITAT.—District of Columbia, Virginia, and Florida.

Types in Coll. Ashmead.

Diapria agromyzæ Fitch.

Second N. Y. Report, p. 303; Cress. Syn. Hym. N. A., p. 251.

"The Wheat Mow Fly's Parasite. They measure 0.06ᵐᵐ in length, and to the tip of the closed wings, 0.08ᵐᵐ. They are black and shining, with shanks thickened toward their tips, the hind pair very long, and the legs are pale yellowish, with the thighs and the thickened ends of the shanks black. The abdomen is elliptic. The antennæ in the males are thread-like and nearly as long as the body, composed of 14 joints, which are very distinct, equal, oval, a third longer than broad, the apical one being a little longer and egg-shaped, and the basal one club-shaped, and thrice as long, but scarcely thicker, than the following ones. In the female they are shorter and composed of 12 joints, which are compacted together, the three last enlarged and forming a kind of knob or club, the last joint nearly as long as the two which precede it, its end bluntly rounded."

HABITAT.—New York and Long Island.

Reared by Dr. Fitch from *Agromyza tritici.*

A single ♀ specimen, doubtfully referred to this species, is in my collection obtained from Mr. Martin Linell. It measures 1.5ᵐᵐ long and is similar to *D. erythropus,* but differs as follows: Antennæ piceous-black; the seventh funiclar joint is only a little longer and stouter than the sixth; the legs are honey-yellow, with the hind coxæ black, and the clavate parts of the femora and tibiæ piceous.

Diapria texana, sp. nov.

♀. Length, 1.8ᵐᵐ. Closely allied to *D. erythropus,* but with the legs more decidedly yellow, the funiclar joints very gradually shortening, the joints after the first only a little longer than thick, the seventh shorter than the sixth and very little stouter, the club much larger. The fovea at base of scutellum is more deeply impressed, the metathorax more pubescent, while the petiole is black; otherwise as in *D. erythropus.*

HABITAT.—Texas.

Type in Coll. Ashmead.

Diapria erythrothorax Ashm.

Can. Ent., XIX, p. 196.

♀. Length, 1.5ᵐᵐ. Head and abdomen black, shining; thorax, reddish-yellow; legs and antennæ, honey-yellow. Antennæ, 14-jointed, the joints pedicellate-nodose, with whorls of bristles; the second flagellar joint the longest, clavate, a little curved, the first, being but two-thirds as long as the second, obconic. The scutellum has a fovea across the base, without distinct lateral impressed lines, the apex slightly ridged or compressed. Collar and petiole, woolly. Wings hyaline, strongly fringed; the nervures pale yellow. Abdomen oval, black, polished, shorter than thorax.

HABITAT.—Jacksonville, Fla.

Type in Coll. Ashmead.

Diapria columbiana, sp. nov.

♀. Length, 2ᵐᵐ. Polished black, impunctured; antennæ, except the slender 4-jointed club, and the legs reddish-yellow. Head sub-globose, narrowed behind the eyes. Antennæ 12-jointed, long and slender, the first flagellar joint twice as long as the pedicel, slender, cylindrical, joints 2 to 6 very gradually decreasing, the second, fully four times as long as thick, the last, a little thicker than the preceding, very little more than twice as long as thick; club 4-jointed, black, the last joint ovate, a little longer and thicker than the preceding, the others oval.   Thorax obliquely narrowed before, the collar produced into a little rounded neck.   Scutellum with a curved impression at base.   Wings clear hyaline, strongly fringed, the nervures pale yellowish.   Hind coxæ large, oblong; all the legs very long, the trochanters long, the femora and tibiæ long, clavate.   Abdomen ovate, black, polished; the petiole twice as long as thick, fluted.

HABITAT.—District of Columbia and Virginia.

Type in Coll. Ashmead.

Diapria armata Ashm.

Can. Ent., xx, p. 53.

♀.   Length, 1.6ᵐᵐ.   Polished black, impunctured, cheeks, collar, metathorax, and petiole, woolly; legs piceous-black, with the trochanters, base of femora, tibiæ, and tarsi rufous.   Antennæ 12-jointed, black, clavate, the club 4-jointed, incrassated, the last joint conic-ovate, stouter and as long as the 3 preceding joints united; first funiclar joint a little longer and slenderer than the pedicel, the following joints shorter, a little longer than thick and very slightly widened at tips, the last joint a little thicker than the preceding.   Scutellum with a transverse fovea at base connected with the lateral impressed lines.   Metathorax with a very sharp, prominent, triangularly acute keel just back of the postscutellum, the sides densely woolly or pubescent.   Wings hyaline, strongly fringed, with a streak across the wing below the stigma. Abdomen ovate, polished black, the petiole a little longer than thick, fluted.

HABITAT.—Ottawa, Canada.

Type in Coll. Ashmead.

A single specimen received from Mr. W. Hague Harrington.

Diapria colon Say.

Psilus colon Say., Bost. Jour., 1, 284; Lec. Edit., ii, p. 729.
Diapria colon Cress., Syn. Hym., p. 251.

♀.   Length, 1.1ᵐᵐ.   Polished black, impunctate; cheeks behind, collar and metapleura woolly; legs black, sutures of trochanters, base of tibiæ, and the tarsi piceous or brown; wings subhyaline, strongly fringed, with a dusky streak below the stigma.   Antennæ 12-jointed, terminating in a 4-jointed club, the last joint oblong, as long as the two preceding joints united, the first and second joints moniliform, the first much smaller than the second, the third quadrate, wider.   Scutellum with a distinct fovea, the fovea itself with 2 punctures at bottom and connected with impressed lines at the sides, which themselves are connected with a transverse punctate line at the apex of the scutellum.

Abdomen oblong-oval, polished black, the petiole a little longer than thick, striated, pubescent.

HABITAT.—California, Indiana.

This species is recognized from a single specimen discovered in a small lot of Hymenoptera, purchased from a collector in California.

Diapria tetraplasta, sp. nov.

♀. Length, 1.6ᵐᵐ. Polished black; antennæ, except the three terminal joints of the club which are black, dark rufous; legs pale rufous or reddish-yellow. Antennæ 12-jointed, terminating in a 4-jointed club, the flagellum two and a half times as long as the scape; pedicel longer and thicker than the first funicular joint, the following joints to the club scarcely longer than thick; the first joint of club small, oval, the second larger, the third still larger, transverse, the last longer, conic, but not thicker than the preceding. Scutellum with a shallow, smooth, transverse fovea at base. Wings fusco-hyaline, pubescent, and with long ciliæ, the marginal vein small, triangular. Abdomen ovate, pointed at apex, polished black, the petiole about twice as long as thick, piceous, and fluted.

HABITAT.—Carolina and Washington, D. C.

Types in Berlin Museum and Coll. Ashmead.

Described from a specimen in the Berlin Museum, labeled "Carolina, Zimmermann," and specimens in my collection from Washington.

Diapria virginica, sp. nov.

♀. Length, 1.5ᵐᵐ. Polished black; antennæ, except the 4-jointed club and the legs, reddish-yellow; collar, metapleura and petiole covered with a cinereous wool. Antennæ 12-jointed; first funicular joint shorter than the pedicel; the second almost as long as the first; joints 3 and 4 very little longer than thick; 5 and 6 rounded; club black, 4-jointed, the first joint not quite as wide, or as long, as the following, rounded behind, cup-shaped, the second and third quadrate, the last ovate, one-half longer than the preceding. Scutellum with a shallow, transverse fovea at base; the impressed lateral lines only indicated posteriorly, entirely wanting anteriorly. Wings clear hyaline, strongly fringed, the stigma yellowish. Abdomen ovate, polished black, the petiole short, densely woolly.

HABITAT.—Arlington, Va.

Types in Coll. Ashmead.

Described from three specimens.

Diapria muscæ, sp. nov.

♂♀. Length, 1.6 to 2ᵐᵐ. Polished black, with some sparse long hairs, the cheeks, collar, metapleura, and petiole covered with a dense, woolly pubescence. Head rounded, when viewed from above, a little longer than wide. Eyes rather small, oval, coarsely facetted. Mandibles and palpi pale brown. Antennæ inserted on a frontal prominence; in ♀ 12-jointed, pale brown, terminating in a black, 4-jointed club; the scape projects far above the ocelli, above dusky, beneath pale; pedicel stouter and slightly longer than the first funiclar joint; the funiclar joints slender, cylindric, the joints very slightly shortened to club; the club joints large, quadrately oval, the last oblong; in the ♂ 14-jointed, longer than the body, pale brown, the flagellar joints nodose-pedicellate with whorls of hairs; the pedicel is about half the length of the first flagellar joint, the following to the last shorter, the last pointed, fusiform. Scutellum with a large rounded fovea at base. Tegulæ honey-yellow. Wings hyaline, with long ciliæ. Legs, including coxæ, reddish-yellow, the tip of the posterior tibiæ slightly dusky, the coxæ more or less pubescent, the legs sparsely pilose. Abdomen ovate, scarcely longer than the thorax, subdepressed, the second segment occupying most of its surface, the following segments very short, the third in ♀ being longer than the others.

HABITAT.—Sacramento County, Cal.

Types in National Museum.

Described from many specimens, reared in September, 1890, by Albert Koebele, from Dipterous puparia found in the ground.

A specimen of the puparium, sent with the parasite, shows it to belong to the large family *Muscidæ*.

Diapria abdominalis Say.

Psilus abdominalis Say, Lec. Ed. Say's Works, II, p. 729.
Platymischus abdominalis Ashm., Can. Ent., XIX, p. 195.
Diapria abdominalis Cr., Syn. Hym., p. 251.

Antennæ clavate, as long as the body; black, abdomen whitish.
Inhabits Indiana.

Body black; antennæ broken at the second joint; first joint one-fourth the whole length, whitish; second joint obconic; terminal joint ovate-fusiform, longer than the three preceding joints together; wings very deeply ciliated; abdomen whitish, particularly at base; tarsi white.

Length about one-fourth of an inch. (Say.)

Diapria obtusa Say.

Psilus obtusus Say, Lec. Ed. Say's Works, I, p. 383.
Galesus obtusus Ashm., Can. Ent., XIX, p. 195.
Diapria obtusa Cr., Syn. Hym., p. 251.

Black; feet whitish, thighs black in the middle.
Inhabits Indiana.

Body black, polished; antennæ fuscous; anterior wings white and very obtuse, finely ciliated; ciliæ very short; feet whitish; thighs black, white at base and tip; coxæ black.

Length nearly one-twentieth of an inch. (Say.)

Diapria ciliata Say.

Psilus ciliatus Say, Lec. Ed. Say's Works, I. p. 333.
Galesus ciliatus Ashm., Can. Ent., XIX, p. 195.
Diapria ciliata Cr., Syn. Hym., p. 251.

Black; feet whitish; hairs of the wings elongated.
Inhabits Indiana.

Body black, polished; antennae with an oblong acute club, at base honey-yellow; petiole of the abdomen and feet honey-yellow; wings deeply ciliated, the hairs longer than the transverse diameter of the wings.

Length less than one-twentieth of an inch.   (Say.)

This appears to me to be a *Cosmocoma* in the family *Mymaridæ*.

Diapria apicalis Say.

Psilus apicalis Say., Bost. Jour., I. p. 283; Lec. Ed. Say's Works, II, p. 729.
Aneurhynchus apicalis Ashm., Can. Ent., XIX. p. 195.
Diapria apicalis Cres., Syn. Hym., p. 251.

Antennae at the tip of the head, which is a little prominent.
Inhabits Indiana.

Body black, polished; antennae as long as the body, fuscous, with subquadrately moniliform joints; basal joint honey-yellow; terminal joint not much longer than the preceding one; inserted at the tip of the head; beneath the antennae is rather a broad prominence; costal nervure but little less than half the length of the wing, triangular and black at its tip; feet honey-yellow; petiole distinct.

Length one-twenty-fifth of an inch.   (Say.)

CERATOPRIA Ashm., gen. nov.

(Type *C. longiceps* Ashm.)

Head globose or oblong, with a frontal impression, the occiput faintly margined, the cheeks with a tuft of wool behind; ocelli 3, very small; eyes rounded.

Antennæ inserted on a frontal ledge, in ♀ 12-jointed, terminating in an abrupt 3-jointed club, the pedicel much longer than the first funiclar joint, the first funiclar joint only a little longer than thick; in ♂ 14-jointed, filiform, the pedicel small, rounded, the first funiclar joint shorter than the second, the second stouter, the joints beyond, except the last, submoniliform, bristly.

Maxillary palpi 4-jointed, stout, the last joint terminating in 2 long bristles.

Mandibles short bifid.

Thorax ovoid, the collar usually woolly, the mesonotum smooth, without furrows, scutellum most frequently bifoveated at base, with lateral grooved lines; metathorax short, woolly, emarginated, bicarinated above, the angles usually acute.

Front wings pubescent, the submarginal vein ending in a small punctiform marginal vein at about one-third the length of the wing; no basal nervure.

Abdomen oblong-oval, with a very short, thick, woolly petiole, the second segment very large, occupying more than two-thirds its whole surface, the following segments very short.

Legs clavate, pilose, the basal joint of posterior tarsi about twice as long as the second.

Very closely allied to *Loxotropa* Förster, but the absence of the basal nervure at once separates it.

The following table will aid in determining the species:

TABLE OF SPECIES.

FEMALES.

Scutellum with a single fovea at base, head globose................................... 3
Scutellum with an arcuate impressed line at base, head oblong................. 2
Scutellum with two small foveæ at base, head oblong.
  Club abrupt, 3-jointed, black, the joints nearly equal.
    Antennæ and legs yellow.....................................C. LONGICEPS, sp. nov.
  Club not so abrupt, 3-jointed, black, the last joint very large, about as long as
    the two preceding together.
    Antennæ wholly black, or piceous-black, legs yellow.
                                C. MEGAPLASTA, sp. nov.
    Antennæ and legs yellow.........................C. BIFOVEOLATA, sp. nov.
2. Club abrupt, 3-jointed, black, the first joint not half the length of the second, the
    third one-third larger than the second; rest of antennæ and legs
    yellow...C. PUSILLA, sp. nov.
3. Club 3-jointed, incrassated.
  Antennæ piceous or black, legs rufous, coxæ and the swollen portion of the
    femora and tibiæ piceous.
    Last joint of club not much larger than the penultimate; first funiclar joint
      thrice as long as thick.....................C. INFUSCATIPES, sp. nov.
    Last joint of club greatly enlarged, as large as the two preceding united;
      first funiclar joint twice as long as thick......C. FLORIDANA Ashm.

Ceratopria longiceps, sp. nov.

♀. Length, 1.4ᵐᵐ. Polished black, impunctured; cheeks behind, collar and metathorax pubescent; antennæ, except club, and the legs yellow. Head oblong, a little longer than wide, with some sparse long hairs. Antennæ 12-jointed, ending in an abrupt 3-jointed club; pedicel longer than the first and second funiclar joints united and much stouter; funiclar joints 2 to 5 not longer than thick; joints 6 and 7 transverse; club 3-jointed, black, the first and second joints quadrate, distinctly separated, the third or last oval, closely joined to the second. Scutellum flat, triangular, with two small shallow foveæ at base. Metathorax tricarinate, the surface channeled between the carinæ, pubescent. Wings clear hyaline, fringed, but the fringe short; stigma yellowish. Abdomen oblong oval, polished black, the petiole stout, as wide as long.

The ♂ differs in having 14-jointed, filiform-moniliform antennæ, the first flagellar joint being only two-thirds as long as the second, the second being the longest and the stoutest joint and slightly curved; the following joints are moniliform, except the last, which is ovate; all the flagellar joints have rather short stiff bristles; the head is globose, while the scutellar fovea is a little more deeply impressed than in the female.

HABITAT.—Arlington, Va.

Types in Coll. Ashmead.

Ceratopria megaplasta, sp. nov.

♀. Length, 1.1^{mm}. Polished black, impunctured; cheeks behind, collar and metatborax woolly; antennæ wholly black, the scape piceous beneath; legs yellow. Head oblong, very little wider before than behind, covered with some long, sparse hairs. Antennæ 12-jointed, clavate; first funiclar joint about as long as the pedicel but more slender, cylindrical, the following joints moniliform, very slightly increasing in size to the club; club 3-jointed, the first and second joints quadrate-moniliform, the last oblong, a little larger than the 2 preceding joints together. Scutellum flat, with 2 small foveæ at base. Wings hyaline, fringed, the stigma yellowish. Abdomen ovate, black, polished, the petiole short, stout, rugose, fully as wide as long.

HABITAT.—Ottawa, Canada.

Type in Coll. Ashmead.

A single specimen, received from Mr. W. Hague Harrington, differs from the preceding species in the color of the antennæ, the proportions of the antennal joints and in the shape of the abdomen.

Ceratopria bifoveolata, sp. nov.

♀. Length, 1^{mm}. Agrees in shape and general appearance with C. megaplasta, but with the following differences: The antennæ, except the club and the legs, are reddish-yellow; the first funiclar joint is a little shorter than the pedicel; the abdomen oblong-oval, obtusely rounded at tip; while the petiole is short and so woolly that its sculpture can not be made out; otherwise it agrees perfectly with C. megaplasta.

HABITAT.—Harpers Ferry, W. Va.

Type in Coll. Ashmead.

A single specimen. Taken by Mr. E. A. Schwarz, June 19, 1891.

Ceratopria pusilla, sp. nov.

♀. Length, 0.8^{mm}. Polished black, impunctate; antennæ, except the club and the legs, honey-yellow. Head a little longer than wide. Antennæ 12-jointed; the club abrupt, 3-jointed, black, the first joint not half the length of the second and much narrower, the last ovate, longer than the preceding joint; first funiclar joint a little longer than thick, the following to last scarcely as long as thick, the last transverse. Scutellum with a curved, slightly impressed line at base. Wings hyaline, pubescent, with short cilia. Abdomen oblong, black, polished, truncate behind, the petiole small and short, woolly, the second segment with 2 foveolæ at base.

HABITAT.—Arlington, Va.

Type in Coll. Ashmead.

A single specimen.

Ceratipria infuscatipes, sp. nov.

(Pl. XVIII, Fig. 3, ♀.)

♀. Length, 1.2ᵐᵐ. Polished black, impunctured; antennæ piceous black; legs rufous, coxæ and the swollen parts of femora and tibiæ, piceous or dusky. Head globose, with long, sparse hairs. Antennæ 12-jointed; the club 3-jointed, incrassated, the first joint rounded, a little, smaller than the second, the second quadrate, the last ovate twice as long as the preceding; first funicular joint cylindrical, as long as the pedicel, funicular joints 2 to 6 shortening, but all much longer than thick, the seventh oval. Collar woolly, metathorax and petiole densely pubescent. Scutellum with a rather large, deep fovea at base. Wings hyaline, fringed, the stigma piceous. Abdomen ovate, black polished, the petiole not longer than thick.

HABITAT.—Ottawa, Canada.

Type in Coll. Ashmead.

A single specimen received from Mr. W. Hague Harrington. In the shape of the head, the large scutellar fovea, length of the funicular joints, and the ovate abdomen, this species is quite distinct from those previously described, and its position in this genus is doubtful.

Ceratopria floridana Ashm.

Cephalonomia floridana Ashm., Can. Ent., XIX, p. 196.

♀. Length, 1.1ᵐᵐ. Polished black, impunctured; trochanters, base of tibiæ, and tarsi honey-yellow. Antennæ 12-jointed, ending in a 3-jointed, incrassated club, the last joint being oblong, much stouter than the penultimate, and as long as the two preceding joints together, the first club joint being smaller than the second; first funicular joint cylindrical, about twice as long as thick, the following very slightly widened to the club, a little wider than long. Scutellum with a moderately deep fovea at base. Metathorax and petiole covered with a cinereous pubescence. Wings hyaline, strongly fringed, the stigma brown. Abdomen oblong-oval, black, polished.

HABITAT.—Jacksonville, Fla.

Types in Coll. Ashmead.

TRICHOPRIA Ashmead, gen. nov.

(Type *T. pentaplasta* Ashm.)

Head subglobose, without a frontal impression, the occiput rounded, cheeks woolly; ocelli 3 small; eyes rounded.

Antennæ, in ♀ 12-jointed, the club gradually incrassated, not abrupt, 4 or 5 jointed; in ♂ 14-jointed, filiform, covered with short, sparse hairs;

the pedicel is small, oblong, or rounded; the second funicular joint
shorter than the first, slightly emarginate at base, angulated towards
one side at tip, the joints beyond the third oval-moniliform, only slightly
longer than thick.

Maxillary palpi 5-jointed.

Mandibles short, bifid.

Thorax ovoid, the prothorax appearing above as a transverse ridge,
the sides always woolly; mesonotum smooth, without furrows; scutellum with a single shallow fovea at base, very rarely with two small
shallow foveae, and without lateral impressed lines; metathorax short,
bicarinated above, hardly emarginate, the posterior angles not acute or
prominent, and always woolly or covered with a dense appressed pubescence.

Front wings pubescent, ciliated, the submarginal vein terminating
in a small, triangular marginal vein before attaining one-third the
the length of the wing; no basal, or other nervures.

Abdomen oblong-oval, the petiole longer than thick, fluted, woolly
above and beneath; the second segment occupies most of the remaining surface, with a depression or sulcus at base above, the following
segments very short.

Legs clavate, pilose, the posterior tarsi long, slender, the basal joint
about twice as long as the second, claws long, curved.

The males in this genus are quite distinct from *Diapria* in the antennal characters, while the females, as before remarked, are quite
similar and difficult to separate. The antennae, however, have always
a 4- or 5-jointed club; the scutellum has a more shallow fovea, or two
small foveae at base, and is without the lateral grooved lines which are
always present in *Diapria;* the metathorax has rarely the acute prominent ridge at base; while the abdomen is more truncate at apex and
less pointed than in *Diapria*.

These characters, I believe, are constant and justify me in creating
a new genus, and, with a little study, the student will soon be able to
distinguish them at a glance.

As a rule the species are smaller than in *Diapria*, and the following
table shows that the genus is well represented in North America:

TABLE OF SPECIES.

FEMALES.

1. Antennal club 4-jointed 2
   Antennal club 5-jointed.
     Antennae not entirely black, the club alone black or fuscous,
       Scutellum with a single fovea at base.
         Pedicel shorter than the first funicular joint.
           Legs, reddish-yellow; pedicel twice as long as thick
                                            T. PENTAPLASTA, sp. nov.
           Legs yellow, the femora and tibiae fuscous; pedicel not longer than thick.
                                            T. ZIMMERMANNI, sp. nov.

Scutellum with 2 small foveæ at base.
Pedicel about as long as the first funiclar joint.
Legs rufous ....T. RUFIPES, sp. nov.
Antennæ entirely black.
Pedicel shorter but thicker than the first funiclar joint.
Legs pale rufous, base of coxæ and the thickened portions of the femora and
tibiæ more or less piceous ....T. CAROLINENSIS, sp nov.
2. Antennæ not wholly black 3
Antennæ wholly black.
Pedicel thicker and a little longer than the first funiclar joint.
Last joint of club large, twice the length of the penultimate, first joint
small.
Legs black, the trochanters, base of tibiæ and tarsi pale.
T. POPENOEI Ashm.
3. Pedicel longer and thicker than the first funiclar joint.
Antennæ fuscous or dark brown, club slender, the joints a little longer than
wide, the 6 funiclar joints slender, cylindric, twice as long as thick.
Legs rufous, the coxæ, trochanters and knees yellow..T. HARRINGTONII Ashm.
Antennæ fuscous only towards the tip, rufous basally, the club stout, the joints,
except the first, as broad as long.
Legs reddish-yellow, the femora and tibiæ tinged with rufous.
T. PACIFICA, sp. nov.
Legs dark rufous; club abrupt, fuscous, four jointed, the joints except the
last quadrate...............................T. HIRTICOLLIS Ashm.
Pedicel stouter but not longer than the first funiclar joint.
Club black, rest of antennæ yellow, the funiclar joints, except the last, which
is moniliform, are longer than thick.
Legs yellow...T. FLAVIPES, sp. nov.

Trichopria pentaplasta, sp. nov.

♀. Length, 1.5ᵐᵐ. Black, polished, impunctured; antennæ, except the club and legs, reddish-yellow. Head rounded. Antennæ 12-jointed, long, terminating in a 5-jointed club; first funiclar joint a little longer than the pedicel; the following joints subequal, about three times as long as thick, the last, thickened; club black, the first joint oblong, the second, third, and fourth quadrate, equal, the last, conic, longer but not thicker than the penultimate. Scutellum with a distinct fovea at base. Metathorax carinated, pubescent. Wings hyaline, strongly fringed; the stigma yellowish. Abdomen ovate, black, polished, the petiole 1½ times as long as thick, fluted, pubescent, the second segment truncate at base and slightly overlapping the base of the petiole.

The ♂ has 14-jointed, filiform-moniliform antennæ, the first flagellar joint twice as long as the pedicel and a little longer than the second; the second stouter, a little curved and dilated into an acute point outwardly at tip; the following joints moniliform, the last conical, all the joints with stiff bristles. Scutellum with a deep fovea at base and a slight ridge toward tip; otherwise similar to the ♀.

HABITAT.—Washington, D. C.; Arlington, Va.; and Riley County, Kans.

Types in Coll. Ashmead.

Trichopria Zimmermanni, sp. nov.

♀. Length, about 2ᵐᵐ. Polished black; collar, metapleura, and petiole covered with a dense griseous wool; legs yellowish, the clavate parts of femora and tibiæ and the last tarsal joint fuscous. Antennæ 12-jointed, nearly as long as the body, fuscous; flagellum more than thrice the length of the scape; scape yellowish; pedicel not longer than thick; funiclar joints cylindric, about 2½ times as long as thick and a little paler than the club; club brown-black, the first joint narrower than the following, a little longer than wide, the three following joints equal, quadrate, the last conic-ovate. Wings hyaline with a slight fuscous tinge, strongly fringed.

HABITAT.—Carolina.

Type in Berlin Museum.

Described from specimens in Berlin Museum labeled "Carolina, Zimmermann."

Trichopria rufipes, sp. nov.

(Pl. XVIII, Fig. 4, ♀; a, ♂ antenna.)

♀. Length, 2 to 2.2ᵐᵐ. Polished, black, with long, sparse hairs; the antennæ, except the 4 apical joints of club, and the legs rufous; cheeks, collar, metathorax, and petiole woolly. Antennæ 12-jointed, the flagellum slightly more than twice the length of the scape; pedicel about as long as the first funiclar joint, but stouter; the third, fourth, and fifth joints equal, scarcely longer than thick; the first joint of the club is only slightly thicker than the last funiclar joint, moniliform; the second is wider, the two following as wide as long, the last short, conic. Scutellum with 2 minute, almost obsolete, foveæ at base. Wings hyaline, fringed.

The male differs in having 14-jointed filiform-submoniliform antennæ, the first flagellar joint being four times as long as the pedicel, subclavate, the second, stouter and two-thirds as long as the first, slightly curved and dilated toward tip, the following oval-moniliform, the last conical, twice as long as the penultimate; all with sparse, bristly hairs.

HABITAT.—District of Columbia and Virginia.

Types in Coll. Ashmead.

Trichopria carolinensis, sp. nov.

♀. Length, about 2ᵐᵐ. Polished black, impunctured, sparsely pubescent; head rounded; antennæ 12-jointed, black, terminating in a 5-jointed club, the joints of which increase gradually in size; the pedicel is shorter but thicker than the first funiclar joint; the funiclar joints to the club are slender, cylindric, the first about 2½ times as long as thick, the following very slightly shorter, the last a little thicker than the preceding; the first joint of the club is oval, the second round, the third and fourth submoniliform, slightly wider than long, the last longer, conic-ovate. Scutellum with a shallow, transverse foveæ at base.

Legs pale rufous, the base of the coxæ and the clavalte portion of the
femora and tibiæ slightly infuscated; sometimes the base of the tibiæ
is yellow. Wings hyaline, strongly iridescent, with long ciliæ, the
marginal vein short, triangular, brown.

HABITAT.—Carolina and Pennsylvania.

Types in Berlin Museum.

Described from specimens labeled as having been received from
Dr. Zimmermann.

Trichopria Harringtonii Ashm.

Loxotropa Harringtonii Ashm., Can. Ent., xx, p. 53.

♀. Length, 1ᵐᵐ. Black, polished, impunctured, with sparse, long
hairs; antennæ dark red or rufo-piceous; legs, including coxæ, entirely
rufous; collar, metathorax, and petiole with a dense cinereous pubescence; wings hyaline, faintly tinged, strongly fringed, the stigma
brown; there is also a brownish streak across the wing from the tip of
the stigma.

Antennæ 12-jointed, ending in a long 4-jointed club, the joints nearly
equal in size, 1½ times as long as thick; the funicler joints are slender,
cylindrical, the first a little shorter than the pedicel; joints 2 to 4
twice as long as thick, 5 and 6 stouter. Abdomen ovate, pointed at
tip, the ovipositor slightly exserted; petiole longer than thick, woolly
or pubescent.

HABITAT.—Ottawa, Canada.

Type in Coll. Ashmead.

A single specimen received from Mr. W. H. Harrington.

Trichopria Popenoei, sp. nov.

♀. Length, 1.5ᵐᵐ. Entirely black, smooth, shining; the trochanters,
base of tibiæ and tarsi alone pale rufous or piceous; scutellum with a
single fovea at base; metathorax carinated, sparsely pubescent; petiole
woolly, not or scarcely longer than thick; abdomen oblong-oval, a
little pointed at tip, smooth and shining; wings hyaline, strongly
fringed, the stigma piceous; there is a yellowish streak across the
wing just beyond the tip of the stigma.

The antennæ are 12-jointed, moniliform; the club 4-jointed, incrassated; the first joint small, rounded; the second larger, cup-shaped; the
fourth quadrate; the last still larger, oblong.

The ♂ agrees with the ♀ except that the antennæ are 14-jointed, filiform, and piceous; the first flagellar joint is less than thrice as long as
the pedicel or a little longer than the second, the latter dilated beneath
a little beyond the middle, the following oval-moniliform.

HABITAT.—Riley County, Kans.

Types in Coll. Ashmead.

My specimens were received from Prof. E. A. Popenoe.

Trichopria pacifica, sp. nov.

♀. Length, 1.1ᵐᵐ. Polished black, impunctured; antennæ, except club and the legs, reddish-yellow; the clavate parts of the femora and tibiæ piceous. Antennæ 12-jointed, ending in a 4-jointed fuscous club, the first joint longer than thick and narrower than the preceding, the second and third, quadrate, the last conic, one-half larger than the preceding; first and second funiclar joints about equal, 1½ times as long as thick, cylindrical, the following moniliform. Scutellum with a single fovea at base; collar, metapleura, and petiole woolly. Wings hyaline, strongly fringed, the stigma yellow. Abdomen elliptic-oval, black, polished, the petiole very short.

HABITAT.—Santa Cruz Mountains, California.

Type in National Museum.

Trichopria hirticollis Ashm.

Cephalonomia hirticollis Ashm., Can. Ent., XIX, p. 195.

♀. Length, 1.9 to 2ᵐᵐ. Robust, polished black, impunctured; antennæ, except club, and the legs dark rufous. Head thick, globose, with some sparse long hairs. Antennæ 12-jointed, ending in a 4-jointed fuscous club, the joints, except the last, quadrate-moniliform, the last conic; funicle slender, cylindrical, the first joint twice as long as thick, a little shorter than the pedicel, the following joints subequal, the last two joints not longer than thick. Collar densely woolly at sides. Scutellum with a large fovea at base. Metathorax tricarinated, the pleura sparsely pubescent. Wings hyaline, the ciliæ not especially long, the stigma honey-yellow. Abdomen ovate, black, polished, the petiole short, thick, scarcely as long as wide, pubescent.

HABITAT.—Jacksonville, Fla.

Types in Coll. Ashmead.

Trichopria flavipes, sp. nov.

♀. Length, 1.1ᵐᵐ. Slender, polished black, impunctured; antennæ, except club, and legs yellow; collar, metathorax, and petiole woolly. Antennæ 12-jointed, long; club 4-jointed, fuscous, the joints, except the last, quadrate-moniliform, the first the smallest, terminal joint ovate. Scutellum with a moderate sized fovea at base. Wings hyaline, fringed, the stigma yellowish. Abdomen ovate, as long as the thorax, polished black, the petiole short and densely covered with wool.

HABITAT.—Arlington, Va.

Types in Coll. Ashmead.

PHÆNOPRIA Ashmead, gen. nov.

(Type *D. minutissima* Ashm.)

Head globose, a little broader than the thorax, the frons not impressed; ocelli 3, small; eyes rounded and broadly oval.

Antennæ inserted on a slight frontal prominence; in ♀ 12-jointed, the pedicel always much larger and stouter than the first funiclar joint, the first funiclar joint only a little longer than wide, the 3 or 4 terminal joints enlarged, forming a club; in ♂ 14-jointed, filiform-moniliform, finely pubescent, the first and second funiclar joints elongated, about equal or the second very slightly the shorter.

Maxillary palpi (?).

Mandibles short, bifid.

Thorax ovoid, the prothorax short, woolly; mesonotum smooth, without furrows; scutellum subconvex, rounded behind, without a trace of the lateral impressed lines or a fovea at base; metathorax short, woolly.

Front wings pubescent, strongly ciliated, the submarginal vein terminating in a punctiform marginal vein at about one-third the length of the wing; no other veins; occasionally apterous.

Abdomen oval, slightly pointed at tip, the petiole short, woolly, the second segment occupying most of the surface, the following segments very short.

Legs clavate, pilose or pubescent, the basal joint of posterior tarsi twice as long as the second.

This genus is readily distinguished from all other genera in the *Diapriinæ*, except *Monelata* Förster, by the non-foveated scutellum, and from *Monelata*, which also has no fovea at the base of the scutellum, by having 12-jointed, not 13-jointed, antennæ.

The species are all very small, and evidently parasitic on Muscid larvæ, *P. hæmatobiæ*, having been reared by Dr. Riley from the Horn-fly, *Hæmatobia serrata*.

Our species may be distinguished by the aid of the following table:

TABLE OF SPECIES.

FEMALES.

Wingless forms... 3
Winged.
   Antennæ with a 4-jointed club... 2
   Antennæ with a 3-jointed club, the last joint of which is enlarged, ovate.
      Antennæ black or brown-black, funiclar joints moniliform, not longer than thick.
         Legs rufous, the coxæ, femora, and tibiæ fuscous.. P. MINUTISSIMA, sp. nov.
         Legs entirely reddish-yellow...................... P. HÆMATOBIÆ, sp. nov.
      Antennæ reddish-yellow, the club black.
         Legs reddish-yellow .... P. VIRGINICA, sp. nov.
2. Club black or fuscous.
   Legs yellow or reddish-yellow.
      All the funiclar joints at least twice as long as thick... P. SCHWARZII, sp. nov.
3. Antennæ black or brown-black, gradually incrassated.
   Legs rufo-piceous, the trochanters, base of tibiæ, and tarsi pale.
      Abdomen long-oval, 2¼ times as wide as the thorax......... P. APTERA, sp. nov.
   Antennæ reddish-yellow, with a distinct 4-jointed black club.
     Legs reddish-yellow.
      Abdomen long-oval, only twice as wide as the thorax..... P. AFFINIS, sp. nov.

MALES.

Wingless forms 2
Winged.
  Legs yellow or reddish-yellow.
    Scape yellow; flagellum pale brown.
      First funiclar joint a little shorter than the second, less than thrice as long
        as thick, the second strongly excised at base, the following joints
        long-oval, fully twice as long as thick, the last conic, more than
        one-third longer than the penultimate .... P. VIRGINICA, sp. nov.
2. Legs rufous, the femora and tibiae piceous.
  Antennae honey-yellow.
    First and second funiclar joints nearly equal, more than thrice as long as thick,
      the joints beyond to the last oval-moniliform, less than twice as long
      as thick, the last conical, one-third longer than the penultimate.
                                            P. PARVA, sp. nov.
  Antennae piceous-black.
    First and second funiclar joints not equal, the second not thrice as long as thick,
      angulate towards one side, the joints beyond to the last long-oval,
      fully twice as long as thick .... P. MONTANA.

Phænopria minutissima, sp. nov.

♀. Length, 0.9ᵐᵐ. Polished black, impunctured; antennae piceous, the scape rufous; legs rufous, the coxae and clavate part of femora and tibiae piceous; wings hyaline, strongly fringed, the stigma piceous; collar at sides, metathorax, and petiole with a fine, cinereous pubescence. The antennae 12-jointed, ending in a 3-jointed club, the last joint oblong, as long as the two preceding joints united; funicle moniliform, the first joint a little longer than thick, the following moniliform, not longer than thick, very slightly increasing in size to the club. Scutellum convex without a trace of a fovea at base. Abdomen ovate, pointed at tip, the petiole very short, not as long as thick.

HABITAT.—Jacksonville, Fla.

Type in Coll. Ashmead.

Phænopria hæmatobiæ, sp. nov.

♀. Length, 0.8ᵐᵐ. Very close to *P. minutissima* but differs as follows: The cheeks have a tuft of pubescence not present in the former; antennae a little stouter, the pedicel larger, oblong; mesonotum a little shorter and not so much narrowed anteriorly as in *P. minutissima;* the petiole a little longer, while the legs are uniformly reddish-yellow.

HABITAT.—Arlington, Va.

Types in National Museum.

The species was reared by Dr. Riley from the larva of *Hæmatobia serrata*.

Phænopria virginica, sp. nov.

(Pl. XVIII. Fig. 5, ♂ ; *a*, ♀ antenna.)

♀. Length, 1.1ᵐᵐ. Polished black, impunctured; antennae, except club, and legs reddish-yellow or yellowish; cheeks behind, collar, and petiole woolly; metathorax pubescent; wings hyaline, fringed, the stigma long, brownish; head thick, globose.

The antennæ are 12-jointed, rather long, the club black or fuscous, 3-jointed, the last joint ovate, about one-half longer than the preceding, the first a little smaller than the second; funicle cylindrical, moniliform, the first joint two-thirds the length of the scape, the second subequal, the following joints not longer than thick. Abdomen ovate, pointed at tip, polished black, the petiole short and very woolly.

In the ♂ the antennæ are 14-jointed, as long as the body, pilose, the second flagellar joint a little longer than the first, slightly curved, and dilated toward apex, the following elliptic-oval, all except the last, about of an equal length, the last conical, much longer than the penultimate. The antennæ, legs, and petiole are yellow.

HABITAT.—District of Columbia and Arlington, Va.

Types in Coll. Ashmead.

Phænopria Schwarzii, sp. nov.

♀. Length, 1 to 1.2ᵐᵐ. Polished black, impunctured; antennæ, except club, and legs reddish-yellow or yellowish; club 4-jointed, fuscous; cheeks behind, collar, metathorax, and petiole woolly. Antennæ about as long as the body, 12-jointed, ending in a 4-jointed club, the joints, except the last, oval-rotund, the last conic, one-half longer than the preceding; funicle very slightly incrassated toward the club, the first joint about as long as the pedicel, but not so thick, the second and following subequal, very slightly increasing in width, the last being only a little narrower than the first club joint. Wings hyaline, fringed, the stigma yellowish. Abdomen ovate, polished black, the petiole thick, as wide as long, the second segment occupying most of the surface of the abdomen, its base inclosing the tip of the petiole.

HABITAT.—Jacksonville, Fla., District of Columbia, Maryland, and Virginia.

Types in Coll. Ashmead and National Museum.

Described from many specimens. The species is dedicated to Mr. E. A. Schwarz, to whom I am indebted for many specimens taken in and around Washington.

Phænopria aptera, sp. nov.

♀. Length, 1.4ᵐᵐ. Apterous; polished black, impunctured. Antennæ black or brown-black, gradually incrassated, submoniliform, the three joints preceding the last very briefly pedicellate, the last conic, not quite as long as the two preceding joints united, the penultimate joint quadrate. Head globose, much wider than the thorax. Cheeks, collar, and metathorax pubescent or woolly. Legs rufo-piceous, the trochanters, base of femora and tibiæ, and the tarsi yellowish. Abdomen oblong-oval, much broader than the thorax, polished black, with sparse hairs at apex, the petiole wider than long, rugose.

HABITAT.—Ottawa, Canada.

Types in Coll. Ashmead.

Described from 8 specimens received from Mr. W. Hague Harrington. The species bears no resemblance to any other species placed in this genus, but closely resembles *Loxotropa pezomachoides* Ashm., with which it was confused in my collection. It is, however, readily separated from that species by the entire absence of a fovea at base of scutellum and the slight difference in the shape of the antennæ.

Phænopria affinis, sp. nov.

♀. Length, 1.2ᵐᵐ. Apterous; polished black, impunctured; antennæ, except the 4-jointed club, and the legs reddish-yellow.

Antennæ 12-jointed, ending in a 4-jointed club, the joints of the club very gradually increasing in size; funicle subcylindrical, the first joint about as long as the pedicel, the following joints very gradually shortening, the last two a little thickened, and none less than twice as long as thick. Cheeks behind pubescent; collar, metathorax, and petiole woolly. Abdomen oblong-oval, wider than the thorax, the second segment overlapping the apex of the petiole, the latter scarcely longer than thick.

HABITAT.—District of Columbia.

Type in Coll. Ashmead.

A single specimen received from Mr. E. A. Schwarz.

Smaller and quite distinct from *P. aptera* in the color of the antennæ and legs and in having different shaped antennæ.

Phænopria parva, sp. nov.

♂. Length, 0.8ᵐᵐ. Apterous; polished black, impunctured; antennæ, coxæ, trochanters, base of tibiæ, and tarsi honey-yellow; rest of the legs rufous. The antennæ are long, 14-jointed, sparsely pubescent, the first and second flagellar joints long, subequal, the second slightly the shorter, a little curved and dilated toward apex; the following joints, except the last, oval-moniliform, nearly equal, 1½ times as long as thick, the last conical, longer than the preceding.

HABITAT.—District of Columbia.

Type in Coll. Ashmead.

A single specimen from Mr. E. A. Schwarz, and possibly the opposite sex of *P. affinis.*

Phænopria montana, sp. nov.

♂. Length, 1ᵐᵐ. Differs from *P. parva* as follows: The antennæ are piceous or black, the first and second funiclar joints long, subequal, the second, angulately produced toward apex, the following joints elliptic-oval, fully twice as long as thick; metapleura and petiole yellowish, but still woolly, while the coxæ are rufous, not honey-yellow.

HABITAT.—Santa Cruz Mountains, California.

Type in Coll. Ashmead.

A single specimen obtained by purchase. The difference in the length of the flagellar joints and the color of the metapleura and petiole will readily distinguish the species.

MONELATA Förster.

Hym. Stud., II, p. 123 (1856).

Corynopria, Hal. Nat. Hist. Rev., IV, p. 170 (1857).

(Type *Diapria parvula* Nees.)

Head rounded, slightly wider than the thorax, with a slight frontal impression, the occiput slightly emarginated and rounded; cheeks pubescent; ocelli 3, small, placed in a triangle anteriorly; eyes rounded.

Antennæ inserted on a slight frontal ledge, in ♀ 13-jointed, clavate, the last joint, which constitutes the club, being abnormally enlarged, several times larger than the preceding; scape cylindrical, reaching considerably beyond the ocelli; pedicel large, a little elongate, much stouter than the funiclar joints, and as long as the first two or three together; in ♂ 13-jointed, filiform.

Maxillary palpi 5-jointed.

Mandibles bifid.

Thorax ovoid, the prothorax visible from above as a transverse woolly line; mesonotum a little longer than wide at base, smooth, without furrows; scutellum subconvex without a trace of a fovea at base and at the most separated from the mesonotum by a faint transverse line; metathorax emarginated posteriorly, sparsely woolly, the angles more or less acute.

Front wings pubescent, strongly ciliated, the costal cell scarcely distinct, the submarginal vein reaching the costa a little before half the length of the wing, marginal vein punctiform, the basal vein wanting.

Abdomen oblong-oval, subtruncate posteriorly, the petiole short, woolly, strigose, the second segment very large, without a basal sulcus, beneath at base woolly, the following segments all very short.

Legs moderate, pubescent, the femora clavate, the tibiæ subclavate, the tibial spurs not very prominent, hind tarsi as long as their tibiæ, slender, basal joint much longer than the second.

This genus could only be confused with *Phænopria*, but the enormously enlarged terminal antennal joint in the female and the 13-jointed antennæ in both sexes sufficiently differentiate the two.

Three species are known in Europe, while but two species have been discovered in our fauna, which may be thus distinguished.

Black; antennæ, except the large terminal joint, rufous or yellow.
  Legs yellow.
    Collar and mesopleura piceous or pale; antennæ yellow.
                                 M. MELLICOLLIS Ashm.
    Collar and mesopleura black; antennæ rufous...........M. HIRTICOLLIS Ashm.

Monelata mellicollis Ashm.

(Pl. XVIII, Fig. 6, ♀; *a*, ♂ antenna.)

Can. Ent., XIX, p. 197.

♀. Length, 1ᵐᵐ. Polished black; antennæ, except the abnormally enlarged terminal joint, collar, and legs, honey-yellow; mesopleura, piceous. Flagellum scarcely twice the length of the scape; pedicel long, stouter than funicle; first funiclar joint twice as long as thick; the following all moniliform, the two or three joints preceding the club a little transverse; the club consists of one enormously enlarged joint, fuscous. Abdomen ovate, polished black, the petiole 1½ times as long as thick. yellowish, densely woolly, the wool beneath extending on to the base of the second ventral segment.

HABITAT.—Jacksonville, Fla.

Type in Coll. Ashmead.

Monelata hirticollis Ashm.

Can. Ent., xx, p. 54.

♀. Length, 1.1ᵐᵐ. Differs from the species just described in being slightly larger, with the antennæ dark rufous, the club black, the legs yellow or reddish-yellow; while the collar, metapleura, and petiole are black and densely woolly.

HABITAT.—Ottawa, Canada.

Type in Coll. Ashmead.

The species was described from a specimen received from Mr. W. Hague Harrington.

BASALYS Westwood.

Phil. Mag., 1833, p. 343.

(Type B. fumipennis Westw.)

Head subglobose, the occiput margined, the clypeus entirely separated, the cheeks woolly; ocelli 3, close together in a triangle; eyes rounded.

Antennæ inserted on a frontal prominence, in ♀ 13-jointed, either moniliform, clavate or gradually incrassated toward tips, the scape very long, reaching far beyond the ocelli; in ♂ 14-jointed, filiform, the pedicel usually smaller than the first flagellar joint, the first flagellar joint not more than half the length of the second, the second the longest joint and always excised outwardly at base and dilated at apex, the following joints, except the last, always shorter than the second.

Maxillary palpi short, 5-jointed, the third and fourth joints small, equal, the last elongate, pilose.

Mandibles bifid at tips.

Thorax ovoid, the collar woolly at sides: mesonotum without furrows (except sometimes indications of them posteriorly in the ♀), but prominently elevated in the middle anteriorly; scutellum with a profound fovea at base or with 2 confluent foveæ; metathorax emarginated, with acute angles and a prominent acute median ridge above.

Wings in ♂ large, broad, pubescent; the submarginal vein attains the costa at about one-third the length of the wing or a little beyond; the costal cell closed; marginal vein short, triangular, usually with a stump of a stigmal vein that has a short backward directed branch; ♀ apterus.

Abdomen pointed, ovate, distinctly petiolated, the second segment very large, with 2 sulci at base.

Legs clavate, pilose, the basal joint of posterior tarsi about one-third the length of the second.

Westwood characterized the genus from the male sex alone, and the difference noticed in the antennæ sufficiently distinguishes it from all other genera with non grooved mesonotum and 14-jointed antennæ.

Thomson characterized the genus as having 12-jointed antennæ, in ♀, but he confused *Loxotropa* Förster with *Basalys*, and all his females are now relegated to that genus.

The genus is undoubtedly closely allied to *Loxotropa* and the females are probably always apterous.

What I take to be the female of *Basalys* has 13-jointed antennæ and agrees in all essential characters with the male. Whether I am right or wrong is, however, questionable, and can only be definitely settled when specimens in both sexes have been reared, or the sexes are captured *in coitu*.

The species in our fauna known to me may be distinguished by the aid of the following table:

TABLE OF SPECIES.

Females 2
Males.
  Scutellum with an oblong fovea across the base.
    Antennæ and legs rufous; wings fuscous.
      First flagellar joint oblong-oval, less than half the length of the second, the second very stout, a little narrowed and curved towards the base, the following joints shorter, fully thrice as long as thick, the last acuminate, 1¼ times as long as the penultimate.
                      B. FUSCIPENNIS, sp. nov.
    Scape, pedicel, and femora piceous, rest of antennæ and legs pale rufous, or brownish-yellow.
      First flagellar joint longer than thick, about half the length of second, the second stouter towards apex, narrowed and curved towards base, the following joints shorter, not more than 2½ times as long as thick, the last much longer than the penultimate ......B. PICIPES, sp. nov.
    Scape polished black, pedicel and flagellum brown.
      First flagellar joint oblong-oval, less than half the length of the second, which is dilated outwardly at apex, the following joints equal, 2½ times as long as thick......................B. UTAHENSIS, sp. nov.
2. Winged; antennæ and legs brownish-yellow, the former gradually incrassated and dusky towards tips; pointed tip of abdomen rufous. B. ANALIS, sp. nov
  Winged; antennæ and legs piceous, the former very short, filiform; bases of tibiæ honey-yellow................................B. BREVICORNIS, sp. nov.
  Apterous; legs and funicle honey-yellow; scape and club brown-black, the latter abrupt, 4-jointed .... B. CALIFORNICA, sp. nov·

Basalys fuscipennis, sp. nov.

♂. Length, 2.7 ᵐᵐ. Polished black, impunctured; cheeks behind and collar with a dense cinereous pubescence; antennae, legs, and petiole brownish-yellow; wings fuscous. The antennae are very long, 14-jointed, tapering toward tips; pedicel rounded, first flagellar joint about twice as long as the pedicel, or less than half the length of the second flagellar joint, the latter the longest joint except the last, very stout and a little curved, the following joints to the last shorter, cylindrical, all about of an equal length, fully thrice as long as thick, the last joint longer than the penultimate. Thorax without furrows, but convexly swollen medially anteriorly, the scutellum with a transverse fovea at base, the sides straight, the postscutellum tricarinated, while the metathorax has the angles lobed and a prominent, blunt, median carina. Wings fuscous, pubescent; the submarginal nervure reaches the costa a little beyond one-third the length of the wing; the marginal vein triangular, piceous, with a cloud below its tip. Abdomen oval, black, shining; the petiole brownish-yellow, only a little longer than thick, fluted, and covered with a fine griseous pubescence.

HABITAT.—Washington, D. C.

Type in Coll. Ashmead.

Described from a single specimen taken by Mr. E. A. Schwarz.

Basalys picipes, sp. nov.

♂. Length, 2ᵐᵐ. Polished black, impunctured; scape, pedicel, and legs piceous; flagellum brownish or fuscous; base of tibiae, trochanters, and tarsi paler or brownish-yellow; wings subfuscous. The antennae are 14-jointed, longer than the body, the first flagellar joint about half as long as the second, the second the longest and stoutest joint except the last, thicker toward apex than at base, a little curved, the following joints not more than two and one-half times as long as thick, cylindrical, the last joint about twice as long as the penultimate, but slenderer. Metathorax covered with a rather dense cinereous pubescence, rugulose. Abdomen entirely black, the petiole fluted, pubescent.

HABITAT.—Washington, D. C.

Type in Coll. Ashmead.

Described from a specimen given me by Mr. O. Heidemann. This species differs from *B. fuscipennis*, in its much smaller size, color of antennae and legs, the shorter flagellar joints, more pubescent metathorax, and its wholly black abdomen.

Basalys utahensis, sp. nov.

♂. Length, 2ᵐᵐ. Polished black, impunctured; scape black, shining; pedicel and flagellum brown-black; legs piceous, the trochanters, knees, base and tips of tibiae and tarsi, rufous. The antennae are 14-jointed, stouter than in the preceding species, the first flagellar joint less than half as long as the second, the latter one-third longer than the third, and a little dilated toward tip, joints 3 to 5 about twice as long as thick, the joints beyond to the last about two and one-half times as long as thick, the last one-half longer than the penultimate. The scutellum has a large, deep quadrate fovea at base, the sides parallel, keeled; the metathorax rugose, with the usual central carina above, prominent posterior angles and almost devoid of pubescence. Wings subhyaline, pubescent, the venation pale, the stigma brown, with a slight cloud beneath the tip. Abdomen oblong-oval, black, shining, the petiole very stout, not longer than thick, rugose, pubescent above and beneath.

HABITAT.—City Canyon, Utah.

Type in Coll. Ashmead.

A single specimen given me by Mr. Schwarz. Resembles *B. picipes*, but the antennae are shorter and stouter, wings paler, while the abdomen is differently shaped.

Basalys analis, sp. nov.

♀. Length, 2.5ᵐᵐ. Polished black; antennae and legs brownish-yellow, the former dusky toward the tips; collar, metathorax and petiole covered with a white woolly pubescence; scutellum bifoveated at base; metathorax with a pyramidal prominence at base.

Antennae 13-jointed, very long, the scape unusually long, more than half the length of the flagellum, the flagellum gradually incrassated towards the tip; pedicel stouter but not longer than the first flagellar joint; the flagellar joints to fifth gradually shortening, but widening, the following to the last, submoniliform, the last, enlarged, ovate, about thrice as large as the penultimate.

Wings subhyaline, pubescent, the marginal vein thickened and truncate at apex. Abdomen conic-ovate, one and a half times as long as the head and thorax together, polished black, the pointed tip reddish.

HABITAT.—Carolina and Texas.

Types in Berlin Museum and Coll. Ashmead.

Described from two specimens. The specimen in the Berlin Museum is labeled "Carolina, Zimmermann."

Basalys ruficornis Prov.

Nat. Can., XII, p. 261; Faun. Ent. Can., II, p. 560.

♂. Length, 0.15 inch. Black, polished, shining; the mandibles, with the legs, rust red; the palpi pale yellow. Antennae ferruginous, inserted on a prominence of the face. Metathorax rugose, short, with two diverging keels. Wings subhyaline, the 2 subcostal nervures, with the stigma, black. Coxae black at the base. Abdomen with a grooved petiole, margined at the sides, the rest polished, shining, the extremity punctured.

HABITAT.—Canada.

Unknown to me.

Basalys brevicornis, sp. nov.

♀. Length, 3ᵐᵐ. Black, polished; antennæ and legs piceous; the scape beneath, the pedicel and three basal joints of the funicle showing more or less yellow; tegulæ, anterior femora beneath, the slender portion of all the tibiæ and the tarsi, honey-yellow. Mesonotum with two very short grooved lines posteriorly. Scutellum with two oblique foveæ at base; while the metanotum is rugose with a distinct median carina. Abdomen oblong-oval, highly polished, and at base overlapping the short, thick, fluted petiole. Antennæ very short, 13-jointed, subfiliform, extending scarcely to the tegulæ; the flagellar joints after the first wider than long, the terminal joint oblong. Wings subhyaline, a distinct costal cell, a basal cell, and rather large, triangular stigma, the latter with a distinct backward directed vein.

HABITAT.—Colorado.

Type in Coll. Ashmead.

Remarkable for the brevity and shape of the antennæ.

Basalys californica, sp. nov.

(Pl. XVIII, Fig. 7, ♂; a, ♀ antenna.)

♀. Length, 1.6ᵐᵐ. Apterous; polished black, impunctured, sparsely pubescent; funicle and legs, including coxæ, honey-yellow or reddish-yellow; scape and club brown-black. Head globose, the frons slightly impressed, abrupt. Antennæ 13-jointed, terminating in an abrupt, 4-jointed club; funicle slender, filiform, the joints after the first moniliform; club abruptly enlarged, the first joint rounded basally, the two middle joints quadrate, the last, oblong. Thorax subovate, the collar woolly at the sides, the mesonotum flat on disk, with indications of two very short grooved lines posteriorly just in front of the scutellum, the latter with a small fovea at base, the metanotum with a delicate median carina. Abdomen oblong-oval, polished, the petiole about twice as long as thick, striated.

HABITAT.—Santa Cruz Mountains, California.

Types in Coll. Ashmead.

Unfortunately, in making my drawings of this species and before I had drawn up my description, I knocked off the head of the male with my pocket lens; it fell to the floor and could not be found. It is now in too poor a condition for description, but I hope it will be recognized by the figure.

MYRMECOPRIA Ashm., gen. nov.

(Type M. mellea, Ashm. ♀.)

Head globose, without a frontal sulcus; the occiput small, convex; ocelli 3, small, close together in a triangle; eyes round, rather coarsely facetted.

Antennæ inserted on a frontal prominence; in ♀ 14-jointed, gradually incrassated, moniliform, with long, bristly hairs; the scape is long,
clavate, the first flagellar joint twice as long as the pedicel and one-
third longer than the second, both narrowed toward base, the following joints round-moniliform, gradually increasing in size.

Maxillary palpi very short.

Mandibles small, bifid.

Thorax ovoid, rounded before, the prothorax not at all visible from
above, the sides bare, flat; mesonotum flat, smooth, without furrows;
scutellum subconvex, with a slight median ridge, separated from the
mesonotum by a transverse furrow all across the base; metathorax not
very short, sloping off posteriorly, the angles not at all prominent; a
carina originates just back of the scutellum, forks and extends obliquely
on each side to the base of the hind coxæ.

Front wings very large and broad, with long pubescence; the submarginal vein reaches the costa before one-third the length of the wing,
ending in a small triangular marginal vein; basal cells two, subequal.

Abdomen short, oval; the petiole scarcely longer than thick, separated from the second abdominal segment by a strong constriction,
dilated beneath, and in structure very similar to the nodes in certain
ants.

Legs long, pilose, the femora clavate, the tibiæ very long, subclavate,
the posterior tarsi thick, somewhat dilated, the basal joint twice as long
as the second, the three following joints subequal, the last longer than
the second, all very hairy.

By far the most remarkable Diapriid yet discovered, and exhibiting
a most remarkable resemblance to certain ants. In venation, the 14-
jointed clavate-moniliform antennæ, and in its metathoracic and abdominal characters, it is quite distinct from all other genera in the group,
and requires no special comment at my hands, as it could not be confounded with any other genus.

Myrmecopria mellea Ashm.

(Pl. XVIII, Fig. 8, ♀.)

Loxotropa mellea Ashm., Can. Ent., XIX, p. 196.

♀. Length, 2.3ᵐᵐ. Honey-yellow, polished, sparsely pilose; eyes
and tip of abdomen brown. Antennæ 14-jointed, reaching to the middle
of the abdomen; scape long, clavate, the length of the first three joints of
flagellum united; pedicel half the length of the first flagellar joint, the
joints beyond the second, moniliform, subpedicellate, gradually increasing in size. Thorax flat above, without grooves, narrowed before; the
prothorax not visible from above; sides flat, but not impressed, bare.
Wings very large, broad, and hairy; the submarginal vein attains the
costa before one-third the length of the wing; marginal vein short, with

a blunt stump of a stigmal vein; beneath the marginal vein is a narrow, dusky streak; veins yellow.

HABITAT.—Jacksonville, Fla.

Type in Coll. Ashmead.

The species was originally described in the genus *Loxotropa*, but the characters are too entirely different from that genus to permit its remaining there.

The following species of *Pantoclis*, described by Abbé Provancher, and overlooked by me, should have followed *P. floridana* on p. 372:

Pantoclis inermis Prov.

Aneurhynchus inermis Prov. Add. à la Faune., p. 179.
Pantoclis inermis Prov. Add., p. 405.

♀. Long. 15 pce. Noir, poli brillant; les mandibles jaunâtres. Antennes longues jaune-brunâtre, le premier article le plus long, avec long poils peu denses, le 2e le plus court, les autres allongés. Tête grosse en carré transversal, épaisse en arrière des yeux, à face renflée en tubercule pour l'insertion des antennes. Thorax poli, brillant; l'écusson soulevé mais inerme. Ailes hyalines iridescentes, frangées, à stigmal, la radiale en forme de triangle allongé, ouverte en arrière, ni cubitales, ni discoidales fermées. Pattes jaune-pâle avec les hanches noires, les cuisses renflées en massue à l'extrémité. Abdomen en ovale, poli, brillant, à pédicule fort, poilu, du tiers de sa longueur environ, le 2e segment fusiforme, noir, le reste d'un jaune plus ou moins noirâtre. (Provancher.)

HABITAT.—Cap Rouge, Ottawa, Canada.

Type in Coll. Provancher.

Unknown to me.

A TABULAR VIEW OF THE BRED NORTH AMERICAN PROCTOTRYPIDÆ.

[The hosts in these tables followed by a query (?) indicate that the records are of questionable accuracy or that the host is doubtful.]

I. HYMENOPTERA.

Parasites.	*Hosts.*
Cephalonomia cynipiphila Ashm	Cynipid gall, *Holcaspis omnivora* (?).
gallicola Ashm	Cynipid gall, *Andricus foliatus* (?).
hyalinipennis Ashm	Cynipid gall, *Amphibolips cinerew* (?).
Isobrachium mandibulare Ashm	Nests of *Camponotus pennsylvanicus*.
montanum Ashm	Nests of *Formica rufibarbis*.
myrmecophilum Ashm	Nests of *Formica rufibarbis*.
rufiventre Ashm	Nests of *Formica obscuripes*.
Polygnotus emuræ Ashm	*Emura s. nodus* on willows (?).

II. DIPTERA.

Aphanogmus floridanus Ashm	*Cecidomyia* sp. feeding on Red-spider.
Lygocerus californicus Ashm	*Cecidomyia* sp. in galls on *Larrea mexicana*.
triticum Tayl	*Cecidomyia destructor* Say.
Inostemma Horni Ashm	Cecidomyiid gall in blossoms of *Vernonia noveboracensis*.
californica Ashm	Cecidomyiid gall on *Telypodium integrifolium*.
Amblyaspis minutus Ashm	*Cecidomyia* sp. in Squash.
Isorhombus arizonensis Ashm	Cecidomyiid gall on an unknown plant in Arizona.
Polymecus lupinicola Ashm	Cecidomyiid gall on *Lupinus arboreu*.
cornicola Ashm	Cecidomyiid gall on *Cornus paniculata*.
Synopeas antennariæ Ashm	*Cecidomyia antennariæ* Whlr.
Anopedias error Fitch	*Diplosis tritici* on wheat.
Trichacis rubicola Ashm	Cecidomyiid stem-gall on Blackberry; ditto on *Vernonia noveboracensis*.
Platygaster Herrickii Pack	*Cecidomyia destructor* Say.
caryæ Ashm	Cecidomyiid gall on Hickory.
Polygnotus alnicola Ashm	Cecidomyiid gall from flower bud of alder.
pinicola Ashm	*Cecidomyia pini-inopis* O. S. on *Pinus inops*.
atriplicis Ashm	Cecidomyiid gall on *Atriplex canescens*.
huachucae Ashm	Cecidomyiid pod-like gall on an unknown plant in Arizona; Cecidomyiid stem-gall on sunflower, and a Cecidomyiid gall on sensitive plant.
diplosidis Ashm	*Diplosis* sp. on pine.
actinomeridis Ashm	Cecidomyiid gall on *Actinomeris* sp.
californicus Ashm	Cecidomyiid gall on *Baccharis pilularis*.

Parasites.	Hosts.
Polygnotus astericola Ashm	Cecidomyiid gall on aster.
asynaptæ Ashm	Cecidomyiid gall *Asynapta* sp.
hiemalis Forbes	*Cecidomyia destructor* Say, on wheat.
baccharicola Ashm	Cecidomyiid gall on *Baccharis halimifolia.*
Platygaster aphidis Ashm	Aphis sp. on *Chenopodium album* (?).
Eritrissomerus cecidomyiæ Ashm	Cecidomyiid gall on Hickory.
Polygnotus striaticeps Ashm	Cecidomyiid gall on *Begeloria* or *Artemisia* sp.
salicicola Ashm	Cecidomyiid gall on midrib of leaves of Willow.
proximus Ashm	*Cecidomyia ananassæ* Riley.
eurotiæ Ashm	Cecidomyiid gall on *Eurotia canata.*
solidaginis Ashm	Cecidomyiid gall on Solidago.
viticola Ashm	Cecidomyiid gall on Grape and *Artemisia.*
cynipicola Ashm	Cecidomyiid in Cynipid gall *Neuroterus batatus.*
coloradensis Ashm	Cecidomyiid gall on Sage bush.
utahensis Ashm	Cecidomyiid gall on *Artemisia tridentata.*
rubi Ashm	*Cecidomyia farinosa* on Blackberry.
vernoniæ Ashm	Trypeta gall on *Vernonia noveboracensis.*
tumidus Ashm	*Cecidomyia symmetrica* O. S. on Oak.
Tropidopria conica Fabr	*Eristalis tenax.*
Diapria muscæ Ashm	Dipterous puparia found in the ground.
Phænopria hæmatobiæ Ashm	*Hæmatobia serrata.*
Diapria agromyzæ Fitch	*Agromyza americana* Say.

III.—Lepidoptera.

Perisemus prolongatus Prov	*Crambus caliginosellus.*
Goniozus Hubbardii How	*Platynota rostrana* and *Platacticus glorerii.*
cellaris Say	Geometrid larva on wheat stubble.
foveolatus Ashm	Tineid larva in dry fungus.
Lygocerus 6-dentatus Ashm	*Sarrothripa rawayana* (?).
Telenomus ichthyuræ Ashm	Eggs *Ichthyura inclusa.*
graptæ How	Eggs *Grapta interrogationis.*
	Eggs *Grapta progue.*
	Eggs *Vanessa antiopa.*
	Eggs *Chrysophanes hypophlæas.*
	Eggs *Pamphila cernes.*
gnophælæ Ashm	Eggs *Gnophæla hopferi.*
spilosomatis Ashm	Eggs *Spilosoma virginica.*
arzamæ Riley	Eggs *Arzama densa.*
heliothidis Ashm	Eggs *Heliothis armigera.*
lavernæ Ashm	Eggs *Laverna luciferella.*
Rileyi How	Eggs *Apatura clyton.*
clisiocampæ Riley	Eggs *Clisiocampa americana.*
bifidus Riley	Eggs *Hyphantria textor.*
Koebelei Ashm	Eggs unknown, Lepidopteron on Solidago.
gossypiicola Ashm	Eggs of a Lepidopteron on cotton.
geometræ Ashm	Eggs unknown, Geometrid on wild cherry.
californicus Ashm	Eggs *Orgyia* sp.
sphingis Ashm	Eggs *Sphinx carolina.*

IV. COLEOPTERA.

Parasites.	Hosts.
Apenesia coronata Ashm	Catogenus rufus under bark of Cercis.
Cephalonomia hyalinipennis Ashm	Hypothenemus eruditus in fig twigs.
Anoxas Chittendenii Ashm	From Cis fusipes living in fungus.
Ateleopterus tarsalis Ashm	Silvanus surinamensis in raisins and in stored grain.
Laelius trogodermatis Ashm	Larva of Trogoderma tarsale.
Proctotrypes obsoletus Say	Stelidota strigosa.
Macroteleia floridana Ashm	From stems of timothy infested with Languria (?)
Prosacantha caroborum Riley	Eggs of Chlaenius impunctifrons.
Ceraphron salicicola Ashm	Bred from old willow wood which was partially covered with fungus and infested with Coleopterous larvæ (?).
Trichacis rufipes Ashm	From acorns infested with Balaninus nasicus (?).

V. ORTHOPTERA.

Caeus œcanthi Riley	Eggs Œcanthus niveus and Œ. latipennis.
Baryconus œcanthi Riley	Eggs Œcanthus niveus.
Scelio ovivora Riley	Eggs Dissosteira carolina.
calopteni Riley	Eggs Caloptenus atlantis.
Luggeri Riley	Eggs Caloptenus sp.
Macroteleia floridana Ashm	Eggs Orchelimum glaberrimum.

VI. HEMIPTERA.

Labeo typhlocybae Ashm	Typhlocyba sp. on Celtis and Elm.
Lygocerus niger How	Siphonophora avenae.
floridanus Ashm	Lachnus australis.
Aphanogmus floridanus Ashm	From twigs containing eggs of Cicada septendecim (?).
Phanurus ovivorus Ashm	Heteropterous eggs.
Telenomus podisi Ashm	Eggs Podisus modestus.
Hubbardii Ashm	Eggs of a Reduviid.
persimilis Ashm	Eggs of an unknown Hemipteron.
Trissolcus podisi Ashm	Eggs Podisus spinosus.
euschisti Ashm	Eggs Euschistus servus.
brochymenae Ashm	Eggs Brochymena arborea.
murgantiae Ashm	Eggs Murgantia histrionica.
Hadronotus floridanus Ashm	Eggs Metapodius femoratus.
leptocorisae How	Eggs Zelus bilobus.
largi Ashm	Eggs Largus succinctus.
rugosus How	Eggs Euthoctha galeator.
anasae Ashm	Eggs Anasa tristis.
Amitus aleurodinis Hald	Aleurodes aceris and A. corn

VII. NEUROPTERA.

Telenomus chrysopae Ashm	Eggs Chrysopa sp.

VIII. ARACHNIDA.

Acoloides saitidis How	Eggs Saitis pulex.
	Eggs Phydippus morsitans.
Emertonii How	Eggs of an unknown spider.
Acolus Zabriskiei Ashm	Eggs of an unknown spider.
Baeus americanus How	Eggs of an Epeirid.

LITERATURE AND ABBREVIATIONS.

A.

Ann. Soc. Ent. Fr.—Annales de la Société Entomologique de France, Paris, 1832 *et seqq.*

Ann. and Mag.— The Annals and Magazine of Natural History, London, 1841 *et seqq.*

Ann. Mus. Nat. Hist.—Annales du Muséum d'Histoire Naturelle, 20 vols., Paris, 1802–'13.

Ashm., Bull. No. 1, Col. Biol. Ass.—Ashmead (W.H.). Bulletin No. 1 of the Colorado Biological Association. Washington, D. C., 1890.

Ashm., Bull. No. 14, U. S. Dept. Agric.—Bulletin No. 14, U. S. Department of Agriculture, Division of Entomology. Washington, D. C., Government Printing Office, 1887.

Ashm., Fla. Agric., IV.—Florida Agriculturist. Vol. IV, 1881, De Land, Fla.

Ayers (H.).—Memoirs of the Boston Society of Natural History, vol. III, 1884, No. 8, Boston, Mass.

B.

Balf., Comp. Embry.—Balfour (E. M.). Comparative Embryology. London, 1880.

Berl. ent. Zeit.—Berliner entomologische Zeitschrift, Berlin, 1857 *et seqq.*

Boh.—Boheman (C. H.). *See* Serials.

Bouché, Naturg.—Bouché (P. F.). Naturgeschichte der Insecten, besonders in Hinsicht ihrer ersten Zustände als Larven und Puppen. Berlin, 1834.

Bost. Journ. N. H.—Boston Journal of Natural History, vol. I [1836] *et seqq.*

Brullé, Hym.—Brullé (A.). Histoire Naturelle des Insectes, par M. le Comte Amedée Lepelletier de Saint-Fargeau. Hyménopteres, par M. Aug. Brullé. 4 vols. Paris, 1837–'46.

C.

Can. Ent.—Canadian Entomologist, 24 vols., London, Ontario, Canada.

Cameron.—Cameron (P.).

Cress., Syn. Hym.—Cresson (E. T.). Synopsis of the Families and Genera of the Hymenoptera of America, North of Mexico, together with a catalogue of the described species, and bibliography. American Entomological Society, Philadelphia, 1887.

Comp.-rend.—Comptes-rendus de l'Académie des sciences à l'Institut de France, Paris, 1835 *et seqq.*

Curt., Brit. Ent.—Curtis (J.). British Entomology, 16 vols., London, 1823–'40.

Curt., Farm Ins.—Farm Insects: being the Natural History and Economy of the Insects injurious to the Field Crops of Great Britain and Ireland. London, 1860.

Curt., McIntosh, Book of the Gard.—Entomological articles in McIntosh's Book of the Garden. 2 vols. Edinburgh and London, 1853–'55.

D.

Dalm., An. Ent.—Dalman (J. W.). Analecta Entomologica. Stockholm, 1823.

E.

Ent. Amer.—Entomologica Americana. A Monthly Journal of Entomology. Edited by John B. Smith. 5 vols., 1885 *et seqq.* Published by Brooklyn Entomological Society.

Ent. News.—Entomological News, etc. Edited by Henry Skinner, Md. Philadelphia, Vol. I, 1890 *et seqq.*

Ent. Mo. Mag.—The Entomologists' Monthly Magazine, London, 1864 *et seqq.*

Enc. Brit.—Encyclopaedia Britannica. Ed. VIII. Vol. IX. Article Entomology. Edinburgh, 1855.

Enc. Méth.—Encyclopédie Méthodique. 10 vols. Paris, 1789-1825.

Ent. Mag.—The Entomological Magazine. 5 vols. London, 1833-'38.

Entom.—Newman's Entomologist, London. 1840-'42 and 1864, *et seqq.*

F.

Fab.,Ent. Sys.—Fabricius (J. C.). Entomologica Systematica. 4 vols. Copenhagen, 1792-'94. Supplement, 1798.

Fab., Piez.—Fabricius (J. C.). Systema Piezatorum. Brunswick, 1804.

Fab., Sys. Ent.—Fabricius (J. C.). Systema Entomologica. Flensburg and Leipzig, 1775.

Först., Beitr.—Förster (A.). Beiträge zur Monographie der Pteromalinen. Aix-la-Chapelle, 1841.

Först., Hym. Stud.—Förster (A.). Hymenopterologische Studien. Heft II. Chalcidiae und Proctotrupii. Aix-la-Chapelle, 1856.

Först., Kl. Monog.—Kleine Monographie, 1878.

G.

Ganin, Ueber d. Embry.- Ganin (M.). Ueber die Embryonalhülle der hymenopt. und lepidopt. Embryonen. St. Petersburg, 1869.

Gard. Chron.—Gardeners' Chronicle, London, 1841 *et seqq.*

Geer, Mém.—Geer (C. de). Mémoires pour servir à l'Histoire des Insectes. 7 vols. Stockholm, 1752-'78.

Germ., Fn. Ins. Eur.—Germar (E. F.). Fauna Insectorum Europae. Heft I and II, by A. Ahrens. Continued by E. F. Germar. 24 pts. Halle, 1812-'48.

H.

Haid., Berichte.—Berichte über d. Mittheilungen v. Freunden d. Naturwissensch., herausgegeben v. W. Haidenger. 7 vols. Vienna, 1847-'51.

Hal., Hym. Brit.—Haliday (A. H.). Hymenoptera Britannica. Oxyura. Fasciculus I. London, 1839. *See* Serials.

How., Scud. But.—Howard (L. O.) in Scudder's Butterflies of the Eastern United States and Canada. Cambridge, Mass., 1889.

Hub., Ins. Aff. Orange.—Hubbard (H. G.). Insects Affecting the Orange. Report on the insects affecting the culture of the orange and other plants of the Citrus family, with practical suggestions for their control or extermination, made under direction of the Entomologist. Washington, D. C., Government Printing Office, 1885.

I.

Ins. Life.—Insect Life, etc. Edited by C. V. Riley and L. O. Howard. Vol. I. 1888 *et seqq.* Washington, D. C., Government Printing Office.

J.

Jour. Acad. Nat. Sci. Phil.—Journal of the Academy of Natural Sciences of Phila-
   delphia, Vol. I. [1817] et seqq.
Jurine, Hym.—Jurine (L.).   Nouvelle Méthode de classer les Hyménoptères et les
   Diptères. Geneva and Paris, 1807.

K.

Kirch., Cat. Hym. Eup.—Kirchner (L.).   Catalogus Hymenopterorum Europæ, 1867.

L.

Lam. Syst.—Monet de Lamarck (J. B. P. A. de).   Système des animaux sans verté-
   bres. Paris, 1801.
Latr., Cuv. Reg. An.—Latreille (P. A.).   Articles in Cuvier's Regne Animal.   Ed. I.
   3 vols. Paris, 1817.
Latr., Gen. Crust. et Ins.—Latreille (P. A.).   Genera Crustaceorum et Insectorum
   secundum ordinem naturalem in familias disposita. 4 vols. Paris and
   Strasburg, 1806-'9.
Latr., Nat. Hist.—Latreille (P. A.).   Histoire Naturelle générale et particuliere des
   Crustacés et des Insectes. 14 vols. Paris, 1802-'5.
Latr., Préc.—Latreille (P. A.).   Précis des Caracteres génériques des Insectes.
   Brive, 1796.
Linn. Ent.—Linnaea Entomologica. 16 vols.   Berlin and Leipsic, 1846-'66.
Linn., Fn. Suec.—Linné (C. von).   Fauna Suecica. Ed. II. Stockholm, 1761.
Linn., Sys. Nat.—Linné (C. von).   Systema Naturæ. Ed. XII. Stockholm, 1766-'68.
Lond. Mag.—London's Magazine of Natural History. 9 vols. London, 1829-'36.

M.

Marsh.—Marshall (T. A.).   In serials.
Marsh., Cat. Brit. Oxy.—Marshall (T. A.).   A Catalogue of the British Hymenop-
   tera Oxyura. London, 1873.
Mém. Acad. St.-Pétersbourg.—Mémoires de l'Académie de St.-Pétersbourg, 1869-'70.
Mem. Accad. Tor.—Memorie della Reale Accademia delle Scienze di Torino, serie
   seconda, Tom. XIII. 1853.
Mem. Soc. Manch.—Memoirs and Proceedings of the Manchester Literary and
   Philosophical Society, 4th series.
Mik.—Mik (J.).   In serials.
Mocs., Magyar Fn.—Mocsary (A.).   Magyar Fauna másnejii darasai (Heterogynidæ
   Faunæ Hungaricæ). Term. Közlem. XVII. pp. 1-93, pls. II.
Müll., Fn. Fridr.—Müller (O. F.).   Fauna Insectorum Fridrichsdalina. Copenha-
   gen and Leipsic, 1761.
Müll., Naturs. Linn.—Müller. (P. L. S.).   Vollständiges Natursystem des C. v. Linné,
   mit einer Erklärung. 6 vols. and supplement. Nuremberg, 1773-'76.

N.

Nat. His. Rev.—The Natural History Review: a quarterly journal of science, con-
   ducted by Haliday and others. 5 vols. Dublin, 1854-'58.
Nees, Monog.—Nees von Esenbeck (C. G.).   Hymenopterorum Ichneumonibus af-
   finium Monographiæ, Genera Europæa et Species illustrantes. 2 vols.
   Stuttgardt and Tübingen, 1834.
Nouv. Dict.—Nouveau Dictionnaire d'Histoire Naturelle. 36 vols. Paris, 1816.

O.

Öfv.—Öfversigt af Kongliga Svenska Vetenskaps-Akademiens Förhandlingar,
   Stockholm, 1815 et seqq.

P.

Pack., Guide.—Packard (A. S.). Guide to Study of Insects, and a treatise on those injurious and beneficial to crops, for the use of colleges, farm-schools, and agriculturalists, by A. S. Packard, jr., M. D. Seventh edition. New York, 1880.

Panz., Fn. Germ.—Panzer (G. W. F.). Faunae Insectorum Germanicae initia. 109 pls. Nuremberg, 1792–1810.

Panz., Krit. Revis.—Panzer (G. W. F.). Kritische Revision der Insecktenfaune Deutschlands, nach dem System bearbeitet. 2 vols. Nuremberg, 1805–'06.

Perris, Ann. Soc. Linn.—Perris (ed.). Annales de la Société Linn. Lyon, série 2. Tom. IV.

Phil. Mag.—The London and Edinburgh Philosophical Magazine and Journal of Science. 16 vols. London, 1832–'40.

Proc. Ent. Soc. Lond.—Proceedings of the Entomological Society of London, 1834 et seqq.

Proc. Ent. Soc. Wash.—Proceedings of the Entomological Society of Washington, vol. I et seqq.

Proc. U. S. N. Mus.—Proceedings U. S. National Museum, Vol. VIII.

Prov., Faun., Hym.—Provancher (Abbé Léon). Petite Faune Entomologique du Canada, etc. Vol. II. Hyménoptères. Quebec, 1883.

Prov., Add. Faun. Hym.—Additions et corrections au volume II de la Faune Entomologique du Canada traitant des Hyménoptères. Quebec, 1889.

R.

Ratz., Ichn. d. Forst.—Ratzeburg (J. T. C.). Die Ichneumonen der Forstinsecten in forstlicher und entomologischer Beziehung. 3 vols. Berlin, 1844–'52.

S.

Say, Lec. Ed.—Say (T.). Complete Writings of Thomas Say on the Entomology of North America. Edited by John L. Leconte, M. D. 2 vols. New York, 1859.

Schäff., Forts. Germ.—Herrich-Schäffer (G. A. W.). Fortsetzung von Panzer, Faunae Insectorum Germanicae initia. Ratisbon, 1829–'44.

Scudd., But.—Scudder (Saml.). Butterflies of the Eastern United States and Canada. Cambridge, Mass., 1889.

Spin., Ins. Lig.—Spinola (Marquis M.). Insectorum Liguriae species novae aut rariores. 2 vols. Genoa, 1806–'8.

Stefano, Nat. Sic.—Stefano. Il Naturalista Siciliano: Giornale delle Scienze Naturali, Ragusa, Palermo.

Step., Ill. Brit. Ent.—Stephens (J. F.). Illustrations of British Entomology. 11 vols. London, 1828–'46.

Stett. Zeit.—Stettiner entomologische Zeitung, Stettin, 1840 et seqq.

St.-Farg.—St.-Fargeau (A. L. M. Le Peletier, Comte de). See serials.

Sv. Ak. Handl.—Kongliga Svenska Vetenskaps-Akademiens Handlingar, Stockholm, 1780 et seqq.

T.

Tasch., Naturg. wirb. Thiere.—Taschenberg (E. L.). Naturgeschichte der in Deutschland, Preussen und Posen den Culturpflanzen schädlichen wirbellosen Thiere. Leipsic, 1869.

Trans. Ent. Soc. Lond.—The Transactions of the Entomological Society of London, 1834 et seqq.

Trans. Linn. Soc.—The Transactions of the Linnaean Society of London, 1791 et seqq.

Turt., Sys. Nat.—Turton (W.). A General System of Nature; translated from Gmelin, Fabricius, etc., Animal Kingdom, vols. II and III. London, 1806.

V

Verh. Pr. Rheinl.—Verhandlungen des naturhistorischen Vereins der preussischen
Rheinlande und Westphalens. Bonn, 1844 *et seqq.*
Vill., Linn. Ent.—Villers (C. J. de). Caroli Linnæi Entomologia. Leyden, 1789.

W.

Walck., Fn. Paris.—Walckenaër (Baron C. A. de). Faune Parisienne. Histoire
Abrégée des Insectes des Environs de Paris. 2 vols. Paris. 1802.
Web. und Mohr, Beitr.—Archiv für die systematische Naturgeschichte. Edited by F.
Weber and W. H. Mohr. Leipsic, 1804. Continued under the title: Beiträge
zur Naturkunde, etc. 2 vols. Kiel, 1805 and 1810.
Westw., Intr.—Westwood (J. O.). An Introduction to the Modern Classification of
Insects. 2 vols. London, 1839–'40.
Westw., Thes. Ent. Ox.—Thesaurus Entomologicus Oxoniensis, etc. Oxford, 1874.
Wien. Ent. Zeit.—Wiener entomologische Zeitung.

Z.

Zett., Ins. Lap.—Zetterstedt (J. W.). Insecta Lapponica descripta. Leipzig, 1840.
Zeits. f. wiss. Zool.—Zeitschrift für wissenschaftliche Zoologie. Leipzig, Bd. xix,
1869.
Zoöl.—The Zoölogist. London, 1843 *et seqq.*

PLATE I.

Fig. 1. *Epyris grandis.*—H. *Head:* v. vertex; o, ocelli; oc, occiput; g, gena, or cheek; f, face; e, eye; cl, clypeus; a, antenna; sc, scape; p, pedicel; fl, flagellum; mxp, maxillary palpi; lbp, labial palpi; m, mandibles. T. *Thorax:* pt, prothorax; c, collar; pn, pronotum; p, propleuron; pstm, prosternum; sp, spiracles; ms, mesoscutum; pf, parapsidal furrows; p, parapsides, or scapulæ; s, mesoscutellum or scutellum; ax, axilla; mps, mesopostscutellum; tg, tegula, or wing scale; mp, mesopleuron; mstm, mesosternum; ep, episternum; mtt, metathorax; mn, metanotum; sp, spiracles; mtp, metapleuron; ihn, insertion of hind wings. A. *Abdomen:* 1 2, 3, 4, 5, 6, 7, and 8, dorsal segments or tergites, number 1 usually designated as the petiole; 2, 3, 4, 5, and 6 urites, or ventral segments; t or sh, sheaths conjoined, forming a tube for the terebra (tba), or ovipositor proper; spc, spiculæ; cx, cx, cx, coxæ; tr, tr, tr, trochanters (1-jointed); fr, fr, fr, femora; ta, ta, ta, tibiæ; tbs, tibial spur, the middle and posterior tibiæ being 2-spurred; ts, tarsi; cl, claws; pr, pulvillus.

2. *Maxilla:* mx, maxilla; s, stipes; c, cardo; pfr, palpifer; mxp, maxillary palpi; ga, galea; la, lacinia.

3. *Labium:* plg, palpiger; lp, labial palpi.

4. *Front wing of Pristocera atra Klug.*—C, costal nervure; sc, subcostal or submarginal nervure; m, median nervure; sm, submedian or anal nervure; st, stigma, or marginal nervure *stigmated:* r, radius, or radial nervure; cbt, cubital nervure; sd, subdiscoidal nervure; b, basal nervure; tm, transverse median nervure; d, first recurrent nervure; sr, second recurrent nervure; ftc, first transverse cubital nervure; stc, second transverse cubital nervure. *Cells:* 1, costal; 2, first basal; 3, second basal; 4, anal; 5, marginal or radial; 6, first discoidal; 7, second discoidal; 8, third discoidal; 9, first submarginal; 10, second submarginal; 11, third submarginal; 12, first apical; 13, second apical.

5. *Hind wing of Belyta:* sc, costal and subcostal united at base; m, median nervure; st, stigma with hooks; b, basal nervure; cbt, cubital. *Cells:* 1, costal cell (open); 2, basal; 3, anal.

6. *Ovipositor of Proctotrypes caudatus* Say: shs, sheaths conjoined, very long and curved at tip; tba, terebra, or ovipositor proper; bp, basal plates; spc, spiculæ; spl, basal lobes of spiculæ.

7. *Ovipositor of Epyris grandis (external view):* shs, sheaths; tba, terebra, or ovipositor proper, with the two spiculæ (spc).

8. *The same (internal view):* shs, sheaths or tube; tba, ovipositor proper; spc, spiculæ, base and apex; bp, basal plates.

9. *Cross section of same:* A, a section from near the base; B, from near the tip

10. *Male genitalia of Epyris carbonarius:* A, viewed from above; p, penal sheath, strongly exserted; o, orifice of penis; as, upper sheath or plate; ls, lower sheath. B, viewed from beneath: as, upper sheath; ls, lower sheath; p, penis; pc, penal claspers; bl, swollen basal lobes or plates, to which the sheaths are attached.

459

Fig. 11. *Male genitalia of Proctotrypes caudatus (side view, with right ventral spine removed)*: *rs,* left ventral spine, or outer sheath; *us,* upper sheath; *ls,* lower sheath.

12. *Male genitalia of Scleroderma cylindricus* Westw. *(after Westwood)*: *us,* upper sheaths; *ls,* lower sheaths.

13. *Ovipositor of Scleroderma ephippium (after Westwood)*: *dd,* spiculae; *b,* recurved bases of spiculae; *ff,* muscular angulated lobe or catch; *cc,* membranous plates.

PLATE II.

Fig. 1. *Pristocera atra* Klug ♂.
2. *Pristocera atra* Klug ♀.
3. *Isobrachium myrmecophilum* Ashm. ♀.
4. *Isobrachium myrmecophilum* Ashm. ♀.
5. *Scleroderma* sp. ♂.
6. *Scleroderma macrogaster* Ashm. ♀.
7. *Dissomphalus xanthopus* Ashm. ♂.
8. *Dissomphalus xanthopus* Ashm. ♀.

PLATE III.

Fig. 1. *Ateleopterus virginiensis* Ashm. ♂.
2. *Ateleopterus virginiensis* Ashm. ♀.
3. *Apenesia coronata* Ashm. ♂.
4. *Apenesia amazonica* Westw. ♀.
5. *Cephalonomia hyalinipennis* Ashm. ♂.
6. *Cephalonomia gallicola* Ashm. ♀.
7. *Laelius nigrispilosus* Ashm. ♀.
8. *Bethylus pedatus* Say ♀.

PLATE IV.

Fig. 1. Front wing of *Eupsenella agilis* Westw. ♂.
2. Front wing of *Sierola* (?) *ambigua* Ashm. ♂
3. *Calyoza staphylinoides* Westw. ♂.
4. *Epiris columbianus* Ashm. ♂.
5. *Mesitius rancourrensis* Ashm. ♀.
6. *Anoxus Chittendenii* Ashm. ♂.
7. *Perisemus floridanus* Ashm. ♀.
8. *Goniozus platynota* Ashm. ♀.

PLATE V.

Fig. 1. *Ampulicomorpha confusa* Ashm. ♂.
2. *Embolemus Raddii* Westw. ♂.
3. *Dryinus americanus* Ashm. ♀.
4. *Gonatopus flavifrons* Ashm. ♀.
5. *Labeo typhlocybae* Ashm. ♂.
6. *Bocchus flavicollis* Ashm. ♀.
7. *Phorbas laticeps* Ashm. ♂.

PLATE VI.

Fig. 1. *Chelogynus canadensis* Ashm. ♀.
2. *Anteon politus* Ashm. ♀.
3. *Aphelopus melaleucus* Dalm. ♀.
4. *Habropelte armatus* Say ♂; *a,* head of ♀.

Fig. 5. *Trichosteresis floridanus* Ashm. ♀ .
6. *Eumegaspilus erythrothorax* Ashm. ♀ ; *a*, head of ♂ .
7. *Megaspilus striatipes* Ashm. ♀ ; *a*, head of *M. californicus* ♂ .
8. *Lygocerus 6-dentatus* Ashm. ♂ ; *a*, head of ♀ .
9. *Atritomus americanus* Ashm. ♂ .
10. *Lagynodes minutus* Ashm. ♀ ; *a*, head of ♂ .

PLATE VII.

Fig. 1. *Aphanogmus varipes* Ashm. ♀ ; *a*, head of *A. niger* Ashm. ♂ .
2. *Ceraphron punctatus* Ashm. ♀ ; *a*, head of ♂ .
3. *Neoceraphron macroneurus* Ashm. ♂ ; *a*, head of ♀ .
4. *Trimorus americanus* Ashm. ♂
5. *Phanurus floridanus* Ashm. ♀ .
6. *Trissolcus brochymenae* Ashm. ♀ .
7. *Telenomus sphingis* Ashm. ♀ ; *a*, ♂ antenna.
8. *Dissolcus nigricornis* Ashm. ♀ .
9. *Aradophagus fasciatus* Ashm. ♀ .

PLATE VIII.

Fig. 1. *Pentacantha canadensis* Ashm. ♀ ; *a*, abdomen, viewed from the side.
2. *Trissacantha americana* Ashm. ♂ .
3. *Xenomeras ergenna* Walk. ♂ .
4. *Prosacantha caraborum* Riley ♀ ; *a*, ♂ antenna.
5. *Teleas pallidipes* Ashm. ♂ ; *a*, ♂ antenna of *T. coxalis* Ashm.
6. *Hoplogryon longipennis* Ashm. ♀ ; *a*, ♂ antenna of *H. solitarius* Ashm.
7. *Gryon borealis* Ashm. ♀ .
8. *Ceratobaeus cornutus* Ashm. ♀ .
9. *Baeus americanus* How.: *a*. ♀ ; *b*. winged ♂ .

PLATE IX.

Fig. 1. *Caloteleia Heidemanni* Ashm. ♀ ; *a*, ♂ antenna.
2. *Baryconus acanthi* Riley ♀ .
3. *Chromoteleia semicyanea* Ashm. ♀ ; *a*, ♂ antenna.
4. *Opisthacantha mellipes* Ashm. ♀ ; *a*, side view of body; *b*, ♂ antenna.
5. *Baeoneura bicolor* Ashm. ♀ .
6. *Macroteleia floridana* Ashm. ♀ ; *a*, ♂ antenna of *M. macrogaster* Ashm.
7. *Calliscelio luticinctus* Ashm. ♀ ; *a*, ♂ antenna.
8. *Lapitha spinosa* Ashm. ♂ .

PLATE X.

Fig. 1. *Hoploteleia floridana* Ashm. ♂ .
2. *Anteris nigriceps* Ashm. ♀ ; *a*, ♂ antenna.
3. *Cremastobaeus bicolor* Ashm. ♀ ; *a*, ♂ antenna.
4. *Hadronotus insularis* Ashm. ♀ ; *a*, ♂ antenna.
5. *Idris aenea* Ashm. ♀ .
6. *Acanthoscelio americanus* Ashm. ♂ .
7. *Sparasion pilosum* Ashm. ♀ .
8. *Sceliomorpha longicornis* Ashm. ♂ .
9. *Scelio hyalinipennis* Ashm. ♀ ; *a*, ♂ antenna; *o*, wing of *S. fuscipennis* Ashm.

PLATE XI.

Fig. 1. *Iphetrachelus americanus* Ashm. ♂ .
2. *Allotropa americana* Ashm. ♀ ; *a*, ♂ antenna.
3. *Metaclisis belonocnema* Ashm. ♀ .

Fig. 4. *Monocrita carinata* Ashm. ♀ ; *a, ♂* antenna.
     5. *Isostasius musculus* Ashm. ♀.
     6. *Inostemma Cressoni* Ashm. ♀.
     7. *Acerota cargo* Ashm. ♂ ; *a, ♀* antenna.
     8. *Piestopleura catillus* Walk. ♀.
     9. *Xestonotus andriciphilus* Ashm. ♀.
    10. *Amblyaspis longipes* Ashm. ♂ ; *a, ♀* antenna.

PLATE XII.

Fig. 1. *Leptacis rugiceps* Ashm. ♀ ; *a, ♂* antenna.
     2. *Isorhombus hyalinipennis* Ashm. ♂ ; *a, ♀* antenna.
     3. *Polymecus pallipes* Ashm. ♀ ; *a. ♂* antenna.
     4. *Sactogaster Howardii* Ashm. ♀ ; *a, ♂* antenna of *S. anomaliventris* Ashm.
     5. *Synopeas rufipes* Ashm. ♀ ; *a. ♂* antenna.
     6. *Calopelta mirabilis* Ashm. ♂.
     7. *Anopedias error* Fitch ♀ ; *a, ♂* antenna.
     8. *Amitus aleurodinis* Hald. ♀ ; *a. ♂* antenna.
     9. *Trichacis rufipes* Ashm. ♀ ; *a. ♂* antenna.
    10. *Hypocampsis pluto* Ashm. ♀.

PLATE XIII.

Fig. 1. *Eritrissomerus cecidomyiae* Ashm. ♀ ; *a, ♂* antenna.
     2. *Polygnotus baccharicola* Ashm. ♀ ; *a, ♂* antenna.
     3. *Platygaster floridensis* Ashm. ♀ ; *a, ♂* antenna.
     4. *Isocybus pallipes* Say ♀ ; *a, ♂* antenna.
     5. *Helorus paradoxus* Prov. ♀
     6. *Disogmus areolator* Hal. ♀.
     7. *Proctotrypes caudatus* Say ♀ ; *a, ♀* abdomen, viewed from the side; *b, ♂* abdomen; *c, ♂* antenna.
     8. *Codrus apterogyus* Hal. ♀.

PLATE XIV.

Fig. 1. *Leptorhaptus conicus* Ashm. ♀ ; *a, ♂* abdomen; *b, ♂* antenna
     2. *Miota glabra* Ashm. ♀.
     3. *Acropiesta flavicauda* Ashm. ♀.
     4. *Belyta frontalis* Ashm. ♀.
     5. *Oxylabis spinosus* Prov. ♀.
     6. *Cinctus mellipes* Say ♀.
     7. *Xenotoma mandibularis* Ashm. ♀.

PLATE XV.

Fig. 1. *Zelotypa longicornis* Ashm. ♂.
     2. *Pantoclis analis* Ashm. ♀ ; *a, ♂* antenna of *P. californica* Ashm.
     3. *Zygota americana* Ashm. ♂ ; *a.* front tibia.
     4. *Aclista rufopetiolata* Ashm. ♀.
     5. *Psilomma colambianum* Ashm. ♀ ; *a, ♂* antenna.
     6. *Ismarus rugulosus* Först. ♂ ; ♀ antenna.
     7. *Acolus Zabriskiei* Ashm. ♂.
     8. *Aneetata hirtifrons* Ashm. ♀.

PLATE XVI.

Fig. 1. *Pantolyta brunnea* Ashm. ♀.
     2. *Polypeza Pergandei* Ashm. ♀.
     3. *Thoron pallipes* Ashm. ♀.

Fig. 4. *Hemiteles subemarginatus* Ashm. ♀ ; *a, ♂* antenna.
5. *Paramesius pallidipes* Ashm. ♀.
6. *Cacus œcanthi* Riley ♀.
7. *Spilomicrus armatus* Ashm. ♀ ; *a, ♂* antenna of *S. flavicornis.*
8. *Hemilexodes floridana* Ashm. ♀ ; *a, ♂* antenna of another species.

PLATE XVII.

Fig. 1. *Glyptonota nigriclavata* Ashm. ♀.
2. *Tropidopsis clavatus* Ashm. ♀ ; *a, ♂* antenna.
3. *Idiotypa pallida* Ashm. ♀ ; *a, ♂* antenna.
4. *Aneurhynchus floridanus* Ashm.; *a, ♂* antenna.
5. *Labolips innupta* Hal. ♀.
6. *Galesus 6-punctatus* Ashm. ♀ ; *a, ♂* antenna.
7. *Platymischus dilatatus* Westw. ♂.
8. *Loxotropa ruficornis* Ashm. ♀ ; *a,* head and antennæ of *♂.*

PLATE XVIII.

Fig. 1. *Tropidopria conica* Latr. ♀.
2. *Diapria californica* Ashm. ♀ ; *a, ♂* antenna.
3. *Ceratopria infuscatipes* Ashm. ♀.
4. *Trichopria rufipes* Ashm. ♀ ; *a, ♂* antenna.
5. *Phænopria virginica* Ashm. ♂ ; *a,* ♀ antenna.
6. *Monelata mellicollis* Ashm. ♀ ; *a, ♂* antenna.
7. *Basalys californica* Ashm. ♂ ; *a,* ♀ antenna.
8. *Myrmecopria mellea* Ashm. ♀.

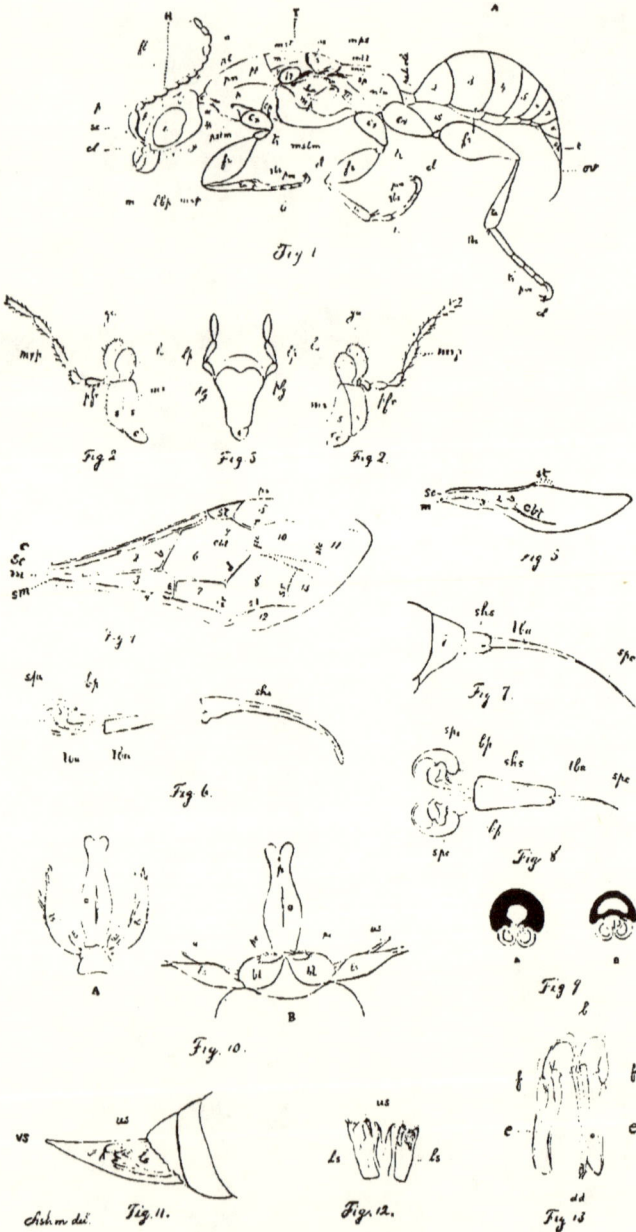

NORTH AMERICAN PROCTOTRYPIDÆ.

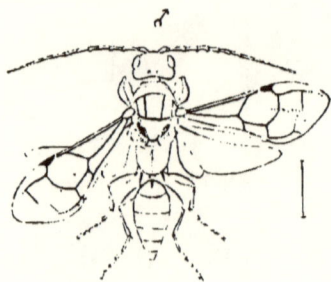

1   PRISTOCERA-Klug

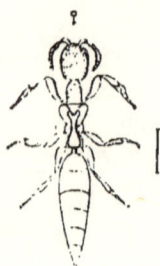

2   PRISTOCERA-Klug

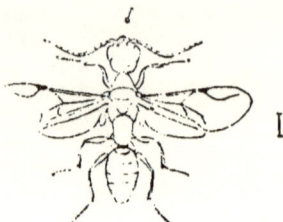

3   ISOBRACHIUM-Forst

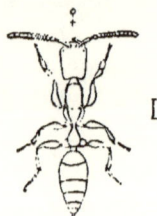

4   ISOBRACHIUM-Forst

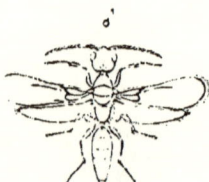

5   SCLERODERMA-Westw

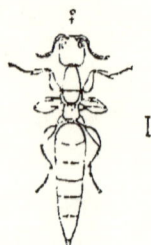

6   SCLERODERMA-Westw

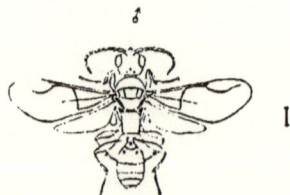

7   DISSOMPHALUS-Ashm

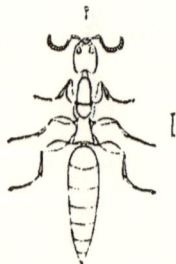

8   DISSOMPHALUS-Ashm

Ashm del

NORTH AMERICAN PROCTOTRYPIDÆ

1 ATELEOPTERUS—Forst

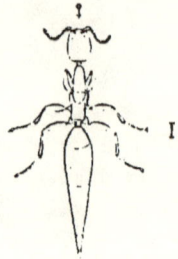

2 ATELEOPTERUS—Forst

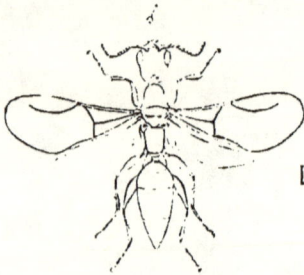

3 APENESIA Westw

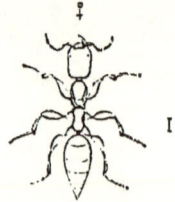

4 APENESIA—Westw

5 CEPHALONOMIA Westw

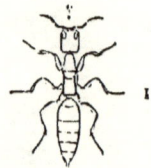

6 CEPHALONOMIA Westw

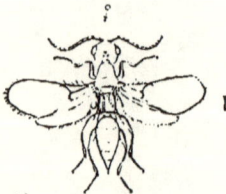

Ashm. del

7 LÆLIUS—Ashm

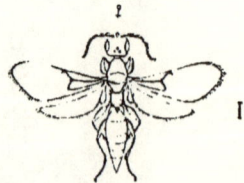

8 BETHYLUS—Latr

NORTH AMERICAN PROCTOTRYPIDÆ.

1   EUPSENELLA-WEST~

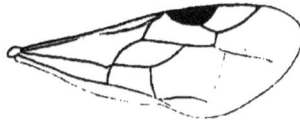

2   SIERDLA-CAN ♀

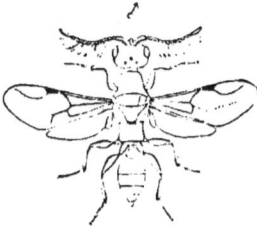

3   CALYOZA-WEST~

4   EPVRIS-WEST~

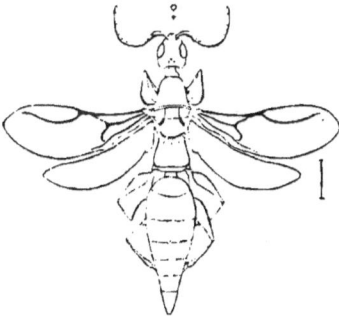

5   VESTIUS-WEST~

6   ANORUS-THONS

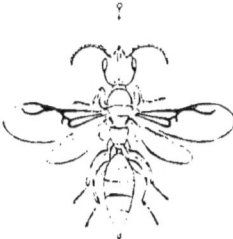

7   PER.SEVLS-FORS~

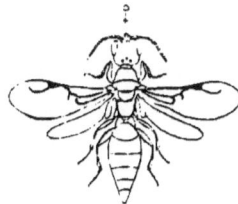

8   GONIOZUS-FORS~

Ashin del

NORTH AMERICAN PROCTOTRYPIDÆ.

www.ingramcontent.com/pod-product-compliance
Lightning Source LLC
Chambersburg PA
CBHW020905210326
41598CB00018B/1775